Advances in Experimental Medicine and Biology

For further volumes:
http://www.springer.com/series/5584

Abdeslem El Idrissi · William J. L'Amoreaux

Editors

Taurine 8

Volume 1: The Nervous System, Immune System, Diabetes and the Cardiovascular System

 Springer

Editors

Abdeslem El Idrissi
City University of New York
Center for Developmental Neuroscience
College of Staten Island
Staten Island, NY, USA

William J. L'Amoreaux
City University of New York
Department of Biology
College of Staten Island
Staten Island, NY, USA

ISSN 0065-2598
ISBN 978-1-4614-6129-6   ISBN 978-1-4614-6130-2 (eBook)
DOI 10.1007/978-1-4614-6130-2
Springer New York Heidelberg Dordrecht London

Library of Congress Control Number: 2012953700

Printed on acid-free paper

Springer is part of Springer Science+Business Media (www.springer.com)

Preface

The organizing committee wishes to thank all attendees of the 18th International Taurine Meeting that took place in Marrakesh, Morocco, from April 7th to 13th. This year, the conference highlighted the *"Mystique of Taurine."* Taurine investigators have had the privilege of attending these scientific meetings on three continents: Asia, Europe, and North America. This marked the first time that our conference was held in Africa. As a result, we present here the data from investigators from five of the six continents (sadly taurine research has yet to hit Antarctica). With this geographical expansion, the interest in taurine research has exponentially grown. This international meeting was attended by approximately 120 scientists. We present here information on the roles of taurine in a variety of organ systems, from the brain to the reproductive system and every system in between. As you are keenly aware, there is certainly a mystique to taurine. Is it beneficial or harmful? Does it protect cells or induce cell death? Can it be used in conjunction with another molecule to benefit health or cause death? The answer (or at least a hint to the answer) to these and other questions lies within this body of works. Of course, not all questions were answered but there were many discussions that generated numerous new ideas that will be taken home and tested in the laboratory.

This meeting was also unique in that many undergraduate and graduate students from the College of Staten Island/CUNY attended and presented their research as part of a study abroad program. This opportunity represented the first time that most of these students attended an international conference. More importantly, it served to stimulate interest in taurine research and recruit future taurine researchers. We are greatly appreciative for the overwhelming support of the College of Staten Island's administration, particularly Dr. Deborah Vess, Associate Provost for Undergraduate Studies and Academic Programs; Dr. William Fritz, the provost; Renee Cassidy, study abroad advisory from The Center for International Service; Debra Evans-Greene, Director of the Office of Access and Success Programs; and Dr. Claude Braithwaite of the City College of New York and the Louis Stokes Alliance for Minority Participation.

The abstracts of the conference were published in the journal "Amino Acids" (Vol. 42, Issue 4). We thank Drs. Lubec and Panuschka for making this possible.

Because of the success of this meeting, the organizing committee wishes to gratefully acknowledge the following:

- Taisho Pharmaceutical Co., Ltd., Tokyo Japan for their generous financial support.
- Professor Dr. Gert Lubec, FRSC (UK), Medical University of Vienna and Editor in Chief of AMINO ACIDS.
- Dr. Claudia Panuschka, Springer Wien, New York, Senior Editor Biomedicine/ Life Sciences.
- Dr. Portia E. Formento, Editor, Biomedicine, Springer US.
- Dr. Melanie Tucker (Wilichinsky) Editor, Genetics and Systems Biology, Springer US.

On behalf of the organizing committee, I thank all the attendees of the 18th international Taurine Meeting and the sponsors that made this meeting possible.

Staten Island, NY, USA

Abdeslem El Idrissi

Contents

Part I
Taurine and Its Actions
on the Nervous System

Chapter 1
Neuropsychopharmacological Actions of Taurine

Shailesh P. Banerjee, Andre Ragnauth, Christopher Y. Chan, Mervan S. Agovic, Vincent Sostris, Iman Jashanmal, Louis Vidal, and Eitan Friedman

Abstract Taurine, an endogenous amino sulfonic acid, exhibits numerous neuropsychopharmacological activities. Previous studies in our laboratory have shown that it is an effective anti-cataleptic and neuro-protective agent. Current investigations show that acute or chronic administration of psychotropic drug cocaine may increase extracellular release of endogenous taurine which may protect

S.P. Banerjee (✉) • C.Y. Chan • V. Sostris • E. Friedman
Department of Physiology, Pharmacology and Neuroscience,
Sophie Davis School of Biomedical Education at CCNY, City University of New York,
Harris Hall Rm. 203, 160 Convent Ave, New York, NY, USA

Neuroscience Subprogram, Doctoral Programs in Biology,
Graduate Center of the City University of New York, New York, NY, USA
e-mail: Banerjee@med.cuny.edu

A. Ragnauth • I. Jashanmal
Department of Physiology, Pharmacology and Neuroscience,
Sophie Davis School of Biomedical Education at CCNY, City University of New York,
Harris Hall Rm. 203, 160 Convent Ave, New York, NY, USA

Neuroscience Subprogram, Doctoral Programs in Psychology,
Graduate Center of the City University of New York, New York, NY, USA

L. Vidal
Department of Physiology, Pharmacology and Neuroscience,
Sophie Davis School of Biomedical Education at CCNY, City University of New York,
Harris Hall Rm. 203, 160 Convent Ave, New York, NY, USA

M.S. Agovic
Department of Physiology, Pharmacology and Neuroscience,
Sophie Davis School of Biomedical Education at CCNY, City University of New York,
Harris Hall Rm. 203, 160 Convent Ave, New York, NY, USA

Neuroscience Subprogram, Doctoral Programs in Biology,
Graduate Center of the City University of New York, New York, NY, USA

Department of Biology, Bronx Community College, The City University
of New York, New York, NY, USA

A. El Idrissi and W.J. L'Amoreaux (eds.), *Taurine 8*, Advances in Experimental
Medicine and Biology 775, DOI 10.1007/978-1-4614-6130-2_1,
© Springer Science+Business Media New York 2013

against deleterious effects of the substances of abuse. Taurine administration was found to prevent cocaine-induced addiction by suppressing spontaneous locomotor activity and conditioned place preference. Taurine markedly delayed tail-flick response in rats which was significantly different from that in the group of animal receiving the same volume of saline, thereby indicating that taurine is a potentially valuable analgesic agent. Both taurine and endomorphin-1 were found to suppress the delayed broad negative evoked field potentials in anterior insular cortex (upper layer 5) by partially inhibiting NMDA receptor system. Thus, taurine is a unique psychopharmacological compound with potential for a variety of therapeutic uses including as a neuro-protective, anti-cataleptic, anti-addicting, and analgesic agent.

Abbreviations

NMDA	N-Methyl-D-aspartate
GABA	Gama-amino-butyric acid
APV	DL-2-Amino-5-phosphonopentanoic acid
CPP	Conditioned place preference

1.1   Introduction

We have focused our attention on the neuropsychological actions of taurine over the last several years. In this chapter, rather than trying to survey all known psychopharmacological activities of taurine, we will restrict our discussion to investigations conducted in our laboratories. Previously, we reported that therapeutic actions of typical and atypical antipsychotic drugs are mediated, in part, by partial agonistic activity on the NMDA receptor subtype glutamatergic transmission (Banerjee et al. 1995; Lidsky and Banerjee 1993, 1996; Lidsky et al. 1997). On the other hand, antipsychotic-induced side effects, such as catalepsy, were shown to occur as a result of complex changes in a variety of neurotransmitter functions, including glutamatergic and dopaminergic neuronal pathways in the striatum (Agovic et al. 2008). It was proposed that haloperidol-induced catalepsy occurs due to augmentation of NMDA-mediated glutamatergic transmission, inhibition of dopamine D_2 receptor-mediated transmission, and unchanged dopamine D_1 receptor-mediated transmission in the basal ganglia following chronic treatment of haloperidol (Agovic et al. 2008). In order to investigate if haloperidol-induced catalepsy is mediated by the degeneration of dopaminergic neurons and/or changes in dopamine- and glutamate-mediated transmissions, we studied the effects of chronic haloperidol administration on the dopaminergic neurons in the basal ganglia. Daily administration of haloperidol for 3 weeks caused a significant diminution of endogenous dopamine levels, and tyrosine hydroxylase activities as well as a significant increase in the densities of dopamine D_2 receptor in the basal ganglia (Lidsky et al. 1994 and Lidsky et al. 1995). These observations indicate that chronic haloperidol treatment may cause degeneration of dopaminergic neurons in the basal ganglia as well as catalepsy. Interestingly, rats that were previously primed with a week of treatment of taurine and then given both haloperidol and taurine for 3 weeks, showed no development of catalepsy or significant changes in the levels of dopamine or tyrosine hydroxylase activities or in the dopamine D_2 receptor densities in the basal ganglia (Lidsky et al. 1995). Taurine therefore appears to prevent both haloperidol-induced neurodegeneration of dopamine neurons in the basal ganglia and the development of catalepsy in rats. The probable mechanisms for the neuro-protective and anti-cataleptic actions of taurine was proposed to be mediated by either its antagonism to glutamate-induced excitotoxicity and/or by its functioning as a partial agonist at the $GABA_A$ receptor (Quinn and Miller 1992). In another study, we found that both pre- and postnatal exposure to cocaine-induced neurotoxicity retards the development of dopamine neurons in the striatum. This was prevented by simultaneous administration of relatively high doses of the NMDA antagonist, clozapine (Lidsky and Banerjee 1992; Lidsky et al. 1993), suggesting that cocaine-induced neurotoxicity may occur due to the development of glutamate-mediated excitotoxicity (Yablonsky-Alter et al. 2005). Since exogenous taurine appears to oppose the neurotoxicity mediated by psychotropic drugs, we wondered whether endogenous taurine levels would be altered as a consequence of chronic substances of abuse exposure. Therefore, we measured extracellular taurine levels in the striatum following chronic cocaine treatment. Chronic cocaine administration significantly increased the extracellular levels of taurine in the striatum which further significantly increased following cocaine challenge to rats that previously had received chronic cocaine (Yablonsky-Alter et al. 2009). Therefore, it appears that endogenous taurine may oppose neurotoxicity induced by glutamate-mediated excitotoxicity, caused by the administration of exogenous psychotropic drugs. Taurine may open chloride channel through activation of a specific taurine receptor which is independent of $GABA_A$ or strychnine-sensitive glycine receptors to oppose glutamate-induced excitotoxicity (Yarbrough et al. 1981; Okamoto et al. 1983a, b). Alternatively, taurine may directly interact with the NMDA receptor to suppress its activity. This possibility was investigated by adopting electrophysiological and receptor binding studies. Taurine inhibited glutamate-induced evoked field potential that is mediated by NMDA receptor system in vitro in the rat prefrontal cortex in the presence of picrotoxin used to block the taurine and GABA-sensitive chloride channel. Also, taurine reduced by at least ten-fold, the apparent affinity of glycine, as well as partially inhibited the polyamine-activated calcium channel opening in NMDA receptor system as analyzed by measuring specific (^{3}H)-MK-801 binding to NMDA receptor in the rat cortical membrane preparations (Chan et al. 2012).

Thus, taurine may inhibit NMDA-mediated glutamatergic transmission by two independent mechanisms. First, it may open chloride channel on postsynaptic neurons to prevent depolarization and activation of NMDA receptor (Yarbrough et al. 1981; Okamoto et al. 1983a). Second, by interacting directly with the NMDA receptor system, taurine has been reported to partially inhibit APV-sensitive glutamate cell firing and polyamine-activated (^{3}H)MK-801 binding (Chan et al. 2012). Since a number of NMDA receptor antagonists have been shown to exhibit a variety of neuropsychopharmacological activities, as exemplified by acamprosate showing anti-addicting action (Heilig and Egli 2006) and ketamine exhibiting antidepressant and analgesic actions (Autry et al. 2011; Sinner and Graf 2008), we decided to investigate possible neuropharmacological effects of taurine and these are described below.

1.2 Methods

1.2.1 Experimental Procedures

All the experimental procedures adopted in the current studies have been thoroughly described previously. Electrophysiological methods are described in another chapter by Chan et al. (2012). For this study, rat slices containing the rostral agranular insular cortex were prepared. Microdialysis assays to measure extracellular amino acids using high-pressure liquid chromatography (HPLC) have been previously described (Yablonsky-Alter et al. 2009). Extracellular dopamine levels were collected by microdialysis and quantified by using HPLC (Yablonsky-Alter et al. 2005). The methods for behavioral experiments including tail-flick test, spontaneous motor activity, and conditioned place preference have been previously described (Rodriguiz et al. 2008; Spinella et al. 1999; Ragnauth et al. 2000, 2001, 2005).

1.2.2 Statistic Analysis

Data were presented as mean ± S.E.M. Statistical significance of group means difference was measured by one-way analysis of variance (ANOVA), followed by Bonferroni's post hoc analysis. The threshold for statistical significance was assumed at $p < 0.05$. All statistical tests were calculated using GraphPad Prism (GraphPad Software, San Diego California, USA).

1.3 Results

1.3.1 Effects of Acute and Chronic Cocaine Treatment on Striatal Amino Acid Release

1.3.1.1 Effects of Acute Cocaine Treatment

The basal extracellular levels of glutamate, glycine, and taurine in striata of control rats were found to be similar in range (Fig. 1.1a). In contrast, the basal extracellular level of glutamine was approximately two to four times that of other three

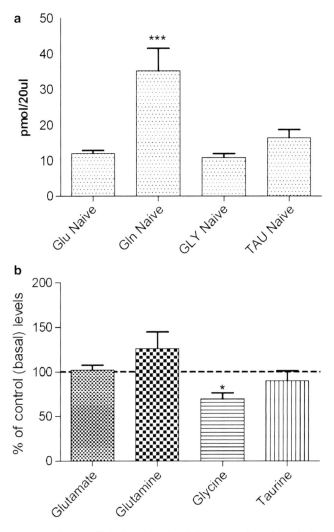

Fig. 1.1 Comparison of extracellular basal levels of glutamate, glutamine, glycine, and taurine in striata of (**a**) normal and (**b**) acute cocaine-treated rats. (**a**) Two groups consisted of six rats in each group. Control group received saline, while acute group received 10 mg/kg of intraperitoneal cocaine 30 min before microdialysis procedure. Glutamine in the control group was severalfold higher compared to other amino acids; no significant difference was observed between other amino acids. One-way ANOVA: $n = 6$; ***$p < 0.01$; Error bars, S.E.M. (**b**) Six, otherwise untreated rats were injected intraperitoneally with 10 mg/kg cocaine 30 min before the microdialysis collection. Only glycine extracellular concentrations were decreased significantly, while glutamine showed slight, not significant, increase. One-way ANOVA: $n = 6$; *$p = 0.0167$; Error bars, S.E.M

amino acids. Administration of 10 mg/kg of cocaine failed to alter the extracellular concentrations of glutamate and taurine (Fig. 1.1b). While acute cocaine significantly decreased the extracellular levels of glycine, it increased striatal extracellular levels of glutamine; however, this change was not statistically significant (Fig. 1.1b).

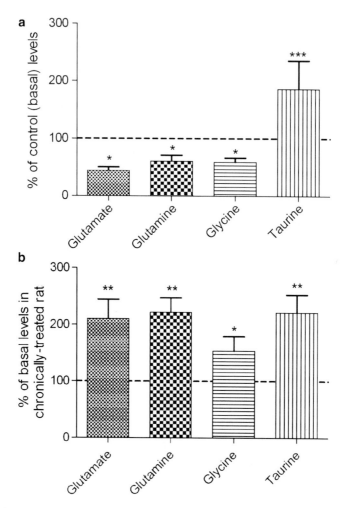

Fig. 1.2 Extracellular release of glutamate, glutamine, glycine, and taurine in striatum after (**a**) chronic cocaine treatment and (**b**) cocaine challenge to/in chronic cocaine-treated rats. (**a**) A group of six rats were injected intraperitoneally with 10 mg/kg cocaine, 6 days each week for 3 weeks and microdialysis was performed 24 h after the last injected dose. Compared to the control group (Fig 1.1a), there was a significant decrease in glutamate, glutamine, and glycine, while taurine was significantly increased. One-way ANOVA: $n=6$; *$p=0.3108$; ***$p=0.0020$; Error bars, S.E.M. (**b**) A group of six rats that was previously treated with cocaine for 3 weeks was challenged 24 h after the last dose with a single dose of 10 mg/kg cocaine, 30 min before microdialysis procedure. An increase in all four neurochemicals was recorded. One-way ANOVA: $n=6$; *$p=0.3108$; **$p<0.001$; Error bars, S.E.M

1.3.1.2 Effects of Chronic Cocaine Treatment on Basal Amino Acids Levels

The effects of chronic cocaine administration on basal extracellular concentrations of the four amino acids are shown in Fig. 1.2a. Basal extracellular concentrations of glutamate, glutamine, and glycine were significantly decreased as compared to basal control levels, whereas, interestingly, the concentration of taurine was significantly increased when measured 24 h following the last daily injection of cocaine.

1.3.1.3 Effects of a Cocaine Challenge in Chronic Cocaine-Treated Rats

A single 10 mg/kg injection of cocaine given to rats which had previously received 3 weeks of chronic cocaine treatment, which was then withdrawn for 24 h, increased extracellular striatal levels of all amino acids including taurine (Fig. 1.2b).

In conclusion, potential cocaine-induced neurotoxicity following acute or chronic administration may be opposed by two separate endogenous mechanisms. First, acute effects are probably mitigated by diminution of glycine release that would reduce over-activation of NMDA receptor function. Second, chronic effects of the psychomotor stimulant may be opposed by endogenous release of taurine.

1.3.2 Effects of Taurine on Cocaine-Induced Locomotor Sensitization and Conditioned Place Preference

Additional endogenous taurine release in response to chronic cocaine treatment led to us to investigate whether pharmacological administration of taurine would prevent cocaine-induced addiction. This was tested by studying the effects of on cocaine-induced sensitization of locomotor activity and development of conditioned place preference (CPP). The effect of taurine on cocaine-induced augmentation of locomotor activity is shown in Fig. 1.3a. These results indicate that taurine is a suppressor of cocaine-induced locomotor sensitization, suggesting that it may oppose psychomotor stimulant's chronic effects including perhaps addiction. Next we investigated the influence of taurine on cocaine-induced place preference acquisition (Fig. 1.3b). Cocaine in a daily intraperitoneal (IP) dose of 15 mg/kg induced significant place preference acquisition after 8 days of habituation protocol. The psychomotor stimulant when co-administered with taurine failed to develop conditioned place preference. Thus, taurine may be effective in blocking cocaine-induced addiction.

1.3.3 Effect of Taurine on Tail-Flick Response in Rats

The NMDA receptor antagonist, ketamine, has been shown to be an effective analgesic agent and prevents hyperalgesia (Mathisen et al. 1995; Sinner and Graf 2008; Tverskoy et al. 1994). Therefore, we investigated the potential value of taurine in the management of pain by using tail-flick response in rats treated with either taurine or saline. Taurine (100 mg/kg) or equal volume of saline was administered IP for 9 days to rats before subjecting them to tail-flick tests. Taurine significantly delayed the tail-flick response as compared to saline-treated animals (Table 1.1),

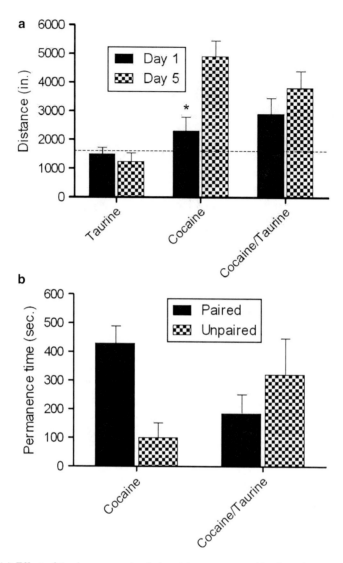

Fig. 1.3 (**a**) Effect of taurine on cocaine-induced locomotor sensitization. Average locomotion during habituation phase did not vary significantly between any of the groups tested and is indicated by the *dashed line*. Daily taurine injections (100 mg/kg) failed to induce significant sensitization after 5 days. When preceded by 7 days of daily 100 mg/kg of taurine preloading and daily injections of 15 mg/kg cocaine-induced sensitization as indicated by a significant increase in distance traveled at day 5 vs. day 1 (* = t test; $p < 0.05$). When co-administered with taurine (100 mg/kg), cocaine failed to induce significant sensitization in animals that were primed with taurine after 5 days. (**b**) Effect of taurine on cocaine-induced place preference acquisition. Cocaine (dose = 15 mg/kg) injections induced significant place preference acquisition following an 8 day habituation protocol (*t* test; $p < 0.05$). When preceded by 7 days of taurine (daily 100 mg/kg) preloading and co-administered with taurine (dose = 100 mg/kg), cocaine (dose = 15 mg/kg) failed to induce significant place preference acquisition

Table 1.1 Effect of taurine with and without naltrexone on tail-flick response

Condition	Mean tail-flick latency (s)	SE	Significance
Saline	11.32	±1.72	ns
Saline+NTX	10.76	±2.31	ns
Taurine	21.87	±1.62	s
Taurine+NTX	11.78	±3.07	ns

suggesting that it may be a useful analgesic agent. Interestingly, rats that received naltrexone (68 μg/kg) 15 min before the final dose of taurine showed no significant difference in the duration of time needed to exhibit tail-flick response as compared to saline- and naltrexone-treated animals (Table 1.1). The ability of naltrexone to at least partially block taurine-mediated analgesia indicates that taurine may relieve pain by activation of endogenous morphine pathways. Alternatively, opioids and taurine may diminish pain signals by the inhibition of glutamatergic transmission mediated by the NMDA receptor on CNS loci in pain pathways.

1.3.4 Effect of Taurine and Endomorphin-1 Glutamate Evoked Response in Anterior Insular Cortex

To seek support for the above possibility, we studied the effects of taurine and endomorphin-1 on evoked field potentials that were evoked in the upper layer-5 of rat rostral insular cortical slices by single pulses (0.04 Hz; 0.03 ms duration, 1.3× threshold current) delivered from concentric bipolar electrodes placed medial to the nucleus accumbens. Its response typically consisted of multiple wavelets, including a large negativity (Fig. 1.4), which is sensitive to the NMDA receptor antagonist APV (unpublished data). Bath application of either 2 mM taurine or 50 μM endomorphin-1 showed specific inhibition of this presumed NMDA-receptor-mediated response (Fig. 1.4a, b, respectively). These results indicate that both taurine and endomorphin-1 may induce analgesia by the inhibition of glutamate transmission mediated by the NMDA receptor. The precise mechanisms for taurine- or endomorphin-1-mediated inhibition of the NMDA receptor in the CNS pain pathway including the dorsal horn in the spinal cord remain to be elucidated.

1.4 Discussion

Taurine, which is an endogenous amino sulfonic acid, has been shown to be an inhibitory neuromodulator. Our previous studies have shown it to be an effective neuro-protective and anti-cataleptic agent (Lidsky et al. 1995). Taurine exhibits

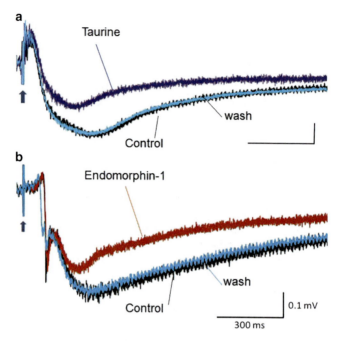

Fig. 1.4 Effect of taurine and of endomorphin-1 on NMDA-receptor-mediated evoked response in rat anterior insular cortical slices. (**a**) Superimposed upper layer-5 rostral agranular insular cortical field potential responses to ventral medial cortical stimulation by single electrical pulses (*arrow head*) recorded from a slice superfused with artificial cerebral spinal fluid (ACSF: control), 2 mM taurine, and ACSF wash. The taurine effect was restricted to the latter part of the negative wave, which is sensitive to NMDA receptor antagonists (unpublished data). (**b**) Experimental conditions were similar to those described in (**a**) except 50 nM endomorphine-1 instead of taurine was used. The endomorphine-1 effect was also mainly restricted to the latter part of the evoked response. Calibration scales for (**b**) apply to (**a**)

neuro-protection by physiological and pharmacological mechanisms. Chronic exposure to psychotropic drugs such as cocaine-induced body insult is opposed by additional extracellular release of taurine (Yablonsky-Alter et al. 2009; Fig. 1.2a). This may be considered as a physiological mechanism for counteracting the adverse effects of chronic intake of substances of abuse. Another possible physiological mechanism for protecting the central nervous system has been identified in this investigation. Acute administration of cocaine was found to suppress extracellular release of glycine (Fig. 1.1b). Since glycine is a co-transmitter with glutamate in the activation NMDA receptor (Lidsky and Banerjee 1993), diminution of extracellular glycine release would be expected to reduce glutamatergic transmission and subsequent possible excitotoxicity. Therefore, two separate physiological mechanisms may be available to counteract adverse effects of substances of abuse. The mechanisms for development for such counteractive protective mechanisms remain to be elucidated. Our and other studies indicate that an increase in either glutamate release or receptor sensitization may often be associated with enhancement of taurine release, but how these different changes may be connected to each other is not known.

Although observed physiological mechanisms involving an extracellular increase in taurine and an extracellular decrease in glycine may potentially oppose deleterious effects of substances of abuse, these are not sufficient to prevent the development of addiction. Drug addiction, however, may be opposed and possibly prevented by taurine-mediated pharmacological mechanisms. Our studies show that cocaine-induced drug addiction as assessed by observing increase in locomotor activity and conditioned place preference may be prevented when animals are primed with taurine for several days followed by simultaneous chronic cocaine and additional taurine administration (Fig. 1.2a, b). Several mechanisms may be considered for pharmacological anti-addicting activity of taurine. Substances of abuse are believed to co-opt synaptic plasticity mechanisms in brain circuits that are involved in reinforcement and reward processing, as well as those which are responsible for learning and memory (Hyman et al. 2006). Although a variety of neuronal systems play a role in the development of addiction, dopamine is recognized to play a leading role in reinforcement and reward processes, while the glutamatergic system is involved in learning and memory (Kauer and Malenka 2007). Taurine may reduce glutamatergic activity by two separate mechanisms. First, taurine has been shown to open chloride channels which are independent of GABA and inhibitory glycine receptors to decrease depolarization state of the target cells (Yarbrough et al. 1981; Okamoto et al. 1983a, b). Second, taurine may directly interact at the glutamate NMDA receptor to suppress glutamatergic transmission (Chan et al. 2012). In addition, preliminary microdialysis studies in our laboratory show that chronic taurine treatment is effective in decreasing extracellular basal levels of dopamine, and it prevents acute cocaine-induced increase in the synaptic levels of dopamine in the nucleus accumbens (data not shown). Thus, taurine may interfere with dopamine-mediated reinforcing and rewarding processes to block drug addiction.

Pain stimuli are believed to originate in the primary afferent neuron at the periphery and then carried to the dorsal horn in the spinal cord, where it synapses with glutamate as well as other neuropeptide transmitters within the secondary neuron. Opioids have been shown to suppress release of glutamate and neuropeptides at the presynaptic sites in the dorsal horn and reverse or oppose postsynaptic depolarization by stimulating potassium efflux to attenuate transmission of pain signals (Schumacher et al. 2012). Clearly, NMDA receptor antagonists may exhibit analgesic activity, as has been shown for ketamine (Mathisen et al. 1995; Sinner and Graf 2008). Therefore, we wondered if taurine would function as a pain-relieving agent. Interestingly, in the tail-flick assay, taurine was found to be an effective analgesic agent (Table 1.1). Surprisingly, the opioid antagonist naltrexone markedly inhibited taurine-induced analgesia (Table 1.1), suggesting that either taurine may influence the opioid receptor system either directly or by co-opting its intracellular signaling mechanisms involved in opioid-induced analgesia. Notably, the NMDA receptor antagonist, ketamine, is known to enhance opioid-induced analgesia (Mathisen et al. 1995) and prevent hyperalgesia (Tverskoy et al. 1994), and ketamine has recently been shown to enhance the opioid-induced extracellular signal-regulated kinase 1/2 (ERK1/2) (Gupta et al. 2011). Since taurine is a NMDA receptor antagonist (Chan et al. 2012) it is possible that taurine-mediated analgesic effect may be mediated by augmentation of ERK1/2 phosphorylation. Effects of taurine on opioid receptor system and opioid-induced ERK1/2 phosphorylation or mitogen-activated kinase are not known. Alternatively, taurine may not influence opioid receptor nor opioid-induced intracellular signaling, and it may, instead, potentiate endogenous opioid-induced inhibition of NMDA receptor activation by different independent mechanisms. Taurine has been shown to directly interactat the NMDA receptor complex to inhibit it at the postsynaptic site (Chan et al. 2012) and suppress glutamate release perhaps by opening chloride channel at the presynaptic site (Mochanova et al. 2007) to prevent depolarization at the presynaptic neurons.

Different mechanisms are involved in opioid-induced inhibition of glutamatergic transmission, such as opening of potassium channels at the postsynaptic site and closing of voltage-gated calcium channels at presynaptic site to suppress glutamate release (Fig. 1.5).

We propose that heat-induced pain leads to release of endogenous morphine peptides that may inhibit NMDA receptor system but fails to achieve the threshold level to cause prolongation of tail-flicking duration. The action of pre-administrated taurine either sums with or potentiates the inhibitory effect of endogenous opioids on the NMDA-receptor-induced response to raise the inhibition past the required level for inducing analgesia. By the same token, the observed reversal of this analgesic effect by naltrexone can be accounted for by a selective removal of the inhibition contributed by endogenous opioids such that the suprathreshold level of inhibition necessary for analgesia no longer exists (Table 1.1). Consistent with this hypothesis, we found that both taurine and endomorphin-1 inhibit glutamate-induced evoked response (Fig. 1.4a, b).

Finally, both opioids and taurine function as analgesic agents either share similar mechanisms or act by different modes of action to achieve the same goal of diminishing NMDA-mediated glutamatergic transmission. Taurine and opioids, however, exhibit opposite effects on drug-induced addiction. How do opioids cause addiction and how does taurine block this effect? It is believed that opioids disinhibit dopamine neurons in the ventral tegmental area by inhibiting GABA neurons at presynaptic site by preventing release of GABA through inhibition of voltage-gated calcium channels and postsynaptic site in dendrites by activating potassium channels to suppress depolarization (Fig. 1.6; Luscher 2012).

Disinhibition of dopamine neurons in the ventral tegmental area would be expected to stimulate dopamine release to reinforce reward system (Fig. 1.6; Hyman et al. 2006; Kauer and Malenka 2007). In contrast, taurine has been shown to suppress dopamine release in the nucleus accumbens by microdialysis in our laboratory (unpublished results). Thus, taurine opposes substances of abuse-induced drug addiction while opioids sustain it.

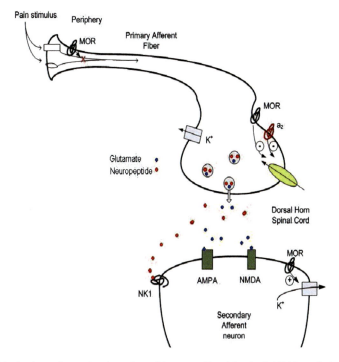

Fig. 1.5 Mechanisms for analgesic action of drugs mediated by the inhibition of glutamatergic transmission via NMDA receptor. The primary afferent neuronal pathway located in the periphery carries pain signals to the dorsal horn of the spinal cord, where it synapses with the secondary afferent neuron via glutamate and neuropeptide transmitters. An opioid agonist may inhibit the pain signal at the periphery or by its action at the dorsal horn cell. In the primary afferent neuron, it may inhibit voltage-gated calcium channel at the presynaptic site to suppress glutamate release and increase potassium efflux to diminish depolarization of the postsynaptic neuron to oppose activation of NMDA receptor and glutamatergic transmission. On the other hand, taurine may open the chloride channel at the presynaptic site to suppress release of glutamate and directly interacts with NMDA receptor at the postsynaptic site to inhibit glutamatergic transmission by different set of mechanisms than those utilized by opioids (Based on Schumacher et al. 2012)

1.5 Conclusion

Our investigations have identified diverse neuropsychopharmacological actions of taurine and these include neuro-protection, anti-cataleptic actions, anti-addiction actions, and analgesic activity. In addition, we found that endogenous taurine is released in the extracellular space perhaps to oppose harmful effects of substances of abuse.

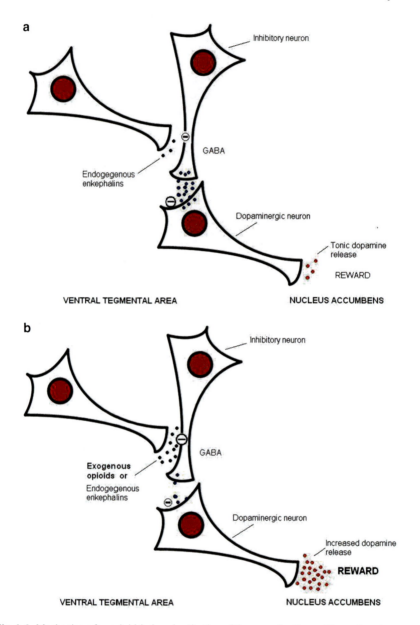

Fig. 1.6 Mechanisms for opioid-induced activation of the reward pathway. Dopaminergic neurons arising at the ventral tegmental area are believed to be involved in activating brain reward pathway in the nucleus accumbens and these are tonically inhibited by GABA interneurons. Opioids may inhibit GABA neuronal activity by suppressing GABA release to disinhibit dopaminergic neurons in activating the reward pathway. An opioid agonist may be either an endogenous substrate (**a**) or an exogenous drug (**b**). Interestingly taurine may have an opposite effect as unpublished studies in our laboratory indicate that taurine may suppress dopamine release in the nucleus accumbens, perhaps by opening chloride channel at the presynaptic site of dopaminergic neuron (Based on Martin et al. 2012)

Acknowledgments This work was supported by DAO18055 grant from NIDA and by PSC-CUNY 41-582 grant.

References

Agovic MS, Yablonsky-Alter E, Lidsky TI, Banerjee SP (2008) Mechanisms for metoclopramide-mediated sensitization and haloperidol-induced catalepsy in rats. Eur J Pharmacol 587:181–186

Autry AE, Adachi M, Nosyreva E, Na ES, Los MF, Cheng PF, Kavalali ET, Monteggia LM (2011) NMDA receptor blockade at rest triggers rapid behavioral anti depressants responses. Nature 475:91–95

Banerjee SP, Zuck LG, Yablonsky-Alter E, Lidsky TI (1995) Glutamate agonist activity: implications for antipsychotic drug action and schizophrenia. Neuroreport 6:2500–2504

Chan CY, Sun HS, Shah SM, Agovic MS, Ho I, Friedman E, Banerjee SP (2012) Direct interaction of taurine with the NMDA glutamate receptor sub-type via multiple mechanisms. 18th annual Taurine meeting. Marrakesh, Morocco

Gupta A, Devi LA, Gomes I (2011) Potentiation of μ-opioid receptor-mediated signaling by ketamine. J Neurochem 119:294–302

Hyman SE, Malenka RC, Nestler EJ (2006) Neural mechanisms of addiction: the role of reward-related learning and memory. Annu Rev Neurosci 29:565–598

Heilig M, Egli M (2006) Pharmacological treatment of alcohol dependence: target symptoms and target mechanisms. Pharmacol Ther 111:855–876

Kauer JA, Malenka RC (2007) Synaptic plasticity and addiction. Nat Rev Neurosci 8:844–858

Lidsky TI, Banerjee SP (1992) Clozapine's mechanisms of action: non-dopaminergic activity rather than anatomical selectivity. Neurosci Lett 139:100–103

Lidsky TI, Banerjee SP (1993) Acute administration of haloperidol enhances dopaminergic transmission. J Pharmacol Exp Ther 265:1193–1198

Lidsky TI, Banerjee SP (1996) Contribution of glutamatergic dysfunction to schizophrenia. Drug News & Perspectives 9:453–459

Lidsky TI, Yablonsky-Alter E, Zuck LG, Banerjee SP (1993) Anti-glutamatergic effects of clozapine. Neurosci Lett 163:155–158

Lidsky TI, Schneider JS, Zuck LG, Yablonsky-Alter E, Banerjee SP (1994) GM1 ganglioside attenuates changes in neurochemistry and behavior caused by repeated haloperidol administration. Neurodegeneration 3:135–140

Lidsky TI, Schneider JS, Yablonsky-Alter E, Zuck LG, Banerjee SP (1995) Taurine prevents haloperidol-induced changes in striatal neurochemistry and behavior. Brain Res 686:104–106

Lidsky TI, Yablonsky-Alter E, Zuck LG, Banerjee SP (1997) Antipsychotic drug effects on glutamatergic activity. Brain Res 764:46–52

Luscher C (2012) Drugs of abuse. In: Katsung BG et al (eds) Basic and clinical pharmacology. The McGraw Hill Companies Inc., New York, pp 565–580

Martin RP, Patel S, Swift MR (2012) Pharmacology of drugs of abuse. In: Golan DE, Tashjian HA, Armstrong JE, Armstrong WA (eds) Principles of pharmacology. Lippincott Williams & Wilkins, Philadelphia, PA, pp 284–309

Martin RP, Patel S, Swift MR (2012) Pharmacology of drugs of abuse. In: Golan DE, Tashjian HA, Armstrong JE, Armstrong WA (eds) Principles of pharmacology. Lippincott Williams & Wilkins, Philadelphia, PA, pp 284–309

Mathisen LC, Skjelbred P, Skoglund LA, Oye I (1995) Effect of ketamine, an NMDA receptor inhibitor, in acute and chronic orofacial pain. Pain 61:215–220

Mochanova SM, Oja SS, Saransaari P (2007) Inhibitory effect of taurine on veratridine-evoked D-[3H]aspartate release from murine corticostriatal slices: involvement of chloride channels and mitochondria. Brain Res 1130:95–102

Okamoto K, Kimura H, Sakai Y. (1983) Taurine-induced increase of the Cl-conductance of cerebellar Purkinje cell dendrites in vitro. Brain Res.259:319–23.

Okamoto K, Kimura H, Sakai Y (1983b) Evidence for taurine as an inhibitory neurotransmitter in cerebellar stellate interneurons: selective antagonism by TAG (6-aminomethyl-3-methyl-4H,1,2,4-benzothiadiazine-1,1-dioxide). Brain Res 265:163–168

Quinn MR, Miller CL (1992) Taurine allosterically modulates flunitrazepam binding to synaptic membranes. J Neurosci Res 33:136–141

Ragnauth A, Znamensky V, Moroz M, Bodnar RJ (2000) Analysis of dopamine receptor antagonism upon feeding elicited by mu and delta opioid agonists in the shell region of the nucleus accumbens. Brain Res 877:65–72

Ragnauth A, Schuller A, Morgan M, Chan J, Ogawa S, Pintar J, Bodnar RJ, Pfaff DW (2001) Female preproenkephalin-knockout mice display altered emotional responses. Proc Natl Acad Sci U S A 198:1958–1963

Ragnauth AK, Devidze N, Moy V, Finley K, Goodwillie A, Kow LM, Muglia LJ, Pfaff DW (2005) Female oxytocin gene-knockout mice, in a semi-natural environment, display exaggerated aggressive behavior. Genes Brain Behav 4:229–239

Rodriguiz RM, Gadnidze K, Ragnauth A, Dorr N, Yanagisawa M, Wetsel WC, Devi LA (2008) Animals lacking endothelin-converting enzyme-2 are deficient in learning and memory. Genes Brain Behav 7:418–426

Schumacher MA, Basbaum AI, Way WL (2012) Opioid analgesics and antagonists. In: Katzung BG et al (eds) Basic and clinical pharmacology. The McGraw Hill Companies Inc., New York, pp 543–564

Sinner B, Graf BM (2008) Ketamine. Handb Exp Pharmacol 182:313–333

Spinella M, Znamensky V, Moroz M, Ragnauth A, Bodnar RJ (1999) Actions of NMDA and cholinergic receptor antagonists in the rostral ventromedial medulla upon beta-endorphin analgesia elicited from the ventrolateral periaqueductal gray. Brain Res 829:151–159

Tverskoy M, Oz Y, Isakson A, Finger J, Bradley EL Jr, Kissin I (1994) Preemptive effect of fentanyl and ketamine on postoperative pain and wound hyper-algesia. Anesth Analg 78:205–209

Yablonsky-Alter E, Gashi E, Lidsky TI, Wang HY, Banerjee SP (2005) Clozapine protection against gestational cocaine-induced neurochemical abnormalities. J Pharmacol Exp Ther 312:297–302

Yablonsky-Alter E, Agovic MS, Gashi E, Lidsky TI, Friedman E, Banerjee SP (2009) Cocaine challenge enhances release of neuroprotective amino acid taurine in the striatum of chronic cocaine treated rats: a micro-dialysis study. Brain Res Bull 79:215–218

Yarbrough GG, Singh DK, Taylor DA (1981) Neuropharmacological characterization of a taurine antagonist. J Pharmacol Exp Ther 3:604–613

Chapter 2
Taurine and Its Neuroprotective Role

Neeta Kumari, Howard Prentice, and Jang-Yen Wu

Abstract Taurine plays multiple roles in the CNS including acting as a neuro-modulator, an osmoregulator, a regulator of cytoplasmic calcium levels, a trophic factor in development, and a neuroprotectant. In neurons taurine has been shown to prevent mitochondrial dysfunction and to protect against endoplasmic reticulum (ER) stress associated with neurological disorders. In cortical neurons in culture taurine protects against excitotoxicity through reversing an increase in levels of key ER signaling components including eIF-2-alpha and cleaved ATF6. The role of communication between the ER and mitochondrion is also important and examples are presented of protection by taurine against ER stress together with prevention of subsequent mitochondrial initiated apoptosis.

2.1 Introduction

Taurine, or 2-aminoethanesulfonic acid, is a sulfonic acid which is derived from cysteine and it is one of the few naturally occurring sulfonic acids. Taurine is widely distributed in animal tissues and one of the most abundant amino acid in mammals. Taurine plays several crucial roles including modulation of calcium signaling, osmoregulation, and membrane stabilization. However, despite extensive study, the mechanisms of action of taurine are not well understood. Based on past studies taurine has appeared as a promising agent for treating several neurological disorders including Alzheimer's disease, Huntington's disease, and stroke because of its ability to prevent apoptosis and its capacity to act as an antioxidant.

N. Kumari • H. Prentice (✉) • J.-Y. Wu (✉)
Department of Biomedical Science, Charles E. Schmidt College of Medicine,
Florida Atlantic University, 777 Glades Road, Boca Raton, FL 33431, USA
e-mail: neetazephyr@gmail.com; hprentic@fau.edu; jwu@fau.edu

A. El Idrissi and W.J. L'Amoreaux (eds.), *Taurine 8*, Advances in Experimental
Medicine and Biology 775, DOI 10.1007/978-1-4614-6130-2_2,
© Springer Science+Business Media New York 2013

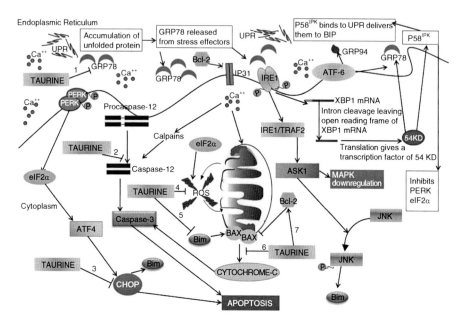

Fig. 2.1 Diagram showing the mode of action of taurine in alleviating the apoptosis induced by ER stress and mitochondrial dysfunction. The sites of action of taurine regulation are indicated a follows: (1) decreased Grp78, (2) decreased caspase 12, (3) decreased CHOP, (4) decreased ROS levels, (5) decreased Bim, (6) decreased cytochrome C release, and (7) increased Bcl-2

In this chapter, we will focus on the neuroprotective role of taurine. There has been extensive research demonstrating that taurine has a unique protective role and that it can downregulate several stress-associated proteins and increase neuronal survival under conditions of glutamate-induced cytotoxicity, mitochondrial stress, and endoplasmic reticulum (ER) stress (Fig. 2.1).

2.2 Taurine and Its Receptors

The taurine-synthesizing enzyme cysteine sulfinic acid decarboxylase (CSAD) in the brain was first identified and purified (Wu 1982) and then localized in the hippocampus (Taber et al. 1986), cerebellum (Chan-Palay et al. 1982b; Chan-Palay et al. 1982a), and the retina (Chan-Palay et al. 1982b; Chan-Palay et al. 1982a; Wu et al. 1985). Taurine fulfills most of the criteria as a neurotransmitter as the molecule is released from neurons in a calcium-dependent manner and binds to specific receptors postsynaptically (Lin et al. 1985a; Lin et al. 1985b; Wu and Prentice 2010; Wu et al. 1985). Taurine is of great interest as a potential neuroprotectant preventing excitotoxicity caused by glutamate which is a major excitatory neurotransmitter in the CNS. Part of the effect of taurine in neuroprotection involves preventing calcium entry into neurons through its action on L, P/Q, and N-type calcium channels and NMDA-R calcium channels (Wu et al. 2005). The mechanisms by which taurine modulates voltage-gated calcium channels may involve binding to GABA/glycinergic receptors resulting in hyperpolarization. Under conditions of excessive calcium entry taurine can downregulate this process. Taurine acts as an agonist of GABA and glycine receptors increasing the duration of chloride channel conductance (for review, Wu and Prentice 2010). Besides taurine binding to GABAergic or glycinergic receptors, we have previously reported that there are also taurine-specific receptors because blocking agents specific to GABA and glycine receptors were found to exert only a minimal effect on taurine receptor binding in neurons (Wu et al. 1992). Taurine is also important for regulating osmotic stress, a cellular response that is reduced though taurine's action on blocking sodium/calcium exchangers, K(ATP) channels, voltage-gated calcium channels, and fast acting sodium channels (Takatani et al. 2004a).

2.3 Role of Taurine in Mitochondrial Dysfunction

Mitochondria are highly sensitive to oxidative damage and there is much evidence of a cytoprotective role of taurine towards mitochondria in addition to other organelles within the cell. In mitochondria, taurine has been shown to be a key regulator of levels of superoxide production and of oxidative phosphorylation since taurine deficiency results in oxidative stress in mitochondria through respiratory chain impairment. Many taurine-conjugated products are functionally involved in energy metabolism and cholesterol metabolism (Schuller-Levis and Park 2003; Yokogoshi et al. 1999). One of the key products is 5-taurinomethyluridine-tRNA leu (UUR) which is involved in stabilization of U-G pairing in an anticodon loop transfer RNA (tRNA) responsible for efficient decoding of UUG (Kurata et al. 2008). Taurine deficiency lowers the taurinomethyluridine-tRNA leu encoded protein synthesis disabling efficient assembly of respiratory chain components (Schaffer et al. 2009).

Superoxide generation is the result of the diversion of electrons to acceptor oxygen from the respiratory chain (Jong et al. 2011a; Jong et al. 2011b). Mitochondrial DNA mutations can cause mitochondrial dysfunction which has been found to be the primary cause of several kinds of mitochondrial disease. Some key studies have demonstrated that these mutated mt tRNAs are the result of the absence of posttranscriptional taurine-dependent modifications leading to the molecular pathogenesis of mitochondrial myopathy, encephalopathy, lactic acidosis, and stroke-like episodes (MELAS) and a second disease, myoclonus epilepsy, associated with ragged red fibers (MERRF) (Suzuki et al. 2002). In an investigation of the protective role of taurine in a rat model of stroke it was shown that taurine can preserve mitochondrial function and prevent the cell death mediated by the mitochondrial pathway of apoptosis (Sun et al. 2011).

2.4 Role of Taurine in Endoplasmic Reticulum Stress

The accumulation of misfolded proteins leading to endoplasmic reticulum stress (ER stress) interferes with neuronal signaling and induces neuronal cell death. Under such conditions of neuronal stress the unfolded protein response (UPR) becomes activated to restore normal cellular function by activating three signaling systems: PERK (PKR-like endoplasmic reticulum kinase), IRE 1 (inositol-requiring enzyme 1), and ATF 6 (activation transcription factor 6) which then regulate protein synthesis during ER stress at both the transcriptional and posttranscriptional levels (Harding et al. 2003; Kaufman 1999). In addition to misfolding of proteins, calcium overload and oxidative stress all can lead to endoplasmic reticulum stress (Lai et al. 2007). The accumulation of unfolded proteins within the ER lumen leads to dissociation of GRP 78 (glucose-regulated protein 78) from PERK, IRE1, and ATF 6, respectively, which is the initiating step in activation of these signaling molecules and their downstream signaling counterparts (Bertolotti et al. 2000; Shen et al. 2002). The primary function of PERK, IRE1, and ATF 6 is to activate signaling events to overcome ER stress but under severe conditions of stress when the UPR system fails to restore correct protein folding and processing capacity PERK, IRE 1, and ATF 6 can also activate apoptosis, not directly, but by activating downstream pro-death components including CHOP, JNK, and caspases (Anand and Babu 2012; Higo et al. 2010; Sokka et al. 2007) (Fig. 2.1).

Taurine is effective in reducing ER stress in a number of neural systems including PC12 cell cultures, primary neuronal cultures, and human neuroblastoma cell lines. Taurine exerts its neuroprotective function in part by restoring the integrity of the structure and function of the ER. Treatment with taurine under conditions of oxidative stress, excitotoxicity, or hypoxic stress results in a decrease in levels of expression of a number of ER stress proteins including Grp 78, CHOP/GADD153, p-IRE and p-eI0046–2 alpha protein, and caspase-12, as well as a decreased ratio of cleaved ATF6 and full-length ATF6 and a decreased ratio of Bax/Bcl2 (Pan et al. 2011; Pan et al. 2010) (Fig. 2.1).

2.5 Role of Taurine in Apoptosis

Taurine has the ability to downregulate several of the molecules that trigger apoptosis. In a recent study on neuronal cell cultures taurine treatment resulted in decreased levels of the Bax to Bcl-2 ratio after exposure to glutamate (Leon et al. 2009). Taurine was able to prevent the decline in Bcl-2 expression in these cultures in the presence of glutamate (Leon et al. 2009). A major route by which taurine can regulate apoptosis is by decreasing intracellular free calcium through inhibiting different types of calcium channels (Leon et al. 2009) and also by increasing of Ca^{2+} levels in mitochondria (Taranukhin et al. 2010) (Fig. 2.1).

Furthermore beyond its reported antioxidant roles there is substantial evidence that taurine can act on apoptotic components and will help to restore the cellular

Bcl-2 pool. Previous studies have shown that taurine can prevent release of cyto-chrome c from mitochondria and also suppress the assembly of the Apaf 1/cas-pase-9 apoptosome complex preventing caspase-9 activation (Takatani et al. 2004b). Further studies investigating this pathway revealed that taurine regulates the interaction of Apaf 1 and caspase-9 through Akt (Takatani et al. 2004a).

In an investigation of morphine-induced toxicity in C6 glioma cells taurine reversed the depletion of Bcl-2 levels in conjunction with increasing the activities of superoxide dismutase, catalase, and glutathione peroxidase (Taranukhin et al. 2010). Thus taurine contributed to preventing the oxidative insult that resulted in morphine-induced apoptosis (Zhou et al. 2011).

2.6   Role of Taurine in Neurological Diseases

In stroke the loss of blood supply leads to ischemic stress which is characterized by an increase in intracellular free calcium, elevated reactive oxygen species, and the development of acidosis. Previous studies have shown the ability of taurine to maintain neuronal calcium homeostasis and to prevent neuronal cell death occurring through necrosis or apoptosis as well as through ER stress (Mantopoulos et al. 2011; Zhang et al. 2010). In stroke patients it is reported that plasma concentrations of taurine are increased (Ghandforoush-Sattari et al. 2011). Further studies are examining the extent to which taurine levels may be a biomarker for recovery in stroke. In experimental stroke administration of taurine can protect in a dose-dependent manner through mechanisms that include up-regulation of calpastatin and down-regulation of calpain and caspase-3 (Sun et al. 2009). In a recent analysis of the effects of taurine on inflammatory markers 22 h after a 2 h transient brain ischemia it was shown that both poly (ADP-ribose) polymerase (PARP)- and nuclear-factor-kappa-B (NF-kB)-driven expression of inflammatory mediators was suppressed by taurine (Sun et al. 2012). Specifically taurine administration resulted in decreased levels of tumor necrosis factor-alpha, inter-leukin-i-beta, inducible nitric oxide synthase, and intracellular adhesion molecule-1.

A rat model of Huntington's disease resulting from striatal lesions induced by the mitochondrial toxin 3-NP has been employed for an investigation of the protective effects of taurine. It was found that pretreatment with taurine significantly protected against the behavioral deficits in this model and increased locomotor activity (Tadros et al. 2005).

In spinal cord injury, the neutrophils that migrate to the site of injury have been shown to contain high taurine concentrations. Using a spinal cord compression model, treatment with taurine was shown to inhibit expression of the pro-inflammatory cytokine IL-6 and to decrease phosphorylation of STAT3 and expression of COX2. In the taurine-fed mice there was a reduced accumulation of neutrophils in addition to recovery of function of the mouse hind-limb (Nakajima et al. 2010).

In epilepsy an imbalance in amino acid content in epileptic foci is found which is characterized by low concentrations of glutamate and taurine and high levels of glycine (Guilarte 1989). Administration of taurine rectifies this imbalance of amino acids through alterations in membrane fluidity and activation of membrane enzymes and transporters including the sodium/calcium exchanger (Jong et al. 2011b). Several studies report that taurine can reverse epileptic symptoms (Junyent et al. 2011; Junyent et al. 2010) and in the KA experimental model increased taurine levels have been reported (Baran 2006). Using a KA mouse model of epilepsy it was recently demonstrated that administration of taurine 12 h before KA administration elicited a reduction or even a disappearance of cellular and molecular KA-derived effects.

Elevated calcium in neurons is strongly linked to seizure activity and a recent study addressed the effects of taurine treatment in mice on the expression of proteins in the hippocampus associated with calcium regulation. Taurine inhibited CaMKII activity in hippocampus which may be related to its neuroprotective effect. Other calcium-binding proteins including calbindin-D28k, calretinin, and parvalbumin were also increased in expression within the same time frame as the previously reported anticonvulsant effect of taurine (Junyent et al. 2010).

2.7 Communication Between the ER and Mitochondrion

Communication between the ER and mitochondrion may play an important role in regulating intracellular free calcium levels. In stressed conditions the ER can trigger signaling events that result in cytochrome C release from the mitochondrion. An integral protein of the ER membrane BAP 31 (B-cell-associated protein 31) is a caspase cleavage product which has been shown to induce mitochondrial fission through ER-derived calcium signals that enhance cytochrome C release (Rutter and Rizzuto 2000). Once released, cytochrome C can translocate to ER where it binds with IP3R and causes a sustained increase in cytosolic calcium levels (Boehning et al. 2003; Wang and El-Deiry 2004). ER stress resulting from the accumulation of unfolded proteins is associated with several neurological diseases including Alzheimer's disease, Parkinson's disease, and cerebral stroke. It is likely that in these disease conditions an increase in intracellular free calcium can lead to ER stress and that ER stress may also be responsible for triggering mitochondrial dysfunction and subsequent apoptosis. Therapeutic interventions controlling ER stress may therefore have potential for preventing apoptotic responses. In a study on the effect of taurine against transient focal cerebral ischemia one proposed protective mechanisms of taurine against ischemia was through blocking the mu-calpain and caspase-3-mediated apoptotic cell death pathways (Sun and Xu 2008). Taurine has been shown in a number of studies to diminish the damaging effects of ER stress. In a recent study on C. elegans exposed to the ER stress inducer tunicamycin it was found that taurine treatment was able to enhance longevity, mobility, and fecundity of the organism (Kim et al. 2010).

2.8   Conclusion

In summary, taurine exerts its neuroprotective function minimally through its action at both the mitochondrial and ER levels by decreasing the expression and/or the activity of Grp78, caspase-12, CHOP, ROS levels, Bim, and cytochrome C release and increasing the level of Bcl-2. The details of the mechanism at each step need further investigation.

References

Anand SS, Babu PP (2012) Endoplasmic reticulum stress and neurodegeneration in experimental cerebral malaria. Neurosignals. doi:10.1159/000336970

Baran H (2006) Alterations of taurine in the brain of chronic kainic acid epilepsy model. Amino Acids 31:303–307

Bertolotti A, Zhang Y, Hendershot LM et al (2000) Dynamic interaction of BiP and ER stress transducers in the unfolded-protein response. Nat Cell Biol 2:326–332

Boehning D, Patterson RL, Sedaghat L et al (2003) Cytochrome c binds to inositol (1,4,5) trisphosphate receptors, amplifying calcium-dependent apoptosis. Nat Cell Biol 5:1051–1061

Chan-Palay V, Lin CT, Palay S et al (1982a) Taurine in the mammalian cerebellum: demonstration by autoradiography with [3H]taurine and immunocytochemistry with antibodies against the taurine-synthesizing enzyme, cysteine-sulfinic acid decarboxylase. Proc Natl Acad Sci USA 79:2695–2699

Chan-Palay V, Palay SL, Wu JY (1982b) Sagittal cerebellar microbands of taurine neurons: immunocytochemical demonstration by using antibodies against the taurine-synthesizing enzyme cysteine sulfinic acid decarboxylase. Proc Natl Acad Sci USA 79:4221–4225

Ghandforoush-Sattari M, Mashayekhi SO, Nemati M, Ayromlou H (2011) Changes in plasma concentration of taurine in stroke. Neurosci Lett 496:172–175

Guilarte TR (1989) Regional changes in the concentrations of glutamate, glycine, taurine, and GABA in the vitamin B-6 deficient developing rat brain: association with neonatal seizures. Neurochem Res 14:889–897

Harding HP, Zhang Y, Zeng H et al (2003) An integrated stress response regulates amino acid metabolism and resistance to oxidative stress. Mol Cell 11:619–633

Higo T, Hamada K, Hisatsune C et al (2010) Mechanism of ER stress-induced brain damage by IP(3) receptor. Neuron 68:865–878

Jong CJ, Azuma J, Schaffer S (2011a) Mechanism underlying the antioxidant activity of taurine: prevention of mitochondrial oxidant production. Amino Acids 42:2223–2232

Jong CJ, Azuma J, Schaffer SW (2011b) Role of mitochondrial permeability transition in taurine deficiency-induced apoptosis. Exp Clin Cardiol 16:125–128

Junyent F, Porquet D, de Lemos L et al (2011) Decrease of calbindin-d28k, calretinin, and parvalbumin by taurine treatment does not induce a major susceptibility to kainic acid. J Neurosci Res 89:1043–1051

Junyent F, Romero R, de Lemos L et al (2010) Taurine treatment inhibits CaMKII activity and modulates the presence of calbindin D28k, calretinin, and parvalbumin in the brain. J Neurosci Res 88:136–142

Kaufman RJ (1999) Stress signaling from the lumen of the endoplasmic reticulum: coordination of gene transcriptional and translational controls. Genes Dev 13:1211–1233

Kim HM, Do C-H, Lee DH (2010) Taurine reduces ER stress in C. elegans. J Biomed Sci 17(Suppl 1): S1–S26

Kurata S, Weixlbaumer A, Ohtsuki T et al (2008) Modified uridines with C5-methylene substituents at the first position of the tRNA anticodon stabilize U.G wobble pairing during decoding. J Biol Chem 283:18801–18811

Lai E, Teodoro T, Volchuk A (2007) Endoplasmic reticulum stress: signaling the unfolded protein response. Physiology (Bethesda) 22:193–201

Leon R, Wu H, Jin Y et al (2009) Protective function of taurine in glutamate-induced apoptosis in cultured neurons. J Neurosci Res 87:1185–1194

Lin CT, Song GX, Wu JY (1985a) Is taurine a neurotransmitter in rabbit retina? Brain Res 337:293–298

Lin CT, Song GX, Wu JY (1985b) Ultrastructural demonstration of L-glutamate decarboxylase and cysteinesulfinic acid decarboxylase in rat retina by immunocytochemistry. Brain Res 331: 71–80

Mantopoulos D, Murakami Y, Comander J, Thanos A, Roh M, Miller JW, Vavvas DG (2011) Tauroursodeoxycholic acid (TUDCA) protects photoreceptors from cell death after experimental retinal detachment. PLoS One 6(9)

Nakajima Y, Osuka K, Seki Y et al (2010) Taurine reduces inflammatory responses after spinal cord injury. J Neurotrauma 27:403–410

Pan C, Giraldo GS, Prentice H, Wu J-Y (2010) Taurine protection of PC12 cells against endoplasmic reticulum stress induced by oxidative stress. J Biomed Sci 17(Suppl 1):S1–S17

Pan C, Prentice H, Price AL, Wu J-Y (2011) Beneficial effect of taurine on hypoxia- and glutamate-induced endoplasmic reticulum stress pathways in primary neuronal culture. Amino Acids 43:845–855. doi:10.1007/s00726-011-1141-6

Rutter GA, Rizzuto R (2000) Regulation of mitochondrial metabolism by ER Ca2+ release: an intimate connection. Trends Biochem Sci 25:215–221

Schaffer SW, Azuma J, Mozaffari M (2009) Role of antioxidant activity of taurine in diabetes. Can J Physiol Pharmacol 87:91–99

Schuller-Levis GB, Park E (2003) Taurine: new implications for an old amino acid. FEMS Microbiol Lett 226:195–202

Shen J, Chen X, Hendershot L, Prywes R (2002) ER stress regulation of ATF6 localization by dissociation of BiP/GRP78 binding and unmasking of Golgi localization signals. Dev Cell 3:99–111

Sokka A-L, Putkonen N, Mudo G et al (2007) Endoplasmic reticulum stress inhibition protects against excitotoxic neuronal injury in the rat brain. J Neurosci 27:901–908

Sun M, Gu Y, Zhao Y et al (2011) Protective functions of taurine against experimental stroke through depressing mitochondria-mediated cell death in rats. Amino Acids 40:1419–1429

Sun M, Xu C (2008) Neuroprotective mechanism of taurine due to up-regulating calpastatin and down-regulating calpain and caspase-3 during focal cerebral ischemia. Cell Mol Neurobiol 28(4):593–611

Sun M, Zhao Y, Gu Y, Xu C (2009) Inhibition of nNOS reduces ischemic cell death through down-regulating calpain and caspase-3 after experimental stroke. Neurochem Int 54(5–6): 339–346

Sun M, Zhao Y-M, Gu Y, Xu C (2012) Therapeutic window of taurine against experimental stroke in rats. Transl Res 160:223–229. doi:10.1016/j.trsl.2012.02.007

Suzuki T, Suzuki T, Wada T et al (2002) Taurine as a constituent of mitochondrial tRNAs: new insights into the functions of taurine and human mitochondrial diseases. EMBO J 21:6581–6589

Taber KH, Lin CT, Liu JW et al (1986) Taurine in hippocampus: localization and postsynaptic action. Brain Res 386:113–121

Tadros MG, Khalifa AE, Abdel-Naim AB, Arafa HMM (2005) Neuroprotective effect of taurine in 3-nitropropionic acid-induced experimental animal model of Huntington's disease phenotype. Pharmacol Biochem Behav 82:574–582

Takatani T, Takahashi K, Uozumi Y et al (2004a) Taurine prevents the ischemia-induced apoptosis in cultured neonatal rat cardiomyocytes through Akt/caspase-9 pathway. Biochem Biophys Res Commun 316:484–489. doi:10.1016/j.bbrc.2004.02.066

Takatani T, Takahashi K, Uozumi Y et al (2004b) Taurine inhibits apoptosis by preventing formation of the Apaf-1/caspase-9 apoptosome. Am J Physiol Cell Physiol 287:C949–C953

Taranukhin AG, Taranukhina EY, Saransaari P et al (2010) Neuroprotection by taurine in ethanol-induced apoptosis in the developing cerebellum. J Biomed Sci 17(Suppl 1):S1–S12

Wang S, El-Deiry WS (2004) Cytochrome c: a crosslink between the mitochondria and the endoplasmic reticulum in calcium-dependent apoptosis. Cancer Biol Ther 3:44–46

Wu H, Jin Y, Wei J et al (2005) Mode of action of taurine as a neuroprotector. Brain Res 1038: 123–131

Wu J-Y, Prentice H (2010) Role of taurine in the central nervous system. J Biomed Sci 17(Suppl 1):S1–S6

Wu JY (1982) Purification and characterization of cysteic acid and cysteine sulfinic acid decarboxylase and L-glutamate decarboxylase from bovine brain. Proc Natl Acad Sci USA 79:4270–4274

Wu JY, Lin CT, Thalmann R et al (1985) Immunocytochemical and physiological identification of taurine neurons in the mammalian CNS. Prog Clin Biol Res 179:261–270

Wu J-Y, Tang XW, Tsai WH (1992) Taurine receptor: kinetic analysis and pharmacological studies. Adv Exp Med Biol 315:263–268

Yokogoshi H, Mochizuki H, Nanami K et al (1999) Nutrient interactions and toxicity dietary taurine enhances cholesterol degradation and reduces serum and liver cholesterol concentrations in rats fed a high-cholesterol diet. J Nutr 129:1705–1712

Zhang B, Yang X, Gao X (2010) Taurine protects against bilirubin-induced neurotoxicity in vitro. Brain Res 1320:159–167

Zhou J, Li Y, Yan G et al (2011) Protective role of taurine against morphine-induced neurotoxicity in C6 cells via inhibition of oxidative stress. Neurotox Res 20:334–342

Chapter 3
Antidepressant-Like Effect of Chronic Taurine Administration and Its Hippocampal Signal Transduction in Rats

Atsushi Toyoda and Wataru Iio

Abstract Taurine is one of the most abundant amino acids in the central nervous system, and it has various important functions as a neuromodulator and antioxidant. Taurine is expected to be involved in the mental disorders such as depression; however, knowledge of its function in relation to depression is limited. In this research, we tried to elucidate the effects of taurine supplementation on antidepressant-like behaviors in rats and depression-related signal transduction in the hippocampus. In behavioral tests, rats fed a high taurine (HT: 45 mmol/kg taurine) diet for 4 weeks (HT4w) showed decreased immobility in the forced swim test (FS) compared to controls. On the other hand, rats fed a low taurine (LT: 22.5 mmol/kg taurine) diet for 4 weeks or an HT diet for 2 weeks (HT2w) did not show a significant difference in FS compared to controls. In western blot analyses, the expression of glutamic acid decarboxylase (GAD) 65 and GAD67 in the hippocampus was not affected by taurine supplementation. However, the phosphorylation levels of extracellular signal-regulated kinase1/2 (ERK1/2), protein kinase B (Akt), glycogen synthase kinase3 beta (GSK3β), and cAMP response element-binding protein (CREB) were increased in the hippocampus of HT4w and HT2w rats. Phosphorylated calcium/calmodulin-dependent protein kinase II (CaMKII) was increased in the hippocampus of HT4w rats only. Moreover, no significant changes in these molecules were observed in the hippocampus of rats fed an HT diet for 1 day. In conclusion, our discoveries suggest that taurine supplementation has an antidepressant-like effect and an ability to change depression-related signaling cascades in the hippocampus.

A. Toyoda (✉) • W. Iio
College of Agriculture, Ibaraki University, Ami, Ibaraki 300-0393, Japan

United Graduate School of Agricultural Science, Tokyo University of Agriculture and Technology, Fuchu-city, Tokyo 183-8509, Japan
e-mail: atoyoda@mx.ibaraki.ac.jp

A. El Idrissi and W.J. L'Amoreaux (eds.), *Taurine 8*, Advances in Experimental Medicine and Biology 775, DOI 10.1007/978-1-4614-6130-2_3,
© Springer Science+Business Media New York 2013

Abbreviations

HT	High taurine
LT	Low taurine
FS	Forced swim test
GAD	Glutamic acid decarboxylase
ERK1/2	Extracellular signal-regulated kinase1/2
Akt	Protein kinase B
GSK3β	Glycogen synthase kinase3 beta
CREB	cAMP response element-binding protein
CaMKII	Calcium/calmodulin-dependent protein kinase II

3.1 Introduction

Depression is one of the serious mental disorders and it affects approximately 20% of the population in the world (Berton and Nestler 2006). To treat depression and other mental disorders, antidepressants including selective serotonin reuptake inhibitors (SSRIs) have been widely used in many countries. Possibly, SSRIs mediate their antidepressant effects via serotonergic systems in the brain; however, the functions and effects of SSRIs on the brain are not largely elucidated. It has been reported that SSRIs have several aversive side effects including nausea and anorexia (Vaswani et al. 2003). Recently, Kobayashi et al. described that chronic treatment of fluoxetine, one of the major SSRIs, induces the immaturation of dentate gyrus matured neurons in mouse hippocampus. Dentate gyrus immaturity has been shown to be involved in abnormal physiological properties in the hippocampus and to psychiatric disorder-like behaviors (Kobayashi et al. 2010; Kobayashi et al. 2011). Therefore, antidepressants without any harmful side effects should be developed to improve quality of life for depressive patients. Possibly, diet is one candidate for the prevention and treatment of depressive disorders. Nutrients should be widely screened for antidepressant-like activity. However, the antidepressant-like effects of nutrients have not yet been clarified. The identification of such nutrients could raise the possibility of treating or preventing depression through dietary regimens.

Taurine is one of the most abundant amino acids in the brain (Hussy et al. 2000). Taurine transporter (TAUT) is expressed in various tissues including those of the blood–brain barrier (Tamai et al. 1995). Taurine is synthesized from cysteine and is known to have many important physiological functions such as membrane stabilization, osmoregulation, and neuroprotection (Hussy et al. 2000; Timbrell et al. 1995; Tanabe et al. 2010). Moreover, taurine acts as an agonist for glycine and gamma-aminobutyric acid (GABA) receptors (Albrecht and Schousboe 2005; del Olmo et al. 2000a). Additionally, taurine modulates intracellular calcium and calcium signaling molecules in the brain (Wu and Prentice 2010). Furthermore, taurine plays a key role in development, especially brain development. Mice whose mother was infected with human influenza virus during the gestational period suffer from brain atrophy and show a decreased concentration of taurine in the brain (Fatemi et al. 2008). Down syndrome patients have an abnormal concentration of taurine in the frontal cortex (Whittle et al. 2007).

Previous papers demonstrated that taurine concentration in the plasma of depressive patients is increased, while its concentration in the cerebrospinal fluid of schizophrenic patients is decreased (Altamura et al. 1995; Do et al. 1995). Perry et al. described that families that suffer from a hereditary taurine deficiency have a tendency to develop depression (Perry et al. 1975). These observations show that taurine is related to depression and other mental disorders. Furthermore, acute stress using FS induced to increase the plasma concentration of taurine in mice (Murakami et al. 2009). Taurine administration showed anxiolytic-like effects in mice and rats (Chen et al. 2004; Kong et al. 2006). Murakami et al. reported that ICR mice fed a taurine-supplemented diet for 4 weeks showed antidepressant-like behaviors (Murakami and Furuse 2010). On the other hand, Whirley et al. presented inconsistent results that daily intraperitoneal injections of taurine for short term in C57BL/6 mice had neither antidepressant-like nor anxiolytic-like effects (Whirley and Einat 2008). Thus, more precise researches about the relation between taurine and depression are needed.

Depression is developed by various molecular changes in the brain. For example, amino acids are involved in depression. Tryptophan and arginine were reported to have an antidepressant effect on FS (Inan et al. 2004; Wong and Ong 2001). Moreover, neurotrophic factors have been related to depression. Brain-derived neurotrophic factor (BDNF) has been implicated in the pathophysiology of stress-related mood disorders, and BDNF expression decreases after exposure to various stresses, including social defeat stress (Krishnan and Nestler 2008). Conversely, infusion of BDNF protein into the hippocampus has shown antidepressant effects (Shirayama et al. 2002). Murakami et al. reported that hippocampal BDNF expression is not influenced by chronic administration of taurine (Murakami and Furuse 2010); however, the intracellular downstream of BDNF cascades and other extracellular signals related to depression have not been characterized. Depression is related to various molecules in the brain including extracellular signal-regulated kinase1/2 (ERK1/2), protein kinase B (Akt), glycogen synthase kinase3 beta (GSK3β), calcium/calmodulin-dependent protein kinase II (CaMKII), and cAMP response element-binding protein (CREB) (Barbiero et al. 2007; Iio et al. 2011; Karege et al. 2007; Krishnan and Nestler 2008). These signaling molecules play pivotal roles in depression and other psychiatric disorders.

Tomida et al. reported that mice with a gene mutation in the ubiquitin-specific peptidase 46 (Usp46) show prolonged mobility in FS (Tomida et al. 2009). An association between the Usp46 gene and major depressive disorder was observed in a haplotype analysis of the Japanese population (Fukuo et al. 2011). Because the 67-kDa isoform of glutamic acid decarboxylase (GAD) 67 was decreased in the hippocampus of the Usp46 mutant mice, hippocampal GAD67 is thought to be involved in behavioral despair and antidepressant-like behaviors in FS.

In this study, we observed the effects of oral taurine supplementation on body weight, food intake, behavioral tests, and the expression and phosphorylation of depression-related proteins in the hippocampus.

3.2   Methods

3.2.1   Animals

Five-week-old male Wistar rats were obtained from Charles River (Yokohama, Japan) and housed individually at room temperature ($22 \pm 1 °C$), with lights on from 6:00 to 18:00 with ad libitum access to food and water. All experimental procedures followed the guidelines of the Animal Care and Use Committee of Ibaraki University.

3.2.2   Experimental Design and Drugs

Animals were divided into six groups: control, HT4w, HT2w, HT1d, LT4w, and LT2w. The control group was fed a normal powder diet (MF; Oriental Yeast, Tokyo, Japan). The HT4w group was fed a high taurine (HT) diet (45 mol taurine/kg diet) for 4 weeks. The HT2w and HT1d groups were fed an HT diet for 2 weeks and 1 day, respectively. The LT4w group was fed a low taurine (LT) diet (22.5 mmol taurine/kg diet) for 4 weeks, and the LT2w group was fed an LT diet for 2 weeks.

3.2.3   Behavioral Tests

All behavioral tests in this study were performed between 13:00 and 17:00 at a room temperature of $22 \pm 1 °C$ and a light intensity of 70 lx.

Open-field test (OF): The method for OF was described previously (Iio et al. 2011). Each subject was placed in the same corner of the open-field apparatus. The total distance traveled (in cm), time spent in the center area (in s), and average speed (in cm/s) during the 10-min session were recorded, and the results were analyzed on a Windows computer using Image J XX (O'Hara & Co., Ltd.), a modified software program based on the public domain Image J program.

Forced swim test (FS): The method for FS was described previously (Iio et al. 2011). Each rat was placed into an acrylic cylinder filled with water ($24 \pm 1 °C$) to a height of 18 cm. After 15 min, the animal was transferred to a 35°C environment for another 15 min (pretest). Twenty-four hours later, the subject was placed into the cylinder again for 5 min (test). Prior to each test, the cylinder was cleaned and filled with fresh water.

3.2.4 Body Weight and Food Intake

Body weight and food intake were measured at the end of the acclimation phase (baseline) and at weekly intervals during taurine administration. Body weight gain was calculated by subtracting baseline body weight from weight at the end of each week.

3.2.5 Protein Preparation and Western Blotting

After behavioral tests, all animals were subjected to biochemical analysis. Upon sacrifice, the rat's brains were rapidly removed and chilled on ice and the hippocampi were dissected out. The tissue was homogenized in ice-cold RIPA buffer (50 mM Tris–HCl pH 7.4, 150 mM NaCl, 1% NP-40, 0.75% sodium deoxycholate, 1 mM EDTA, 100 mM NaF, 2 mM Na_3VO_4, and a protease inhibitor mix (GE Healthcare)) with a Polytron homogenizer. The homogenate was centrifuged at $800 \times g$ for 15 min at 4°C and the supernatant was collected. Protein concentration was determined using the BCA method (Thermo). The method of western blotting followed the protocols of the ECL plus western blotting detection reagents (GE Healthcare), except for the incubation time of the primary antibody. We changed the incubation time of the primary antibody from 1 h to overnight. Detection was performed using the ECL plus western blotting detection reagents and LAS-3000 mini (FUJIFILM). The results of western blotting were quantitatively analyzed using Image J.

3.2.6 Statistical Analysis

Data were analyzed using Excel Toukei 2006 for Windows (Social Survey Research Information Co., Ltd. Tokyo, Japan). The western blotting data were analyzed using Student's t-tests. The behavioral tests were analyzed using one-way ANOVAs and Bonferroni post hoc analysis. Body weight gain and food intake were analyzed using two-way repeated measures ANOVAs.

3.3 Results

3.3.1 The Effects of Chronic Taurine Supplementation on Body Weight Gain and Food Intake

Chronic taurine supplementation did not affect body weight gain. Before taurine supplementation, the body weights of both HT4w and control rats were similar (175.0 ± 1.9 g vs. 171.0 ± 2.1 g, $P = 0.1780$). A two-way repeated measures ANOVA revealed that taurine supplementation (F (1, 72)=0.026, P=0.8748) and taurine supplementation×time interaction (F (4, 72)=0.042, P=0.9966) had no significant effect on body weight gain. Furthermore, taurine supplementation did not affect food intake. A two-way repeated measures ANOVA revealed that taurine supplementation (F (1, 72)=0.023, P=0.8805) and taurine supplementation×time interaction (F (4, 72)=1.283, P=0.2846) had no significant effect on food intake. HT4w rats were fed taurine at 462.8±12 mg/kg body weight/day during this study.

3.3.2 The Effects of Chronic Taurine Supplementation on Behavior

We observed a concentration-dependent effect of taurine supplementation on rat behavior. HT4w rats showed decreased duration of immobility in FS compared to controls (control, 199.9±7.8 s vs. HT, 143.2±15.9 s, P=0.0430). However, LT4w rats did not show any significant differences in FS compared to control or HT4w rats (control vs. LT4w, P=1.0000; LT4w vs. HT4w, P=0.2441). Furthermore, different taurine concentrations in the diet did not affect activity in any of these parameters of OF: total distance traveled (control vs. LT4w, P=1.0000; control vs. HT4w, P=0.4966; LT4w vs. HT4w, P=1.0000), time spent in the center area (control vs. LT4w, P=1.0000; control vs. HT4w, P=0.1728; LT4w vs. HT4w, P=0.2994), and average speed (control vs. LT4w, P=1.0000; control vs. HT4w, P=0.5172; LT4w vs. HT4w, P=0.9981). Then, we observed the effect of 2 weeks of taurine supplementation on rat behavior. Rats in all groups did not show any significant difference in immobility in FS (control vs. LT2w, P=1.0000; control vs. HT2w, P=1.0000; LT2w vs. HT2w, P=1.0000). Moreover, 2 weeks of taurine supplementation had no significant effect on these parameters of OF: total distance traveled (control vs. LT2w, P=0.6185; control vs. HT2w, P=1.0000; LT2w vs. HT2w, P=0.6946), time spent in the center area (control vs. LT2w, P=1.0000; control vs. HT2w, P=1.0000; LT2w vs. HT2w, P=1.0000), and average speed (control vs. LT2w, P=0.6416; control vs. HT2w, P=1.0000; LT2w vs. HT2w, P=0.6736).

3.3.3 The Effects of Chronic and Acute Taurine Supplementation on Hippocampal Protein Expression and Phosphorylation

We evaluated the effect of taurine supplementation on the expression of both GAD65 and GAD67 in the hippocampus. We defined HT1d, HT2w, and HT4w rats as acute-, subchronic-, and chronic-supplemented rats, respectively. In our results, taurine supplementation did not affect the expression of either GAD65 or GAD67 in the hippocampus (4 weeks GAD65, P=0.3506; GAD67, P=0.9759; 2 weeks GAD65, P=0.9434; GAD67, P=0.7155; 1 day GAD65, P=0.8896; GAD67, P=0.1817). Then we observed the effects of taurine supplementation on the expression and phosphorylation of significant depression-related molecules in the hippocampus (Table 3.1).

Table 3.1 Phosphorylation of signaling molecules in the hippocampus

	1 day	2 weeks	4 weeks
ERK1	⇔	⇑	⇑
ERK2	⇔	⇑	⇑
Akt	⇔	⇑	⇑
GSK3β	⇔	⇑	⇑
CaMKII	⇔	⇔	⇑
CREB	⇔	⇑	⇑

First, we checked the expression and phosphorylation of ERK1/2 and CREB. The mitogen-activated protein kinase (MAPK)-CREB cascade plays an important role in depression (Berton and Nestler 2006; Iio et al. 2011). In this study, the ratios of phospho-ERK1/ERK1, phospho-ERK2/ERK2, and phospho-CREB/CREB were increased in the hippocampus of HT4w rats compared to controls (phospho-ERK1/ERK1, $P=0.0032$; phospho-ERK2/ERK2, $P=0.0066$; phospho-CREB/CREB, $P=0.0008$). Similar results were obtained in the hippocampus of HT2w rats (phospho-ERK1/ERK1, $P=0.0343$; phospho-ERK2/ERK2, $P=0.0168$; phospho-CREB/CREB, $P=0.0002$). However, the MAPK-CREB cascade was not changed in the hippocampus of HT1d rats (phospho-ERK1/ERK1, $P=0.7448$; phospho-ERK2/ERK2, $P=0.9737$; phospho-CREB/CREB, $P=0.9938$). In the next step, we observed the expression and phosphorylation of Akt and GSK3β in the hippocampus. The phosphatidylinositol 3-kinase (PI3K)-Akt-CREB cascade in the hippocampus is also implicated in mood disorders (Karege et al. 2007). In our results, the ratios of phospho-Akt/Akt and phospho-GSK3β/GSK3β were increased in the hippocampus of HT4w rats compared to controls (phospho-Akt/Akt, $P=0.0010$; phospho-GSK3β/GSK3β, $P=0.0093$). Similar results were obtained in the hippocampus of HT2w rats (phospho-Akt/Akt, $P=0.0187$; phospho-GSK3β/GSK3β, $P=0.0123$). However, the PI3K-Akt cascade was not changed in the hippocampus of HT1d rats compared to controls (phospho-Akt/Akt, $P=0.5117$; phospho-GSK3β/GSK3β, $P=0.6296$). Moreover, we checked the expression and phosphorylation of CaMKII, which is affected by antidepressant treatment (Barbiero et al. 2007). In our results, the ratio of phospho-CaMKII/CaMKII was increased in the hippocampus of HT4w rats compared to controls (phospho-CaMKII/CaMKII, $P=0.0052$). However, the ratio of phospho-CaMKII/CaMKII was not changed in the hippocampus of HT2w or HT1d rats compared to controls (2 weeks phospho-CaMKII/CaMKII, $P=0.8850$; 1 day phospho-CaMKII/CaMKII, $P=0.6176$). The protein expression levels of the aforementioned molecules were not changed between control and taurine-supplemented rats.

3.4 Discussion

In this study, we investigated the function of taurine supplementation on behaviors and signal transduction in the rat hippocampus. Especially, we focused on characterizing the antidepressant-like effects of taurine using behavioral and biochemical approaches.

A previous report showed that taurine has an antidepressant-like effect because mice fed a taurine-supplemented diet for 4 weeks displayed decreased immobility in FS (Murakami and Furuse 2010). However, another report indicated that mice given short-term taurine injections did not show antidepressant-like behaviors (Whirley and Einat 2008). These studies were carried out using different experimental methods; therefore, the inconsistency of the results was dependent on differences in experimental animal strains, methods, and durations of taurine supplementation, among other factors. In our study, we tried to confirm an antidepressant-like effect of taurine using our original experimental methods using Wistar rats that were reared individually. Murakami et al. described that mice were reared in pairs to observe an antidepressant-like effect of taurine because housing mice individually decreased antidepressant sensitivity in FS (Murakami and Furuse 2010). However, in our study, we observed an antidepressant-like effect of a taurine-containing diet (45 mmol taurine/kg diet for 4 weeks) in FS using Wistar rats, although we adjusted the taurine intake per body weight of Wistar rats to the mice study by Murakami et al. (Murakami and Furuse 2010). Therefore, the duration of taurine supplementation and taurine concentration in the diet may be essential for revealing an antidepressant-like effect of taurine in FS. We observed the effects of chronic taurine intake on body weight gain and food intake in rats for the following reasons. If rats dislike eating taurine-containing diets and their food intake is decreased, the total calories that rats take from foodstuff are decreased. Previous reports described that calorie-restricted mice show antidepressant-like behaviors (Lutter et al. 2008) and that this antidepressant-like effect of caloric restriction was dependent on neuropeptides related to feeding behaviors, such as orexin and ghrelin (Berton and Nestler 2006; Lutter et al. 2008). The taurine-containing diet (45 mmol taurine/kg diet) did not affect body weight gain or food intake in rats; thus, the results indicate that our behavioral and biochemical studies using taurine-fed rats are not influenced by the antidepressant-like effects of caloric restriction.

Next, we observed the effects of chronic taurine supplementation on rat behavior. The taurine-supplemented diet did not influence behaviors in OF. HT4w rats showed antidepressant-like behavior in FS, while HT2w rats did not. These results indicate that the antidepressant-like effects of taurine are not revealed after two successive weeks of supplementation but that four successive weeks of supplementation are needed to reveal an antidepressant-like effect of taurine. Moreover, because LT4w rats did not show antidepressant-like behavior in FS, four successive weeks of taurine supplementation are insufficient to reveal antidepressant-like activity in rats. An HT diet may be necessary to obtain antidepressant-like effects of taurine in rats because the HT diet contained taurine at 45 mmol taurine/kg diet. Taurine supplementation is known to increase exercise performance. Rats fed with taurine at 100 mg taurine/kg body weight/day for 2 weeks increased exercise performance on a treadmill (Miyazaki et al. 2004; Yatabe et al. 2003). In our study, the rats fed with taurine at approximately 400–500 mg taurine/kg body weight/day for 4 weeks did not change their locomotor activity or average speed in OF. Therefore, OF was not influenced by the ability of taurine to improve physical endurance. Our FS data were also not influenced by this ability of taurine because taurine improves physical endurance with a dose less than that of a taurine-supplemented diet (Miyazaki et al. 2004; Yatabe et al. 2003). Therefore, we think that chronic supplementation of taurine shows antidepressant-like effects in rats. The dose of taurine and duration of its supplementation are critical to reveal the antidepressant-like effects of taurine. Similarly, chronic exposure to antidepressants is required for clinical utility in depressive patients, while acute exposure is insufficient for recovery from depression. The action mechanisms of antidepressants are not yet completely understood. A previous report described that chronic imipramine (a tricyclic antidepressant) modifies the chromatin structure of the BDNF gene promoter, enhances BDNF expression in the hippocampus, and reverses depression-like behaviors in socially defeated mice (Tsankova et al. 2006).

Because TAUT is highly expressed in hippocampal CA3, taurine is implicated in significant functions in the hippocampus (Sergeeva et al. 2003). The antidepressant actions of chronic taurine may be based on long-term modifications in neurotransmitters and/or signal transduction in the hippocampus. Due to these estimations, we focused on signal transduction and enzymes related to depression in the hippocampus.

We observed the effects of chronic taurine supplementation on the GABAergic system in the hippocampus. Taurine acts as a neuromodulator and functions as an agonist for glycine and GABA receptors (Albrecht and Schousboe 2005; del Olmo et al. 2000a). Tomida et al. reported that Usp46 mutant mice show negligible immobility in FS and the tail suspension test. These anti-immobile behaviors in Usp46 mutant mice were dependent on the decreased expression of GAD67 in the hippocampus (Tomida et al. 2009). Furthermore, the Usp46 gene is implicated in major depression (Fukuo et al. 2011). Because immobility in FS has been linked to the regulation of GABA action in the hippocampus, we observed the expression of both GAD65 and GAD67 in the hippocampus of rats that were chronically administered taurine. Chronic taurine supplementation did not affect the expression of either GAD65 or GAD67 in the hippocampus. Thus, the antidepressant-like effects of taurine are not based on the downregulation of GAD65 and GAD67 in the hippocampus.

Chronic taurine supplementation affected the phosphorylation of several key molecules related to depression in the hippocampus. In both HT4w and HT2w rats, increased phosphorylation of ERK1/2, Akt (Thr-308), GSK3β (Ser-9), and CREB (Ser-133) was observed in the hippocampus compared to control rats. Moreover, phosphorylation of CaMKII (Thr-286) was increased in HT4w rats. However, no significant change in the phosphorylation of these molecules was observed in HT1d rats. Acute oral supplementation of taurine could not induce profound effects on signal transduction in the hippocampus, but the long-term administration of taurine induced several profound changes in depression-related signal transduction. CREB, which is one of the molecules downstream of MAPK, serotonin (5-HT), and BDNF, plays a pivotal role in depression (Tsankova et al. 2006). Phospho-ERK1/2 and phospho-CREB were increased in the hippocampus after 2 weeks of taurine administration. A previous report showed that hippocampal ERK1/2 and CREB are activated by treatment with an antidepressant (Qi et al. 2006). Also, the expression of

BDNF in the hippocampus is upregulated with antidepressant treatment, as described above (Tsankova et al. 2006). Furthermore, acute infusion of BDNF in the hippocampus induced antidepressant effects in a behavioral model of depression (Shirayama et al. 2002). BDNF also facilitates the PI3K-Akt cascade in the hippocampus (Zheng and Quirion 2004). Akt is activated by PI3K with its phosphorylation of Thr-308 and Ser-473, and then phospho-Akt phosphorylates the Ser-133 of CREB and the Ser-9 of GSK3β (Du and Montminy 1998). Phospho-CREB (Ser-133) is an active transcriptional form. Phospho-GSK3β (Ser-9) is an inactive form of the kinase, and dephosphorylated GSK3β phosphorylates the Ser-129 of CREB. Finally, phospho-CREB (Ser-129) has an attenuated DNA-binding activity, and its transcriptional activity is decreased (Grimes and Jope 2001). In our observations, both phospho-Akt (Thr-308) and phospho-GSK3β (Ser-9) were increased in the hippocampus of rats administered taurine for 2 weeks. Phospho-Akt (Ser-473) was also increased in the hippocampus with 2 weeks of taurine administration (data not shown). However, these molecules might not play an essential role in the antidepressant-like effect of taurine, because the phosphorylation of Akt, GSK3β, and CREB with taurine administration was observed in HT2w rats, which did not show any significant change in antidepressant-like behavior compared to controls. Moreover, Murakami et al. reported that the expression of the BDNF protein in the hippocampus of mice fed a taurine-containing diet for 4 weeks was not altered (Murakami and Furuse 2010). Taurine could possibly activate MAPK and PI3K cascades via a BDNF-independent pathway in the hippocampus, although these cascades may be not necessary to reveal an antidepressant-like effect of taurine.

CaMKII is an abundant serine/threonine protein kinase in the brain. The kinase is activated by the binding of the calcium/calmodulin complex that generates calcium-dependent enzymatic activity. CaMKII plays pivotal roles in synaptic plasticity, the process underlying learning and memory in the hippocampus (Silva et al. 1992a; Silva et al. 1992b). In this study, we found that phospho-CaMKII was increased only in HT4w rats that revealed an antidepressant-like behavior in FS, while it was not observed in other groups of rats fed a taurine-supplemented diet. Therefore, the increase in phospho-CaMKII in the hippocampus may be critical to the antidepressant-like effect of taurine. CaMKII is also implicated in the pathophysiology and pharmacology of psychiatric disorders, because postmortem brain studies of patients with bipolar or unipolar depression indicate significantly reduced CaMKII mRNA levels in certain brain regions (Xing et al. 2002). α-CaMKII-deficient mice exhibit abnormal behaviors resembling schizophrenia and other human psychiatric disorders (Yamasaki et al. 2008). Thus, CaMKII has been found to be one of the target molecules for antidepressants. Chronic treatment with antidepressants increased CaMKII activity in the hippocampus, but acute treatments did not induce any change in the kinase (Barbiero et al. 2007; Tiraboschi et al. 2004). Also, chronic treatment with antidepressants increased the phosphorylation of CaMKII (Thr286) in neuronal cell bodies in the hippocampus (Tiraboschi et al. 2004). However, chronic treatment with antidepressants downregulated the phosphorylation of CaMKII (Thr286) in synaptic terminals and synaptic membranes in the hippocampus. The decrease in CaMKII phosphorylation reduced its interaction with syntaxin-1, thereby changing protein–protein interactions at glutamatergic pre-synaptic terminals and reducing depolarization-evoked glutamate release (Barbiero et al. 2007; Bonanno et al. 2005). However, a previous report indicated that short-term treatment with taurine inhibited CaMKII activity in the hippocampus (Junyent et al. 2010). Therefore, there may be some different physiological mechanisms between the short- and long-term applications of taurine. Recently, Han et al. reported that chronic treatment with nefiracetam, a prototype cognitive enhancer, significantly improved depression-like behaviors in olfactory bulbectomized (OBX) mice, one of the popular animal models of depression. The improvement of depression-like behaviors was associated with activation of CaM kinases, including CaMKII, in the hippocampus, amygdala, and prefrontal cortex. In addition to CaMKII auto-phosphorylation, CaMKI and CaMKIV may be required to counter-act depressive behaviors through CREB phosphorylation in OBX mice (Han et al. 2009). Furthermore, CaMKIV knockout mice and calcineurin knockout mice showed symptoms like mood disorders (Miyakawa et al. 2003; Takao et al. 2010). Taurine plays a crucial role in cellular calcium homeostasis (Junyent et al. 2010), and chronic taurine administration may have antidepressant-like activities due to activation of CaMKII in the hippocampus.

We found that chronic taurine administration has a strong ability to induce modifications in various signaling cascades in the hippocampus, as described above, although the precise mechanisms remain unclear. Because taurine acts as an agonist for glycine and GABA receptors, neuronal activities in the hippocampus and in other brain regions may be changed by oral taurine administration (del Olmo et al. 2000a). Taurine also induces a long-lasting potentiation of excitatory synap-tic potentials in hippocampal slices, which is related to the intracellular accumula-tion of taurine (del Olmo et al. 2000b; del Olmo et al. 2003; Galarreta et al. 1996). Taurine-induced synaptic potentiation requires calcium influx and shares some common mechanisms with tetanus-induced long-term potentiation (del Olmo et al. 2000b). Taurine potentiates presynaptic NMDA receptors in hippocampal Schaffer collateral axons (Suárez and Solís 2006). Chronic taurine administration may induce the intracellular accumulation of taurine in hippocampal neurons and mod-ify intracellular calcium concentration and activate CaMKII. Because chronic, but not acute, taurine application was needed to facilitate the phosphorylation of ERK1/2, Akt, GSK3β, CREB, and CaMKII in the hippocampus, intracellular and/ or extracellular accumulation of taurine in the brain may be essential to reveal antidepressant-like actions. Additionally, some reports have indicated that taurine acts as a trophic factor in neuronal tissues and nonneuronal tissues (Hernández-Benítez et al. 2010; Jeon et al. 2007; Lima and Cubillos 1998). However, the pre-cise function of taurine as a trophic factor in the brain remains to be elucidated. In future studies, we need to investigate the relationship between taurine supplemen-tation and the expression of neurotrophic factors in the brain, such as nerve growth factor, BDNF, and neurotrophin 3, and especially we should focus on the transcrip-tional regulation and epigenetics of these genes. Moreover, there is a possibility

Table 3.2 Effects of HT on behaviors and hippocampal signal transduction

	1 day	2 weeks	4 weeks
Body weight gain	NT	NT	ND
Food intake	NT	NT	ND
Immobility of FS	NT	ND	Down
GAD 65 and 67	ND	ND	ND
MAPK cascade	ND	Up	Up
IP3K-Akt cascade	ND	Up	Up
CaMKII	ND	ND	Up

NT not tested, ND no difference

that intracellular taurine uptaken by the taurine transporter modifies signal transduction in the hippocampus. Oral taurine administration induces an increase in taurine concentration in various tissues including cerebral cortex and hypothalamus (Miyazaki et al. 2004; Murakami and Furuse 2010). However, hippocampal taurine concentration of taurine-fed rats has not been analyzed in this study. And oral taurine administration did not affect the concentration of serotonin in cerebral cortex and hypothalamus, but the serotonin synthesis and release in the hippocampus was not elucidated (Murakami and Furuse 2010). Thus, the effects of oral taurine administration on hippocampal taurine metabolism and serotonergic system should be investigated in the future studies. Intracellular taurine accumulation may be essential for inducing the modification of signal transduction in the hippocampus, although the precise mechanism should be investigated.

3.5 Conclusion

We found that chronic taurine supplementation at 45 mmol taurine/kg diet for 4 weeks induces antidepressant-like effects in rats (Table 3.2). The beneficial effects of chronic taurine supplementation in rats might be mediated by phosphorylation of CaMKII in the hippocampus. In addition to CaMKII phosphorylation, the increase in hippocampal phospho-ERK1/2, phospho-Akt, phospho-GSK3β, and phospho-CREB may be required to reveal the antidepressant-like activities of taurine. Although taurine has various physiological effects on human health, its antidepressant action may be useful for combating depression. Further clinical study is required to evaluate the application of these findings for human health.

Acknowledgments We would like to thank Dr. Hiroko Toyoda (Ibaraki University) for helpful comments regarding the manuscript. Also we thank Dr. Mitsuhiro Furuse and Dr. Shozo Tomonaga (Kyushu University) for fruitful discussion of this study. This research was supported in part by Grants-in-Aid for Scientific Research (C) from the Ministry of Education, Culture, Sports, Science and Technology of Japan.

References

Albrecht J, Schousboe A (2005) Taurine interaction with neurotransmitter receptors in the CNS: an update. Neurochem Res 30:1615–1621

Altamura C, Maes M, Dai J, Meltzer HY (1995) Plasma concentrations of excitatory amino acids, serine, glycine, taurine and histidine in major depression. Eur Neuropsychopharmacol 5(Suppl):71–75

Barbiero VS, Giambelli R, Musazzi L, Tiraboschi E, Tardito D, Perez J, Drago F, Racagni G, Popoli M (2007) Chronic antidepressants induce redistribution and differential activation of alphaCaM kinase II between presynaptic compartments. Neuropsychopharmacology 32:2511–2519

Berton O, Nestler EJ (2006) New approaches to antidepressant drug discovery: beyond monoamines. Nat Rev Neurosci 7:137–151

Bonanno G, Giambelli R, Raiteri L, Tiraboschi E, Zappettini S, Musazzi L, Raiteri M, Racagni G, Popoli M (2005) Chronic antidepressants reduce depolarization-evoked glutamate release and protein interactions favoring formation of SNARE complex in hippocampus. J Neurosci 25: 3270–3279

Chen SW, Kong WX, Zhang YJ, Li YL, Mi XJ, Mu XS (2004) Possible anxiolytic effects of taurine in the mouse elevated plus-maze. Life Sci 75:1503–1511

del Olmo N, Bustamante J, del Río RM, Solís JM (2000a) Taurine activates GABA(A) but not GABA(B) receptors in rat hippocampal CA1 area. Brain Res 864:298–307

del Olmo N, Galarreta M, Bustamante J, Martín del Rio R, Solís JM (2000b) Taurine-induced synaptic potentiation: role of calcium and interaction with LTP. Neuropharmacology 39: 40–54

del Olmo N, Handler A, Alvarez L, Bustamante J, Martín del Río R, Solís JM (2003) Taurine-induced synaptic potentiation and the late phase of long-term potentiation are related mechanistically. Neuropharmocology 44:26–39

Do KQ, Lauer CJ, Schreiber W, Zollinger M, Gutteck-Amsler U, Cuénod M, Holsboer F (1995) Gamma-Glutamylglutamine and taurine concentrations are decreased in the cerebrospinal fluid of drug-naive patients with schizophrenic disorders. J Neurochem 65:2652–2662

Du K, Montminy M (1998) CREB is a regulatory target for the protein kinase Akt/PKB. J Biol Chem 273:32377–32379

Fatemi SH, Reutiman TJ, Folsom TD, Huang H, Oishi K, Mori S, Smee DF, Pearce DA, Winter C, Sohr R, Maternal JG (2008) Maternal infection leads to abnormal gene regulation and brain atrophy in mouse offspring: implications for genesis of neurodevelopmental disorders. Schizophr Res 99:56–70

Fukuo Y, Kishi T, Kushima I, Yoshimura R, Okochi T, Kitajima T, Matsunaga S, Kawashima K, Umene-Nakano W, Naitoh H, Inada T, Nakamura J, Ozaki N, Iwata N (2011) Possible association between ubiquitin-specific peptidase 46 gene and major depressive disorders in the Japanese population. J Affect Disord 133:150–157

Galarreta M, Bustamante J, Martin del Río R, Solís JM (1996) Taurine induces a long-lasting increase of synaptic efficacy and axon excitability in the hippocampus. J Neurosci 16:92–102

Grimes CA, Jope RS (2001) CREB DNA binding activity is inhibited by glycogen synthase kinase-3 beta and facilitated by lithium. J Neurochem 78:1219–1232

Han F, Nakano T, Yamamoto Y, Shioda N, Lu YM, Fukunaga K (2009) Improvement of depressive behaviors by nefiracetam is associated with activation of CaM kinases in olfactory bulbectomized mice. Brain Res 1265:205–214

Hernández-Benítez R, Pasantes-Morales H, Saldaña IT, Ramos-Mandujano G (2010) Taurine stimulates proliferation of mice embryonic cultured neural progenitor cells. J Neurosci Res 88:1673–1681

Hussy N, Deleuze C, Desarménien MG, Moos FC (2000) Osmotic regulation of neuronal activity: a new role for taurine and glial cells in a hypothalamic neuroendocrine structure. Prog Neurobiol 62:113–134

Iio W, Matsukawa N, Tsukahara T, Kohari D, Toyoda A (2011) Effects of chronic social defeat stress on MAP kinase cascade. Neurosci Lett 504:281–284

Inan SY, Yalcin I, Aksu F (2004) Dual effects of nitric oxide in the mouse forced swimming test: possible contribution of nitric oxide-mediated serotonin release and potassium channel modulation. Pharmacol Biochem Behav 77:457–464

Jeon SH, Lee MY, Kim SJ, Joe SG, Kim GB, Kim IS, Kim NS, Hong CU, Kim SZ, Kim JS, Kang HS (2007) Taurine increases cell proliferation and generates an increase in [Mg2+]i accompanied by ERK 1/2 activation in human osteoblast cells. FEBS Lett 581:5929–5934

Junyent F, Romero R, de Lemos L, Utrera J, Camins A, Pallàs M, Auladell C (2010) Taurine treatment inhibits CaMKII activity and modulates the presence of calbindin D28k, calretinin, and parvalbumin in the brain. J Neurosci Res 88:136–142

Karege F, Perroud N, Burkhardt S, Schwald M, Ballmann E, La Harpe R, Malafosse A (2007) Alteration in kinase activity but not in protein levels of protein kinase B and glycogen synthase kinase-3beta in ventral prefrontal cortex of depressed suicide victims. Biol Psychiatry 61:240–245

Kobayashi K, Ikeda Y, Sakai A, Yamasaki N, Haneda E, Miyakawa T, Suzuki H (2010) Reversal of hippocampal neuronal maturation by serotonergic antidepressants. Proc Natl Acad Sci USA 107:8434–8439

Kobayashi K, Ikeda Y, Suzuki H (2011) Behavioral destabilization induced by the selective serotonin reuptake inhibitor fluoxetine. Mol Brain 4:12

Kong WX, Chen SW, Li YL, Zhang YJ, Wang R, Min L, Mi X (2006) Effects of taurine on rat behaviors in three anxiety models. Pharmacol Biochem Behav 83:271–276

Krishnan V, Nestler EJ (2008) The molecular neurobiology of depression. Nature 455:894–902

Lima L, Cubillos S (1998) Taurine might be acting as a trophic factor in the retina by modulating phosphorylation of cellular proteins. J Neurosci Res 53:377–384

Lutter M, Krishnan V, Russo SJ, Jung S, McClung CA, Nestler EJ (2008) Orexin signaling mediates the antidepressant-like effect of calorie restriction. J Neurosci 28:3071–3075

Miyakawa T, Leiter LM, Gerber DJ, Gainetdinov RR, Sotnikova TD, Zeng H, Caron MG, Tonegawa S (2003) Conditional calcineurin knockout mice exhibit multiple abnormal behaviors related to schizophrenia. Proc Natl Acad Sci USA 100:8987–8992

Miyazaki T, Matsuzaki Y, Ikegami T, Miyakawa S, Doy M, Tanaka N, Bouscarel B (2004) Optimal and effective oral dose of taurine to prolong exercise performance in rat. Amino Acids 27:291–298

Murakami T, Furuse M (2010) The impact of taurine- and beta-alanine-supplemented diets on behavioral and neurochemical parameters in mice: antidepressant versus anxiolytic-like effects. Amino Acids 39:427–434

Murakami T, Yamane H, Tomonaga S, Furuse M (2009) Forced swimming and imipramine modify plasma and brain amino acid concentrations in mice. Eur J Pharmacol 602:73–77

Perry TL, Bratty PJ, Hansen S, Kennedy J, Urquhart N, Dolman CL (1975) Hereditary mental depression and Parkinsonism with taurine deficiency. Arch Neurol 32:108–113

Qi X, Lin W, Li J, Pan Y, Wang W (2006) The depressive-like behaviors are correlated with decreased phosphorylation of mitogen-activated protein kinases in rat brain following chronic forced swim stress. Behav Brain Res 175:233–240

Sergeeva OA, Chepkova AN, Doreulee N, Eriksson KS, Poelchen W, Mönnighoff I, Heller-Stilb B, Warskulat U, Häussinger D, Haas HL (2003) Taurine-induced long-lasting enhancement of synaptic transmission in mice: role of transporters. J Physiol 550:911–919

Shirayama Y, Chen AC, Nakagawa S, Russell DS, Duman RS (2002) Brain-derived neurotrophic factor produces antidepressant effects in behavioral models of depression. J Neurosci 22:3251–3261

Silva AJ, Paylor R, Wehner JM, Tonegawa S (1992a) Impaired spatial learning in alpha-calcium-calmodulin kinase II mutant mice. Science 257:206–211

Silva AJ, Stevens CF, Tonegawa S, Wang Y (1992b) Deficient hippocampal long-term potentiation in alpha-calcium-calmodulin kinase II mutant mice. Science 257:201–206

Suárez LM, Solís JM (2006) Taurine potentiates presynaptic NMDA receptors in hippocampal Schaffer collateral axons. Eur J Neurosci 24:405–418

Takao K, Tanda K, Nakamura K, Kasahara J, Nakao K, Katsuki M, Nakanishi K, Yamasaki N, Toyama K, Adachi M, Umeda M, Araki T, Fukunaga K, Kondo H, Sakagami H, Miyakawa T (2010) Comprehensive behavioral analysis of calcium/calmodulin-dependent protein kinase IV knockout mice. PLoS One 5:e9460

Tamai I, Senmaru M, Terasaki T, Tsuji A (1995) Na(+)- and Cl(-)-dependent transport of taurine at the blood-brain barrier. Biochem Pharmacol 50:1783–1793

Tanabe M, Nitta A, Ono H (2010) Neuroprotection via strychnine-sensitive glycine receptors during post-ischemic recovery of excitatory synaptic transmission in the hippocampus. J Pharmacol Sci 113:378–386

Timbrell JA, Seabra V, Waterfield CJ (1995) The in vivo and in vitro protective properties of taurine. Gen Pharmacol 26:453–462

Tiraboschi E, Giambelli R, D'Urso G, Galietta A, Barbon A, de Bartolomeis A, Gennarelli M, Barlati S, Racagni G, Popoli M (2004) Antidepressants activate CaMKII in neuron cell body by Thr286 phosphorylation. Neuroreport 15:2393–2396

Tomida S, Mamiya T, Sakamaki H, Miura M, Aosaki T, Masuda M, Niwa M, Kameyama T, Kobayashi J, Iwaki Y, Imai S, Ishikawa A, Abe K, Yoshimura T, Nabeshima T, Ebihara S (2009) Usp46 is a quantitative trait gene regulating mouse immobile behavior in the tail suspension and forced swimming tests. Nat Genet 41:688–695

Tsankova NM, Berton O, Renthal W, Kumar A, Neve RL, Nestler EJ (2006) Sustained hippocampal chromatin regulation in a mouse model of depression and antidepressant action. Nat Neurosci 9:519–525

Vaswani M, Linda FK, Ramesh S (2003) Role of selective serotonin reuptake inhibitors in psychiatric disorders: a comprehensive review. Prog Neuropsychopharmacol Biol Psychiatry 27:85–102

Whirley BK, Einat H (2008) Taurine trials in animal models offer no support for anxiolytic, antidepressant or stimulant effects. Isr J Psychiatry Relat Sci 45:11–18

Whittle N, Sartori SB, Dierssen M, Lubec G, Singewald N (2007) Fetal Down syndrome brains exhibit aberrant levels of neurotransmitters critical for normal brain development. Pediatrics 120:e1465–e1471

Wong PT, Ong YP (2001) Acute antidepressant-like and antianxiety-like effects of tryptophan in mice. Pharmacology 62:151–156

Wu JY, Prentice H (2010) Role of taurine in the central nervous system. J Biomed Sci 17(Suppl 1):S1

Xing G, Russell S, Hough C, O'Grady J, Zhang L, Yang S, Zhang LX, Post R (2002) Decreased prefrontal CaMKII alpha mRNA in bipolar illness. Neuroreport 13:501–505

Yamasaki N, Maekawa M, Kobayashi K, Kajii Y, Maeda J, Soma M, Takao K, Tanda K, Ohira K, Toyama K, Kanzaki K, Fukunaga K, Sudo Y, Ichinose H, Ikeda M, Iwata N, Ozaki N, Suzuki H, Higuchi M, Suhara T, Yuasa S, Miyakawa T (2008) Alpha-CaMKII deficiency causes immature dentate gyrus, a novel candidate endophenotype of psychiatric disorders. Mol Brain 1:6

Yatabe Y, Miyakawa S, Miyazaki T, Matsuzaki Y, Ochiai N (2003) Effects of taurine administration in rat skeletal muscles on exercise. J Orthop Sci 8:415–419

Zheng WH, Quirion R (2004) Comparative signaling pathways of insulin-like growth factor-1 and brain-derived neurotrophic factor in hippocampal neurons and the role of the PI3 kinase pathway in cell survival. J Neurochem 89:844–852

Chapter 4
Direct Interaction of Taurine with the NMDA Glutamate Receptor Subtype via Multiple Mechanisms

Christopher Y. Chan, Herless S. Sun, Sanket M. Shah, Mervan S. Agovic, Ivana Ho, Eitan Friedman, and Shailesh P. Banerjee

Abstract Taurine has neuroprotective capabilities against glutamate-induced excitotoxicity through several identified mechanisms including opening of the Cl^- channel associated with $GABA_A$ and glycine receptors, or a distinct Cl^- channel. No existing work has however shown a direct interaction of taurine with the glutamate NMDA receptor. Here we demonstrate such direct interactions using electrophysiological and receptor binding techniques on rat medial prefrontal cortical (mPFC) slices and well-washed rat cortical membrane. Electrically evoked field potential responses were recorded in layer 4/5 of mPFC in the presence of picrotoxin to prevent opening of Cl^- channels gated by GABA or taurine. Applied taurine markedly diminished evoked-response amplitude at the peak and latter phases of the response. These phases were predominantly

C.Y. Chan (✉) • E. Friedman • S.P. Banerjee
Department of Physiology, Pharmacology & Neuroscience, Sophie Davis School
of Biomedical Education at CCNY, CUNY, Harris Hall Rm. 203,
160 Convent Ave, New York, NY 10031, USA

Neuroscience Subprogram, Doctoral Programs in Biology,
Graduate Center of CUNY, New York, NY, USA
e-mail: cyc@med.cuny.edu

H.S. Sun • S.M. Shah • I. Ho
Department of Physiology, Pharmacology & Neuroscience, Sophie Davis School
of Biomedical Education at CCNY, CUNY, Harris Hall Rm. 203,
160 Convent Ave, New York, NY 10031, USA

M.S. Agovic
Department of Physiology, Pharmacology & Neuroscience, Sophie Davis School
of Biomedical Education at CCNY, CUNY, Harris Hall Rm. 203,
160 Convent Ave, New York, NY 10031, USA

Department of Biology, Bronx Community College, CUNY, New York, NY, USA

A. El Idrissi and W.J. L'Amoreaux (eds.), *Taurine 8*, Advances in Experimental
Medicine and Biology 775, DOI 10.1007/978-1-4614-6130-2_4,
© Springer Science+Business Media New York 2013

sensitive to the NMDA antagonist, MK-801, but not the AMPA/kainate receptor antagonist CNQX. Furthermore, this taurine effect was blocked by APV pretreatment. Taurine (0.1 mM) decreased spermine-induced enhancement of specific (^{3}H) MK-801 binding to rat cortical membrane in the presence of glycine, though it was ineffective in the absence of spermine. Our preliminary work shows that taurine diminished the apparent affinity of NMDA receptor to glycine in the presence of spermine. These results indicate that taurine may directly interact with the NMDA receptor through multiple mechanisms.

Abbreviations

AMPA	(RS)-α-Amino-3-hydroxy-5-methyl-4-isoxazolepropionic acid
ACSF	Artificial cerebral spinal fluid
APV	DL-2-Amino-5-phosphonopentanoic acid
CNQX	6-Cyano-7-nitroquinoxaline-2,3-dione
GABA	Gamma-aminobutyric acid
mPFC	Medial prefrontal cortex
NMDA	N-Methyl-D-aspartate
TAG	6-Aminomethyl-3-methyl-4H-1,2,4-benzothiadiazine 1,1-dioxide

4.1 Introduction

It is well established that excessive stimulation of glutamate receptors can cause cell damage and cell death, i.e., excitotoxicity (Coyle and Puttfarcken 1993; Ikonomidou et al. 1999; Besancon et al. 2008). Taurine may be neuroprotective against glutamate-induced excitotoxicity by decreasing the intracellular levels of free calcium (El Idrissi and Trenkner 1999; Chen et al. 2001; Saransaari and Oja 2000; Lidsky et al. 1995). Several mechanisms have been demonstrated to underlie this function, including diverse modes of interactions of taurine with intracellular or membrane transporters, as well as Ca^{2+} or Cl^- channels (El Idrissi and Trenkner 2003; Wu et al. 2005). Notably, taurine has inhibitory properties through gating a Cl^- channel shared with GABA and glycine, thereby reducing Ca^{2+} influx through voltage-gated Ca^{2+} channels or the NMDA receptor (Belluzzi et al. 2004). In addition, a similar indirect effect can alternatively be mediated by a taurine-specific Cl^- channel. This channel is exclusively sensitive to 6-aminomethyl-3-methyl-4H-1,2,4-benzothiadiazine 1,1-dioxide (TAG, Yarbrough et al. 1981; Frosini et al. 2003a,b), though interestingly it is also sensitive to picrotoxin, the $GABA_A$ receptor noncompetitive antagonist (Molchanova et al. 2007). No previous work has indicated a direct interaction of taurine with the glutamate NMDA receptor. We aimed to examine the merits of this possible mechanism electrophysiologically, using a saturating dose of picrotoxin in all control and drug solutions so as to isolate any observed taurine effects from those requiring Cl^- channel activation. We also performed (^{3}H) MK-801 (10 nM) binding studies to evaluate direct taurine interactions with the NMDA receptor.

4.2   Methods

4.2.1   Rat Brain Slice Preparation and Field Potential Recording

DL-APV was purchased from Tocris Bioscience (Minneapolis, MO). Male Sprague Dawley rats of 4–6 weeks of age were purchased from Taconic Farms (Albany, NY) and housed in the City College of New York animal facilities. All procedures for care and use of the rats adhered to the protocols approved by the City College Institutional Animal Care and Use Committee. We prepared 400-µm brain slices in ice-cold artificial cerebral spinal fluid (ACSF) from adult male rats by slicing at 30° to the coronal plane, with the dorsal edge of the cutting plane being most anterior, as previously described (Orozco-Cabal et al. 2006) using a Vibratome (Technical Products International). Standard procedures for electrophysiological recording from brain slices were followed, including the makeup of the ACSF. Compounds dissolved in ACSF were applied through the chamber as needed from a side tube in the superfusion line. We stimulated the tissue with single negative current pulses (0.1 ms, 0.04 Hz; current about 1.5×threshold), delivered to a ventral cortical location 500–600 µm from the medial edge and about 1 mm ventral to the recording location in the infralimbic area of mPFC via a concentric electrode. D.C.-amplified signals were digitally averaged over six frames.

4.2.2   Neurochemistry

Rat cortical tissue for receptor binding assay was purchased from Pel-Freeze Biologicals, (Rogers, AR). (^{3}H) MK801 (24 Ci/mmol) was purchased from PerkinElmer Life and Analytical Sciences (Boston, MA). Standard receptor binding techniques were used (Banerjee et al. 1995). The rat cortical tissue was homogenized and centrifuged at $49,000 \times g$ twice, frozen and thawed and washed thrice. Receptor binding assays were carried out by incubating 800 µl of the crude fractions containing 0.8 mg tissue each with radiolabeled (^{3}H) MK-801 for 45 min at 25°C and then filtered. The filters were washed and processed for conventional scintillation counting. To determine nonspecific binding, a parallel set of tubes was incubated with a large excess of 100 µM nonradioactive MK-801 ligand. Specific binding is determined as the difference between total binding and nonspecific binding.

We stimulated specific (^{3}H) MK-801 (10 nM) binding with glycine (30 µM). The data were fitted to the one-site binding or one-site competition equation using Prism-3 software (Graphpad, La Jolla), based on nonlinear least-square regression.

4.3   Results

4.3.1   Taurine Inhibited Evoked Responses in Layers 4/5 of Rat mPFC

The population response to single-pulse electrical stimulation was characterized using MK-801and CNQX. 10 μM MK-801, the NMDA receptor antagonist, drastically reduced the amplitude of the major negative (excitatory) wave in the evoked response. The peak as well as latter parts of this response was most affected (middle trace, Fig. 4.1a). When 10 μM CNQX, the AMPA/kainate antagonist, was subsequently co-applied with APV to the same slice, the remaining response further reduced in amplitude, mainly in the initial part of the response (top trace). The differential inhibition by these two antagonists thus showed that the earlier part of the evoked response is mediated by the AMPA/kainate receptor, whereas the middle (peak) and the long trailing parts are predominantly mediated by NMDA receptors.

We tested the effect of taurine on the evoked response in the mPFC. Bath applied taurine (0.1–10 mM) caused inhibition of this response. A typical response in the presence of 2 mM taurine (six slices) is shown in Fig. 4.1b. Taurine robustly and selectively reduced the response around the peak and latter phases of the response, and spared the initial phase. This phase-specific characteristic was shared by the action of MK801 but not that of CNQX, indicating a taurine effect on the NMDA receptor-mediated response component. The presence of a saturating dose of picrotoxin in all control and drug solutions precluded the possibility that taurine's inhibition of the NMDA receptor was mediated by involvement of one of the two Cl^- channels known to be associated with taurine.

4.3.2   The Inhibition of the Evoked Response by Taurine Was Blocked by Co-Applied APV

Another NMDA antagonist, APV, was bath applied to confirm the above finding. APV (100 μM) also inhibited the peak and latter part of the evoked response in all slices tested. In many (seven) but not all slices, the initial phase of the large negative wave was enhanced rather than inhibited (Fig. 4.2). This was likely due to disinhibition of major neuronal elements through inhibition by APV of GABA release from interneurons, which would have stimulated the $GABA_B$ receptor even as $GABA_A$ receptors had been functionally blocked by picrotoxin in the recording solution. We tested the effect of taurine on the evoked response that had been pretreated with 100 μM APV. In all four pretreated slices tested, 2 mM taurine failed to cause inhibition of the evoked response (Fig. 4.2). This result reinforces the above finding that taurine interacts with the NMDA receptor.

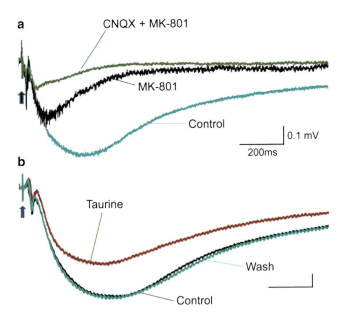

Fig. 4.1 Action of taurine on NMDA-receptor-mediated evoked response in rat medial prefrontal cortical slices. (**a**) Postsynaptic response of electrically evoked field potential in drug-free ACSF (Control) is superimposed on a response evoked by an identical stimulus (*arrow head*) when the slice was superfused with ACSF containing the NMDA-receptor antagonist, MK-801 (10 μM), and on a response evoked in the presence of MH-801 and the AMPA-receptor antagonist, CNQX (10 μM), indicating that the peak and subsequent parts of the typical population neuronal response in layer 4/5 of the mPFC evoked by stimulating at the medial ventral cortical area is predominantly mediated by the NMDA receptor. The response level recovered toward the control level after 30 min washing with ACSF (trace omitted for clarity). (**b**) Evoked field-potential responses in a different slice recorded in drug-free control ACSF and in 2 mM taurine, showing inhibition by taurine of the peak and latter parts of the population response evoked by the same stimulus strength (at *arrow head*). The inhibitory effect of taurine was completely reversed by a 40 min wash with ACSF (Wash). Calibration scale same as in (**a**)

4.3.3 Glycine-Stimulated (^{3}H)MK-801 Binding Was Affected by Taurine

Our preliminary data show that 100 μM spermine enhanced glycine-stimulated (^{3}H) MK-801 binding, and also that taurine had no effect on glycine-stimulated (^{3}H) MK-801 binding in the absence of spermine. These results prompted us to examine in the present study the displacement of specific (^{3}H) MK-801 binding in the presence of 100 μM spermine by eight different concentrations of taurine (0.1 nM to 1 mM). Taurine dose dependently displaced the spermine-enhanced (^{3}H) MK-801 binding with an IC_{50} of around 20 μM. We are in the process of studying the specific binding of (^{3}H) MK-801 to NMDA receptor either in the presence of 100 μM spermine or a

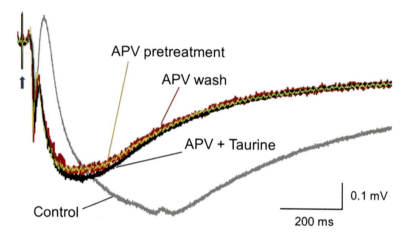

Fig. 4.2 The selective NMDA antagonist, APV, blocked the inhibitory action of taurine on layer 4/5 mPFC-evoked response. Superimposed evoked neuronal population responses of layer 4/5 mPFC to stimulation of the ventral medial cortical area recorded from the same slice superfused first with drug-free ACSF (Control), and then with 100 μM DL-APV (APV pretreatment), followed by 2 mM taurine in the presence of 100 μM DL-APV (APV + Taurine). The traces show an absence of any effect of taurine in the presence of APV. The apparent enhancement by APV at the initial part of the population response was possibly due to disinhibition of an inhibitory pathway involving GABA$_B$ receptors. Trace for wash back toward control level is omitted for clarity

mixture of spermine and taurine, in each case also with one of eight different concentrations of glycine (0.01 nM to 0.1 mM). Preliminary results shows a rightward shift of the dose–response curve of glycine-activated (^{3}H) MK-801 binding in this condition (i.e., enhanced by spermine) when taurine was present, suggesting an allosteric inhibition by taurine of the spermine site of the NMDA receptor.

4.4 Discussion

Taurine has not been thought to directly interact with the glutamate NMDA receptor. In the present study, we have identified at least two different mechanisms by which taurine exerts direct inhibition on the NMDA receptor in rat cerebral cortex. These are selective inhibition of the MK-801-sensitive (i.e., NMDA-receptor mediated) parts of the evoked neuronal population response, and partial inhibition of the enhancing effect of spermine on glycine-stimulated NMDA receptor activation. In addition, our ongoing work also suggests a third mechanism, i.e., diminution of the apparent affinity of the NMDA receptor for glycine.

Taurine is previously found to open a Cl$^-$ channel causing reduced neuronal membrane excitability, and that action is not gated by GABA or glycine (Yarbrough et al. 1981; Molchanova et al. 2007). This Cl$^-$ channel gating taurine binding site is blocked by TAG (Frosini et al. 2003a,b), and appears to play a major role in the expression of taurine's neuronal inhibitory activity. It is possible that taurine inhibits voltage-gated Ca^{2+} channels through interacting with the TAG-sensitive taurine receptor (see also Yarbrough et al. 1981). In this report, we have identified and demonstrated the actions of a TAG-insensitive taurine receptor that is specifically located on the glutamate NMDA receptor and may function to counteract the excitotoxicity of glutamate. Because of its being situated on the NMDA receptor, this receptor is expected to be more selectively located among regions of the CNS than the TAG-sensitive taurine receptor, and may therefore present itself as a more selective therapeutic target. The relative CNS distribution of these two receptors should be determined in future studies. Currently, we have no information on whether the TAG-insensitive effects of taurine on the NMDA receptor system represent the actions of a single receptor site or multiple receptor subtypes.

Interestingly, an earlier electrophysiological study has shown that in a subpopulation of striatal neurons, the synthetic taurine analog, acamprosate, partially inhibits the spermine-induced potentiation of NMDA receptor activity (Popp and Lovinger 2000), agreeing with the finding from the present binding study. This mode of action would allow taurine to suppress or prevent over-excitation of glutamatergic transmission without significantly interfering with the physiological functions of glutamate. This feature may also translate into significant potential therapeutic advantage.

4.5   Conclusion

The present study provided data to demonstrate the existence of a functional taurine receptor that interacts with the glutamate NMDA receptor directly, i.e., independently of involvement of Cl^- channel activation. This novel TAG-insensitive taurine receptor activates multiple mechanisms to reduce the NMDA-receptor-mediated response. The possibility that there are subtypes of this receptor is a subject for future studies.

Acknowledgments This work was supported by DA018055 grant from NIDA and by a PSCCUNY-award 63679-0041.

References

Banerjee SP, Zuck LG, Yablonsky-Alter E, Lidsky TI (1995) Glutamate agonist activity: implications for antipsychotic drug action and schizophrenia. Neuroreport 6:2500–2504

Belluzzi O, Puopolo M, Benedusi M, Kratskin I (2004) Selective neuroinhibitory effects of taurine in slices of rat main olfactory bulb. Neuroscience 124:929–944

Besancon E, Guo S, Lok J, Tymianski M, Lo EH (2008) Beyond NMDA and AMPA glutamate receptors: emerging mechanisms for ionic imbalance and cell death in stroke. Trends Pharmacol Sci 29:268–275

Chen WQ, Jin H, Nguyen M, Carr J, Lee YJ, Hsu CC, Faiman MD, Schloss JV, Wu JY (2001) Role of taurine in regulation of intracellular calcium level and neuroprotective function in cultured neurons. J Neurosci Res 66:612–619

Coyle JT, Puttfarcken P (1993) Oxidative stress, glutamate, and neurodegenerative disorders. Science 262:689–695

El Idrissi A, Trenkner E (1999) Growth factors and taurine protect excitotoxicity by stabilizing calcium homeostasis and energy metabolism. J Neurosci 19:9459–9468

El Idrissi A, Trenkner E (2003) Taurine regulates mitochondrial calcium homeostasis. Adv Exp Med Biol 526:527–536

Frosini M, Sesti C, Dragoni S, Valoti M, Palmi M, Dixon HB, Machetti F, Sgaragli G (2003a) Interactions of taurine and structurally related analogues with the GABAergic system and taurine binding sites of rabbit brain. Br J Pharmacol 138:1163–1171

Frosini M, Sesti C, Saponara S, Ricci L, Valoti M, Palmi M, Machetti F, Sgaragli G (2003b) A specific taurine recognition site in the rabbit brain is responsible for taurine effects on thermoregulation. Br J Pharmacol 139:487–494

Ikonomidou C, Bosch F, Miksa M, Bittigau P, Vöckler J, Dikranian K, Tenkova TI, Stefovska V, Turski L, Olney JW (1999) Blockade of NMDA receptor, apoptotic neurodegeneration in developing brain. Science 283:70–74

Lidsky TI, Schneider JS, Yablonsky-Alter E, Zuck LG, Banerjee SP (1995) Taurine prevents haloperidol-induced changes in striatal neurochemistry and behavior. Brain Res 686:104–106

Molchanova SM, Oja SS, Saransaari P (2007) Inhibitory effect of taurine on veratridine-evoked D-[3H] aspartate release from murine corticostriatal slices: Involvement of chloride channels and mitochondria. Brain Res 1130:95–102

Orozco-Cabal L, Pollandt S, Liu J, Vergara L, Shinnick-Gallagher P, Gallagher JP et al (2006) A novel rat medial prefrontal cortical slice preparation to investigate synaptic transmission from amygdala to layer V prelimbic pyramidal neurons. J Neurosci Methods 151:148–158

Popp LR, Lovinger DM (2000) Interaction of acamprosate with ethanol and spermine on NMDA receptors in primary cultured neurons. Eur J Pharmacol 394:221–231

Saransaari P, Oja SS (2000) Taurine and neural damage. Amino Acids 19:509–526

Wu H, Jin Y, Wei J, Jin H, Sha D, Wu JY (2005) Mode of action of taurine as a neuroprotector. Brain Res 1038:123–131

Yarbrough GG, Singh DK, Taylor DA (1981) Neuropharmacological characterization of a taurine antagonist. J Pharmacol Exp Ther 219:604–613

Chapter 5
The Modulatory Role of Taurine in Retinal Ganglion Cells

Zheng Jiang, Simon Bulley, Joseph Guzzone, Harris Ripps, and Wen Shen

Abstract Taurine (2-aminoethylsuphonic acid) is present in nearly all animal tissues, and is the most abundant free amino acid in muscle, heart, CNS, and retina. Although it is known to be a major cytoprotectant and essential for normal retinal development, its role in retinal neurotransmission and modulation is not well understood. We investigated the response of taurine in retinal ganglion cells, and its effect on synaptic transmission between ganglion cells and their presynaptic neurons. We find that taurine-elicited currents in ganglion cells could be fully blocked by both strychnine and SR95531, glycine and $GABA_A$ receptor antagonists, respectively. This suggests that taurine-activated receptors might share the antagonists with GABA and glycine receptors. The effect of taurine at micromolar concentrations can effectively suppress spontaneous vesicle release from the presynaptic neurons, but had limited effects on light-evoked synaptic signals in ganglion cells. We also describe a metabotropic effect of taurine in the suppression of light-evoked response in ganglion cells. Clearly, taurine acts in multiple ways to modulate synaptic signals in retinal output neurons, ganglion cells.

Z. Jiang • S. Bulley • J. Guzzone • W. Shen (✉)
Department of Biomedical Science, Charles E Schmidt College of Medicine,
Florida Atlantic University, 777 Glades Road, Boca Raton, FL 33431, USA
e-mail: zjiang7@jhmi.edu; sbulley@uthsc.edu; wshen@fau.edu

H. Ripps
Department of Ophthalmology and Visual Sciences, University of Illinois
College of Medicine, Chicago, IL, USA

Marine Biological Laboratory, Woods Hole, MA, USA

A. El Idrissi and W.J. L'Amoreaux (eds.), *Taurine 8*, Advances in Experimental
Medicine and Biology 775, DOI 10.1007/978-1-4614-6130-2_5,
© Springer Science+Business Media New York 2013

Abbreviations

OPL   Outer plexiform layer
IPL   Inner plexiform layer
ONL   Outer nuclear layer
INL   Inner nuclear layer
GCL   Ganglion cell layer
CNS   Central nervous system

5.1   Introduction

In the retina, taurine is found primarily in glutamatergic neurons, i.e., in photoreceptors and bipolar cells in the eyes of goldfish, amphibians, murines, and cynomolgus monkeys (Marc et al. 1995; Kalloniatis et al. 1996; Omura and Inagaki 2000; Militante and Lombardini 2002). However, in early development other retinal cells, such as amacrine and ganglion cells, Muller cells, and pigmentary epithelium cells, have also been shown to take up taurine (Orr et al. 1976; Kennedy and Voaden 1976; Lake et al. 1978; Pow and Crook 1994). There is a strong link between taurine deficiency and visual dysfunction in retinal development, disorders that can be reversed through taurine dietary supplementation (Lombardini 1991). The previous study showed that application of taurine to cultured rat retinas promotes rod photoreceptor production (Altshuler et al. 1993). However, application of glycine or GABA to the culture media did not have the same effect as taurine to promote photoreceptor growth, although the molecular structures of taurine, glycine, and GABA are remarkably similar, suggesting that taurine activated different receptors (Rentería et al. 2004). Recent studies indicate that the ability of taurine to promote rod photoreceptor differentiation could be through the activation of the glycine receptor subtype GlyRα2, although glycine seems unlikely to be the ligand for triggering the events in rod photoreceptor development (Young and Cepko 2004). These studies, while demonstrating the importance of taurine in retinal neurodevelopment, pose an unsolved question as to whether taurine activates a specific receptor other than the glycine and GABA receptors.

The molecular structures of taurine, glycine, and GABA are similar and all capable of activating ionotropic receptors that are permeable to Cl$^-$. In each case it results in an inhibitory neuronal response. Because glycine and GABA are widely accepted as major inhibitory neurotransmitters in retinas, less attention has been paid to taurine, despite the fact that the endogenous taurine levels in retinas are much higher than either GABA or glycine. In fact, both glycine and GABA receptors have been cloned and their receptor pharmacology is well defined. Two ionotropic (GABA$_A$ and GABA$_C$) and one metabotropic (GABA$_B$) receptors have been characterized in the retina, with distinctive pharmacology and dose-dependent response properties. Moreover, the strychnine-sensitive ionotropic glycine receptor has been

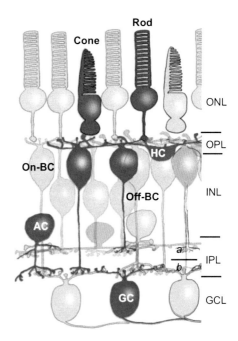

Fig. 5.1 Retinal structure and cellular organization. *Highlighted* with *dark* coloring are the five major cell types: photoreceptors (*rods* and *cones*), bipolar cells (On-BC and Off-BC), horizontal cells (HC), amacrine cells (AC), and ganglion cells (GC). These neurons synapse within two synaptic layers: the outer plexiform layer (OPL) and inner plexiform layer (IPL)

well studied in the CNS, and the use of specific agonists and antagonists of these receptors has enabled study of their function and mechanism of action in the retina. In contrast, neither the pharmacology nor molecular evidence of a taurine-specific receptor has been elucidated. Thus, the role of taurine in neurotransmission and modulation is not well understood.

At this juncture it is important to consider the structural and functional properties of the vertebrate retina, essentially an outgrowth of the CNS that is responsible for detecting environmental light and translating the visual scene into a series of electrochemical signals. The structural components and cellular organization of the retina are highly conserved among vertebrate species. In general, retinal neurons are classified into five major types (including subtypes): photoreceptors (rods and cones), bipolar cells (On-bipolar cells and Off-bipolar cells), horizontal cells, amacrine cells, and ganglion cells. They communicate within two synaptic layers: the outer plexiform layer (OPL) and inner plexiform layer (IPL), as shown in the schematic image of Fig. 5.1. The laminar organization of the retina generates two streams of visual information: a vertical pathway from photoreceptors to ganglion cells via bipolar cells, and a lateral pathway that comprises local feedback from horizontal cells and amacrine cells in the OPL and IPL, respectively. The vertical pathway is directly involved in sending signals to the brain, whereas the feedback circuits adjust the gains of pre- and postsynaptic neurons to optimize signal transmission within the vertical pathway. Receiving the integrated signals from bipolar and amacrine cells are the ganglion cells, output neurons that process the information and relay it to their first station (lateral geniculate nucleus) in the CNS. Because ganglion cells are the sole output neurons in the retina, their activity represents a critical stage in the visual pathway.

The amphibian retina is often used as a model system for studying retinal physiology, since this tissue has been well characterized as to both its neural structure and its signal pathway (Fig. 5.1). More importantly, the amphibian retina shares the same functions and mechanisms of neurotransmissions with higher vertebrate animals, but is accessed much more easily for electrophysiological recordings owing to the fact that its cells are larger in size and the overall structure of the network is more compact. Our previous study of the amphibian retina has shown that taurine is primarily present in photoreceptors and bipolar cells (Bulley and Shen 2010), and a similar finding has been reported in a study of the mammalian retina (Pow et al. 1994). Photoreceptors and bipolar cells are both glutamatergic neurons, and it is likely that glutamate and taurine are released from both cell types to regulate the activities of second-order (horizontal and bipolar cells) and third-order neurons (amacrine and ganglion cells). Indeed, we found that taurine suppresses glutamate-induced $[Ca^{2+}]_i$ changes in amacrine and ganglion cells (Bulley and Shen 2010). In this study we examined the regulatory effects of taurine on the spontaneous firing of ganglion cells, and on the light-evoked synaptic transmission in the neuronal network.

5.2   Methods

All procedures were performed in accordance with the guidelines of the National Institutes of Health Guide for the Care and Use of Laboratory Animals, and approved by the University's Animal Care Committee.

5.2.1   Retinal Slice Preparation

Larval tiger salamanders (*Ambystoma tigrinum*), purchased from Kons Scientific (Germantown, WI) and Charles Sullivan (Nashville, TN), were used in this study. The animals were kept in aquaria at 13°C under a 12-h dark–light cycle with continuous filtration. The retinas were collected from animals kept at least 6 h in the darkness. Briefly, the animals were decapitated and double-pithed and the eyes were enucleated. Retinal slices were prepared in a dark room under a dissecting microscope equipped with powered Night-Vision scopes (BE Meyer Co., Redmond, WA), an infrared illuminator (850 nm), an infrared camera, and a video monitor. The retina was removed from an eyecup in Ringer's solution and mounted on a piece of microfilter paper (Millipore, Bedford, MA), with the ganglion cell layer downward. The filter paper with retina was vertically cut into 250 nm slices using a tissue slicer

(Stoelting Co, IL). A single retinal slice was mounted in a recording chamber and superfused with oxygenated Ringer's solution, consisting of (in mM) NaCl (111), KCl (2.5), $CaCl_2$ (1.8), $MgCl_2$ (1.0), HEPES (5.0), and dextrose (10), pH = 7.7. The recording chamber was placed on an Olympus BX51WI microscope equipped with a CCD camera linked to a monitor.

5.2.2 Cell Dissociation

Retinas were removed from the eyecups into a Ringer's solution. The tissue was then digested in a freshly prepared enzymatic solution containing 50 µl papain (12 U/ml) in 0.5 ml of Ringer's solution to which was added 5 mM L-cysteine and 1 mM EDTA (pH = 7.4), and bathed for 20–35 min at room temperature. The papain-treated retinas were washed and mechanically dissociated through a flame polished Pasteur pipette into the standard Ringer's solution. The dissociated cells were seeded on 18 mm glass coverslips freshly coated with lectin and allowed to set for 20 min before use. The coverslip was then placed in a recording chamber on an Olympus BX51WI microscope equipped for electrophysiological recording, and viewed with a CCD camera linked to a monitor. After superfusion with Ringer's solution, recordings were made at room temperature and within a few hours after cell dissociation.

5.2.3 Electrophysiological Recording

Whole-cell recordings were performed on amacrine and ganglion cells (third-order retinal neurons) in dark-adapted retinal slices or after dissociation. Patch electrodes (5–8 MΩ) were pulled with an MF-97 microelectrode puller (Sutter Instrument Co.), and filled with a high K^+ solution containing (in mM) K-gluconate (100), $MgCl_2$, (1), EGTA (5) and HEPES (5), and an ATP-regenerating cocktail consisting of (in mM) ATP (20), phosphocreatine (40), and creatine phosphokinase (2); (pH = 7.4). An EPC-10 amplifier and HEKA Pulse software (HEKA Co., Germany) were used for data acquisition, and analysis was performed with Igor Pro software (WaveMetrics).

A gravity-driven perfusion system was used to superfuse all external solutions in the retinal slice preparation. The perfusion tube was placed 3 mm away from the retinal slice and was manually controlled for delivering drugs during the experiments. All recordings were performed under 850 nm infrared illumination to avoid exposure to visible light. A red LED light source (660 nm peak emission) was focused directly upon the retinal slice, and the brief (3 s) light stimuli were controlled by the output of the HEKA amplifier.

The isolated cells were constantly superfused with Ringer's solution or drug solutions via a DAD-VM automated superfusion system (ALA Scientific Instruments, Farmingdale, NY). All the chemicals were purchased from Tocris Bioscience (Minneapolis, MN) and Sigma-Aldrich Co. (St. Louis, MO).

5.2.4   Cell Selection

The third-order neurons in retinal slices were identified by their localization and their light-evoked response patterns. On the other hand, isolated cells often were not so readily identified. During the isolation process, many cells lost their typical morphological features, although dissociated ganglion cells could frequently be identified by the long axonal processes that extended from their cell somas. However, amacrine and bipolar cells often were not distinguishable, and it was necessary to rely on physiological criteria, i.e., their distinctive transient Na^+ currents in whole-cell recording. Unlike bipolar and horizontal cells, which show extremely large inward rectifier currents, ganglion cells typically generate large Na^+ currents (exceeding 500 pA), and display very small inward rectifier currents at negative voltages, whereas amacrine cells have relatively small transient Na^+ currents as well as small inward rectifier currents. Because there are many different types of amacrine cells in salamander retina, these criteria are less than ideal, but did not significantly influence the results.

5.3   Results

5.3.1   Pharmacology of the Ionotropic Taurine Response in Ganglion Cells

We noted earlier that taurine activates ionotropic receptors permeable to Cl^-. In that case, the reversal potential for taurine-generated currents would be expected to be near the Cl^- equilibrium potential (E_{Cl}). We used a low Cl^- and high K^+ intracellular solution to measure the E_{Cl} in ganglion cells with whole-cell voltage-clamp recording. Figure 5.2 shows the current responses to a puff of 350 µM taurine when a ganglion cell (in a retinal slice) was voltage-clamped at various potentials (−80, −70, −65, and −60 mV) near the E_{Cl}. Taurine generated inward and outward currents that reversed at or near −70 mV, indicating that taurine activated Cl^--permeable receptors in ganglion cells with a reversal potential at the E_{Cl}.

To determine the sensitivity of retinal ganglion cells to taurine we determined their dose–response function. Taurine was applied at various concentrations (10 µM, 100 µM, 500 µM, 1 mM, 2 mM, and 5 mM) to ganglion cells in the retinal slice preparation, and taurine-elicited currents were recorded at −60 mV (Fig. 5.2b). The taurine dose–response curve shows that on average the maximum current response was generated by 2 mM taurine, and the EC_{50} (concentration producing a half maximal response) was around 350 µM ($n = 7$). Note that 10 µM taurine did not produce a measurable current in ganglion cells ($n = 21$, Fig. 5.2b, lower panel). Similar dose-dependent responses could be obtained from isolated ganglion cells (unpublished data).

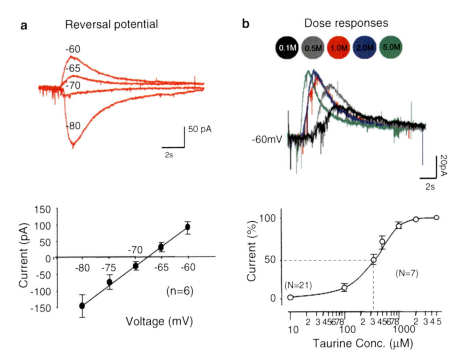

Fig. 5.2 Whole-cell recording of the taurine current reversal potential and the taurine dose response in retinal ganglion cells. (**a**) 350 μM taurine elicited currents at various voltages of −80, −70, −65, and −60 mV (*upper*), and the current reversal potential (*lower*) was near −70 mV (*n* = 6). (**b**) Taurine dose-dependent currents recorded from a single ganglion cell (*upper*) and the dose–response curve generated from a number of cells (*in parentheses*)

Several studies of neurons in the CNS have reported that taurine acts like a weak agonist on GABA and glycine receptors based on the observation that antagonists for GABA and glycine receptors can block the taurine response. It is unknown whether the ganglion cell response to taurine is also sensitive to GABA and glycine receptor antagonists. To test this we applied 350 μM taurine (the EC_{50} concentration) to isolated ganglion cells in which retinal network inputs were eliminated. Ganglion cells were voltage-clamped at a potential negative to the E_{Cl} and the inward current was elicited in response to the taurine (Fig. 5.3a, b). The taurine current was fully blocked by 2 μM strychnine, a glycine receptor-specific antagonist (Fig. 5.3a). Taurine-elicited currents were also sensitive to the $GABA_A$ receptor-specific inhibitor, 10 μM SR95531 (Fig. 5.3b). These results suggest that taurine-activated receptors might share the same receptor antagonists as those of $GABA_A$ and glycine receptors. It seems less likely that taurine is activating both GABA and glycine receptors, since the effect of taurine could be separately blocked by either the $GABA_A$ antagonist or the glycine receptor antagonist.

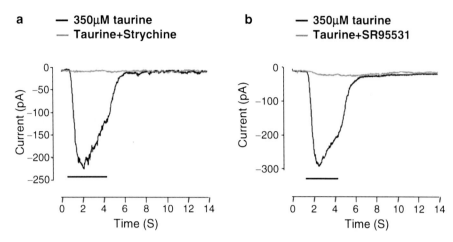

Fig. 5.3 Taurine-elicited currents were blocked by glycine and GABA receptor antagonists. 350 µM taurine elicited currents that were fully blocked by either 2 µM strychnine, a specific glycine receptor antagonist (**a**), or 10 µM SR95531, a potent GABA$_A$ receptor antagonist (**b**)

5.3.2 The Effects of Micromolar Taurine on Spontaneous and Light-Evoked Responses in Ganglion Cells

Ganglion cells receive glutamate inputs from bipolar cells, and GABA and glycine inputs from amacrine cells. In the dark, spontaneous vesicular release of glutamate, GABA, and glycine from the presynaptic neurons produces spontaneous excitatory postsynaptic currents (sEPSCs) and inhibitory postsynaptic currents (sIPSCs) in ganglion cells. To determine whether taurine affects these local vesicle release events, we recorded sEPSCs and sIPSCs from ganglion cells in dark-adapted retinal slices. To minimize the direct ionotropic effect of taurine caused membrane conductance changes in ganglion cells, we used a sufficiently low concentration of taurine (10 µM) that did not significantly change the membrane conductance in ganglion cells as shown in the dose–response curve in Fig. 5.2b.

sEPSCs in ganglion cells were recorded at −70 mV where GABAergic and glycinergic currents were close to zero. Figure 5.4a shows an example of sEPSCs that were continuously recorded from ganglion cells in dark-adapted retinal slices, first in control, then in the presence of taurine, and lastly after washout of taurine. In darkness, there is a high-frequency discharge with a combination of large and small amplitudes sEPSCs, which reflect the rate of vesicle release events from bipolar cell terminals. Application of taurine largely reduced the frequency and amplitude of the sEPSCs, effects that were totally eliminated within a minute after withdrawal of taurine.

The effect of taurine on the amplitude and frequency of sEPSCs in ganglion cells was analyzed with the Minianalysis Program (Synaptosoftware). Figure 5.4b shows quantitative analysis of sEPSCs frequency and amplitude changes in control, with taurine and after taurine washout, sampled from the recordings in Fig. 5.4a. The histograms indicate that taurine effectively eliminated the sEPSCs with

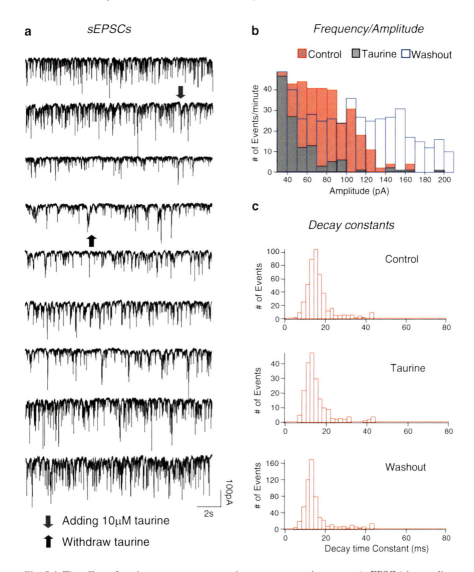

Fig. 5.4 The effect of taurine on spontaneous excitatory postsynaptic currents (mEPSCs) in ganglion cells in dark-adapted retinal slices. mEPSCs were recorded from a ganglion cell at −70 mV. (**a**) Continuously recording of sEPSCs in control, with 10 μM taurine and washout of taurine in a ganglion cell. (**b**) Histogram of the frequencies (time of the events) and amplitudes of the sEPSCs and (**c**) histogram of the decay time constants of the spontaneous synaptic events with and without taurine

amplitudes greater than 100 pA and frequencies less than 30 events/min, but it had limited effect on the high-frequency sEPSCs with amplitudes less than 40 pA. Also, taurine largely reduced the frequencies of sEPSCs that had amplitudes ranging from 40 to 100 pA. The analysis showed that the average frequencies of the sEPSCs were 7.63 Hz in control, 3.35 Hz with taurine, and 9.51 Hz after washout. The large reduction in the frequencies of sEPSCs implies that taurine suppressed the presynaptic release events. The effect of taurine on sEPSC amplitude was further analyzed from measurements of the decay time constants of a series of single events, since the decay time constants of sEPSCs may represent the properties of glutamate receptors on postsynaptic neurons, ganglion cells. Figure 5.4c shows the quantitative analyses of the decay time constants of individual sEPSCs in the control, in the presence of taurine, and after taurine withdrawal. The decay phases of each sEPSCs with an amplitude greater than 20 pA were fitted with a double-exponential curve, and the decay time constants were obtained by calculating the time to decay to 37% of the peak amplitudes. The histogram bars show that the distribution patterns of decay time constant of sEPSCs with and without taurine. The average decay time constant for sEPSCs was around 17 ms in control, and was not affected by taurine, although the number of the events/min (frequency) was reduced by taurine. This suggests that taurine may affect the spontaneous presynaptic release of glutamate in the bipolar cells; but had a much lesser effect on the decay time constants of glutamate receptors on the ganglion cells. Taken together, taurine at this low concentration reduced the frequencies of large sEPSCs in ganglion cells; we speculate that this effect might result from the suppression of presynaptic sites on bipolar cell terminals. Note that the frequencies of the large sEPSCs were increased in ganglion cells after taurine was withdrawn. This was commonly observed in ganglion cells, and could be due to an accumulation of synaptic vesicles in the release pools of bipolar cell terminals during taurine application, and their discharge once the suppressive effect of taurine was removed.

Taurine also strongly suppressed sIPSCs in ganglion cells. Figure 5.5 shows sIPSCs recorded from the same ganglion cell as in Fig. 5.4, but with the cell voltage-clamped at 0 mV (the reversal potential for glutamate currents). At this voltage, sIPSCs in ganglion cells produced by GABA and glycine synaptic inputs were outward currents in the dark-adapted retina. The sIPSCs were almost abolished when taurine was applied, and they fully recovered within a minute after return to the control Ringer's solution. In general, the effect of taurine (10 μM) was stronger on suppression of sIPSCs than sEPSCs in ganglion cells. We postulated that the strong suppression of sIPSCs by taurine could be due to the effects of both pre- and postsynaptic regulation by limitation of the presynaptic spontaneous releases of GABA and glycine, as well as down-regulation of GABA and glycine receptor activity in the postsynaptic ganglion cells.

Ganglion cells receive light-evoked signals from bipolar cells and amacrine cells in the network. Photic stimulation generates excitatory and inhibitory neurotransmitter release from bipolar and amacrine cells, which can be recorded from ganglion cells in retinal slice preparation. Figure 5.6a shows typical light-evoked excitatory currents in ganglion cells in 3 s duration (dark trace). These large EPSCs, elicited at the onset and offset of the light stimulus, are the result of glutamatergic inputs from presynaptic On- and Off-bipolar cells. Taurine (10 μM) had little effect on the amplitude of these light-evoked currents. However, the miniature events riding on the light-evoked currents and after light offset were significantly reduced by taurine (Fig. 5.6a, light trace). On average, taurine suppressed the peak currents evoked by the onset and offset of the light stimulus by $7.8 \pm 3\%$ and $8.2 \pm 4\%$ ($n = 18$), respectively (Fig. 5.6b).

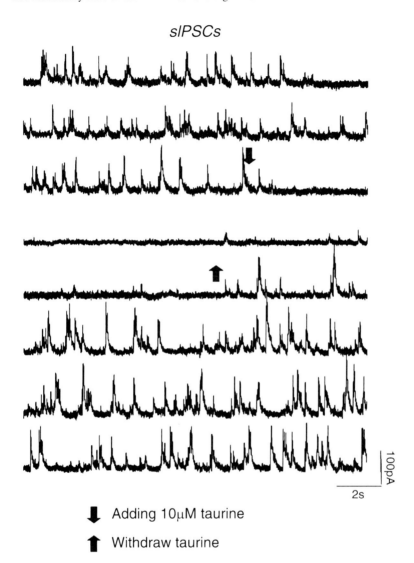

Fig. 5.5 Taurine suppresses spontaneous inhibitory postsynaptic currents (sIPSCs) in ganglion cells. sIPSCs were recorded from a ganglion cell held at 0 mV in a dark-adapted retinal slice. The *arrows* indicate the time points for application and withdrawal of 10 μM taurine during the recording

The effect of taurine on light-evoked IPSCs in ganglion cells is shown in Fig. 5.6c, recorded from the same cell as in Fig. 5.6a. With the cell voltage-clamped at 0 mV, taurine produced a small suppression of light-evoked IPSCs, but significantly reduced the fast frequency spontaneous events. Statistical analysis showed that taurine reduced the light-evoked IPSCs at the onset by $8 \pm 2\%$ and offset by $7.4 \pm 4\%$, ($n = 14$, Fig. 5.6d). It is apparent that taurine reduced the concomitant miniature events on both the light-evoked excitatory and inhibitory responses in ganglion cells.

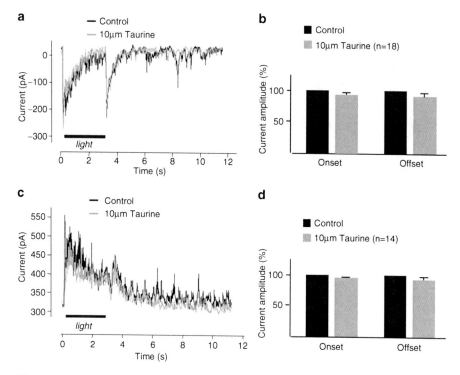

Fig. 5.6 The effects of taurine on light-evoked excitatory and inhibitory responses in ganglion cells. (**a**) Transit currents were evoked by the light onset and termination (offset) in a ganglion cell clamped at −70 mV. Taurine (10 μM) slightly reduced the amplitudes of the light-evoked EPSCs, and more effectively suppressed miniature currents. (**b**) Statistical analysis of the effect of taurine on the light-evoked current amplitudes from the 18 ganglion cells recorded. (**c**) The effect of taurine on light-evoked IPSCs recorded at 0 mV and the (**d**) statistics of the current amplitudes in control and with taurine

In summary, taurine (10 μM) had a limited effect on light-evoked large synaptic signals, but suppressed miniature responses in ganglion cells. It appears that taurine in the IPL may play a role in filtering out the synaptic noise spontaneous miniatures in ganglion cells.

5.3.3 Taurine Suppresses Light-Evoked Responses in Ganglion Cells via a Metabotropic Pathway

To further study the effect of taurine on light-evoked responses in ganglion cells, we used a higher concentration of taurine (100 μM) with picrotoxin (100 μM) and strychnine (5 μM) to block taurine-sensitive, as well as GABA- and glycine-sensitive Cl⁻-permeable receptors in ganglion cells. Meanwhile, picrotoxin and strychnine would increase excitatory synaptic inputs in ganglion cells due to blocking the

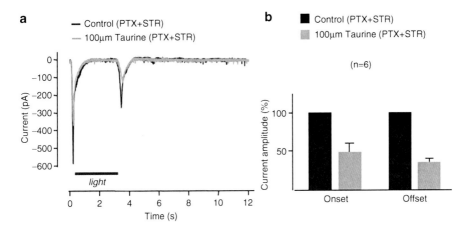

Fig. 5.7 Taurine suppresses light-evoked response in ganglion cells via a metabotropic mechanism. With picrotoxin (100 μM) and strychnine (5 μM) to block Cl⁻-permeable receptors, taurine (100 μM) effectively reduced light-evoked EPSCs in a ganglion cell (**a**); also shown are the statistics of the effect of taurine on suppression of light-evoked currents in ganglion cells (**b**)

inhibitions in the network. In the presence of picrotoxin and strychnine, light-evoked EPSCs were recorded from ganglion cells in dark-adapted retinal slices with the same protocol as used in Fig. 5.6. The addition of taurine (100 μM) reduced the amplitude of the light-evoked onset and offset EPSCs (Fig. 5.7a). The effect was present in all ganglion cells tested ($n = 6$). To quantitatively analyze the effect of taurine on light-evoked onset and offset EPSCs in the six ganglion cells, we measured the amplitudes of the currents from the ganglion cells in the control (with picrotoxin and strychnine) and presence of taurine. On average, with picrotoxin and strychnine, taurine suppressed $48.7 \pm 6\%$ of the onset currents and $41.3 \pm 4\%$ of the offset currents in the ganglion cells, shown in the histogram (Fig. 5.7b). Since taurine-sensitive Cl⁻- permeable receptors were blocked by the antagonists, the taurine regulation of light-evoked EPSCs in ganglion cells should be via a metabotropic pathway. We speculated that the effect of taurine might be due to suppression of presynaptic glutamate release from bipolar cell terminals, since metabotropic regulation of Ca^{2+}-dependent glutamate release is a common mechanism present in neural systems.

5.4 Discussion

5.4.1 Taurine-Sensitive Receptors

Although taurine has often been considered a neurotransmitter or neuromodulator in the CNS, the site of taurine's action is unclear, since a taurine-specific receptor has yet to be identified. One of the difficulties is the lack of a specific antagonist to differentiate the action of taurine from either glycine or GABA. A major question yet to be resolved is whether taurine acts on GABA or glycine receptors, or whether it activates a specific receptor that is also sensitive to the antagonists of GABA and glycine receptors. The evidence from this study is hardly conclusive, but it suggests that taurine-sensitive receptors might share the same receptor antagonists with GABA and glycine receptors.

Since glycine receptors and $GABA_A$ receptors have been considered as the major Cl^--permeable receptors coexisting in amphibian ganglion cells (Lukasiewicz and Werblin 1990), less has been known about taurine-sensitive receptors in the retinal neurons. Strychnine and SR95531 are widely used as the specific antagonists to block glycine and $GABA_A$ receptors, respectively. The concentrations of strychnine (2 μM) and SR95531 (10 μM) used in our study were in the range that has been shown to specifically block each receptor and it should have minimum or no antagonist cross talk to the other receptor. We showed that taurine currents could be completely blocked by either strychnine or SR95531 in isolated ganglion cells (see Fig. 5.3), implying that both antagonists might block the same site that was sensitive to taurine. The previous studies of taurine in photoreceptors show that taurine-mediated responses cannot be mimicked by GABA and glycine (Rentería et al. 2004; Young and Cepko 2004). Our results are consistent with the results of previous studies showing that taurine did not activate the same sites as either GABA or glycine. Although pharmacological studies demonstrate a discrepancy between taurine's action on GABA and glycine receptors in retinal neurons, the answer as to whether taurine activates a specific receptor type still awaits molecular evidence.

Much of what is known about taurine leads to the conclusion that taurine activates Cl^--permeable receptors. With picrotoxin and strychnine to block ionotropic taurine responses, the metabotropic effect of taurine was revealed. This is the first time that a metabotropic effect of taurine has been implicated in regulation of a light-evoked response in ganglion cells, although metabotropic effects of taurine have been reported in brain neurons, where it has been suggested that the effects might be via a $GABA_B$ receptor (Kontro and Oja 1990; Smith and Li 1991). Nevertheless, the actual site of taurine binding and the intracellular metabotropic pathways for the taurine effect still need to be elucidated.

5.4.2 Taurine Modulates Signal Transmission in Ganglion Cells

Taurine is found in the outer nuclear layer (ONL), mainly in photoreceptors (rods and cones) and Off-bipolar cells in the amphibian retina (Bulley and Shen 2010). Seemingly, Off-bipolar cells release glutamate in the IPL, and probably taurine as well. This study shows that taurine in the IPL performs multiple roles in the regulation of spontaneous and light-evoked synaptic transmissions in ganglion cells, which integrate these signals, and convert them into a train of spikes that carry both information and some unwanted noise to the CNS. In low micromolar concentrations, taurine preferentially suppressed spontaneous excitatory and inhibitory synaptic events (noise), but not the light-evoked synaptic signals in ganglion cells.

Since taurine effectively reduced the frequencies of sEPSCs, the suppressive effects of taurine seem to be acting on presynaptic spontaneous vesicle release sites on bipolar cells. These results suggest that endogenous taurine released from bipolar cells may feedback to regulate spontaneous release from bipolar cells. The fact that the suppressive effect of taurine was seemingly much stronger on sIPSCs suggests that, in addition of reducing spontaneous vesicle releases from amacrine cells, taurine might also interact with GABA and glycine receptors and suppress these receptors on ganglion cells. Unquestionably, the effect of taurine on GABA and glycine receptors needs to be studied further.

5.5 Conclusion

We find that taurine activates Cl^--permeable receptors that may also be sensitive to strychnine and SR95331, specific antagonists for glycine and $GABA_A$ receptors, respectively. Taurine applied in low micromolar concentrations reduced spontaneous vesicle release (noise), but not light-evoked signals (information) in ganglion cells, thereby acting as a negative control to enhance the signal/noise ratio of signal transmission. The light-evoked responses of ganglion cells were suppressed by taurine via a metabotropic pathway that has yet to be identified. Thus, taurine serves as a neuromodulator that regulates synaptic transmission at multiple sites in the inner retina.

Acknowledgements This work was supported by research grants to WS from NSF (1021646) and NIH (EY14161).

References

Altshuler DM, Lo Turco JJ, Rush J, Cepko C (1993) Taurine promotes the differentiation of a vertebrate retinal cell type *in vitro*. Development 119:1317–1328

Bulley S, Shen W (2010) Reciprocal regulation between taurine and glutamate response via Ca^{2+} dependent pathways in retinal third-order neurons. J Biomed Sci 17(Suppl 1):5

Kalloniatis M, Marc RE, Murry RF (1996) Amino acid signatures in the primate retina. J Neurosci 16:6807–6829

Kennedy AJ, Voaden MJ (1976) Studies on the uptake and release of radioactive taurine by the frog retina. J Neurochem 27:131–137

Kontro P, Oja SS (1990) Interactions of taurine with GABAB binding sites in mouse brain. Neuropharmacology 29:243–247

Lake N, Marshall J, Voaden MJ (1978) High affinity uptake sites for taurine in the retina. Exp Eye Res 27(6):713–718

Lombardini JB (1991) Taurine: retinal function. Brain Res 16:151–169

Lukasiewicz PD, Werblin FS (1990) The spatial distribution of excitatory and inhibitory inputs to ganglion cell dendrites in the tiger salamander retina. J Neurosci 10(1):210–221

Marc RE, Murry RF, Basinger SF (1995) Pattern recognition of amino acid signatures in retinal neurons. J Neurosci 15(7 Pt 2):5106–5129

Militante JD, Lombardini JB (2002) Taurine: evidence of physiological function in the retina. Nutr Neurosci 5(2):75–90

Omura Y, Inagaki M (2000) Immunocytochemical localization of taurine in the fish retina under light and dark adaptations. Amino Acids 19(3–4):593–604

Orr HT, Al C, Lowry OH (1976) The distribution of taurine in the vertebrate retina. J Neurochem 26(3):609–611

Pow DV, Crook DK (1994) Rapid postmortem changes in the cellular localization of amino acid transmitters in the retina as assessed by immunocytochemistry. Brain Res 653(1–2):199–209

Pow DV, Sullivan R, Reye P, Hermanussen S (1994) Localization of taurine transporters, and H taurine accumulation in the rat retina, pituitary and brain. Glia 37:153–168

Rentería RC, Johnson JDR, Copenhagen DR (2004) Need rods? Get glycine receptors and taurine. Neuron 41:839–841

Smith SS, Li J (1991) GABA$_B$ receptor stimulation by baclofen and taurine enhances excitatory amino acid induced phosphatidylinositol turnover in neonatal rat cerebellum. Neurosci Lett 132:59–64

Young TL, Cepko CL (2004) A role for ligand-gated ion channels in rod photoreceptor development. Neuron 41:867–879

Chapter 6
Taurine Is a Crucial Factor to Preserve Retinal Ganglion Cell Survival

Nicolas Froger, Firas Jammoul, David Gaucher, Lucia Cadetti, Henri Lorach, Julie Degardin, Dorothée Pain, Elisabeth Dubus, Valérie Forster, Ivana Ivkovic, Manuel Simonutti, José-Alain Sahel, and Serge Picaud

Abstract Retinal ganglion cells (RGCs) are spiking neurons, which send visual information to the brain, through the optic nerve. RGC degeneration occurs in retinal diseases, either as a primary process or secondary to photoreceptor loss. Mechanisms involved in this neuronal degeneration are still unclear and no drugs directly targeting RGC neuroprotection are yet available. Here, we show that taurine is one factor involved in preserving the RGC survival. Indeed, a taurine depletion induced by the antiepileptic drug, vigabatrin, was incriminated in its retinal toxicity

N. Froger (✉) • F. Jammoul • D. Gaucher • L. Cadetti • H. Lorach • J. Degardin • D. Pain
• E. Dubus • V. Forster • I. Ivkovic • M. Simonutti
INSERM, U968, Institut de la Vision, Paris, France

UPMC Université Paris 06, UMR_S 968, Institut de la Vision, Paris, France

CNRS, UMR 7210, Institut de la Vision, Paris, France
e-mail: nicolas.froger@inserm.fr

J.-A. Sahel
INSERM, U968, Institut de la Vision, Paris, France

UPMC Université Paris 06, UMR_S 968, Institut de la Vision, Paris, France

CNRS, UMR 7210, Institut de la Vision, Paris, France

Centre Hospitalier National d'Ophtalmologie des Quinze-Vingts, Paris, France

Institute of Ophthalmology, University College of London, London, UK

Fondation Ophtalmologique Adolphe de Rothschild, Paris, France

French Academy of Sciences, Paris, France

S. Picaud
INSERM, U968, Institut de la Vision, Paris, France

UPMC Université Paris 06, UMR_S 968, Institut de la Vision, Paris, France

CNRS, UMR 7210, Institut de la Vision, Paris, France

Fondation Ophtalmologique Adolphe de Rothschild, Paris, France

A. El Idrissi and W.J. L'Amoreaux (eds.), *Taurine 8*, Advances in Experimental
Medicine and Biology 775, DOI 10.1007/978-1-4614-6130-2_6,
© Springer Science+Business Media New York 2013

leading to the RGC loss. Similarly, we showed that RGC degeneration can be induced by pharmacologically blocking the taurine-transporter with the chronic administration of a selective inhibitor, which results in a decrease in the taurine levels both in the plasma and in the retinal tissue. Finally, we found that taurine can directly prevent RGC degeneration, occurring either in serum-deprived pure RGC cultures or in animal models presenting an RGC loss (glaucomatous rats and the P23H rats, a model for *retinitis pigmentosa*). These data suggest that the retinal taurine level is a crucial marker to prevent RGC damage in major retinal diseases.

Abbreviations

RGC(s)	Retinal ganglion cell(s)
GES	Guanidinoethane sulfonate
IOP	Intraocular pressure
VGB	Vigabatrin
Tau-T	Taurine-transporter
DAPI	4′,6-Diamidino-2-phenylindole
DIV	Days in vitro

6.1 Introduction

Taurine is a free amino-sulfonic acid which is present in large amounts in the central nervous system (Brosnan and Brosnan 2006), and the most abundant in the retina where it represents nearly half of the free amino acid content (Macaione et al. 1974) reaching up to 50 μmol/g of wet weight retina (Voaden et al. 1977). Taurine is mainly provided by nutrient intake, although endogenous synthesis occurred in most of species, excepted in cats (MacDonald et al. 1984). Indeed, a taurine-free diet was found to trigger photoreceptor degeneration in cats (Hayes et al. 1975). The effect of the taurine depletion on photoreceptor survival was subsequently confirmed in monkeys (Imaki et al. 1987). The mechanism of this taurine dependence still remains enigmatic, although the requirement for the taurine-transporter (Tau-T) in photoreceptor survival was evidenced by administering an inhibitor or competitive substrate of Tau-T in rats (Pasantes-Morales et al. 1983) or knocking out Tau-T in mice (Rascher et al. 2004). More recently, we reported that the photoreceptor toxicity of the antiepileptic drug, vigabatrin, is caused by taurine depletion in rats, mice and possibly human patients (Jammoul et al. 2009). Since these early discoveries involving taurine in photoreceptor survival, taurine was also reported to prevent neuronal excitotoxicity by reducing the glutamate-induced increase in intracellular calcium and endoplasmic reticulum stress (El Idrissi 2008; Wu and Prentice 2010), suggesting an intracellular action of taurine.

Here, we focused on the taurine effect exerted on retinal ganglion cell (RGC) survival. RGC are spiking neurons which send visual information to the brain (Roska and Werblin 2001). Degeneration of these neurons occurs in different retinal disease, either as primary process like in glaucoma (Quigley 1999) or as secondary to the photoreceptor loss, like in *retinitis pigmentosa* (Humayun et al. 1999). Because the RGC loss leads to blindness, prevention of RGC degeneration is a major challenge to be addressed. Different mechanisms involved in this neuronal degeneration have been reported (Tezel 2006), but no drugs directly targeting RGC neuroprotection are yet available.

As RGC degeneration appears as a primary process in the retinal toxicity of the antiepileptic drug, vigabatrin, we assessed if this RGC degeneration is caused by the taurine depletion we had previously correlated to the retinal toxicity of this drug (Jammoul et al. 2009). To investigate further the molecular mechanisms of this RGC degeneration, we then examined the consequence of a chronic pharmacological Tau-T blockade on RGC survival. Finally, we assessed if taurine can exert a direct action on RGC survival using serum-deprived pure RCG cultures. Then, the role of taurine was evaluated in different animal models of retinal pathologies with primary RGC degeneration, as in glaucoma (Shareef et al. 1995) or secondary RGC degeneration as in *retinitis pigmentosa* (Kolomiets et al. 2006; Garcia-Ayuso et al. 2010). We here described these studies, which support the administration of taurine for the prevention of RGC degeneration in retinal diseases.

6.2 Methods

6.2.1 Long-Term Treatment with Vigabatrin and the Selective Blocker of Taurine-Transporter (GES)

Vigabatrin treatment was administrated to rats by daily intraperitoneal injection (50 mg/Kg) for 25 days (Jammoul et al. 2010). Taurine (420 mg/Kg) was co-administrated by daily intraperitoneal injections at the same time as vigabatrin during the same period. Chronic treatment with guanidinoethane sulfonate (GES), a selective blocker of Tau-T, was performed on mice during 2 months. GES was administrated through the drinking water at the concentration of 1% (Gaucher et al. 2012).

6.2.2 Taurine Supplementation in Animal Models of RGC Degeneration

Animal models of RGC degeneration consist in (1) episcleral vein occlusion in rats (Shareef et al. 1995) considered as a model of glaucoma with increase in intraocular pressure (IOP) and (2) P23H line of rats, a model *of retinitis pigmentosa* in which

RGC degeneration occurred secondary to photoreceptor loss (Kolomiets et al. 2010). Animals were treated with taurine *per os* (through their drinking water) during 3 months (for rats) (Froger et al. 2012).

6.2.3   Taurine Plasmatic Level

Blood samples were collected in hemolysis tubes containing heparin (14 IU/ml) and centrifuged ($2,200 \times g$, 15 min). Plasmatic taurine measurements were performed on each animal by Serba laboratories (Cergy-Pontoise, France) using HPLC technique (Jammoul et al. 2009).

6.2.4   Evaluation of RGC Density on Retinal Sections

At the end of long-term treatments (vigabatrin, GES or taurine; see below), immunostaining with POU4F1, a specific marker for RGCs, was performed on retinal cryosections. RGCs were counted on retinal sections, after fluorescent image acquisition, and their densities were evaluated by reporting the number of cells per length of sections (Froger et al. 2012).

6.2.5   Primary Cultures of Purified RGCs from Adult Rats

RGCs were purified by immunopanning, following the protocol previously described (Barres et al. 1988). Using immunostaining experiments with NF200 and β-III tubulin markers (vs. DAPI counterstaining), RGC purity was estimated at 98% and 92%, respectively. RGCs were cultured in a low nutritive medium composed by Neurobasal-A plus glutamine (without serum). After 6 days in vitro (DIV), alive RGCs, labelled with CalceinAM, were counted from seven fields taken on coverslips to evaluate the RGC survival (Froger et al. 2012).

6.2.6   Statistical Analysis

All data are expressed as means ± SEM. A two-tailed unpaired Student's *t*-test was used to compare means of two groups. For more than two groups compared, a one-way ANOVA was used for variance analysis, followed in case of significance by either a Bonferroni post-hoc test (Gaussian distribution) or a Dunn's post hoc test (no Gaussian distribution) to compare the means of each group. Differences were considered significant at $*p < 0.05$, $**p < 0.01$ and $***p < 0.001$.

6.3   Results

6.3.1   Long-Term Administration of Vigabatrin-Induced Degeneration in Retinal Ganglion Cells

Chronic treatment with vigabatrin (50 mg/Kg/day) in rats induced a retinal toxicity that results in a reduction of RGC density, evaluated by the counting of POU4F1-positive cells in ganglion cell layer along retinal cryosections (Fig. 6.1a, b, e). This

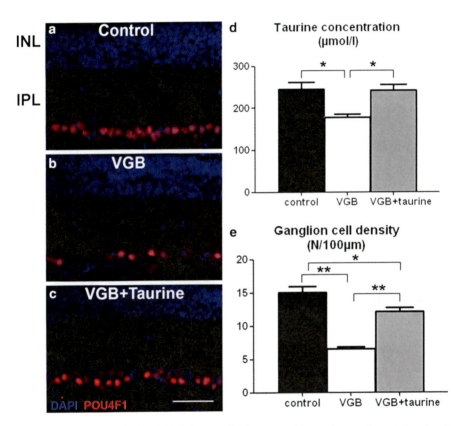

Fig. 6.1 The vigabatrin-induced RGC degeneration is prevented by taurine supplementation. (**a–c**) Representative retinal cryosections showing POU4F1-positive RGC immunolabelling (POU4F1; *red*) and retinal cell nuclei staining (DAPI) in untreated rats (control, **a**), in vigabatrin-treated rats (VGB, **b**) or in vigabatrin plus taurine-treated rats (VBB + taurine, **c**). The scale bar represents 50 μm. *INL* inner nuclear layer, *IPL* inner plexiform layer. (**d**) Taurine plasma levels measured in untreated rats (control), in rats treated with vigabatrin (VGB) or in rats treated with vigabatrin plus taurine (VBG + taurine). (**e**) Quantification of POU4F1-positive RGCs on retinal cryosections from untreated rats (control), in rats treated with vigabatrin (VGB) or in rats treated with vigabatrin plus taurine (VBG + taurine) *$p < 0.05$ and **$p < 0.01$ as compared to indicated groups (one-way ANOVA followed by a Bonferroni post hoc test to compare means) (Adapted from Gaucher et al. 2010)

vigabatrin-induced retinal toxicity is associated with drastic taurine depletion in plasma (Fig. 6.1d). Interestingly, this taurine depletion is directly relied to vigabatrin toxicity on RGCs. Indeed, taurine supplementation, which leads to recovery in taurine plasma levels (Fig. 6.1d) in vigabatrin-treated rats, can prevent the RGC loss induced by vigabatrin (Fig. 6.1c, e). This result indicates that taurine deficiency could be responsible for RGC degeneration and thus that this amino acid could be a key factor for the maintenance of RGC survival.

6.3.2   Long-Term Treatment with a Taurine-Transporter Selective Blocker Leads to RGC Loss in Mice

The major part of taurine is obtained by nutrition. Taurine assimilation from diet is highly dependent upon the Tau-T. Accordingly, long-term treatment in mice with GES, a selective blockade of Tau-T, induced a strong significant depletion of taurine plasmatic level (Fig. 6.2a) from 4-week treatment, and was maintained after 8-week treatment. In addition, taurine amounts were measured in retinal tissue (whole retina) showing from 4-week GES treatment a significant decrease in retinal taurine concentration (Fig. 6.2b). The consequences on RGC survival of the taurine depletion, observed in both plasma and retinal tissue after GES treatment, were evaluated by measuring the RGC density. We found that POU4F1-positive RGC number measured along retinal sections was significantly reduced (−19.3%) in mice subjected to 8-week GES treatment, as compared to untreated mice (Fig. c–e). These results further demonstrated that taurine depletion can cause RGC damage. They also strengthen the crucial role of Tau-T activity in the stability of taurine levels, both in plasma and in tissues, like retina.

6.3.3   Taurine Directly Prevents RGC Degeneration In Vitro

In studies presented above, RGC degeneration was occurring in parallel to photoreceptor degeneration. Therefore, it remained unclear whether the prevention of the RGC degeneration by taurine was indirect. Indeed, it could be consecutive to an effect on photoreceptor or on any other cell types such as glial cells. Therefore, to determine if taurine can affect directly RGC survival, we used purified RGCs from adult rats, cultured under serum-deprivation condition, to mimic ischemic conditions. After 6 days in vitro (DIV), a low density of viable RGCs, revealed by CalceinAM dye, was observed in control untreated conditions (Fig. 6.3a). A positive control was obtained by adding serum (B27 supplement) that increased RGC survival by 190% (Fig. 6.3c). Interestingly, direct application of 1 mM of taurine into the culture medium for the whole period of culture (6 DIV) significantly increased by 68% ($p < 0.001$) the RGC survival, as compared to RGCs cultured in taurine-free medium (Fig. 6.3b, c). Since taurine action requires Na^+-dependent selective uptake to exert its cellular activities, we investigated whether RGCs express the Tau-T.

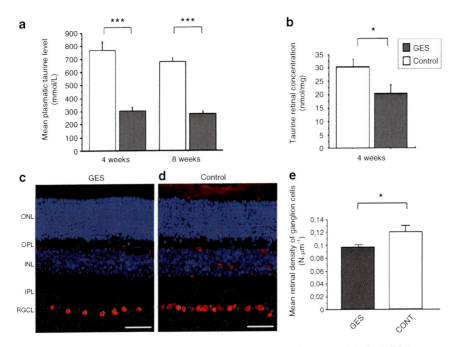

Fig. 6.2 Long-term treatment with GES induced a taurine depletion responsible for RGC damage.
(**a**) Taurine plasmatic levels after 4-week and 8-week treatment with GES, administrated in the drinking water (*grey bars*), as compared to control mice drinking GES-free water (*white bars*). (**b**) Taurine levels in mouse retinal tissues after 4-week GES treatment (*grey bar*) as compared to control mice drinking free-GES water (*white bar*). (**c**, **d**) Representative images showing POU4F1-immunopositive RGCs (*red*) on mice retinal sections after 8-week GES treatment (**c**), as compared to control mice drinking GES-free water (**d**). Counterstaining with DAPI (*blue*) is provided to reveal retinal nuclear layers. (**e**) Quantification of densities of POU4F1-positive RGC in mice treated with GES (*grey bar*) as compared to untreated mice (*white bar*). *ONL* outer nuclear layer, *OPL* outer plexiform layer, *INL* inner nuclear layer, *IPL* inner plexiform layer, *RGCL* retinal ganglion cell layer. Scale bars represent 25 μm (Adapted from Gaucher et al. 2012)

In freshly purified adult RGCs, the Tau-T was detected at the mRNA level. At the protein level, Tau-T was revealed by immunostaining in the cultured RGCs and we found that all βIII-tubulin-positive RGCs (Fig. 6.3d) also expressed Tau-T protein ($n = 3$ independent cultures; Fig. 6.3e, f). This data indicates that RGCs could generate taurine uptake *in vitro*. To assess if taurine uptake could account for the increase in RGC survival, taurine was co-incubated for 6 DIV with a blocker of the Tau-T in RGC cultures. Thus, addition of GES (1 mM) with taurine (1 mM) significantly suppressed the protective effect exerted by taurine on pure RGC cultures ($p < 0.05$, Fig. 6.3g). In this condition, the difference in RGC survival was no longer statistically significant from the control condition (+19%, $p > 0.05$). Similarly, application of GES alone (1 mM) did not significantly modify RGC survival as compared to control conditions ($p > 0.05$, Fig. 6.2e). These results indicated that the protective effect of taurine on RGCs is critically dependent on the Tau-T activity.

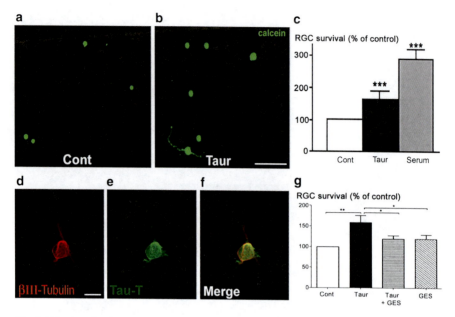

Fig. 6.3 Taurine stimulates the survival of pure adult RGCs in culture, involving the taurine-transporter (Tau-T) activity. (**a, b**) Representative images showing calcein-positive viable RGCs, cultured for 6 days in vitro (DIV) under serum-deprivation, in control untreated condition (Cont; **a**) or following 1 mM taurine application (Taur; **b**). (**c**) Quantification of RGC densities after 6 DIV, either in the control condition (Cont), or with 1 mM taurine application (Taur), or with the B27 supplement, providing a positive control condition (serum). In each experiment, the respective RGC densities were expressed as a percentage of the control condition at 6 DIV. Data are means ± SEM from 21 independent experiments. (**d–f**) Representative confocal images of Tau-T immunolabelling (*green*, **f**) in βIII-tubulin-positive RGCs (*red*, **d**) showing that Tau-T is localized with βIII-tubulin (*Merge*, **f**) after 6 days in culture. (**g**) Quantification of RGC densities after 6 DIV either in control conditions (Cont, $n=11$), in the presence of 1 mM taurine (Taur, $n=10$), in the presence of 1 mM GES (GES, $n=9$; *oblique hatched bar*) or in the presence of both taurine and GES (Taur+GES, $n=11$). Data are means ± SEM from independent experiments. ***$p<0.001$, **$p<0.01$ and *$p<0.05$ as compared to indicated groups, one-way ANOVA followed by a Dunn's post hoc test. The scale bars represent 100 μm in panels (**a, b**) and 10 μm in panels (**d–f**) (Adapted from Froger et al. 2012)

6.3.4   Taurine Prevents RGC Degeneration in an Animal Model of Glaucoma

To investigate the potential role of taurine in retinal diseases with RGC degeneration, taurine supplementation was provided to glaucomatous rats. Their glaucoma was induced by episcleral vein cauterization leading to vein occlusion and a long-term and stable increase in intraocular pressure (IOP) (Mittag et al. 2000). The IOP increase is indeed the main risk factor in glaucoma and all available drugs are currently targeting this IOP increase rather than the RGC degeneration. The consequence of chronic elevated IOP in the cauterized eyes on RGC survival was evaluated by counting the RGC density after POU4F1 immunostaining on retinal sections. The IOP increase that occurred in operated eyes induced a significant 15% RGC loss when compared to unoperated eyes ($p<0.05$; Fig. 6.4a, b, d). Interestingly, taurine supplementation

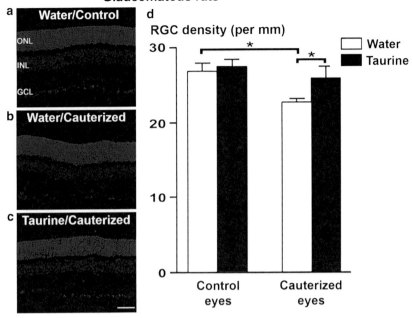

Glaucomatous rats

a **Water/Control**

ONL

INL

GCL

b **Water/Cauterized**

c **Taurine/Cauterized**

d

RGC density (per mm)

□ Water
■ Taurine

e **P23H rats (Retinitis pigmentosa)**

RGC density (per mm)

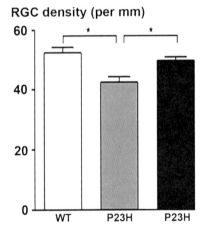

Fig. 6.4 Taurine supplementation prevents the RGC degeneration in animal models for retinal diseases. (**a–c**) Representative confocal images of retinal cryo-sections showing the POU4F1 immunopositive RGC immunolabelling (*red*) and cell nuclei staining (DAPI; *blue*) performed in unoperated eyes (Control, **a**) and cauterized eyes (cauterized; **b, c**) in rats without (water; **a, b**) or with taurine supplementation (taurine; **c**) added to their drinking water. (**d**) Quantification of POU4F1-positive RCG densities in both control eyes and cauterized eyes from rats without (water; *white bars*; mean ± SEM, $n = 11$ and 10 for control and cauterized eyes, respectively) or with tau rine supplementation (taurine; *black bars*; mean ± SEM, $n = 11$ and 9 for control and cauterized eyes, respectively). *$p < 0.05$ and ***$p < 0.001$ as compared to indicated group in (**d, e**), one-way ANOVA followed by a Bonferroni post hoc test. (**e**) Quantification of POU4F1 immunopositive RGCs in retinal cryo-sections from Sprague-Dawley wild-type animals (WT water, *white bar*), from untreated heterozygous P23H rats (P23H water; *grey bar*) and from taurine-supplemented P23H rats (P23H taurine; *black bar*). Data expressed as RGC per mm of retinal section are means ± SEM from $n = 7$ animals for each group. Scale bars represent 50 μm in panels (**a–c**) (Adapted from Froger et al. 2012)

(0.2 M administered through the drinking water for 3 months) significantly prevented 87% of the RGC damage in the cauterized eyes ($p<0.05$; Fig. 6.4c, d). Such taurine supplementation was found to produce a significant increase (by 95%) of the plasma taurine concentration ($p<0.01$ compared to taurine-free water-drinking rats). This RGC prevention was also associated with an increase in the electroretinogram amplitude providing evidence for a functional improvement of the diseased retina. These results indicate that the in vivo taurine administration can prevent the RGC loss in a glaucomatous eye showing an IOP increase.

6.3.5 Taurine Prevents RGC Degeneration in an Animal Model of Retinitis Pigmentosa

To generalize to other pathological conditions the potential interest of taurine in preventing RGC degeneration, we administered taurine in the drinking water of P23H rats, a model of *retinitis pigmentosa*. At 9 months of age, these P23H rats have lost their photoreceptors and they exhibit a secondary RGC degeneration (Garcia-Ayuso et al. 2010; Kolomiets et al. 2010). At 1 year, P23H rats showed a significant 18% loss in RGC density (42.6 ± 4.7 cells/mm), when compared to the age-matched Sprague-Dawley control rats (52.3 ± 5.9 cells/mm) (Fig. 6.4e). Again, the taurine supplementation for 3 months, from 9 to 12 months, prevented significantly 65% of this RGC loss (49.0 ± 2.9 cells/mm; $p<0.05$; Fig. 6.4e). These results indicate that taurine can prevent the secondary RGC loss in this animal model of *retinitis pigmentosa*.

6.4 Discussion

In our studies, we found that taurine is a crucial factor for maintaining RGC survival. In different animal models both *in vitro* and *in vivo*, we demonstrated that taurine deficiency is responsible for RGC degeneration. These results appear as a direct effect of taurine on RGCs because taurine directly prevented RGC degeneration in pure cultures subjected to serum deprivation. This effect of taurine could be important in treating retinal diseases because taurine supplementation prevented RGC loss occurring either as a primary process in glaucomatous rats or secondary to photoreceptor in P23H rats, a model of *retinitis pigmentosa*. The clinical relevance of these studies is provided by the observation of RGC loss in vigabatrin-treated patients exhibiting a taurine depletion.

The first indication of the major role of taurine in RGC survival came from our study on the vigabatrin toxicity. Vigabatrin, an antiepileptic drug effective against infantile spasm and complex partial seizure in adult (Ben-Menachem et al. 1989; Curatolo et al. 2006), was described to cause retinal toxicity both in human (Krauss et al. 1998; Miller et al. 1999; van der Torren et al. 2002; Frisen and Malmgren

2003; Westall et al. 2003; Buncic et al. 2004) and in treated animals (Butler et al. 1987; Duboc et al. 2004). In rats, the retinal toxicity induced by chronic vigabatrin administration is characterized by an alteration in electroretinogram (Duboc et al. 2004), associated with histological disorganization, with cone damage, neuronal plasticity and reactive gliosis (Duboc et al. 2004; Wang et al. 2008). Interestingly, this retinal toxicity was correlated to a taurine deficiency responsible for the cone damage (Jammoul et al. 2009). Such role of taurine in photoreceptor survival was already described, in particular in cats fed with free-taurine diet (Hayes et al. 1975). However, in vigabatrin-treated patients, the primary focus of degeneration appeared located in the RGC layer (Ravindran et al. 2001; Frisen and Malmgren 2003; Kilic et al. 2006). These correlated events led us to demonstrate that taurine depletion can induce RGC degeneration.

Taurine is mainly provided by diet and Tau-T activity constitutes the essential function to provide exogenous taurine to tissues. Indeed, Tau-T are expressed at the level of intestine epithelium (Satsu et al. 1997) to provide exogenous taurine from diet to the blood. In retinal tissue, taurine was provided from blood through the Tau-T expressed in retinal capillary endothelial cells (Tomi et al. 2008). The genetic inactivation of Tau-T in mice (see Heller-Stilb et al. 2002; Ito et al. 2008) produced an alteration in ganglion cell layer, although no specific characterization was provided (Heller-Stilb et al. 2002). Here, the pharmacological inactivation of Tau-T activity by the selective blocker GES induced a strong reduction in taurine level both in plasma and into retinal tissue. This taurine depletion is associated with an RGC loss, while it also affects the photoreceptor survival (Gaucher et al. 2012).

To further demonstrate that taurine exerts a direct action on RGC survival, we examined the effects of taurine treatments on RGC damages occurred in both *in vitro* and *in vivo* models. We provide evidence that taurine can promote the survival of adult RGCs in a pure culture, under serum-deprivation conditions. This result demonstrates a direct neuroprotective action of taurine on RGCs, which is consistent with the taurine-elicited resistance to hypoxia in an immortalized RGC cell line (Chen et al. 2009). As reported in the immortalized RGC cell line, the taurine neuroprotection of RGCs could occur through intracellular pathways, by reducing the intracellular calcium levels and by inhibiting the opening of mitochondrial permeability transition pores (Chen et al. 2009). Taurine was also shown to be essential for the translation of the mitochondrial DNA (Schaffer et al. 2009). The specific expression of the Tau-T in freshly purified RGCs suggests that an intracellular mechanism is involved in taurine protective action. This hypothesis was here validated by the loss of taurine neuroprotection found in presence of the Tau-T inhibitor, GES. The in vitro survival effect reported in this study was obtained at a millimolar concentration that remains in the physiological range since the retinal concentration of taurine was reported as high as 50 mM (Schmidt and Aguirre 1985). The taurine concentration used in our experiments is also similar to those used for in vitro studies on cerebellar neurons (El Idrissi 2008) and on the immortalized RGC line (Chen et al. 2009). Taken together, these results indicate that taurine can directly stimulate the RGC survival through an intracellular mechanism following its uptake by the Tau-T.

In addition, we found that taurine prevents RGC death induced by glutamate excitotoxicity in NMDA-treated retinal explants from adult rats (not shown, see Froger et al. 2012). Taurine was previously reported to prevent glutamate excitotoxicity in embryonic cultured cerebellar neurons (El Idrissi 2008) or mixed brain neurons (Wu and Prentice 2010). The intracellular effects of taurine discussed above could explain this prevention of RGC glutamate excitotoxicity in NMDA-treated retinal explants.

Finally, we assessed the in vivo effect of taurine supplementation on RGC survival in rat animal models of RGC degeneration. The first animal model is glaucomatous rats with an increased intraocular pressure leading to RGC degeneration and the second is a model of *retinitis pigmentosa*, the P23H rat, which exhibits RGC degeneration secondary to a photoreceptor loss. We observed RGC neuroprotection by taurine supplementation in these two models. A similar result was also obtained in another model of glaucoma, the DBA2 mice (Froger et al. 2012). In these models, the taurine neuroprotection suggests that the retinal taurine content may decrease below the optimum level for RGC survival. In fact, the lower ocular perfusion pressure, which is consecutive to the IOP increase and defined as a risk factor for primary open angle glaucoma (Araie et al. 2009), could cause a decrease in the retinal taurine content. In P23H rats, the major vascular atrophy also seen in patients following photoreceptor degeneration (Pennesi et al. 2008) could reduce the taurine retinal intake by reducing the number of capillary endothelial cells, transporting taurine into the retinal tissue (Tomi et al. 2008). Altogether these results suggest that reducing any parameter affecting taurine retinal influx may lead to RGC degeneration: (1) taurine plasma concentrations following vigabatrin (Jammoul et al. 2010) and GES treatments (Gaucher et al. 2012), (2) ocular blood flow as in glaucoma (Flammer et al. 2002; Araie et al. 2009; Leske 2009) and (3) density of the retinal vascular plexus as in *retinitis pigmentosa* (Pennesi et al. 2008).

6.5 Conclusion

Future studies will have to define whether taurine can become a treatment for the prevention of RGC loss in retinal degenerative diseases. In the case of *retinitis pigmentosa*, a decrease in plasma taurine level was described in patients, and a combined treatment with taurine was already found to improve patient vision although this effect was attributed to an improvement in photoreceptor function (Pasantes-Morales et al. 2002). Future epidemiological studies are needed to investigate how the nutritional taurine intake is correlated to the development of RGC degenerative diseases. Indeed, a great variability in nutritional taurine intake was reported among different world populations when correlating taurine levels with heart failure (Yamori et al. 2006). These considerations indicate that a proper diet or taurine supplementation could contribute to the prevention and/or treatment of RGC degeneration in different pathological conditions.

Acknowledgments We would like to thank Bernard Gilly, Pierre Belichard, Didier Pruneau, Annabelle Amiard, Marie-Laure Bouttier (Fovea Pharma), Olivier Lorentz, Katia Marazova (Fondation voir et Entendre), Gordon Fain (University of California Los Angeles), Alain Bron (Ophthalmology department, Dijon) and Jean-François Legargasson (Hôpital Lariboisière) for help and comments. We are grateful to Dr Matthew Lavail for providing the P23H rat line. This work was supported by INSERM, Université Pierre et Marie Curie (Paris VI), Fondation Ophtalmologique A. de Rothschild (Paris), Agence Nationale pour la Recherche (ANR: GLAUCOME), European Community (contrat TREATRUSH no HEALTH-F2-2010-242013), Fédération des Aveugles de France, IRRP, the city of Paris and the Regional Council of Ile-de-France. LC and NF received fellowships from the Fondation pour la Recherche Médicale and Fondation Bailly.

References

Araie M, Crowston J, Iwase A, Tomidokoro A, Leung C, Zeitz O, Vingris A, Schmetterer L, Ritch R, Kook M, Erlich R, Gherghel D, Graham S, Pillunat L, Aung T, Hafez A, Liu J, Harris A (2009) Clinical relevance of ocular blood flow (OBF) measurements including effects of general medications or specific glaucoma treatment. Kugler Publications, Amsterdam, The Netherlands

Barres BA, Silverstein BE, Corey DP, Chun LL (1988) Immunological, morphological, and electrophysiological variation among retinal ganglion cells purified by panning. Neuron 1:791–803

Ben-Menachem E, Persson LI, Schechter PJ, Haegele KD, Huebert N, Hardenberg J, Dahlgren L, Mumford JP (1989) The effect of different vigabatrin treatment regimens on CSF biochemistry and seizure control in epileptic patients. Br J Clin Pharmacol 27(Suppl 1):79S–85S

Brosnan JT, Brosnan ME (2006) The sulfur-containing amino acids: an overview. J Nutr 136:1636S–1640S

Buncic JR, Westall CA, Panton CM, Munn JR, MacKeen LD, Logan WJ (2004) Characteristic retinal atrophy with secondary "inverse" optic atrophy identifies vigabatrin toxicity in children. Ophthalmology 111:1935–1942

Butler WH, Ford GP, Newberne JW (1987) A study of the effects of vigabatrin on the central nervous system and retina of Sprague Dawley and Lister-Hooded rats. Toxicol Pathol 15:143–148

Chen K, Zhang Q, Wang J, Liu F, Mi M, Xu H, Chen F, Zeng K (2009) Taurine protects transformed rat retinal ganglion cells from hypoxia-induced apoptosis by preventing mitochondrial dysfunction. Brain Res 1279:131–138

Curatolo P, Bombardieri R, Cerminara C (2006) Current management for epilepsy in tuberous sclerosis complex. Curr Opin Neurol 19:119–123

Duboc A, Hanoteau N, Simonutti M, Rudolf G, Nehlig A, Sahel JA, Picaud S (2004) Vigabatrin, the GABA-transaminase inhibitor, damages cone photoreceptors in rats. Ann Neurol 55:695–705

El Idrissi A (2008) Taurine increases mitochondrial buffering of calcium: role in neuroprotection. Amino Acids 34:321–328

Flammer J, Orgul S, Costa VP, Orzalesi N, Krieglstein GK, Serra LM, Renard JP, Stefansson E (2002) The impact of ocular blood flow in glaucoma. Prog Retin Eye Res 21:359–393

Frisen L, Malmgren K (2003) Characterization of vigabatrin-associated optic atrophy. Acta Ophthalmol Scand 81:466–473

Froger N, Cadetti L, Lorach H, Martins J, Bemelmans AP, Dubus E, Degardin J, Pain D, Forster V, Chicaud L, Ivkovic I, Simonutti M, Fouquet S, Jammoul F, Leveillard T, Benosman R, Sahel J-A, Picaud S (2012) Taurine provides neuroprotection against retinal ganglion cell degeneration. PLoS One 7(10):e42017

Garcia-Ayuso D, Salinas-Navarro M, Agudo M, Cuenca N, Pinilla I, Vidal-Sanz M, Villegas-Perez MP (2010) Retinal ganglion cell numbers and delayed retinal ganglion cell death in the P23H rat retina. Exp Eye Res 91:800–810

Gaucher D, Arnault E, Husson Z, Froger N, Dubus E, Gondouin P, Dherbecourt D, Degardin J, Simonutti M, Fouquet S, Benahmed MA, Elbayed K, Namer IJ, Massin P, Sahel JA, Picaud S (2012) Taurine deficiency damages retinal neurones: cone photoreceptors and retinal ganglion cells. Amino Acids 43:1979–1993

Hayes KC, Carey RE, Schmidt SY (1975) Retinal degeneration associated with taurine deficiency in the cat. Science 188:949–951

Heller-Stilb B, van Roeyen C, Rascher K, Hartwig HG, Huth A, Seeliger MW, Warskulat U, Haussinger D (2002) Disruption of the taurine transporter gene (taut) leads to retinal degeneration in mice. FASEB J 16:231–233

Humayun MS, Prince M, de Juan E Jr, Barron Y, Moskowitz M, Klock IB, Milam AH (1999) Morphometric analysis of the extramacular retina from postmortem eyes with retinitis pigmentosa. Invest Ophthalmol Vis Sci 40:143–148

Imaki H, Moretz R, Wisniewski H, Neuringer M, Sturman J (1987) Retinal degeneration in 3-month-old rhesus monkey infants fed a taurine-free human infant formula. J Neurosci Res 18:602–614

Ito T, Kimura Y, Uozumi Y, Takai M, Muraoka S, Matsuda T, Ueki K, Yoshiyama M, Ikawa M, Okabe M, Schaffer SW, Fujio Y, Azuma J (2008) Taurine depletion caused by knocking out the taurine transporter gene leads to cardiomyopathy with cardiac atrophy. J Mol Cell Cardiol 44:927–937

Jammoul F, Degardin J, Pain D, Gondouin P, Simonutti M, Dubus E, Caplette R, Fouquet S, Craft CM, Sahel JA, Picaud S (2010) Taurine deficiency damages photoreceptors and retinal ganglion cells in vigabatrin-treated neonatal rats. Mol Cell Neurosci 43:414–421

Jammoul F, Wang Q, Nabbout R, Coriat C, Duboc A, Simonutti M, Dubus E, Craft CM, Ye W, Collins SD, Dulac O, Chiron C, Sahel JA, Picaud S (2009) Taurine deficiency is a cause of vigabatrin-induced retinal phototoxicity. Ann Neurol 65:98–107

Kilic U, Kilic E, Jarve A, Guo Z, Spudich A, Bieber K, Barzena U, Bassetti CL, Marti HH, Hermann DM (2006) Human vascular endothelial growth factor protects axotomized retinal ganglion cells in vivo by activating ERK-1/2 and Akt pathways. J Neurosci 26:12439–12446

Kolomiets B, Sahel JA, Picaud S (2006) Single-unit activity and visual response characteristics in retinal ganglion cells of ex vivo retina recorded using 3D MEA microelectrode array recordings in mice. University College London, London

Kolomiets B, Dubus E, Simonutti M, Rosolen S, Sahel JA, Picaud S (2010) Late histological and functional changes in the P23H rat retina after photoreceptor loss. Neurobiol Dis 38:47–58

Krauss GL, Johnson MA, Miller NR (1998) Vigabatrin-associated retinal cone system dysfunction: electroretinogram and ophthalmologic findings. Neurology 50:614–618

Leske MC (2009) Ocular perfusion pressure and glaucoma: clinical trial and epidemiologic findings. Curr Opin Ophthalmol 20:73–78

Macaione S, Ruggeri P, De Luca F, Tucci G (1974) Free amino acids in developing rat retina. J Neurochem 22:887–891

MacDonald ML, Rogers QR, Morris JG (1984) Nutrition of the domestic cat, a mammalian carnivore. Annu Rev Nutr 4:521–562

Miller NR, Johnson MA, Paul SR, Girkin CA, Perry JD, Endres M, Krauss GL (1999) Visual dysfunction in patients receiving vigabatrin: clinical and electrophysiologic findings. Neurology 53:2082–2087

Mittag TW, Danias J, Pohorenec G, Yuan HM, Burakgazi E, Chalmers-Redman R, Podos SM, Tatton WG (2000) Retinal damage after 3 to 4 months of elevated intraocular pressure in a rat glaucoma model. Invest Ophthalmol Vis Sci 41:3451–3459

Pasantes-Morales H, Quiroz H, Quesada O (2002) Treatment with taurine, diltiazem, and vitamin E retards the progressive visual field reduction in retinitis pigmentosa: a 3-year follow-up study. Metab Brain Dis 17:183–197

Pasantes-Morales H, Quesada O, Carabez A, Huxtable RJ (1983) Effects of the taurine transport antagonist, guanidinoethane sulfonate, and beta-alanine on the morphology of rat retina. J Neurosci Res 9:135–143

Pennesi ME, Nishikawa S, Matthes MT, Yasumura D, LaVail MM (2008) The relationship of photoreceptor degeneration to retinal vascular development and loss in mutant rhodopsin transgenic and RCS rats. Exp Eye Res 87:561–570

Quigley HA (1999) Neuronal death in glaucoma. Prog Retin Eye Res 18:39–57

Rascher K, Servos G, Berthold G, Hartwig HG, Warskulat U, Heller-Stilb B, Haussinger D (2004) Light deprivation slows but does not prevent the loss of photoreceptors in taurine transporter knockout mice. Vision Res 44:2091–2100

Ravindran J, Blumbergs P, Crompton J, Pietris G, Waddy H (2001) Visual field loss associated with vigabatrin: pathological correlations. J Neurol Neurosurg Psychiatry 70:787–789

Roska B, Werblin F (2001) Vertical interactions across ten parallel, stacked representations in the mammalian retina. Nature 410:583–587

Satsu H, Watanabe H, Arai S, Shimizu M (1997) Characterization and regulation of taurine transport in Caco-2, human intestinal cells. J Biochem 121:1082–1087

Schaffer SW, Azuma J, Mozaffari M (2009) Role of antioxidant activity of taurine in diabetes. Can J Physiol Pharmacol 87:91–99

Schmidt SY, Aguirre GD (1985) Reductions in taurine secondary to photoreceptor loss in Irish setters with rod-cone dysplasia. Invest Ophthalmol Vis Sci 26:679–683

Shareef SR, Garcia-Valenzuela E, Salierno A, Walsh J, Sharma SC (1995) Chronic ocular hypertension following episcleral venous occlusion in rats. Exp Eye Res 61:379–382

Tezel G (2006) Oxidative stress in glaucomatous neurodegeneration: mechanisms and consequences. Prog Retin Eye Res 25:490–513

Tomi M, Tajima A, Tachikawa M, Hosoya KI (2008) Function of taurine transporter (Slc6a6/TauT) as a GABA transporting protein and its relevance to GABA transport in rat retinal capillary endothelial cells. Biochim Biophys Acta 1778:2138–2142

van der Torren K, Graniewski-Wijnands HS, Polak BC (2002) Visual field and electrophysiological abnormalities due to vigabatrin. Doc Ophthalmol 104:181–188

Voaden MJ, Lake N, Marshall J, Morjaria B (1977) Studies on the distribution of taurine and other neuroactive amino acids in the retina. Exp Eye Res 25:249–257

Wang QP, Jammoul F, Duboc A, Gong J, Simonutti M, Dubus E, Craft CM, Ye W, Sahel JA, Picaud S (2008) Treatment of epilepsy: the GABA-transaminase inhibitor, vigabatrin, induces neuronal plasticity in the mouse retina. Eur J Neurosci 27:2177–2187

Westall CA, Nobile R, Morong S, Buncic JR, Logan WJ, Panton CM (2003) Changes in the electroretinogram resulting from discontinuation of vigabatrin in children. Doc Ophthalmol 107: 299–309

Wu JY, Prentice H (2010) Role of taurine in the central nervous system. J Biomed Sci 17(Suppl 1):S1

Yamori Y, Liu L, Mizushima S, Ikeda K, Nara Y (2006) Male cardiovascular mortality and dietary markers in 25 population samples of 16 countries. J Hypertens 24:1499–1505

Chapter 7
Taurine Regulation of Voltage-Gated Channels in Retinal Neurons

Matthew J.M. Rowan, Simon Bulley, Lauren A. Purpura, Harris Ripps, and Wen Shen

Abstract Taurine activates not only Cl⁻-permeable ionotropic receptors but also receptors that mediate metabotropic responses. The metabotropic property of taurine was revealed in electrophysiological recordings obtained after fully blocking Cl⁻-permeable receptors with an inhibitory "cocktail" consisting of picrotoxin, SR95531, and strychnine. We found that taurine's metabotropic effects regulate voltage-gated channels in retinal neurons. After applying the inhibitory cocktail, taurine enhanced delayed outward rectifier K^+ channels preferentially in Off-bipolar cells, and the effect was completely blocked by the specific PKC inhibitor, GF109203X. Additionally, taurine also acted through a metabotropic pathway to suppress both L- and N-type Ca^{2+} channels in retinal neurons, which were insensitive to the potent $GABA_B$ receptor inhibitor, CGP55845. This study reinforces our previous finding that taurine in physiological concentrations produces a multiplicity of metabotropic effects that precisely govern the integration of signals being transmitted from the retina to the brain.

M.J.M. Rowan • S. Bulley • L.A. Purpura • W. Shen (✉)
Department of Biomedical Science, Charles E Schmidt College of Medicine, Florida Atlantic University, 777 Glades Road, Boca Raton, FL 33431, USA
e-mail: matt.rowan@mpfi.org; sbulley@uthsc.edu; wshen@fau.edu

H. Ripps
Department of Ophthalmology and Visual Sciences, University of Illinois College of Medicine, Chicago, IL 60612, USA

Whitman Investigator, Marine Biological Laboratory, Woods Hole, MA 02543, USA

A. El Idrissi and W.J. L'Amoreaux (eds.), *Taurine 8*, Advances in Experimental Medicine and Biology 775, DOI 10.1007/978-1-4614-6130-2_7,
© Springer Science+Business Media New York 2013

Abbreviations

AMPA   α-Amino-3-hydroxy-5-methyl-4-isoxazole-propionic acid
GABA   γ-Aminobutyric acid
IPL   Inner plexiform layer
ONL   Outer nuclear layer
INL   Inner nuclear layer
GCL   Ganglion cell layer

7.1   Introduction

Taurine, like GABA and glycine, activates ionotropic receptors that produce an inhibitory effect on neurons by promoting an influx of Cl^-. It is also capable of activating metabotropic receptors that regulate a wide range of mechanisms through intracellular second messenger and G-protein sensitive pathways. Although molecular evidence of a taurine-sensitive metabotropic pathway has not been uncovered, the metabotropic effects of taurine have been reported in brain tissues (Wu et al. 2005). Indeed, taurine has been found to regulate intracellular protein interaction, various aspects of mitochondrial function, and Ca^{2+} release from internal stores, resulting in changing activities of glutamate receptors and Na^+/Ca^{2+} exchange (Li and Lombardini 1991; El Edrissi and Trenkner 1999; Wu and Prentice 2010). Many of these effects are believed to result from activation of metabotropic pathways via receptor-coupled G-proteins (Wu and Prentice 2010).

In the vertebrate retina, taurine acts as a neurotransmitter as well as a modulator of neuronal activity. Many neurotransmitters serve to regulate voltage-gated channels (Parnas and Parnas 2010) and thereby control cell excitability and neurotransmitter release. In a previous study from this laboratory, using Ca^{2+} imaging and whole-cell patch-clamp recordings of retinal neurons, we showed that taurine provides a dose-dependent regulation of Ca^{2+}-permeable glutamate receptors and voltage-gated Ca^{2+} channels. Moreover, we found that this regulatory activity was via CaMKII- and PKA-dependent intracellular pathways (Bulley and Shen 2010). In addition, the regulation of Ca^{2+}-dependent synaptic release by taurine was reported in earlier studies of amphibian, rabbit, and ox retinas (Burkhardt 1970; Cunningham and Miller 1976; DiGiorgio et al. 1977). Currently, we have discovered that taurine also can suppress spontaneous synaptic release from inner retinal neurons. Clearly, taurine plays a key role in retinal synaptic transmission and modulation. In this report we will present evidence showing that taurine regulates voltage-gated K^+ and Ca^{2+} channels in retinal bipolar cells and third-order neurons.

In many earlier studies, the effects of taurine were generated by application in concentrations of considered to be of no relevance physiologically. However, in the course of our studies, we discovered that taurine produces metabotropic regulation at much lower concentrations and that the effects of taurine were insensitive to the antagonists of GABA and glycine receptors. Interestingly, this taurine action leads to a specific modulation of voltage-gated K^+ channels and Ca^{2+} channels, critical components of cell excitability and synaptic transmission.

7.2 Methods

All procedures were performed in accordance with the guidelines of the National Institutes of Health Guide for the Care and Use of Laboratory Animals and approved by the University's Animal Care Committee.

7.2.1 Retinal Slice Preparation

Larval tiger salamanders (*Ambystoma tigrinum*) purchased from the Charles Sullivan (Nashville, TN, USA) and Kons Scientific (Germantown, WI, USA) were used in this study. The animals were kept in aquaria at 13°C under a 12-h dark–light cycle with continuous filtration. The retinas were collected from animals kept at least 6 h in the dark. After the animals were decapitated and double-pithed, the eyes were enucleated. Retinal slices were prepared in a dark room, under a dissecting microscope equipped with powered night-vision scopes (BEMeyer Co., Redmond, WA, USA), an infrared illuminator (850 nm), an infrared camera, and a video monitor. The retina was removed from the eyecup in Ringer's solution consisting of (in mM) NaCl (111), KCl (2.5), $CaCl_2$ (1.8), $MgCl_2$ (1.0), HEPES (5.0), and dextrose (10), pH=7.7. It was then mounted on a piece of microfilter paper (Millipore, Billerica, MA, USA) with the ganglion cell layer downward. The filter paper with retina was vertically cut into 250 μm slices using a tissue slicer (Stoelting Co.,Wood Dale, IL, USA).

7.2.2 Cell Dissociation

Retinas were removed from the eyecup in Ringer's solution. The tissue was then digested in a freshly prepared enzymatic solution containing 50 μl papain (12 U/ml) in 500 μl of Ringer's solution to which was added 5 mM L-cysteine and 1 mM EDTA (pH=7.4), and bathed for 20–35 min at room temperature. The papain-treated retinas were washed and mechanically triturated through a flame-polished Pasteur pipette into the standard Ringer's solution. The dissociated cells were seeded on 18 mm glass coverslips coated with Lectin and allowed to set for 20 min. The coverslip was then moved into the recording chamber and superfused with oxygenated Ringer's solution.

7.2.3 Whole-Cell Patch-Clamp Recording

Tissues (retinal slices or isolated cells) were placed in a recording chamber on the stage of an Olympus BX51WI microscope equipped with a CCD camera linked to a monitor. Whole-cell patch-clamp recordings were performed on bipolar cells, amacrine cells, and ganglion cells in retinal slices as well as isolated cells using an EPC-10 amplifier and Pulse software (HEKA Instruments Inc., Bellmore, NY, USA). Patch electrodes (5–8 MΩ) were pulled with an MF-97 microelectrode puller (Sutter Instrument Co., Novato, CA, USA). Recordings were obtained 5–10 min after membrane rupture in order to allow the cells to stabilize after dialysis of the electrode solution. Data were analyzed and plotted using Pulse and Igor/Excel software (WaveMetrics, Inc., Lake Oswego, OR, USA).

A gravity-driven perfusion system was used to superfuse all external solutions. The perfusion tube was placed 3 mm away from the retinal slice and was manually controlled for delivering drugs during the experiments. All of the chemicals used in this study were purchased from Sigma (St. Louis, MO) and Tocris Bioscience (Ellisville, MO, USA).

7.2.4 Extracellular and Intracellular Solutions

To study the metabotropic effect of taurine on delayed outward K$^+$ (K$_v$) channel currents, the Ringer solution contained an inhibitory "cocktail" of Cd^{2+} (100 μM), picrotoxin (100 μM), and strychnine (10 μM), thereby blocking ionotropic membrane receptors.

To study taurine's metabotropic regulation of voltage-dependent Ca^{2+} channels, the modified Ringer solution contained tetraethyl ammonium chloride (TEA, 40 mM), tetrodotoxin (TTX, 1 μM), and barium chloride (BaCl$_2$, 10 mM). To adjust for the excess concentration of Cl$^-$ ions in the TEA- and Ba^{2+}-containing Ringer's solution, the concentration of NaCl was reduced accordingly.

7.2.5 Identification of Cell Types and Subtypes

On- and Off-bipolar cells in retinal slices were identified by their neuronal morphology after intracellular dialysis with the fluorescent dye Lucifer Yellow, introduced during whole-cell recording. The axonal loci of most On- and Off-bipolar cells are within sublamina b and a of the inner plexiform layer (IPL), respectively. However, many of the Off-bipolar cells were "displaced," with somas in the outer nuclear layer (Maple et al. 2004). Their electrical signature and unique location allowed us to identify the Off-bipolar cells.

Isolated third-order neurons (amacrine and ganglion cells) and bipolar cells often were distinguished based on physiological criteria, i.e., their distinctive transient

Na$^+$ currents in whole-cell recording. Unlike bipolar and horizontal cells, which show extremely large inward rectifier currents, ganglion cells typically generate large Na$^+$ currents (exceeding 500 pA) and display very small inward rectifier currents at negative voltages, whereas amacrine cells have relatively small transient Na$^+$ currents as well as small inward rectifier currents. Because there are many different types of amacrine cells in salamander retina, these criteria are less than ideal but did not significantly influence the results.

7.3 Results

7.3.1 Taurine-Enhanced K_v Currents in Off-Bipolar Cells

Retinal bipolar cells are interneurons that convey photoreceptor signals to amacrine and ganglion cells via mono- or multisynaptic axon terminals located in the synaptic IPL. In general, they can be divided into On- and Off-bipolar cells by their depolarizing and hyperpolarizing responses to photic stimuli. However, as noted above, the two types can also be distinguished morphologically by the location of their terminal endings within the IPL. On-bipolar cells end in the more proximal portion (sublamina *b*) of the IPL, whereas the Off-bipolar cells terminate in the distal portion (sublamina *a*) of the IPL.

Immunocytochemical analysis of the amphibian retina indicated that both taurine and the taurine transporter are present in photoreceptors and Off-bipolar cells (Bulley and Shen 2010), suggesting that these neurons probably release taurine. To study the metabotropic effects of taurine, the tissues were bathed in an extracellular solution that contained an inhibitory "cocktail" consisting of Cd^{2+} (100 μM) which blocks voltage-gated Ca^{2+} channels as well as Ca^{2+}-dependent K$^+$ channels and Cl$^-$ receptors were inhibited by picrotoxin (100 μM) and strychnine (5 μM). This solution is referred to as Cd(I) throughout this chapter.

Voltage-gated channels in bipolar cells in retinal slices were activated by a series of 25 ms voltage steps ranging from −100 to +60 mV. Figure 7.1a shows an example of currents elicited in an On- and Off-bipolar cell by this protocol with the tissue bathed in the Cd(I) solution (black traces). The large-sustained rectifying K$^+$ currents appeared at voltages more positive than 0 mV. Since voltage-gated Ca^{2+} channels in bipolar cells were blocked in Cd(I) solution, it is likely that Ca^{2+}-dependent K$^+$ (K_{Ca}) channel currents were abolished. Therefore, the large-sustained K$^+$ currents were mediated by outward rectifier K_v channels. Interestingly, taurine (80 μM) in Cd(I) solution greatly enhanced K_v current preferentially in Off-bipolar cells; whereas taurine slightly suppressed K_v currents in On-bipolar cells (red traces, On- and Off-bipolar cell). The K_v currents recovered to control levels within 1–2 min after taurine was withdrawn. The enhancement by taurine of K_v currents was likely a metabotrophic effect, since Cd(I) blocked taurine's ionotropic action on Cl$^-$-permeable receptors. Notice that the currents at very negative voltages, which were

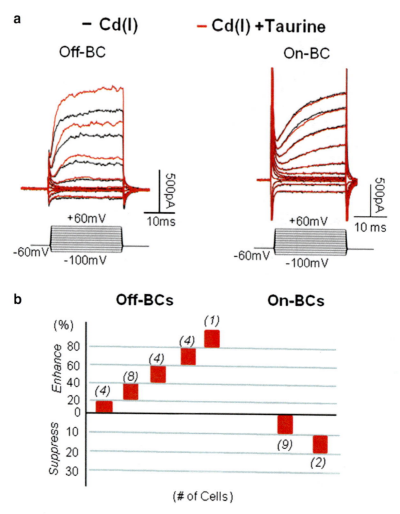

Fig. 7.1 Metabotropic effect of taurine on regulation of K_v currents in bipolar cells studied in whole-cell patch-clamp recording. (**a**) Current responses were recorded from both On- and Off-bipolar cells activated by multiple depolarizing steps from −100 to +60 mV with 10 mV increments. Large K_v currents were generated from bipolar cells at positive voltages in the Cd(I) solution; taurine (80 μM) enhanced K_v currents in Off-bipolar cells, but not On-bipolar cells. (**b**) Taurine-induced percentage increase or decrease of K_v currents in Off- and On-bipolar cells measured at +60 mV

commonly observed in bipolar cells, were not affected by taurine; these currents likely represent inward rectifier K^+ (K_{ir}) channel currents, activated at −100 mV.

The percentage increase and decrease of K_v currents by taurine were quantified at +60 mV from each recorded cell. As depicted in Fig. 7.1b, the statistical analysis indicates that taurine enhancement of K_v currents varied between 10 and 80% from cell to cell. On average, taurine caused $44.37 \pm 8.5\%$ increase in K_v currents in Off-bipolar cells ($n = 21$; $p < 0.001$). Taurine inhibition of K_v currents in On-bipolar cells varied by 10–20% of control with an average inhibition of $8.32\% \pm 2.3$ ($n = 11$;

$p < 0.01$). The differing degree of regulation for K_v channels in On- and Off-bipolar cells suggests that different receptors or different intracellular metabotropic pathways might mediate taurine's actions. Interestingly, we found that the enhancement in Off-bipolar cells by taurine persisted even after the axons were severed, indicating that taurine probably activates receptors in the somatodendritic areas of the cells. Since taurine is naturally found in high concentrations in the outer retina, these findings suggest that taurine could provide metabotropic regulation of Off-bipolar cells.

7.3.2   Role of Protein Kinases in Taurine-Enhanced K_v Currents

To confirm that a metabotropic pathway was involved in taurine regulation of K_v channels in bipolar cells, we tested the effects of specific inhibitors of PKA and PKC, the major intracellular signaling pathways in metabotropic regulation. We recorded K_v current changes with and without taurine using a single voltage step protocol to activate K_v channels in Off-bipolar cells. K_v currents were elicited by a 25 ms voltage step from −60 to +60 mV in control Cd(I) solution, both with and without taurine. As illustrated in Fig. 7.2a, the K_v current was significantly enhanced by taurine, and the histogram in Fig. 7.2b shows that, on average, the K_v currents (measured at the time indicated by the dotted line) were enhanced by approximately 80%. When GF109203X (10 μM), a membrane permeable, PKC-specific inhibitor, was added to the control solution, it suppressed the sustained portion of the K_v currents in Off-bipolar cells (Fig. 7.2c, black trace), and interestingly, it completely suppressed the enhancing effect of taurine on the K_v currents (red trace). When PKC was inhibited pharmacologically, taurine only produced on average a 3.7% increase in the K_v currents of Off-bipolar cells (Fig. 7.2d, $n = 4$). Although this result strongly suggests that the internal pathway activated by taurine is PKC-specific, the result was confounded by the fact that the taurine-enhancing effect was still observed when we dialyzed PKI, a PKA-specific antagonist, into the cell during whole-cell recording (data not shown). It is therefore not possible to conclude that a PKA-mediated pathway is involved in taurine's metabotropic action on Off-bipolar cells.

7.3.3   Taurine Shapes the Voltage Response of Off-Bipolar Cells

A physiological role for the regulation of K_v channels in Off-bipolar cells was addressed by recording bipolar cell membrane changes in response to a transient depolarization pulse that mimicked the physiological response of the cells to light offset. Although bipolar cells do not normally generate spike potentials, Off-bipolar cells typically show a transient "overshoot" depolarization at light offset due to the fact that they receive large, multivesicular phasic signals at the termination of a light pulse. Outward rectifying K_v channels play an important role in the membrane repolarization in "overshoot"; they are activated at positive voltages, which allow the outward flow of K^+ ions, and thus effectively controlling the repolarization rate.

92

M.J.M. Rowan et al.

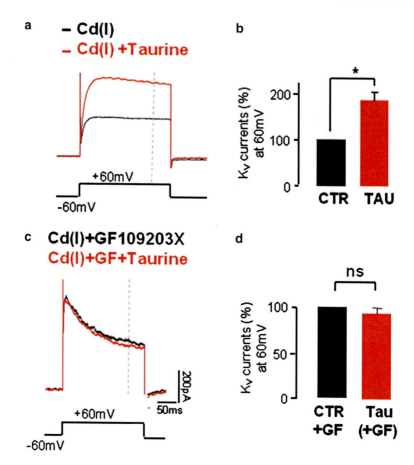

Fig. 7.2 Taurine enhanced K_v currents in Off-bipolar cells via a PKC pathway (**a**) K_v currents in a displaced Off-bipolar cell, elicited by a single voltage step, were significantly enhanced by taurine measured at the dotted line. (**b**) Histogram depicting taurine enhancing K_v current about 80% in Off-bipolar cells. (**c**) K_v current of Off-bipolar cells was suppressed by the specific PKC inhibitor GF109203X (*black trace*), whereas taurine had no effect on the K_v currents in the presence of GF109203X. GF109203X also reduced the sustained K_v currents in the cells. (**d**) The taurine effect in Off-bipolar cells was completely eliminated ($p > 0.05$) compared to control when the PKC inhibitor GF109203X was applied

In current-clamp mode, a protocol was developed to mimic the phasic signal received by Off-bipolar cells. Extracellular Cd(I) was still applied to block the network effects of bath applied taurine, and a brief (5 ms) current injection was used to depolarize the cell to +20 mV from its resting level. After depolarization, the cell repolarized via multiple mechanisms, including K_v channel activation. To confirm that K_v channels are critical for establishing the repolarization rate in Off-bipolar cells, we employed the broad spectrum K^+ channel blocker tetraethylammonium (TEA, 20 mM). As expected, with the cell in control solution, the rate of repolarization following current injection was rapid and described a single exponential decay

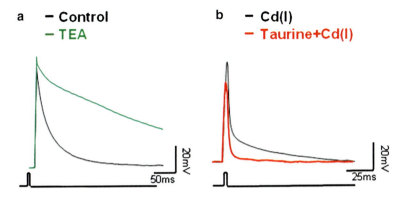

Fig. 7.3 Voltage waveform in Off -bipolar cells is shaped via changes in K⁺ conductance (**a**) Off-bipolar cell under current-clamp conditions was subjected to short 5 ms current injections sufficient to depolarize the cell to +20 mV (*black*). After blocking K⁺ conductance with TEA (20 mM), an identical current injection caused a larger voltage increase with an increase in the time to decay. (**b**) The voltage waveform is reduced in amplitude with taurine (80 μM) with an increase in the decay rate as compared with control

function with a time constant of $\tau = 32$ ms (Fig. 7.3a). This was in stark contrast to the slower repolarization rate ($\tau = 272$ ms) when TEA was used to block K⁺ channels (including K_v). This is not an ideal state in which to encode multiple large and transient signals. It seems likely that the long, delayed repolarization with TEA reflects the ability of K_v channels to ease the cell back to its resting level following brief, intense stimuli.

The test was repeated in Off-bipolar cells in Cd(I) control with and without 80 μM taurine. Since the enhancement of K_v currents by taurine would cause more K⁺ efflux and increase the membrane conductance, signals are expected to decay more rapidly as K⁺ efflux increases, and the cell will quickly repolarize following depolarization. As shown in Fig. 7.3b, the control signal (*black trace*) was greater in response to depolarization, and the decay rate was slower compared with the results when taurine was in the bath (*red trace*). Taurine caused a fast signal decay and limited the duration of depolarization, indicating that taurine plays a role in encoding offset light "overshoot" signals in Off-bipolar cells, especially those receiving rapid, transient responses from cones.

7.3.4 Taurine Suppresses Voltage-Gated Ca²⁺ Channels Via a Metabotropic Pathway

To determine whether taurine directly influences synaptic vesicle release in retinal neurons, the effect of taurine was studied on voltage-gated Ca²⁺ channels in isolated neurons free of inputs from the retinal network. All control Ca²⁺ channel currents

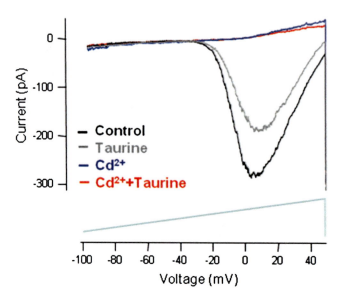

Fig. 7.4 Taurine suppression of voltage-gated Ca⁺ channel currents in third-order neurons recorded in the Ba(I)—modified Ringer solution. Voltage-activated Ca⁺ channel currents were generated with a voltage ramp from −100 to +50 mV within 2 s. Ca^{2+} channel currents were suppressed by taurine (100 μM). Both Ca^{2+} channel currents and taurine inhibition were fully blocked by the voltage-gated Ca^{2+} channel blocker Cd^{2+}

were recorded in a Ba^{2+}- and TEA-containing Ringer's solution, since Ca^{2+} channels are more permeable to Ba^{2+} than to Ca^{2+}, and TEA blocked all voltage-gated K^+ channels. In addition, taurine-sensitive Cl^--permeable channels were blocked with picrotoxin (100 μM) and strychnine (10 μM). TTX (1 μM) was used to block voltage-gated Na^+ channels. The combination of Ba^{2+} and the cocktail of inhibitors served as the control solution referred to as Ba(I) throughout this chapter.

Ca^{2+} channel currents were studied using a ramp protocol with voltages increasing from −100 to +50 mV over a period of 2 s. An inward Ca^{2+} channel current was generated by the ramp in Ba(I), and the addition of taurine (100 μM) caused a reduction in the current (Fig. 7.4); this suppression could be washed away after 1–2 min in the control solution. Ca^{2+} channel currents could be fully blocked by bath application of Cd^{2+} (100 μM), a voltage-dependent Ca^{2+} channel blocker. In the Cd^{2+}-containing Ba(I) solution, taurine had no further effect, indicating that the inhibitory effect of taurine was on voltage-gated Ca^{2+} channels. Taurine regulation of Ca^{2+} channels was further examined in the third-order neurons and some bipolar cells. We found that taurine suppressed Ca^{2+} channel currents rather consistently in most of the cells recorded, but the degree of suppression varied among the different types of neurons.

Several studies have reported that taurine activates metabotropic $GABA_B$ receptors (Kontro and Oja 1990; Smith and Li 1991; Nicoll 2004). To investigate whether the $GABA_B$ receptor was involved in mediating taurine regulation of Ca^{2+} channels,

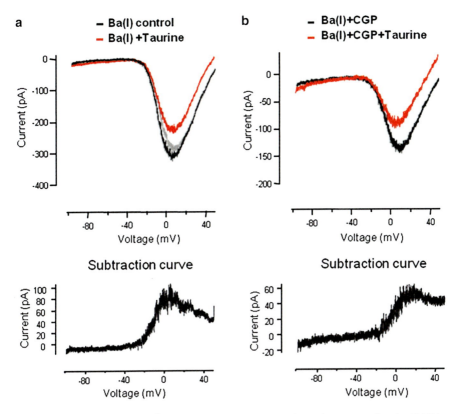

Fig. 7.5 Taurine suppressed Ca^{2+} channels via a receptor or pathway that was unrelated to $GABA_B$ receptors. (**a**) Taurine reduced voltage-gated Ca^{2+} channel currents in control with external Ba(I), and the suppression could be washed away (the *light trace*); the suppression was estimated by subtraction of taurine-regulated Ca^{2+} current curve from control curve. (**b**) With CGP55845 applied, taurine still suppressed Ca^{2+} currents; the effect of taurine was derived by subtracting the two curves with and without taurine

CGP55845 (10 µM), a potent $GABA_B$ receptor inhibitor, was used to block the $GABA_B$ receptors on retinal neurons. We then tested the inhibitory effect of taurine on Ca^{2+} channel currents with and without presence of CGP55845. The results obtained in recordings from amacrine cells (known to express $GABA_B$ receptors), Fig. 7.5a, b show an example of recordings. In the Ba(I) control, taurine suppressed Ca^{2+} channel currents, and the effect was recovered after taurine was removed (Fig. 7.5a). Then, CGP55845 was applied in Ba(I). We observed that CGP55845 itself had a suppressive effect on Ca^{2+} channel currents in some cells (Fig. 7.5b). With CGP55845 in the solution, taurine still suppressed Ca^{2+} channel currents. Subtraction of the Ca^{2+} channel activation curves with and without taurine during control and in CGP55845 conditions revealed that the same amount of current was suppressed by taurine (see the lower panels in Fig. 7.5a, b), supporting that CGP55845 did not influence the effect of taurine. This finding indicates that taurine is not acting on the $GABA_B$ receptor in amacrine cells.

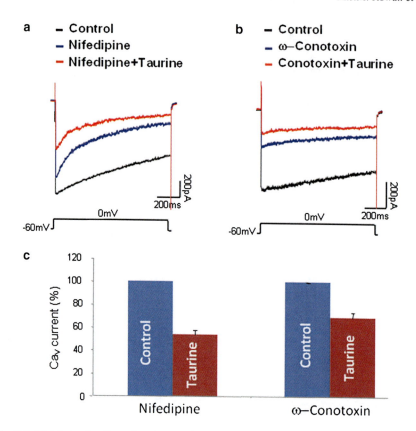

Fig. 7.6 Metabotropic effect of taurine regulates voltage-gated L-type and N-type Ca^{2+} channels. (**a**) Ca^{2+} channel currents were generated in a single voltage depolarization step from -60 to 0 mV in a second. Nifedipine reduced most of sustained Ca^{2+} channel currents and had less effect on the transient Ca^{2+} channel currents, and taurine suppressed Ca^{2+} channel current with nifedipine. (**b**) ω-conotoxin suppressed both sustained and transient Ca^{2+} currents and taurine further reduced Ca^{2+} channel currents. (**c**) With nifedipine and ω-conotoxin, average suppressive effect of taurine on Ca^{2+} channel currents in the third-order neurons ($n=6$)

7.3.5 Both L- and N-Type Ca^{2+} Channel Currents Are Regulated by Taurine

To determine which types of Ca^{2+} channels are regulated by taurine, the specific L-type Ca^{2+} channel blocker, nifedipine (100 μM), and N-type Ca^{2+} channel blocker, ω-conotoxin (1 μM) were used to block the individual Ca^{2+} channel subtype in third-order neurons.

Voltage-gated Ca^{2+} channel currents were generated with a single depolarizing step from -60 to 0 mV, the voltage that elicits a maximum Ca^{2+} channel current in neurons. Figure 7.6a shows that nifedipine partially suppressed the inward Ca^{2+}

channel currents (30% at the beginning of the pulse and 70% near the end of the pulse), and with nifedipine to block L-type channels, taurine reduced approximately 50% of the Ca^{2+} channel current (measured near the end of the pulse) (red trace). On the other hand, application of the N-type channel blocker ω−conotoxin reduced by almost 50% the Ca^{2+} channel currents. And in the presence of ω-conotoxin, taurine suppressed an additional 30% of the Ca^{2+} channel current. These examples show that taurine regulates both the L- and N-type Ca^{2+} channels in the retinal neurons. The histogram bars show that on average taurine suppressed 46% of nifedipine-insensitive Ca^{2+} channel currents ($n=6$) and 24% of ω-conotoxin-insensitive Ca^{2+} channel currents ($n=6$).

7.4   Discussion

We report here the results of a series of experiments which strongly suggest that taurine has a metabotropic effect that is insensitive to GABA and glycine receptor antagonists. In addition, we have shown that taurine regulates voltage-gated K_v channels and Ca^{2+} channels via a metabotropic pathway. Also shown is the more ubiquitous role of taurine in the modification of synaptic transmission and cell excitability by its ability to modulate voltage-gated channels of retinal neurons. These properties may also apply to other brain regions and sensory systems in the CNS.

Although the exact site at which taurine acts in Off-bipolar cells is unknown, a recent study from this laboratory indicates that taurine enhances delayed rectifier K_v channels in third-order neurons by activating the serotonin receptor, $5HT_{2A}$ via a metabotropic pathway (paper in press). It is possible that taurine regulates K_v channels in Off-bipolar cells by the same receptor and same mechanism we find in third-order neurons. Our findings provide new information that (1) a novel metabotropic taurine response in retinal neurons is completely distinct from the earlier described metabotropic $GABA_B$ receptor, (2) the taurine response leads to the downstream activation of K_v channels and appears to act through a PKC-dependent pathway within Off-bipolar cells, (3) this activity is completely different in On-bipolar cells which apparently lack the receptor site for this PKC-dependent effect, and (4) since taurine increased K^+ efflux through K_v channels and increased the rate of cell repolarization, it may possibly shape the rapid "overshoot" signals in Off-bipolar cells. In fact, a transient "overshoot" depolarization at light offset occurs in Off-bipolar cells, but not in On-bipolar cells. Therefore, the effect of taurine on regulation of K_v channels may allow Off-bipolar cells to increase the encoding rate of light-induced signals.

In general, taurine's ability to increase K_v channel activity may serve as a mechanism for protecting cells from over-excitation. As voltage-gated K^+ channels are endogenous suppressors of neuronal excitability, their modulation leads to potential therapeutic benefits by reducing neuronal hyperexcitability in stroke and epileptic patients. Delayed outward rectifier K_v channels are particularly important in somatodendritic excitability in hippocampal and cortical pyramidal neurons (Bekkers 2000; Zhang et al. 2010; Korngreen and Sakmann 2000). Taurine has been shown to exert a neuroprotective role on the brain when administered before or after a middle cerebral artery occlusion (Wang et al. 2007), and it prevents stroke in stroke-prone spontaneously hypertensive rats (Yamori et al. 2010). It is possible that its metabotropic effect is one of the potential mechanisms for neuroprotection.

7.5 Conclusion

We find that taurine at physiological concentrations (μM) generate metabotropic effects that are taurine specific, insensitive to the antagonists of GABA and glycine receptors. The metabotropic effects of taurine play a dual function by enhancing voltage-gated and delayed rectifier K_v channels and suppressing high voltage-activated Ca^{2+} channels in retinal neurons.

Acknowledgments This work was supported by research grants to W.S. from NSF (1021646) and NIH (EY14161).

References

Bekkers JM (2000) Properties of voltage-gated potassium currents in nucleated patches from large layer 5 cortical pyramidal neurons of the rat. J Physiol 525(Pt 3):593–609

Bulley S, Shen W (2010) Reciprocal regulation between taurine and glutamate response via Ca^{2+}-dependent pathways in retinal third-order neurons. J Biomed Sci 17(Suppl I):55

Burkhardt DA (1970) Proxamal negative response of frog retina. J Neurophysiol 33:405–420

Cunningham RA, Miller RF (1976) Taurine; its selective action on neuronal pathways in the rabbit retina. Brain Res 117:341–345

DiGiorgio RM, Macainone S, DeLuca G (1977) Subcellular distribution of hypotaurine oxidase activity in ox retina. Life Sci 20:1657–1662

El Edrissi A, Trenkner E (1999) Growth factors and taurine protect against excitotoxicity by stabilizing calcium homeostasis and energy metabolism. J Neurosci 19:9459–9468

Kontro P, Oja SS (1990) Interactions of taurine with GABAB binding sites in mouse brain. Neuropharmacology 29:243–247

Korngreen A, Sakmann B (2000) Voltage-gated K + channels in layer 5 neocortical pyramidal neurones from young rats: subtypes and gradients. J Physiol 525(Pt 3):621–639

Li YP, Lombardini JB (1991) Inhibition by taurine of the phosphorylation of specific synaptosomal proteins in the rat cortex: Effects of taurine on the stimulation of calcium uptake in mitochondria and inhibition of phosphoinositide turnover. Brain Res 553:89–96

Maple BR, Zhang J, Pang JJ, Gao F, Wu SM (2004) Characterization of displaced bipolar cells in the tiger salamander retina. Vision Res 45:697–705

Nicoll RA (2004) My close encounter with GABAB receptors. Biochem Pharmacol 68:1667–1674

Parnas I, Parnas H (2010) Control of neurotransmitter release: from Ca2+ to voltage-dependent G-protein coupled receptors. Pflugers Arch 460:975–990

Smith SS, Li J (1991) GABAB receptor stimulation by baclofen and taurine enhances excitatory amino acid induced phosphatidylinositol turnover in neonatal rat cerebellum. Neurosci Lett 132(1):59–64

Wang GH, Jiang ZL, Fan XJ, Zhang L, Li X, Ke KF (2007) Neuroprotective effect of taurine against focal cerebral ischemia in rats possibly mediated by activation of both GABAA and glycine receptors. Neuropharmacology 52(5):1199–1209

Wu H, Jin Y, Wei J, Jin H, Sha D, Wu J-Y (2005) Mode of action of taurine as a neuroprotector. Brain Res 1038:123–131

Wu J-Y, Prentice H (2010) Role of taurine in the central nervous system. J Biomed Sci 17(Suppl 1):S1

Yamori Y, Taguchi T, Hamada A, Kunimasa K, Mori H, Mori M (2010) Taurine in health and diseases: consistent evidence from experimental and epidemiological studies. J Biomed Sci 17(Suppl 1):S6

Zhang X, Bertaso F, Yoo JW, Baumgärtel K, Clancy SM, Lee V, Cienfuegos C, Wilmot C, Avis J, Hunyh T, Daguia C, Schmedt C, Noebels J, Jegla T (2010) Deletion of the potassium channel Kv12.2 causes hippocampal hyperexcitability and epilepsy. Nat Neurosci 13(9): 1056–1058

Chapter 8
The Effect of Folic Acid on GABA$_A$-B 1 Receptor Subunit

Kizzy Vasquez, Salomon Kuizon, Mohammed Junaid, and Abdeslem El Idrissi

Abstract Autism contains a spectrum of behavioral and cognitive disturbances of childhood development that is manifested by deficits in social interaction, impaired communication, repetitive behavior, and/or restricted interest. Much research has been dedicated to finding the genes that are responsible for autism, but less than 10% of the cases can be attributed to one gene. Autism prevalence has increased in the last decade and there may be environmental components that are leading to this increase. There are reports of disruption of epigenetic mechanisms controlling the regulation of gene expression as probable cause for autism. Folic acid (FA) is prescribed to women during pregnancy, and can cause epigenetic changes. GABAergic pathway is involved in inhibitory neurotransmission in the central nervous system and plays a crucial role during early embryonic development. Autism may entail defect or deregulation of the GABAergic receptor pathway in the brain. Gamma-aminobutyric acid (type A) beta 1 receptor (GABRB1) disruption has been implicated in autism. In the present study, we investigated GABRB1 expression in response to FA supplementation in neuronal cells. Western blot analysis showed GABRB1 protein levels increased

K. Vasquez(✉)
Department of Biology, College of Staten Island, Staten Island, NY, USA

City University of New York Graduate School, New York, NY, USA

Department of Structural Neurobiology, New York State Institute for Basic Research, Staten Island, NY 10314, USA
e-mail: kizzyv@gmail.com

S. Kuizon • M. Junaid
Department of Structural Neurobiology, New York State Institute for Basic Research, Staten Island, NY 10314, USA

A. El Idrissi
Department of Biology, College of Staten Island, Staten Island, NY, USA

City University of New York Graduate School, New York, NY, USA

A. El Idrissi and W.J. L'Amoreaux (eds.), *Taurine 8*, Advances in Experimental Medicine and Biology 775, DOI 10.1007/978-1-4614-6130-2_8,
© Springer Science+Business Media New York 2013

in the FA-treated cells in a concentration-dependent manner. FA-dependent increased expression of GABRB1 was further confirmed at the mRNA level using quantitative RT-PCR. These results suggest that epigenetic control of gene expression may affect the expression of GABRB1 and disrupt inhibitory synaptic transmission during embryonic development.

Abbreviations

GABRB1	Gamma-aminobutyric acid type A receptor beta 1 subunit
$GABA_A$	Gamma-aminobutyric acid type A receptors
FA	Folic acid
GAD	Glutamic acid decarboxylase

8.1   Introduction

Folic acid (FA) supplementation was recommended and mandated by the US Food and Drug Administration (FDA) (MMWR 1992; USPSTF 2009) to help alleviate the higher incidences of neural tube defects (NTDs) that were prevalent in the human population. Prior to these guidelines, NTDs were on the rise, and FA supplementation of cereals and grains led to over 70% decrease in the incidences of NTDs.

The mandate for FA supplementation and fortification was done to ensure that pregnant women were receiving enough FA during the crucial first few weeks of pregnancy. It was recognized through experimental and observational studies that the FA is necessary for proper neural tube closure and development (Wolff et al. 2009). In addition to fortification, the FDA also recommended a daily intake of 400 μg to 4 mg of FA (dosage depending on the individual risk factor for NTDs) for all women of childbearing age or women with a history of a prior child with NTD, respectively (USPSTF 2009). These FDA guidelines were instrumental in reducing the occurrence of NTDs by more than 70%. Consequently, there have been reports of excessive FA use by pregnant women, and there is a lack of substantiated reports of the long-term effects of a FA-rich diet. Recent epidemiological reports are pointing to excessive FA supplementation with increased incidences of asthma (Bekkers et al. 2012) and autism (Beard et al. 2011) among children.

According to a recent study from our laboratory, DNA microarray analysis revealed that a significant number of genes were either up-regulated or down-regulated in response to folic acid supplementation (Junaid et al. 2011). One of the prominent down-regulated genes was *FMR1*, which has been associated with fragile X mental retardation syndrome. In fragile X syndrome, the expansion of CGG trinucleotide repeats in the promoter region results in decreased synthesis of *FMR* protein (McLennan et al. 2011). In an *fmr1* knockout mouse model, levels of gamma-aminobutyric acid (GABA) pathway were shown to be affected (Zhang et al. 2009). The GABA neurotransmitter system, that includes GABA$_A$ and GABAC classes of ligand-gated ion channels, is the major inhibitory system, and responds to the ligand GABA. Postsynaptic GABA$_A$ receptors are heteropentamer proteins composed of subunits from seven different families (α1–6, β1–3, γ1–3, δ, ε, θ, and ρ1–3) mostly occurring in a 2α:2β:γ stoichiometry that performs a GABA-mediated chloride channel (Olsen and Sieghart 2009). The subunit composition of the GABA$_A$ receptor complex determines many of its functional and pharmacological properties. GABA$_A$ receptor β1 (GABRB1) is a subunit involved in inhibitory affects on neurotransmission. The spatiotemporal patterns of electrical signaling in many brain areas are controlled by the inhibitory neurotransmission mediated by GABA and GABA$_A$ receptor complex (Huang et al. 2007).

In the present study we have investigated the effect of FA supplementation on the expression of GABRB1 at the protein and mRNA levels in SY5Y cells. We found the GABRB1 gene expression is increased both at the mRNA and protein levels by FA in a concentration-dependent manner.

8.2   Methods

8.2.1   Cell Culture and Treatment

Human SH-SY5F neuroblastoma cells were grown in advanced D-MEM/F12 media (GIBCO), supplemented with glutamine and antibiotics (penicillin and streptomycin, 100 μg/ml each), and incubated at 37°C. At 40–50% confluency, cells were differentiated by incubating with 20 μM retinoic acid (Sigma), and after 24 h, FA was added at concentrations between 0 and 250 ng/ml medium. Following incubation with FA, cells were collected after 24 h by centrifugation (800×g for 5 min), and total RNA was prepared immediately by lysing the cells in TRIZOL reagent (Invitrogen Life Technologies, Carlsbad, CA) and further purified using RNeasy kit (Qiagen, Valencia, CA) as described earlier (Junaid et al. 2011). Purity of the RNA fractions was evaluated using agarose-formaldehyde gel. All the RNA samples had a 260/280 nm absorption ratio of 2 or more, and the amount of 28S band was two times greater in area and intensity than 18S band as determined by formaldehyde-agarose gel.

8.2.2   SDS-PAGE and Western Blot

SDS-PAGE and Western blot analysis were done as previously described (Kuizon et al. 2010). Briefly, the cell pellets were prepared by low-speed centrifugation (800×g for 5 min), followed by solubilization in the 2× reducing sample buffer by boiling for 5 min. Proteins were separated on a 10% Tris/HEPES/SDS-polyacrylamide gels (Life Gels), and electrotransferred onto Optitran BA-S83 reinforced nitrocellulose membrane (Whatman). The nonspecific sites on the blots were blocked with 5% blotto, and incubated with the anti-rabbit GABRB1 antibody (cell signaling) at 1:1,000 dilution, overnight at 4°C under constant shaking conditions. The membrane was washed once with TBST (5 min) and twice with TBS (5 min each), and incubated with secondary anti-rabbit IgG coupled with horse raddish alkaline peroxidase (Sigma) for 1 h at room temperature (1:2,000 dilution). The chemiluminescent signal was developed by using the SuperSignal West Pico substrate (Thermo Scientific, Rockford, IL). Band intensities were captured using a UVP Bioimaging System.

8.2.3   Quantitative Reverse Transcriptase Polymerase Chain Reaction

Purified RNA samples were converted to cDNA with the RT2 First Strand cDNA synthesis Kit (Qiagen, Valencia, CA) using oligo-dT primer as described by the manufacturer. These templates were then added to the ready-to-use RT2 SYBR Green qRT-PCR Master Mix (Qiagen, Valencia, CA) in a 96-well plate format. The specific primers obtained from Qiagen, Valencia, CA, were used (PPH01886A for GABRB1 as target gene, and PPH00150A for GAPDH as the housekeeping gene). RT-PCR was performed in an Eppendorf Realplex Mastercycler. Realplex software collected threshold cycle (Ct) values were tabulated for all the genes and triplicate readings were averaged. Calculation of the relative fold changes in gene expression for pair-wise comparison was done by using the $\Delta\Delta$Ct method. The consistency and accuracy of the experiment was evaluated by examination of Ct value for the housekeeping genes.

8.2.4   Statistical Analysis

Each value was expressed as the mean $\pm$ SEM for three independent experiments. Differences were considered statistically significant when the calculated p value was less than 0.05.

8.3   Results

8.3.1   Folic Acid Supplementation Increased the Amount of Gamma-Aminobutyric Acid Type A Receptors Beta 1 Subunit Protein

In our previous study, we investigated the FA supplementation effects in lymphoblastoid cells. This model, however, is not representative of neuronal system. Hence, we evaluated the effect of FA supplementation in a neuronal cell model using the

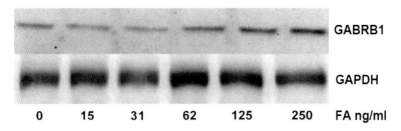

Fig. 8.1 Representative Western blot showing expression of GABRB1 as a function of increasing FA concentration. *Lower panel* shows GAPDH used as a housekeeping control

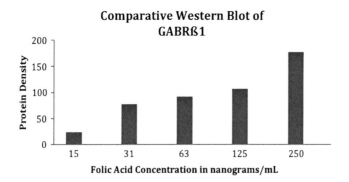

Fig. 8.2 Relative intensities of GABRB1 and GAPDH proteins in response to FA supplementation

SH-SY5Y cell line. Figure 8.1 shows the expression of GABRB1 protein in FA-supplemented SH-SY5Y cells. The Western blot showed increased expression of the protein in response to increasing FA concentration. GAPDH staining was used as a housekeeping control, since expression of this gene did not change in response to FA treatment (Junaid et al. 2011). Figure 8.2 shows the relative band intensities of the GABRB1 and GAPDH stained spots. There was over 176% increase in GABRB1 expression as the concentration of FA was increased to 250 ng/ml.

8.3.2   Folic Acid Supplementation Increases the Gene Expression of GABRβ1 in SH-SY5Y Cells

Table 8.1 shows the quantitative RT-PCR data of the GABRB1 mRNA expression in response to FA supplementation in SH-SY5Y cells. The qRT-PCR experiment was performed three independent times. We have used GAPDH and HPRT1 genes as housekeeping controls for comparison. The results confirmed the GABRB1 protein expression data. FA increased the expression of GABRB1 mRNA in a concentration-dependent manner. We found a significant increase in the GABRB1 expression both at the protein and mRNA levels with FA concentration as little as 15 ng/ml media.

Table 8.1 A comparative RT-PCR assay was used to quantify the expression of $GABA_A$ receptor beta 1 mRNA for the SH-SY5Y folic acid treated cells 0-250 ng/mL, as well as a positive control 18snRNA. A significant difference was found between 0ng/mL, 31ng/mL, 125ng/mL, and 250ng/mL

FA	Gene expression ratio trial 1	Gene expression ratio trial 2	Gene expression ratio trial 3	Average	STD Dev	t-test
0	1.1	1.0	1.0	1		
15 ng	1.8	5.3	1.3	2.8	1.68	0.52
31 ng	4.1	4.1	1.7	3.3	1.65	0.047
63 ng	2.2	4.0	0.4	2.2	1.30	0.324
125 ng	6	11	19.7	12.23	7.54	0.048
250 ng	18.4	12	17	15.8	3.36	0.001

$GABA_A$ receptor subunit ß1 protein levels increased in the folic acid-treated neuronal cells in comparison to control according to our t-test there is a significance difference ($p < 0.05$).

$GABA_A$ receptors subunit ß1 mRNA was up-regulated, and it seems to be folic acid dosage dependent in the neuronal cells. The pathway on which folic acid affects the expression of the gene is not known, but it warrants investigation. There is a possible correlation with $R^2 = 0.76373$, an indication that as the folic acid concentration increase, the expression of the GABRB1 increases as well.

8.4 Discussion

There has been an emphasis in the past decade to increase blood levels of synthetic FA in women, to prevent a deficiency that was resulting in NTDs in newborns. FA supplementation represents by far the most successful effort in preventing debilitating disabilities by simply increasing dietary intake of a water-soluble vitamin. Until now, there are almost no reports of any adverse effects of FA supplementation. Only recently a few epidemiological studies are pointing to the association of asthma and autism with the timing of FA supplementation guidelines.

Recently, we have reported widespread change in gene expression in lymphoblastoid cells, in response to FA supplementation. There is a likelihood that such gene expression changes are also experienced by the developing fetus that has been in an environment of FA supplementation. There is ample evidence of beneficial effects of FA in preventing NTDs. However, unregulated excessive FA supplementation may exert unintended side effects. In the present study, we have found increased expression of GABRB1 with increased FA supplementation in a human neuroblastoma cell line. Concentration of FA reported in human blood was able to cause significant increases in GABRB1 expression both at protein and mRNA levels (Pfeiffer et al. 2004).

How increased expression of one of the subunits of the receptor complex affects the function of $GABA_A$ receptor remains to be determined. Moreover, the effect of FA on other $GABA_A$ receptor subunits needs to be studied. Increases in expression of GABRB1 can result either from direct methylation of the promoter or from a change

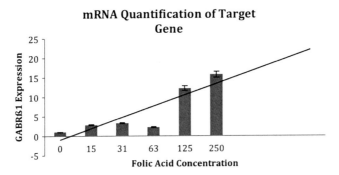

Fig. 8.3 Demonstrating the average values from three qRT-PCRs of the target gene, gamma-aminobutyric acid (type A) beta 1 receptor subunit and the R^2 shows an obvious trend of increased expression with concentration $y = 2.8914 \times -3.8978$; $R^2 = 0.76373$

in expression of another gene which may indirectly control the expression. There are no reports in the literature that the GABRB1 promoter undergoes methylation.

Recent epidemiological reports (Figs. 8.3 and 8.4) are suggesting an increased incidence of autism with the mandate to increase FA supplementation (Beard et al. 2011). While this could be mere coincidence, if proved true, the occurrence of autism may be reduced by regulating the intake of FA to harness the beneficial effects without reaching concentrations that begin to exert unintended changes.

In summary, this study has provided preliminary evidence that FA supplementation of the SH-SY5Y neuroblastoma cells resulted in up-regulation of the gene and the gene product of gamma-aminobutyric acid type A receptor beta 1 subunit. Further investigation is necessary to determine how FA-induced GABRB1 changes affect the neuronal cell function and development, and how this could be leading to a mechanism that may be causing the pathogenesis of autism.

In our research we observed an increase in gamma-aminobutyric acid type A beta 1 receptor subunit protein (Fig. 8.1). There was also an increase in the mRNA as measured by qRT-PCR in Fig. 8.3 with significance difference $p < 0.05$. There is a positive correlation of $R^2 = 0.76373$ with a linear regression formula that is indicative of a dosage-dependent response that was observed in our experiment.

As the number of receptors increases the sensitivity to GABA increases; therefore the cell is more likely to become inhibited, and less excitatory. Alteration in GABAergic receptors has long been suspected to be responsible for the pathogenesis of autism (Fatemi et al. 2010). Gamma-aminobutyric acid type A beta 1 receptor subunit gene has also been implicated with alcoholism, another illness that has been on the rise amount women in the past decade.

8.5   Conclusion

In summary, this study shows that supplementation folic acid in the medium of the SH-SY5Y cells resulted in up-regulation of the gene and the gene product of gamma-aminobutyric acid type A receptor beta 1 subunit. Further investigation is

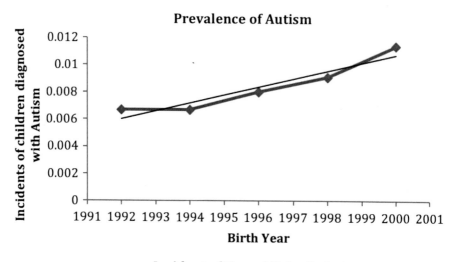

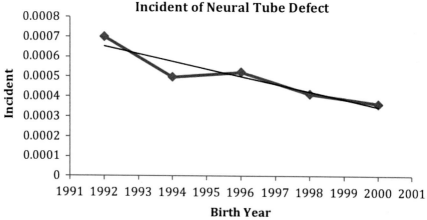

Fig. 8.4 Depicts the trend of neural tube defect decreasing, at the same time the prevalence of autism was increasing

needed to determine how folic acid induces change on the cells, and how this could be leading to an alteration of the $GABA_A$ receptors subunit, a physiological change that has been observed as one of the mechanism that may be causing the pathogenesis of autism.

Acknowledgments This study is supported by funds from the New York State Office for People with Developmental Disabilities, The City University of New York, The College of Staten Island, and Louis Stokes Alliance for Minority Participation a division of National Science Foundation (NSF). Financial assistance to Kizzy Vasquez is gratefully acknowledged.

References

Beard CM, Panser LA, Katusic SK (2011) Is excess folic acid supplementation a risk factor for autism? Med Hypotheses 77(1):15–17

Bekkers MB, Elstgeest LE, Scholtens S, Haveman-Nies A, de Jongste JC, Kerkhof M, Koppelman GH, Gehring U, Smit HA, Wijga AH (2012) Maternal use of folic acid supplements during pregnancy, and childhood respiratory health and atopy. Eur Respir J 39(6):1468–1474

Fatemi SH, Reutiman TJ, Folsom TD, Rooney RJ, Patel DH, Thuras PD (2010) mRNA and protein levels for GABAAalpha4, alpha5, beta1 and GABABR1 receptors are altered in brains from subjects with autism. J Autism Dev Disord 40(6):743–750

Huang ZJ, Di Cristo G, Ango F (2007) Development of GABA innervation in the cerebral and cerebellar cortices. Nat Rev Neurosci 8:673–686

Junaid MA, Kuizon S, Cardona J, Azher T, Murakami N, Pullarkat RK, Brown WT (2011) Folic acid supplementation dysregulates gene expression in lymphoblastoid cells -implications in nutrition. Biochem Biophys Res Commun 412(4):688–692

Kuizon S, DiMaiuta K, Walus M, Jenkins EC Jr, Kuizon M, Kida E, Golabek AA, Espinoza DO, Pullarkat RK, Junaid MA (2010) A critical tryptophan and Ca2+ in activation and catalysis of TPPI, the enzyme deficient in classic late-infantile neuronal ceroid lipofuscinosis. PLoS One 5(8):e11929

McLennan Y, Polussa J, Tassone F, Hagerman R (2011) Fragile x syndrome. Curr Genomics 12(3):216–224

MMWR (1992) Recommendations for the use of folic acid to reduce the number of cases of spina bifida and other neural tube defects. Morb Mortal Wkly Rep 41:1–7

Olsen RW, Sieghart W (2009) GABA A receptors: subtypes provide diversity of function and pharmacology. Neuropharmacology 56(1):141–148

Pfeiffer CM, Fazili Z, McCoy L, Zhang M, Gunter EW (2004) Determination of folate vitamers in human serum by stable-isotope-dilution tandem mass spectrometry and comparison with radio-assay and microbiologic assay. Clin Chem 50:423–432

U.S. Preventive Services Task Force (2009) Folic acid for the prevention of neural tube defects: U.S. Preventive Services Task Force recommendation statement. Ann Intern Med 150:626–632

Wolff T, Witkop CT, Miller T, Syed SB, U.S. Preventive Services Task Force (2009) Folic acid supplementation for the prevention of neural tube defects: an update of the evidence for the U.S. Preventive Services Task Force. Ann Intern Med 150(9):632–639

Zhang A, Shen CH, Ma SY, Ke Y, El Idrissi A (2009) Altered expression of Autism-associated genes in the brain of Fragile X mouse model. Biochem Biophys Res Commun 379(4):920–923

Chapter 9
Taurine Counteracts the Suppressive Effect of Lipopolysaccharide on Neurogenesis in the Hippocampus of Rats

Gaofeng Wu, Takashi Matsuwaki, Yoshinori Tanaka, Keitaro Yamanouchi, Jianmin Hu, and Masugi Nishihara

Abstract Neurogenesis has been generally accepted to happen in the subventricular zone lining the lateral ventricle and subgranular zone (SGZ) in the hippocampus of adult mammalian brain. Recent studies have reported that inflammatory stimuli, such as injection of lipopolysaccharide (LPS), impair neurogenesis in the SGZ. Taurine, a sulfur-containing β-amino acid, is a major free intracellular amino acid in many tissues of mammals and having various supplementary effects on the mammalian body functions including the brain. Recently, it has been also reported that taurine levels in the brain significantly increase under stressful conditions. The present study was aimed to evaluate the possible beneficial effects of taurine on the neurogenesis in the SGZ under the condition of acute inflammatory stimuli by LPS. Adult male rats were intraperitoneally injected with taurine once a day for 39 days. Twenty-four hours before the animals were sacrificed on the last day of taurine treatment, LPS was injected simultaneously with bromodeoxyuridine (BrdU). Immunohistochemistry for BrdU, Ki67, and Iba-1 in the brain was performed, and serum levels of TNF-α and IL-1β 2 h after LPS injection were determined.

G. Wu
Department of Veterinary Physiology, Graduate School of Agricultural and Life Sciences,
The University of Tokyo, 1-1-1 Yayoi, Bunkyo-ku, Tokyo 113-8657, Japan

College of Animal Science and Veterinary Medicine, Shenyang Agricultural University,
Shenyang, Liaoning Province 110866, P.R. China

T. Matsuwaki • Y. Tanaka • K. Yamanouchi • M. Nishihara(✉)
Department of Veterinary Physiology, Graduate School of Agricultural and Life Sciences,
The University of Tokyo, 1-1-1 Yayoi, Bunkyo-ku, Tokyo 113-8657, Japan
e-mail: amnishi@mail.ecc.u-tokyo.ac.jp

J. Hu(✉)
College of Animal Science and Veterinary Medicine, Shenyang Agricultural University,
Shenyang, Liaoning Province 110866, P.R. China
e-mail: hujianmin59@163.com

A. El Idrissi and W.J. L'Amoreaux (eds.), *Taurine 8*, Advances in Experimental
Medicine and Biology 775, DOI 10.1007/978-1-4614-6130-2_9,
© Springer Science+Business Media New York 2013

The results showed that LPS significantly decreased the number of immunoreactive cells for both BrdU and Ki67 in the SGZ, while increased that for Iba-1, all of which were restored by taurine administration. Meanwhile, the serum concentrations of TNF-α and IL-1β were significantly increased, which were significantly attenuated by taurine administration. These results suggest that taurine effectively maintains neurogenesis in the SGZ under the acute infectious condition by attenuating the increase of microgliosis in the hippocampus as well as proinflammatory cytokines in the peripheral circulation.

Abbreviations

CNS	Central nervous system
SGZ	Subgranular zone
LPS	Lipopolysaccharide
BrdU	Bromodeoxyuridine
BBB	Blood–brain barrier
SVZ	Subventricular zone
DG	Dentate gyrus
TNF-α	Tumor necrosis factor-α
IL-1β	Interleukin-1β

9.1   Introduction

Neurogenesis, which is a process of generating functionally integrated neurons from progenitor cells, was traditionally believed to occur only during embryonic stages in the mammalian central nervous system (CNS) (Ramon y Cajal 1913). Recently it has become generally accepted that new neurons were indeed added in discrete regions of the adult mammalian CNS (Gross 2000; Kempermann and Gage 1999; Lie et al. 2004). In the adult mammalian brains, neural progenitor cells are located in specific sites within the brain including the subventricular zone (SVZ) of the lateral ventricle and in the subgranular zone (SGZ) of the dentate gyrus (DG) in the hippocampus and generate thousands of new neurons each day. Neurogenesis out of these two regions appears to be extremely limited, or nonexistent, in the intact adult mammalian CNS. The role of neurogenesis for brain function is still unclear, but some experimental evidence suggests its involvement in memory formation and mood regulation (Shors et al. 2001). Recent studies have reported that exposure to acute or chronic stressors during prenatal period (Lemaire et al. 2006; Mandyam et al. 2008) or in adulthood (Ekdahl et al. 2003) suppressed both cell proliferation and cell survival in the hippocampus of adult male rat. As a kind of acute stressor, inflammation has been shown to impair neurogenesis in the SGZ. Ekdahl et al. delivered lipopolysaccharide (LPS) into the cortex of young adult rats continuously by an osmotic mini pump, which reduced the number of newborn neurons to 85% (Ekdahl et al. 2003). It has also been reported that LPS could acutely inhibit proliferation of neural precursor cells in the DG in adult rats (Fujioka and Akema 2010).

Taurine, a sulfur-containing β-amino acid is the major free intracellular amino acid presenting in many tissues of human and other mammals. There are a lot of evidence that taurine exerts physiological and pharmacological functions, such as maintaining structure and function normally of the hematological system, immune system, reproductive system, visual system, cardiovascular system, and nervous system. Taurine demonstrates multiple cellular functions including the roles in the CNS as a neurotransmitter and as a trophic factor in CNS development. In addition, taurine maintains the structural integrity of the membrane and regulates calcium transport and homeostasis, acting as an osmolyte, neuromodulator, neurotransmitter, and neuroprotector against L-glutamate induced-neurotoxicity (Wu and Prentice 2010). Clinical and electroencephalographic observations and psychological tests have revealed the beneficial effects of taurine on various neuronal functions (Montanini and Gasco 1974). In addition, El Idrissi (2008) has reported that taurine could improve learning and retention in aged mice possibly through alternations of the GABAergic system, suggesting that taurine may play a vital role in the CNS. Recently, moreover, it was reported that taurine levels in the brain significantly increase under stressful conditions (Wu et al. 1998). The present study was aimed to clarify whether LPS may acutely depress neurogenesis in the DG of hippocampus and the effects of taurine on the decrease of neurogenesis induced by LPS.

9.2 Materials and Methods

9.2.1 Animals

Twenty-four male Wistar-Imamichi rats were obtained from Imamichi Institute for Animal Reproduction (Tsuchiura, Japan). The animals were group housed and given 1 week to acclimate to vivarium conditions before the start of the experiment. They were kept under controlled illumination (lights on, 500–1,900 h) and temperature and given food and water ad libitum. The experiments were conducted according to the Guidelines for the Care and Use of Laboratory Animals, Graduate School of Agriculture and Life Sciences, the University of Tokyo.

9.2.2 Experimental Procedure

LPS (055:B5) and 5-bromo-2-deoxyuridine (BrdU) were purchased from Sigma-Aldrich (St. Louis, Mo) and dissolved in saline and phosphate-buffered saline (PBS), respectively. From day 1 to day 39, the animals were intraperitoneally injected with taurine (200 mg/kg; Sigma-Aldrich; $n = 12$) or saline ($n = 12$) every day. Internal jugular vein cannulation was performed on all the animals on the day 33 or 34. After recovery of 1 week, all the animals were treated with a single injection of BrdU (200 mg/kg). LPS (1 mg/kg) or the same volume of saline was injected simultaneously with the BrdU injection. Blood samples were collected via inserted cannula 2 h after LPS injection for tumor necrosis factor-α (TNF-α) and interleukin-1β (IL-1β) detection. All rats were sacrificed 24 h after LPS injection and brains were collected for BrdU, Ki67, and Iba-1 immunohistochemical examination.

9.2.3 Immunohistochemistry for BrdU, Ki67, and Iba-1

Immunohistochemistry of BrdU, Ki67, and Iba-1 was performed in every six sections (30 μm) through the whole hippocampus. The primary antibodies employed were sheep anti-BrdU (1:300; Exalpha Biologicals, Shirley, MA), rabbit anti-Iba-1 (1:500; Wako, Osaka), and mouse anti-Ki67 (1:25, Dako, Carpinteria, CA). Briefly, the slices mounted on slides were incubated in primary antibody solution overnight at 4 °C, following intervening rinses with 0.03% Triton-X PBS, incubated with secondary antibodies, donkey anti-sheep Alexa 488 and donkey anti-rabbit Alexa 568, both from Molecular Probes (Eugene, OR), and HRP-conjugated goat anti-rat (Simple Stain, Nichirei Bioscience, Tokyo, Japan), the later followed by the common signal amplification method using DAB, peroxide, and NiCl. Positive cells located in the SGZ were counted bilaterally through the DG.

9.2.4 Cytokines Assay

The serum concentrations of TNF-α and IL-1β were measured with ELISA Kit (GE Healthcare, Buckinghamshire, UK). The samples were tested in duplicate. According to the manufacturer, the ELISA was highly specific for TNF-α and IL-1β, with a detection limit of 15 pg/ml for TNF-α and 12 pg/ml for IL-1β. ELISA was performed according to the manufacturer's instructions. Final results were expressed as picograms per milliliter.

9.2.5 Statistical Analysis

Values are mean ± SE. Comparisons were performed by using one-way analysis of variance (ANOVA), followed by Tukey's post hoc test. Differences were considered significant at $P < 0.05$.

9.3   Results

9.3.1   Taurine Counteracted the Effects of LPS on the Numbers of BrdU-, Ki67-, and Iba-1-Positive Cells in the SGZ

Injection of LPS significantly reduced the number of immunoreactive (ir) cells for both BrdU and Ki67 in the SGZ compared with the control group ($P<0.05$). Pretreatment with taurine counteracted this suppressive effect of LPS on the number of BrdU-ir cells partially and that of Ki67-ir cells almost completely, while taurine per se did not affect these numbers in rats without LPS injection (Fig. 9.1). On the other hand, LPS injection significantly increased the number of Iba-1-ir cells in the SGZ, which was restored by taurine pretreatment. The number of Iba-1-ir cells in animals treated with taurine was comparable with that of the control animals. These results indicate that LPS decreases the neurogenesis and increases the microgliosis in the SGZ, which can be restored by taurine pretreatment (Fig. 9.1).

9.3.2   Taurine Restored the Effect of LPS on the Concentrations of TNF-α and IL-1β in the Serum

The serum concentrations of TNF-α and IL-1β were significantly increased after LPS injection ($P<0.05$), but taurine pretreatment inhibited the increase in the concentration of IL-1β almost completely and that of TNF-α at least partially (Fig. 9.2). These results suggest that taurine reduces the peripheral inflammatory responses induced by LPS.

9.4   Discussion

Adult neurogenesis in the DG of the hippocampus undergoes five development stages including proliferation, differentiation, migration, axon or dendrite targeting, and synaptic integration. In the proliferation process, newborn neurons are located within the SGZ in the DG, followed by project through the granular cell layer and short tangential processes that extent along the border of the granule cell layer and hilus, the process of which would happen within 25 h. BrdU is incorporated into newly synthesized DNA at S-phase of cell division and can be detected even after that phase. On the other hand, Ki67, a nuclear protein, is expressed in the cells just undergoing S-phase (9.5 h) at the moment of sampling. In the present study, taurine restored the number of BrdU-ir cells after LPS injection significantly but not completely while it maintained the number of Ki67-ir cells at the control level, suggesting that LPS has negative effects not only on the S-phase in the proliferation period and that taurine can suppress the influence of LPS completely in the S-phase but not in the other phases.

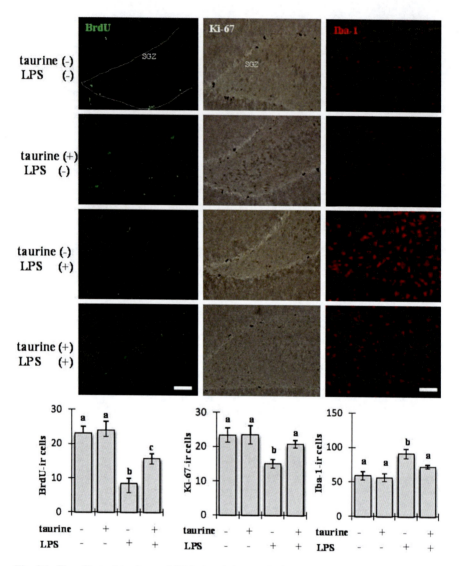

Fig. 9.1 The effect of taurine on LPS-induced changes in the number of BrdU-ir, Ki67-ir, and Iba-1-ir cells. LPS was injected 24 h before sacrifice at the same time of BrdU injection. The sections were cut 30 μm thick, cells located in the SGZ were counted bilaterally through the DG under fluorescence microscope or light microscope in every six sections, and the mean value was calculated. Each column and vertical bar represents the mean ± SEM. Values with different letters are significantly different ($P < 0.05$, ANOVA followed by the Tukey's post hoc test)

The main cell type in the CNS that is responsible for immunity is microglia, which is the main source of inflammatory cytokines and in which Iba-1 a calcium-binding protein is restrictedly expressed. It was reported that 24 h after the injection of LPS (1 mg/kg), Iba-1 immunoactivity was remarkably increased in all the regions

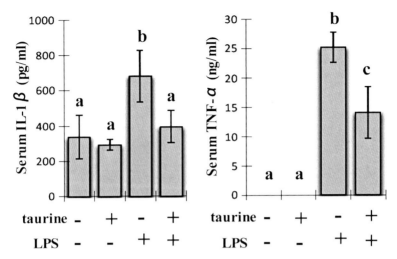

Fig. 9.2 The effect of taurine on LPS-induced increases in the serum levels of IL-1β and TNF-α. LPS was injected 24 h before sacrifice at the same time of BrdU injection. Serum samples were collected 2 h after LPS injection. Each column and vertical bar represents the mean ± SEM. Values with different letters are significantly different ($P < 0.05$, ANOVA followed by the Tukey's post hoc test)

of the hippocampus of mice (Banks et al. 1995). Monje et al. (2003) also found that LPS given systemically caused an increase in microglia activation and a decrease in neurogenesis. The results of the present study are well consistent with those of these previous studies, suggesting that inflammatory signals are propagated across the blood–brain barrier (BBB) to the CNS. On the other hand, taurine administration effectively decreased the number of Iba-1-positive cells, indicating that taurine could play an anti-inflammatory role in the CNS.

The anti-inflammatory effects of taurine have been reported by Sun et al. (2012). We also have previously demonstrated that taurine could act an anti-inflammatory role in alcoholic liver disease in rats, via reversing the elevation of IL-2, IL-6, and TNF-α in the serum caused by alcohol (Wu et al. 2009). Inflammation is an active process with the purpose of removing or inactivating potentially damaging agents and tightly regulates neurogenesis in the SGZ. LPS, a gram-negative bacterial cell surface proteoglycan, is known as a bacterial endotoxin. LPS has been used for a model of the systemic inflammatory response induced by infections. There are contradictory reports about whether LPS induces the disruption of the BBB or not. It was reported that LPS damaged the BBB and made it relatively porous (Gailard et al. 2001; Gailard et al. 2003; Xaio et al. 2001), while it has also been reported that 1 mg/kg LPS did not rupture the BBB directly because of low dosages (Chung et al. 2010). However, systematic administration of LPS could induce the increase of serum cytokines, and it has now become clear that some cytokines such as IL-1α, IL-1β, and TNF-α can directly and rapidly cross the BBB by saturable transport systems to induce inflammation and injury in the CNS (Banks et al. 1995).

In the CNS, cytokines from peripheral tissue stimulate microglia to produce more cytokines that act in autocrine or paracrine manner. It has been reported that peripherally injected LPS could cause significant changes of blood cytokines within 5 h (Kim et al. 2007), and the maximum change of TNF-α and IL-1β was found to be at 1 h and 3 h after LPS injection, respectively (Kuboyama et al. 2007). In the present study, therefore, blood samples were collected 2 h after LPS injection, and inhibition of the LPS-induced increase of TNF-α and IL-1β by taurine pretreatment was observed, indicating that taurine can play an anti-inflammatory role in the LPS-induced inflammatory processes as well as that caused by alcohol. As taurine can be transported into the CNS through BBB, this anti-inflammatory role of taurine can be played both in peripheral and central tissues.

9.5 Conclusion

Taurine may counteract the LPS-induced decrease in neurogenesis in the DG of the hippocampus through the suppression of inflammatory responses in the brain as well as in the peripheral tissue.

Acknowledgments This work was supported in part by the project of JSPS Fellowship for Researcher in JAPAN (2010 long term) to GW and a Grant-in-Aid for Scientific Research (23228004) from the JSPS to MN.

References

Banks WA, Kastin AJ, Broadwell RD (1995) Passage of cytokines across the blood-brain barrier. Neuroimmunomodulation 4:241–248
Chung DW, Yoo KY, Hwang IK, Kim DW, Chung JY, Lee CH, Choi JH, Choi SY, Youn HY, Lee IS, Won MH (2010) Systemic administration of lipopolysaccharide induces cyclooxygenase-2 immunoreactivity in endothelium and increases microglia in the mouse hippocampus. Cell Mol Neurobiol 30:531–541
Ekdahl CT, Claasen JH, Bonde S, Kokaia Z, Lindvall O (2003) Inflammation is detrimental for neurogenesis in adult brain. Proc Natl Acad Sci USA 100:13632–13637
El Idrissi A (2008) Taurine improves learning and retention in aged mice. Neurosci Lett 436:19–22
Fujioka H, Akema T (2010) Lipopolysaccharide acutely inhibits proliferation of neural precursor cells in the dentate gyrus in adult rats. Brain Res 1352:35–42
Gailard PJ, de Boer AB, Breimer DD (2003) Pharmacological investigations on lipopolysaccharide-induced permeability changes in the blood-brain barrier in vitro. Microvasc Res 65:24–31
Gailard PJ, Voorwinden LH, Nielsen JL, Ivanov A, Atsumi R, Engman H, Ringborn C, de Boer AG, Breimer DD (2001) Establishment and functional characterization of an in vitro model of the blood-brain barrier, comprising a co-culture of brain capillary endothelial cells and astrocytes. Eur J Pharm Sci 12:215–222
Gross CG (2000) Neurogenesis in the adult brain: death of a dogma. Nat Rev Neurosci 1:67–73
Kempermann G, Gage FH (1999) New nerve cells for the adult brain. Sci Am 280:48–53
Kim YW, Kim KH, Ahn DK, Kim HS, Kim JY, Lee DC, Park SY (2007) Time-course changes of hormones and cytokines by lipopolysaccharide and its relation with anorexia. J Physiol Sci 57:159–165

Kuboyama N, Sato Y, Saito T, Abiko Y (2007) Stimulation of TNF-α, IL-1β and IL-6 levels in rat serum by actinobacillus actinomycetemcomitans LPS challenge. Med Biol 151:21–26

Lemaire V, Lamarque S, Le Moal M, Piazza PV, Abrous DN (2006) Postnatal Stimulation of the pups counteracts prenatal stress-induced deficits in hippocampal neurogenesis. Biol Psychiatry 59:786–792

Lie DC, Song H, Colamarino SA, Ming GL, Gage FH (2004) Neurogenesis in the adult brain: new strategies for central nervous system diseases. Annu Rev Pharmacol Toxicol 44:399–421

Mandyam CD, Crawford EF, Eisch AJ, Rivier CL, Richardson HN (2008) Stress experienced in utero reduces sexual dichotomies in neurogenesis, microenvironment, and cell death in the adult rat hippocampus. Dev Neurobiol 68:575–589

Monje ML, Toda H, Palmer TD (2003) Inflammatory blockade restores adult hippocampal neurogenesis. Science 302:1760–1765

Montanini R, Gasco P (1974) Taurine in the treatment of diffuse cerebral arteriopathies. Clinical and electroencephalographic observations and psychological tests. Clin Ter 71:427–436

Ramon y Cajal S (1913) Degeneration and regeneration of the nervous system. Oxford Univ. Press, London

Shors TJ, Miesegaes G, Beylin A, Zhao M, Rydel T, Gould E (2001) Neurogenesis in the adult is involved in the formation of trace memories. Nature 410:372–376

Sun M, Zhao Y, Gu Y, Xu C (2012) Anti-inflammatory mechanism of taurine against ischemic stroke is related to down-regulation of PARP and NF-κB. Amino Acids 42:1735–1747

Wu G, Yang J, Sun C, Luan X, Shi J, Hu J (2009) Effect of taurine on alcoholic liver disease in rats. Amino Acid 36:457–464

Wu JY, Prentice H (2010) Role of taurine in the central nervous system. J Biomed Sci 17(Suppl 1):S1

Wu JY, Tang XW, Schloss JV, Faiman MD (1998) Regulation of taurine biosynthesis and its physiological significance in the brain. Adv Exp Med Biol 442:339–345

Xaio H, Banks WA, Niehoff ML, Moreley JE (2001) Effect of LPS on the permeability of the blood-brain barrier to insulin. Brain Res 896:36–42

Chapter 10
Perinatal Taurine Exposure Programs Patterns of Autonomic Nerve Activity Responses to Tooth Pulp Stimulation in Adult Male Rats

Sawita Khimsuksri, J. Michael Wyss, Atcharaporn Thaeomor, Jarin Paphangkorakit, Dusit Jirakulsomchok, and Sanya Roysommuti

Abstract Perinatal taurine excess or deficiency influences adult health and disease, especially relative to the autonomic nervous system. This study tests the hypothesis that perinatal taurine exposure influences adult autonomic nervous system control of arterial pressure in response to acute electrical tooth pulp stimulation. Female Sprague–Dawley rats were fed with normal rat chow with 3% β-alanine (taurine depletion, TD), 3% taurine (taurine supplementation, TS), or water alone (control, C) from conception to weaning. Their male offspring were fed with normal rat chow and tap water throughout the experiment. At 8–10 weeks of age, blood chemistry, arterial pressure, heart rate, and renal sympathetic nerve activity were measured in anesthetized rats. Age, body weight, mean arterial pressure, heart rate, plasma electrolytes, blood urea nitrogen, plasma creatinine, and plasma cortisol were not significantly different among the three groups. Before tooth pulp stimulation,

S. Khimsuksri
Department of Physiology, Faculty of Medicine, Khon Kaen University,
Khon Kaen 40002, Thailand

Department of Oral Biology, Faculty of Medicine, Khon Kaen University,
Khon Kaen 40002, Thailand

J.M. Wyss
Department of Cell, Developmental and Integrative Biology,
School of Medicine, University of Alabama at Birmingham, Birmingham, AL 35294, USA

A. Thaeomor
School of Physiology, Institute of Science, Suranaree University of Technology,
Nakhonratchasima 30000, Thailand

J. Paphangkorakit
Department of Oral Biology, Faculty of Dentistry, Khon Kaen University,
Khon Kaen 40002, Thailand

D. Jirakulsomchok • S. Roysommuti (✉)
Department of Physiology, Faculty of Medicine, Khon Kaen University,
Khon Kaen 40002, Thailand
e-mail: sanya@kku.ac.th

A. El Idrissi and W.J. L'Amoreaux (eds.), *Taurine 8*, Advances in Experimental
Medicine and Biology 775, DOI 10.1007/978-1-4614-6130-2_10,
© Springer Science+Business Media New York 2013

low- (0.3–0.5 Hz) and high-frequency (0.5–4.0 Hz) power spectral densities of arterial pressure were not significantly different among groups while the power spectral densities of renal sympathetic nerve activity were significantly decreased in TD compared to control rats. Tooth pulp stimulation did not change arterial pressure, heart rate, renal sympathetic nerve, and arterial pressure power spectral densities in the 0.3–4.0 Hz spectrum or renal sympathetic nerve firing rate in any group. In contrast, perinatal taurine imbalance disturbed very-low-frequency power spectral densities of both arterial pressure and renal sympathetic nerve activity (below 0.1 Hz), both before and after the tooth pulp stimulation. The power densities of TS were most sensitive to ganglionic blockade and central adrenergic inhibition, while those of TD were sensitive to both central and peripheral adrenergic inhibition. The present data indicate that perinatal taurine imbalance can lead to aberrant autonomic nervous system responses in adult male rats.

Abbreviations

AP	Arterial pressure
C	Control
ETPS	Electrical tooth pulp stimulation
HR	Heart rate
HF	High frequency
LF	Low frequency
MAP	Mean arterial pressure
RA	Renal sympathetic nerve activity
SD	Sprague–Dawley
TD	Taurine depletion
TS	Taurine supplementation

10.1 Introduction

The perinatal environment can greatly influence adult health and disease. Perinatal programming can occur at any time from conception to weaning (i.e., from pregnancy through lactation), and perinatal life encompasses most of the critical periods during which such programming occurs, e.g., embryogenesis, placental development, organogenesis, prepartum maturation of the fetus, and sucking period or lactation (Zhang et al. 2011). Since different organs develop at different rates, timing is important in specifying the organs affected by this programming. Nutritional environment in utero and early life has been widely studied. Maternal undernutrition produces low birth weight and has long-term effects on the offspring (Godfrey and Barker 2001). Adult abnormalities from undernutrition include physiological changes and other diseases, especially coronary heart disease, hypertension, diabetes, impaired glucose tolerance, and renal dysfunction. Some of these changes can be permanent and can transfer to the next generation (primarily via epigenetic mechanisms). In addition, maternal overnutrition causes developmental adaptation and can have long-term effects in adult life (Hanson and Gluckman 2008).

Previous studies suggest that neonatal taurine may program the adult nervous system, particularly the central regulators of the autonomic nervous system. For example, neonatal administration of taurine modifies the hippocampal function and can affect learning and memory in later life (Franconi et al. 2004; Suge et al. 2007). Perinatal taurine depletion impairs autonomic nervous system regulation of arterial pressure in adult rat offspring (Roysommuti et al. 2009a). In such animals, the sympathetic nervous system can be markedly overactive in response to a high sugar diet (Roysommuti et al. 2009a; Thaeomor et al. 2010). Baroreceptor reflex function also decreases in these animals, and this perinatal programming by taurine can lead to other cardiovascular dysfunctions in the adult life, e.g., hypertension and diabetes mellitus (Aerts and Van Assche 2002).

Peripheral autonomic nerve activity is closely regulated by the central autonomic nervous system and can be modulated by external and internal factors via sensory input pathways, e.g., baroreceptors, chemoreceptors, proprioceptors, and nociceptors. The sympathetic nerves differentially supply various organs that have unique functions. In the kidney, peripheral sensory activation may differentially affect sympathetic nerve fibers that supply a specific part of the kidneys (DiBona 2005a). This can occur despite a similar response of whole sympathetic nerve stimulation of the organ. Analysis by nerve recording and power spectral density analysis of integrated sympathetic nerve activity can assist in dissecting the underlying mechanisms of these effects.

Dental pulp pain or toothache is known to induce cardiovascular reflex responses (Sousa and Lindsey 2009). Increases in arterial pressure and heart rate caused by dental pain possibly lead to clinical complications, particularly in subjects prone to develop cardiovascular diseases (Brotman et al. 2007; Montebugnoli et al. 2004). Electrical tooth pulp stimulation can stimulate either sympathetic or parasympathetic nerve activity. Like most pain conveyed by trigeminal nerves, dental pain primarily activates parasympathetic rather than sympathetic vasodilator nerve activity to orofacial vessels (Drummond 1995). Thus, orofacial vasodilation is commonly observed during facial or dental surgery. In addition, orofacial pain may induce vasoconstriction in the nose and finger skin, probably via sympathetic nerve activation (Kemppainen et al. 2001). Taurine can reduce pain (Renno et al. 2008; Terada et al. 2011), but it is still unclear whether it has any physiological function in dental pain responses and what the underlying mechanisms are. This study tests the hypothesis that perinatal taurine exposure influences adult autonomic nervous system control in response to acute electrical tooth pulp stimulation.

10.2   Methods

10.2.1   Animal Preparation and Treatment

Sprague–Dawley (SD) rats were obtained from the animal unit of Faculty of Medicine, Khon Kaen University, and maintained at constant humidity ($60 \pm 5\%$), temperature $(24 \pm 1°C)$, and light cycle (06.00–18.00 h). Female SD rats were fed with normal rat chow and water ad libitum from conception to weaning. They were divided into three groups by being fed with tap water alone for control (C), 3% β-alanine in tap water for taurine depletion (TD), or 3% taurine in tap water for taurine supplementation (TS). After weaning, all male offspring were fed with normal rat chow and tap water alone throughout the experiment. All rats were studied at 8–10 weeks of age.

The experimental protocol was approved by the Animal Ethics Committee of Khon Kaen University (Khon Kaen, Thailand), based on the Ethics of Animal Experimentation of National Research Council of Thailand (Record No. AEKKU 21/2553). The study was conducted in accordance with the US National Institutes of Health Guidelines.

10.2.2 Experimental Protocol

At 8–10 weeks of age, rats were anesthetized by thiopental (50 mg/kg body weight, i.p.), and were tracheostomized, and their femoral artery and vein were inserted with polyethylene tube (PE-50 fused to PE-10), respectively. The arterial catheter was connected to a pressure transducer for continuous recording of arterial pressure pulse (BioPac Systems, Goleta, California, USA), and the venous one was connected to an infusion pump for fluid and drug injection. After 20–30 min recovery, a blood sample (1.0 ml) was drawn from the arterial catheter and the blood loss was replaced with an equal volume of saline. After laparotomy and inserting a stainless steel electrode (12 MΩ, 0.01 Taper, A-M System, Sequim, Washington, USA), renal sympathetic nerve activity was continuously recorded by connecting the electrode to DAM-80 amplifier (World Precision Instruments, Sarasota, Florida, USA) and BioPac Systems, respectively. Multiunit recording of renal nerve activity was conducted only on nerve units that responded to changes in arterial pressure following sodium nitroprusside infusion.

Body temperature was servo control at $37 \pm 0.5°C$ by a rectal probe connected to a temperature regulator controlling an overhead heating lamp. After 20–30 min recovery, the responses of arterial pressure, heart rate, and renal nerve activity to tooth pulp stimulation at different voltages and frequencies were performed before and after adrenergic blockades. The blockades were ordered from hexamethonium (a ganglionic blocker, 10 mg/kg, i.v.), clonidine (a central $α_2$-adrenergic receptor agonist, 5 μg/kg, i.v.), and prazosin (a peripheral $α_1$-adrenergic receptor antagonist, 5 μg/kg, i.v.), respectively. Full recovery periods were allowed between blockades. These doses of drugs were reported to completely block the sympathetic nerve activity in rats (Mozaffari et al. 1996). At the end of experiment, all animals were terminated by a high dose of thiopental anesthesia.

10.2.3   Electrical Tooth Pulp Stimulation

Left lower incisor of the rat was separated using a rubber dam sheet. Then a custom-made silver electrode filmed with electrode gel was placed on the left lower incisor (as a surface electrode) and a ground lead was attached to the left cheek. Before stimulation, resting period was provided to get a steady state of arterial pressure and renal sympathetic nerve activity. The electrical stimulation was applied to the tooth pulp via the surface electrode connected to a current isolator controlled by a Grass stimulator (Grass Technologies, West Warwick, Road Island, USA) at 2 ms pulse duration and 5 s delay. The current strength (0.1–1.5 mA) was adjusted by step-increasing voltage of 10, 20, and 30 V, respectively. At each voltage, the stimulating frequencies were set to 1, 5, 10, 20, and 40 Hz, respectively. Each test was set to 1 min electrical stimulation and 2 min resting.

10.2.4   Data Analyses

All recoded data were analyzed by Acknowledge software version 3.9.1 (BioPac Systems, Goleta, California, USA). The percent power spectral densities of low frequency (0.3–0.5 Hz, LF) or high frequency (0.5–4.0 Hz, HF) to the total power of LF and HF indicate sympathetic and parasympathetic nerve activity, respectively (Roysommuti et al. 2009a; Thaeomor et al. 2010). The patterns of the power spectral densities of renal sympathetic nerve activity and arterial pressure were analyzed individually by using fast Fourier transformation analysis. For each study, the graph of power spectrum was averaged from three to ten rats by using Acknowledge software.

Plasma electrolytes (sodium, potassium, chloride, and bicarbonate), blood urea nitrogen, plasma creatinine, and plasma cortisol levels were analyzed by the Srinagarind Hospital Laboratory Unit (Faculty of Medicine, Khon Kaen University, Khon Kaen, Thailand). Hematocrit was determined by a standard technique and blood sugar by standard glucostrips and a glucometer (Accu-Chek Advantage II, Roche, Indianapolis, Indiana, USA).

10.2.5   Statistical Analyses

All data are expressed as mean ± SEM. Statistical comparisons among groups were tested by using one-way ANOVA and a post hoc Duncan's multiple range test. Student's paired t-test was used to compare values within a group. Significant criteria were p-value of less than 0.05.

Table 10.1 Basic parameters of the rats in all experimental groups

Parameters	C	TD	TS
n	10	10	9
Age (weeks)	9.2 ± 0.3	9.2 ± 0.2	9.1 ± 0.2
Body weight (g)	284 ± 4	284.1 ± 11	277 ± 8
Blood urea nitrogen (mg/dl)	26.31 ± 0.94	22.99 ± 0.80	26.65 ± 2.49
Creatinine (mg/dl)	0.22 ± 0.01	0.20 ± 0.02	0.23 ± 0.02
Sodium (mEq/l)	142.80 ± 1.43	141.00 ± 0.50	142.50 ± 1.40
Potassium (mEq/l)	4.22 ± 0.23	3.87 ± 0.11	4.15 ± 0.23
Bicarbonate (mEq/l)	23.56 ± 0.38	22.84 ± 0.89	21.56 ± 0.37
Chloride (mEq/l)	103.90 ± 1.21	105.11 ± 0.77	105.88 ± 1.63
Hematocrit (%)	47.10 ± 0.46	46.90 ± 0.60	46.44 ± 0.93
Blood sugar (mg/dl)	107.80 ± 3.43	106.11 ± 3.37	95.22 ± 6.18
Plasma cortisol (μg/dl)	1.71 ± 0.19	2.04 ± 0.22	1.94 ± 0.18
MAP (mm Hg)	104 ± 8	103 ± 6	108 ± 6
HR (bpm)	369 ± 14	377 ± 9	377 ± 19
RA (spikes/s)	52.73 ± 8.11	55.87 ± 6.06	58.48 ± 10.38

Values are mean ± SEM. All parameters were not significantly different among groups. *C* control group, *TD* taurine-depleted group, *TS* taurine-supplemented group, *MAP* mean arterial pressure, *HR* heart rate, *RA* renal sympathetic nerve activity

10.3 Results

10.3.1 General Baseline Data

Compared to control treatment, perinatal taurine depletion or supplementation did not alter body weight, blood urea nitrogen, plasma creatinine, plasma electrolytes, hematocrit, blood sugar, plasma cortisol, mean arterial pressure, heart rate, low- and high-frequency power spectral densities of arterial pressure, and renal sympathetic nerve firing rate in adult animals (Table 10.1). Electrical tooth pulp stimulation at different voltages and frequencies had minimal effects on arterial pressure and heart rate in the present study (data not shown). These responses were well preserved in all groups after any adrenergic blockade. In all groups, mean arterial pressures significantly decreased nearly equally after adrenergic drug administration, and the values remained low during tooth pulp stimulation.

10.3.2 Power Spectral Density of Arterial Pressure

Perinatal taurine depletion or supplementation did not affect baseline sympathetic and parasympathetic control of arterial pressure, as indicated by the similar low- and high-frequency power spectral densities of arterial pressure in experimental compared to control rats (Table 10.2). Ganglionic blockade by hexamethonium, central α_2-adrenergic stimulation by clonidine, and peripheral α_1-adrenergic blockade by prazosin each significantly decreased the sympathetic component of

Table 10.2 Power spectral density of arterial pressure before (baseline) and after drug administration

Parameters	C	TD	TS
Baseline			
LF/(LF + HF) (%)	2.6 ± 0.6	4.2 ± 1.3	2.7 ± 0.7
HF/(LF + HF) (%)	97.4 ± 0.6	95.8 ± 1.3	97.3 ± 0.7
Hexamethonium			
LF/(LF + HF) (%)	$0.4 \pm 0.1*$	$0.2 \pm 0.1*$	$0.1 \pm 0.0*$
HF/(LF + HF) (%)	$99.6 \pm 0.1**$	$99.8 \pm 0.1**$	$99.9 \pm 0.0**$
Clonidine			
LF/(LF + HF) (%)	$0.2 \pm 0.0*$	$0.4 \pm 0.2*$	$0.2 \pm 0.0*$
HF/(LF + HF) (%)	$99.8 \pm 0.0**$	$99.6 \pm 0.2**$	$99.8 \pm 0.0**$
Prazosin			
LF/(LF + HF) (%)	$0.3 \pm 0.1*$	$0.5 \pm 0.2*$	$0.3 \pm 0.1*$
HF/(LF + HF) (%)	$99.7 \pm 0.1**$	$99.5 \pm 0.2**$	$99.7 \pm 0.1**$

Values are mean ± SEM. *C* control group, *TD* taurine-depleted group, *TS* taurine-supplemented group, *LF* low frequency (0.3–0.5 Hz); *HF* high frequency (0.5–4.0 Hz); *, ** $P < 0.05$ compared to LF and HF baselines of same groups, respectively

low-frequency power densities to a similar extent, i.e., no significant differences between groups. In contrast, parasympathetic nerve activity was significantly increased by these treatments, as indicated by increases in high-frequency power densities of the three groups.

Tooth pulp stimulation at different voltages and frequencies increased low-frequency (0.3–0.5 Hz) power spectral densities of arterial pressure similarly in all the three groups and high-frequency (0.5–4.0 Hz) power spectral densities were minimally affected (data not shown). The optimal strength of repetitive stimulation was 20 V at 20 Hz. The tooth pulp stimulation also increased power densities of very-low-frequency component (below 0.1 Hz, Fig. 10.1); however, the adrenergic-dependent patterns of very-low-frequency spectral densities differ from those observed in 0.3–4.0 Hz spectral densities (Fig. 10.2). Although acute hexamethonium infusion decreased these densities in the three groups, their frequency patterns were still preserved. Clonidine infusion decreased the very-low-frequency power spectral densities during tooth pulp stimulation much more in TD than in control and TS groups. In addition, prazosin markedly decreased these very-low-frequency densities only in TD during tooth pulp stimulation.

10.3.3   Power Spectral Density of Renal Sympathetic Nerve Activity

Despite similar renal sympathetic nerve activity (Table 10.1), perinatal taurine depletion (but not supplementation) significantly decreased low- and increased high-frequency power spectral densities of integrated renal nerve amplitudes when

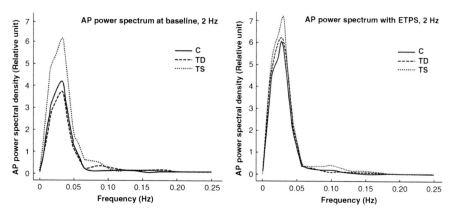

Fig. 10.1 Patterns of arterial pressure variability before (*left*) and after (*right*) tooth pulp stimulation in control (C), taurine-depleted (TD), and taurine-supplemented (TS) groups. Each graph is averaged from nine to ten rats. *AP* arterial pressure, *ETPS* electrical tooth pulp stimulation

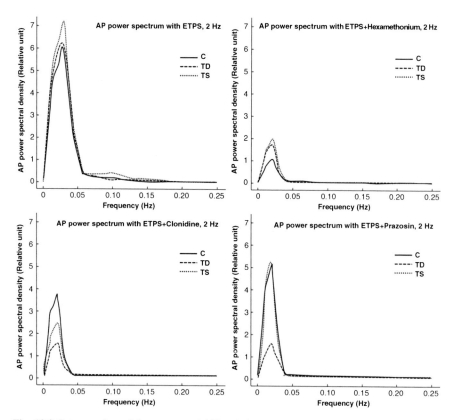

Fig. 10.2 Patterns of arterial pressure variability during tooth pulp stimulation before adrenergic blockades (*top left*), with hexamethonium (*top right*), with clonidine (*bottom left*), or with prazosin (*bottom right*) in control (C), taurine-depleted (TD) and taurine-supplemented (TS) groups. Each graph is averaged from three to ten rats. *AP* arterial pressure, *ETPS* electrical tooth pulp stimulation

Table 10.3 Power spectral densities of renal sympathetic nerve activity before (baseline) and after drug administration

Parameters	C	TD	TS
Baseline			
LF/(LF+HF) (%)	6.6±0.2	0.1±0.2 *	3.4±1.0
HF/(LF+HF) (%)	93.4±0.2	99.0±0.2 *	96.6±1.0
Hexamethonium			
LF/(LF+HF) (%)	4.8±1.7	1.5±0.3	5.8±2.1
HF/(LF+HF) (%)	95.2±1.7	98.5±0.3	94.2±2.1
Clonidine			
LF/(LF+HF) (%)	4.6±1.5	5.6±4.1	2.9±1.1
HF/(LF+HF) (%)	95.4±1.5	94.4±4.1	97.1±1.1
Prazosin			
LF/(LF+HF) (%)	8.1±2.1	4.7±1.0	1.9±1.2
HF/(LF+HF) (%)	91.9±2.1	95.3±1.0	98.1±1.2

Values are mean±SEM. *C* control group, *TD* taurine-depleted group, *TS* taurine-supplemented group, *LF* low frequency (0.3–0.5 Hz); *HF* high frequency (0.5–4.0 Hz); *$P<0.05$ compared to C

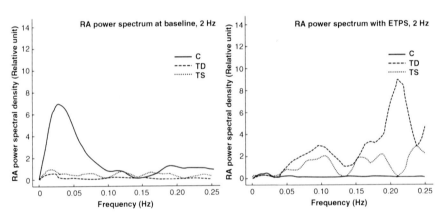

Fig. 10.3 Patterns of renal nerve activity variability before (*left*) and after (*right*) tooth pulp stimulation in control (C), taurine-depleted (TD) and taurine-supplemented (TS) groups. Each graph is averaged from nine to ten rats. *RA* renal sympathetic nerve activity, *ETPS* electrical tooth pulp stimulation

compared to responses in control fed rats (Table 10.3). Unlike arterial pressure power spectrum, renal nerve activity spectral densities at frequencies of 0.3–4.0 Hz were not altered by central and peripheral adrenergic blockade; however, perinatal taurine imbalance (especially depletion) markedly depressed very-low-frequency power densities of renal nerve activity (below 0.1 Hz) when compared to control rats (Fig. 10.3 left).

In the present study, tooth pulp stimulation at different voltages and frequencies had minimal effect on renal sympathetic nerve firing rates (increased by ~5% in all groups; data not shown). In addition, these responses were not modified by adrenergic blockade, but such stimulation altered the patterns of renal sympathetic nerve

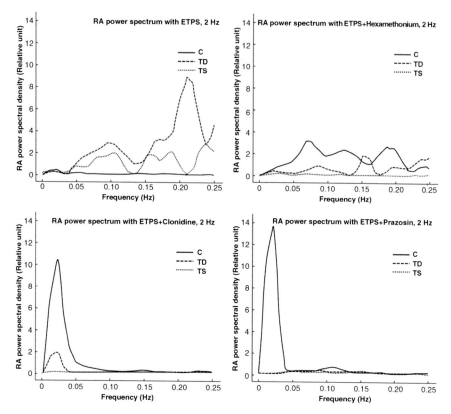

Fig. 10.4 Patterns of renal nerve activity variability during tooth pulp stimulation before adrenergic blockades (*top left*), with hexamethonium (*top right*), with clonidine (*bottom left*), or with prazosin (*bottom right*) in control (C), taurine-depleted (TD) and taurine-supplemented (TS) groups. Each graph is averaged from three to ten rats. *RA* renal sympathetic nerve activity, *ETPS* electrical tooth pulp stimulation

activity (Fig. 10.3). During tooth pulp stimulation, perinatal taurine depletion or supplementation markedly increased power spectral densities at the frequency spectra below 0.2 Hz when compared to control taurine treatment. The highest increase was observed in TD rats and the response was distributed in nearly all frequencies within these spectra. Electrical tooth pulp stimulation increased the TS (vs. TD) power density of integrated renal nerve activity more at lower frequencies (below 0.1 Hz).

During tooth pulp stimulation, the very-low-frequency power spectral densities of control animals were heightened by adrenergic inhibition (Fig. 10.4). In contrast, the power density of TS during tooth pulp stimulation was flattened by any adrenergic blockade, while that of TD was more decreased by prozosin, clonidine, and hexamethonium treatment, respectively.

10.4   Discussion

Perinatal taurine depletion predisposes adult rats to respond to a high sugar diet with sympathetic nerve overactivity and baroreceptor reflex dysregulation (Roysommuti et al. 2009a; Thaeomor et al. 2010). In contrast, perinatal taurine supplementation leads to minimal effects in adults (Roysommuti et al. 2009b; Roysommuti et al. 2011). The present data further indicate that perinatal taurine imbalance alters the pattern of autonomic nervous system function at very-low-frequency components of arterial pressure and renal sympathetic nerve activity. This phenomenon is modified by tooth pulp stimulation, which models pain-induced autonomic nerve responses. In addition, the present study suggests that both central and peripheral adrenergic mechanisms contribute to the effects of perinatal taurine exposure on the autonomic responses to tooth pulp stimulation in adults.

Peripheral autonomic nerve outflow is primarily regulated by the central nervous system, particularly hypothalamus and medulla, and many lines of evidence indicate that the sympathetic nerve response to stimuli is not uniform (Ninomiya et al. 1973; Ninomiya and Irisawa 1975). To test the overall sympathetic and parasympathetic nervous system control of arterial pressure, many indirect noninvasive techniques have been used, including power spectral analysis of arterial pressure pulse and heart rate (Malliani et al. 1994; Parati et al. 1995). In rats, the power spectral density of arterial pressure is closely correlated with the autonomic nerve activity, directly measured in many organs, and in humans, power spectrum of heart rate variability is also closely correlated to autonomic nerve activity. The frequency spectra that coincide to the autonomic control of cardiovascular system are usually within 0.2–8.0 Hz in rats (Persson et al. 1992) and 0.05–0.15 or 0.3–4.0 Hz in humans and dogs (Furlan et al. 1993; Malliani et al. 1994; Pagani et al. 1984).

Previous experiments in our laboratory suggested that frequency of arterial pressure power spectral density between 0.3 and 4.0 Hz is indicative of autonomic nervous system control of arterial pressure in rats (Roysommuti et al. 2009a; Thaeomor et al. 2010). The present study indicates that the arterial pressure power density at frequencies below 0.2 Hz may better model the effect of perinatal taurine exposure on the sympathetic nerve control of arterial pressure. This hypothesis is strengthened by the present finding that either ganglionic blockade by hexamethonium or central sympathetic inhibition by α_2-adrenergic agonist clonidine depresses the power density at this frequency spectrum. This phenomenon was not observed in the frequency spectrum of 0.3–4.0 Hz.

Several lines of evidence indicate that peripheral nociceptive stimulation increases sympathetic activity to heart, blood vessel, and kidney. However, the cardiovascular responses to pain are rather complex. Stimulation of orofacial pain fibers can increase both sympathetic and parasympathetic nerve activities (Andrew and Matthews 2002; Kemppainen et al. 2001). Tooth pulp stimulation predominately activates the parasympathetic vasodilator nerve to orofacial skin and mucosal blood vessels (Kemppainen et al. 2001; Satoh-Kuriwada et al. 2003). Sympathetic nerve activity to the heart is also activated in some cases (Allen and Pronych 1997; Schaller 2004; Sousa and Lindsey 2009). The present study is the first to indicate that electrical tooth pulp stimulation at the strengths and frequencies that have minimal effect on arterial pressure and heart rate does not alter renal sympathetic nerve firing rates but does affect its pattern of response. Our data also suggests that perinatal taurine imbalance increases the synchronization of sympathetic outflow to the kidneys in response to the tooth pulp stimulation. This effect was also reported in rats following heat stimulation of their tails (DiBona and Sawin 1999). The increased response of renal sympathetic nerve to tooth pulp stimulation may confirm previous experiments that indicated high sugar intake induces sympathetic overactivity in perinatal taurine-depleted, adult rats (Roysommuti et al. 2009a; Thaeomor et al. 2010).

In TD compared to TS rats, the power spectral densities of renal sympathetic nerve activity after hexamethonium or clonidine either at baseline or during tooth pulp stimulation were different. TD rats were more sensitive to clonidine while TS rats responded more equally to both hexamethonium and clonidine. Clonidine acts mainly at the rostral ventrolateral medulla (RVLM) to inhibit peripheral sympathetic outflow (Guyenet et al. 2001; Madden and Sved 2003), while hexamethonium blocks both parasympathetic and sympathetic outflows. These data suggest that perinatal taurine depletion has significant effects in the RVLM or related areas.

Several lines of evidence have confirmed that renal sympathetic nerve supplies three main renal structures including renin-releasing juxtaglomerular cells, renal tubules, and renal vessels. Peripheral sensory inputs (i.e., hypotension, pain, or heat) may alter both firing rates and patterns of renal sympathetic nerve activity (DiBona and Sawin 1999). Similar total firing frequency of the renal nerve can bring about different patterns of renal responses, depending on the selective action on different fibers in the nerve. Renin secretion responds actively at very-low-frequency spectrum of renal nerve activity, while renal tubular sodium reabsorption and renal blood flow respond at higher frequency ranges (DiBona 2005b). The present study indicates that perinatal taurine exposure with or without tooth pulp stimulation primarily affects the very-low-frequency component of renal nerve activity spectrum, suggesting that it programs mainly the sympathetic-mediated renal renin secretion. This notation is supported by the previous data from our laboratory indicating that perinatal taurine depletion affects the renin–angiotensin system (Thaeomor et al. 2010).

10.5   Conclusion

In summary, perinatal taurine excess or deficiency can alter autonomic nervous system and renal function in adult animals. In the present study, tooth pulp stimulation at strengths and frequencies that do not affect arterial pressure, heart rate, renal nerve firing rate, or arterial pressure power spectral densities of 0.3–4.0 Hz (the standard measures of autonomic nerve control of arterial pressure) was shown to greatly alter the pattern of very-low-frequency components (i.e., below 0.2 Hz) of both renal nerve activity and arterial pressure. These effects are altered differently by adrenergic blockade. Thus, the present data indicate that perinatal taurine depletion can significantly affect aspects of autonomic nervous system activity in adult animals, and this could significantly alter adult physiological function.

Acknowledgments This study was supported by a grant from the Faculty of Medicine, Khon Kaen University, Khon Kaen 40002, Thailand and by the US National Institutes of Health (NIH) grants AT 00477 and NS057098 (JMW).

References

Aerts L, Van Assche FA (2002) Taurine and taurine-deficiency in the perinatal period. J Perinat Med 30:281–286

Allen GV, Pronych SP (1997) Trigeminal autonomic pathways involved in nociception-induced reflex cardiovascular responses. Brain Res 754:269–278

Andrew D, Matthews B (2002) Properties of single nerve fibres that evoke blood flow changes in cat dental pulp. J Physiol 542:921–928

Brotman DJ, Golden SH, Wittstein IS (2007) The cardiovascular toll of stress. Lancet 370: 1089–1100

DiBona GF (2005a) Dynamic analysis of patterns of renal sympathetic nerve activity: implications for renal function. Exp Physiol 90:159–161

DiBona GF (2005b) Physiology in perspective: the wisdom of the body. Neural control of the kidney. Am J Physiol Regul Integr Comp Physiol 289:R633–R641

DiBona GF, Sawin LL (1999) Renal hemodynamic effects of activation of specific renal sympathetic nerve fiber groups. Am J Physiol 276:R539–R549

Drummond PD (1995) Lacrimation induced by thermal stress in patients with a facial nerve lesion. Neurology 45:1112–1114

Franconi F et al (2004) Taurine administration during lactation modifies hippocampal CA1 neurotransmission and behavioural programming in adult male mice. Brain Res Bull 63:491–497

Furlan R, Piazza S, Dell'Orto S, Gentile E, Cerutti S, Pagani M, Malliani A (1993) Early and late effects of exercise and athletic training on neural mechanisms controlling heart rate. Cardiovasc Res 27:482–488

Godfrey KM, Barker DJ (2001) Fetal programming and adult health. Public Health Nutr 4: 611–624

Guyenet PG, Schreihofer AM, Stornetta RL (2001) Regulation of sympathetic tone and arterial pressure by the rostral ventrolateral medulla after depletion of C1 cells in rats. Ann N Y Acad Sci 940:259–269

Hanson MA, Gluckman PD (2008) Developmental origins of health and disease: new insights. Basic Clin Pharmacol Toxicol 102:90–93

Kemppainen P, Forster C, Handwerker HO (2001) The importance of stimulus site and intensity in differences of pain-induced vascular reflexes in human orofacial regions. Pain 91:331–338

Madden CJ, Sved AF (2003) Rostral ventrolateral medulla C1 neurons and cardiovascular regulation. Cell Mol Neurobiol 23:739–749

Malliani A, Lombardi F, Pagani M (1994) Power spectrum analysis of heart rate variability: a tool to explore neural regulatory mechanisms. Br Heart J 71:1–2

Montebugnoli L, Servidio D, Miaton RA, Prati C (2004) Heart rate variability: a sensitive parameter for detecting abnormal cardiocirculatory changes during a stressful dental procedure. J Am Dent Assoc 135:1718–1723

Mozaffari MS, Roysommuti S, Wyss JM (1996) Contribution of the sympathetic nervous system to hypertensive response to insulin excess in spontaneously hypertensive rats. J Cardiovasc Pharmacol 27:539–544

Ninomiya I, Irisawa A, Nisimaru N (1973) Nonuniformity of sympathetic nerve activity to the skin and kidney. Am J Physiol 224:256–264

Ninomiya I, Irisawa H (1975) Non-uniformity of the sympathetic nerve activity in response to baroreceptor inputs. Brain Res 87:313–322

Pagani M et al (1984) Power spectral density of heart rate variability as an index of sympatho-vagal interaction in normal and hypertensive subjects. J Hypertens Suppl 2:S383–S385

Parati G, Saul JP, Di RM, Mancia G (1995) Spectral analysis of blood pressure and heart rate variability in evaluating cardiovascular regulation. A critical appraisal. Hypertension 25:1276–1286

Persson PB, Stauss H, Chung O, Wittmann U, Unger T (1992) Spectrum analysis of sympathetic nerve activity and blood pressure in conscious rats. Am J Physiol 263:H1348–H1355

Renno WM, Alkhalaf M, Mousa A, Kanaan RA (2008) A comparative study of excitatory and inhibitory amino acids in three different brainstem nuclei. Neurochem Res 33:150–159

Roysommuti S, Suwanich A, Jirakulsomchok D, Wyss JM (2009a) Perinatal taurine depletion increases susceptibility to adult sugar-induced hypertension in rats. Adv Exp Med Biol 643: 123–133

Roysommuti S, Suwanich A, Lerdweeraphon W, Thaeomor A, Jirakulsomchok D, Wyss JM (2009b) Sex dependent effects of perinatal taurine exposure on the arterial pressure control in adult offspring. Adv Exp Med Biol 643:135–144

Roysommuti S, Thaewmor A, Lerdweeraphon W, Khimsuksri S, Jirakulsomchok D, Schaffer SW (2011) Perinatal taurine exposure alters neural control of arterial pressure via the renin-angiotensin system but not estrogen in rats. Amino Acids 41:S84

Satoh-Kuriwada S, Sasano T, Date H, Karita K, Izumi H, Shoji N, Hashimoto K (2003) Centrally mediated reflex vasodilation in the gingiva induced by painful tooth-pulp stimulation in sympathectomized human subjects. J Periodontal Res 38:218–222

Schaller B (2004) Trigeminocardiac reflex. A clinical phenomenon or a new physiological entity? J Neurol 251:658–665

Sousa LO, Lindsey CJ (2009) Cardiovascular and baroreceptor functions of the paratrigeminal nucleus for pressor effects in non-anaesthetized rats. Auton Neurosci 147:27–32

Suge R, Hosoe N, Furube M, Yamamoto T, Hirayama A, Hirano S, Nomura M (2007) Specific timing of taurine supplementation affects learning ability in mice. Life Sci 81:1228–1234

Terada T, Hara K, Haranishi Y, Sata T (2011) Antinociceptive effect of intrathecal administration of taurine in rat models of neuropathic pain. Can J Anaesth 58:630–637

Thaeomor A, Wyss JM, Jirakulsomchok D, Roysommuti S (2010) High sugar intake via the renin-angiotensin system blunts the baroreceptor reflex in adult rats that were perinatally depleted of taurine. J Biomed Sci 17(Suppl 1):S30

Zhang S, Rattanatray L, McMillen IC, Suter CM, Morrison JL (2011) Periconceptional nutrition and the early programming of a life of obesity or adversity. Prog Biophys Mol Biol 106:307–314

Chapter 11
Regulation of Taurine Release in the Hippocampus of Developing and Adult Mice

Simo S. Oja and Pirjo Saransaari

Abstract Taurine release in mouse hippocampal slices is regulated by several neurotransmitter receptor systems. The ionotropic glutamate receptors and the adenosine receptor A_1 are the most effective. The effect of N-methyl-D-aspartate receptors is mediated via activation of the pathway involving nitric oxide and 3',5'-cyclic guanosine monophosphate. The activation of excitatory amino acid receptors causes at the same time an increase in taurine release. The activation of adenosine A_1 receptors also potentiates taurine release. The taurine released may counteract any excitotoxic effects of glutamate, particularly in the developing hippocampus.

Abbreviations

AMPA	2-Amino-3-hydroxy-5-methyl-4-isoxazolepropionate
8-Br-cGMP	8-Bromoguanosine 3',5'-cyclic monophosphate
CACA	Cis-4-aminocrotonate
cGMP	3',5'-Cyclic guanosine monophosphate
CGS 21680	2-p-(2-Carboxyethyl)phenylamino-5'-N-ethylcarboxaminoadenosine hydrochloride
CHA	N^6-Cyclohexyladenosine
CNQX	6-Cyano-7-nitroquinoxaline-2,3-dione
DPCPX	8-Cyclopentyl-1,3-dipropylxanthine
DMPX	3,7-Dimethyl-1-propargylxanthine

S.S. Oja (✉)
Department of Paediatrics, Tampere University Hospital,
Finland
e-mail: simo.oja@uta.fi

P. Saransaari
Medical School, FI-33014 University of Tampere, Finland

A. El Idrissi and W.J. L'Amoreaux (eds.), *Taurine 8*, Advances in Experimental Medicine and Biology 775, DOI 10.1007/978-1-4614-6130-2_11,
© Springer Science+Business Media New York 2013

GABA   γ-Aminobutyrate
NBQX   2,3-Dioxo-6-nitro-1,2,3,4-tetrahydrobenzo(f)quinoxaline-7-sulfonamide
MK-801   [5R,10S]-[+]-5-methyl-10,11-dihydro-5H-dibenzo[a,d]cyclohepten-5,
         10-imine (dizocilpine)
NMDA   N-methyl-D-aspartate
NO   Nitric oxide
R-PIA   R(−)N^6-(2-Phenylisopropyl)adenosine
TACA   Trans-4-aminocrotonate
TPMPA   (1,2,5,6-Tetrahydro-pyridine-4-yl)methylphosphinate
ZAPA   (Z)-3-[(Aminoiminomethyl)thioprop-2-enoate

11.1 Introduction

There obtains a delicate balance between the excitatory and inhibitory processes in the central nervous system. Taurine, an inhibitory agent, participates in the maintenance of this balance. An increase in extracellular taurine upon excessive stimulation of glutamate receptors may serve as an important protective mechanism against excitotoxicity (Trenkner 1990). Taurine is particularly enriched in the developing rodent nervous system, where its concentration exceeds those of all other amino acids (Saransaari and Oja 2008). It may thus be essential particularly in the developing brain.

The major part of excitatory innervation in the hippocampus is glutamatergic, the function of the glutamate neurons being modulated by inhibitory γ-aminobutyrate (GABA)-releasing interneurons (Frotscher et al. 1984; Freund and Buzsáki 1988). Adenosine is an inhibitory neuromodulator which regulates neurotransmitter release in the brain (Cunha 2001). Taurine is also released from interneurons and inhibits firing of the main hippocampal pyramidal neurons, members of the glutamatergic excitatory circuit (Taber et al. 1986). In addition to its other roles, taurine has been thought to function as a regulator of neuronal activity, inducing hyperpolarization and inhibiting firing of central neurons (Oja and Kontro 1983). Furthermore, taurine has a special role in immature brain tissue (Kontro and Oja 1987), being apparently essential for the development and survival of neural cells (Sturman 1993). The following gives an overview of the effects of activation of different neurotransmitter receptors on taurine release in the mouse hippocampus.

11.2 Methodological Comments

All results shown in this article are from experiments on NMRI mice, aged 3 months (adult) and 7 days. Slices 0.4-mm thick weighing 15–20 mg were from the hippocampi. In the experiments with labeled taurine, they were first preloaded for 30 min with 10 μM (50 MBq/l) [^{3}H]taurine in preoxygenated Krebs-Ringer-Hepes-glucose medium under O_2 and then individually superfused for 50 min as described

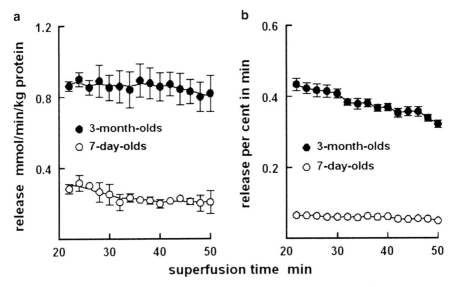

Fig. 11.1 Spontaneous release of endogenous taurine (panel **a**) and preloaded [³H]taurine (panel **b**) from hippocampal slices prepared from 7-day- and 3-month-old mice after a stabilization period of 20 min. Mean values ± SEM of six to eight independent experiments

in detail by Kontro and Oja (1987). The medium was pooled during the first 20 min of superfusion, during which time labeled taurine attached to membranes and retained in the extracellular spaces is washed out. Thereafter, 2-min fractions (0.5 ml) were collected. The effluent samples were counted for radioactivity. The content of endogenous taurine in the slices and its release into the medium were estimated by high-pressure liquid chromatography (Saransaari and Oja 1998). The taurine efflux rate constants (k_2) shown here are for the time interval of 34 to 50 min of superfusion.

11.3   General Properties of Taurine Release

The properties of the release of endogenous and exogenous labeled taurine are largely similar (Saransaari and Oja 1998, 2000a). The basal release is significantly greater from slices from adult as against developing mice (Fig. 11.1). However, K⁺ stimulation and cell-damaging conditions enhance the release markedly more in developing than in adult mice (Kontro and Oja 1987; Saransaari and Oja 1998).

11.4   GABA Receptors

The potentiation of taurine release by GABA is reduced by the $GABA_B$ receptor antagonists phaclofen and saclofen and the $GABA_C$ receptor antagonist (1,2,5,6-tetrahydropyridine-4-yl)methylphosphinate (TPMPA) at both ages

(Saransaari and Oja 2000a). Several GABA$_B$ receptor effectors are also able to inhibit K$^+$-stimulated taurine release in adults, while the GABA$_C$ receptor agonist cis-4-aminocrotonate (CACA) potentiated it. Trans-4-aminocrotonate (TACA) and the other GABA$_C$ receptor agonist (Z)-3-[(aminoiminomethyl)thio]prop-2-enoate (ZAPA), which also activate GABA$_A$ receptors, stimulate the basal release of taurine, particularly in the developing hippocampus. However, CACA and ZAPA can also act as substrates for GABA transport systems in the brain (Allan et al. 1991; Chebib and Johnston 1997). The potentiation of taurine release by these in both the adult and the developing hippocampus may result from the involvement of transporters operating outwards. Such a conception is corroborated by the moderate but significant inhibition of taurine uptake by the same compounds in our experiments (Saransaari and Oja 2000a). Nevertheless, all the effects were relatively minor and GABA receptors do not thus play any major role in taurine release in the hippocampus.

11.5 Adenosine Receptors

The adenosine A$_1$ receptor agonist R(−)N^6-(2-phenylisopropyl)adenosine (R-PIA) potentiated taurine release in developing mice and depressed it in adults in a receptor-mediated manner, since the respective A$_1$ receptor antagonists 8-cyclopentyl-1,3-dipropylxanthine (DPCPX) totally blocked the R-PIA effect in developing mice and attenuated it in adults (Fig. 11.2). The other adenosine A$_1$ receptor agonist N^6-cyclohexyladenosine (CHA) has shown the same effects (Saransaari and Oja 2000b). The neuromodulator adenosine is known to inhibit the release of neurotransmitters acting presynaptically, the adenosine A$_1$ receptors being particularly involved in this regulation (Fredholm and Dunwiddie 1988). The other adenosine receptor A$_{2a}$, which is also abundant in the brain, has exhibited varying effects, both reduction and enhancement of transmitter release or no effects having been observed (Burke and Nadler 1988; Yoon and Rothman 1991; Ribeiro 1999). In the case of taurine release in the mouse hippocampus, adenosine A$_{2a}$ receptors have had only minor effects. The specific adenosine A$_{2a}$ agonist 2-p-(2-carboxyethyl)phenylamino-5'-N-ethylcarboxaminoadenosinehydrochloride (CGS 21680) has slightly inhibited hippocampal taurine release but only in adult mice, while the adenosine A$_{2a}$ receptor antagonist 3,7-dimethyl-1-propargylxanthine (DMPX) has had no marked effects (Saransaari and Oja 2000b).

11.6 Glutamate Receptors

The activation of all three ionotropic glutamate receptors significantly enhanced taurine release in both developing and adult mice (Fig. 11.3). Of the agonists, N-methyl-D-aspartate (NMDA) was the most effective, but also 2-amino-3-hydroxy-5-methyl-4-

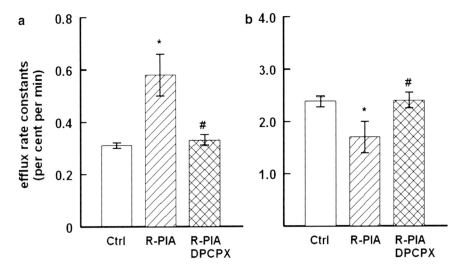

Fig. 11.2 Effects of the adenosine A_1 receptor agonist $R(-)N^6$-(2-phenylisopropyl)-adenosine and its antagonist 8-cyclopentyl-1,3-dipropylxanthine (DPCPX) on taurine release in hippocampal slices from developing (panel **a**) and adult (panel **b**) mice. Mean values $\pm$ SEM of six independent experiments. Significance of the effect of R-PIA: $*P < 0.01$, and the effect of DPCPX on the R-PIA-stimulated release: $^\#P < 0.01$

isoxazolepropionate (AMPA) in developing mice and kainate in adults were markedly effective at a 100-μM concentration. The greater efficacy of NMDA in enhancing taurine release in hippocampal slices has also been reported in other experiments with rats (Magnusson et al. 1991). The antagonists of the NMDA, kainate and AMPA receptors, dizocilpine (MK-801), 6-cyano-7-nitroquinoxaline-2,3-dione (CNQX), and 2,3-dioxo-6-nitro-1,2,3,4-tetrahydrobenzo(f)quinoxaline-7-sulfonamide (NBQX), respectively, totally blocked the effects in developing mice and significantly reduced them in adults (Fig. 11.4).

The excitatory ionotropic glutamate receptors have also enhanced taurine release more efficiently in the cerebral cortex in developing than in adult mice (Saransaari and Oja 1991). In hippocampal slices from aging mice, the response to NMDA is further attenuated (Saransaari and Oja 1997). However, the effects of ionotropic glutamate receptors are not the same in all brain areas and under different experimental conditions. For instance, the ionotropic glutamate receptor agonists have had no effect on taurine release in the immature mouse brain stem under ischemic conditions, whereas in adults the release has been enhanced in a receptor-mediated manner (Saransaari and Oja 2010).

Metabotropic glutamate receptors also participate in taurine release. Group I metabotropic glutamate receptors mostly enhance the release in the adult but inhibit it in the developing hippocampus (Saransaari and Oja 1999a). Group II and III receptors are also involved, but their effects are variable in the developing and adult hippocampus. On the whole, however, the effects of these receptors are relatively minor as compared to those of the ionotropic glutamate receptors.

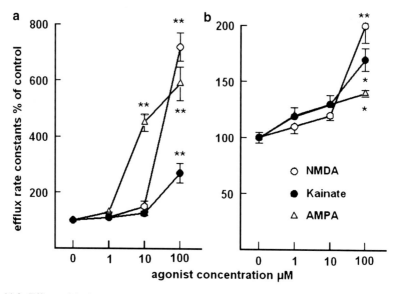

Fig. 11.3 Effects of the ionotropic glutamate receptor agonists on taurine release from hippocampal slices from developing (panel **a**) and adult (panel **b**) mice. The efflux rate constants are from the superfusion period of 34–50 min. Mean values ± SEM of six to eight independent experiments. Significance of differences from values without agonist: $*P<0.05$, $**P<0.01$

11.7 How NMDA Receptors Enhance Taurine Release

Nitric oxide (NO) plays a significant role in signal transduction in the brain (Bruhwyler et al. 1993). The production of NO is linked to the activation of the NMDA receptors (Garthwaite 1991). Glutamate activates postsynaptic NMDA receptors and the receptor-linked Ca^{2+} channels allow Ca^{2+} to enter the cells upon depolarization. NO synthase is a Ca^{2+}-dependent enzyme (Bredt et al. 1992), being activated in the presence of Ca^{2+} and then producing NO. NO stimulates soluble guanylate cyclase and in this manner enhances the production of $3',5'$-cyclic guanosine monophosphate (cGMP) (Schuman and Madison 1994). The levels of cGMP are indeed increased after NMDA receptor activation in hippocampal slices (East and Garthwaite 1991). The release of taurine has been enhanced by the cGMP analog 8-Br-cGMP and the phosphodiesterase inhibitor zaprinast, particularly in the immature hippocampus, this indicating that increased cGMP levels indeed induce taurine release (Saransaari and Oja 2002). Several NO donors have been used to simulate the natural NO production in the brain. Of these, hydroxylamine may best mimic the effect of intracellular generation of NO, since it penetrates into cells easily and is broken down by a catalase-dependent reaction (DeMaster et al. 1989). Hydroxylamine significantly enhances taurine release from hippocampal slices in both developing and adult mice (Fig. 11.5), its effect being clearly concentration dependent. Other NO donors

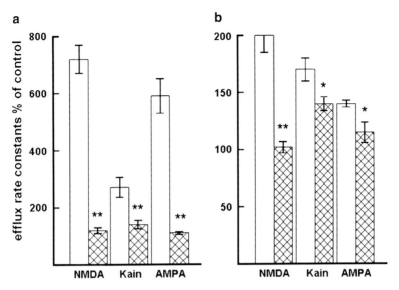

Fig. 11.4 Effects of the antagonists on the ionotropic receptor agonist-enhanced taurine release from hippocampal slices from developing (panel **a**) and adult (panel **b**) mice. The effects of the antagonists of N-methyl-D-aspartate (NMDA), kainate and 2-amino-3-hydroxy-5-methyl-4-isox-azolepropionate (AMPA), dizocilpine (MK-801), 6-cyano-7-nitroquinoxaline-2,3-dione (CNQX), and 2,3-dioxo-6-nitro-1,2,3,4-tetrahydrobenzo(f)-quinoxaline-7-sulfonamide (NBQX), respectively (crosshatched bars). Mean values ± SEM of six independent experiments. Significance of the effects of the antagonists: *$P < 0.05$, **$P < 0.01$

have also affected taurine release from hippocampal slices, but their effects have been less consistent (Saransaari and Oja 1999b).

11.8   Conclusions

Taurine release in hippocampal slices is affected by several neurotransmitter receptors. Of these, the ionotropic glutamate receptors and the adenosine receptor A_1 are the most effectual. The metabotropic glutamate receptors, adenosine A_{2a} receptor, and GABA receptors play minor roles in the regulation of taurine release. The effect of NMDA receptor activation is mediated via the NMDA/NO/cGMP pathway. The release of the inhibitory neuromodulator taurine may maintain homeostasis in the brain, counteracting any excitotoxic effects of glutamate possibly released in excess. Enhanced taurine release could be particularly important in the developing hippocampus. GABA cannot protect against excitotoxicity in the developing hippocampus, since during early development GABA is rather excitatory than inhibitory (Ben-Ari 2002). The rationale is this that the activation of excitatory amino acid receptors causes at the same time an

Fig. 11.5 Effects of different concentrations of hydroxylamine on taurine release from hippocampal slices from developing and adult mice. Mean values ± SEM of six independent experiments. Significance of differences from the efflux without hydroxylamine: *$P<0.05$, **$P<0.01$

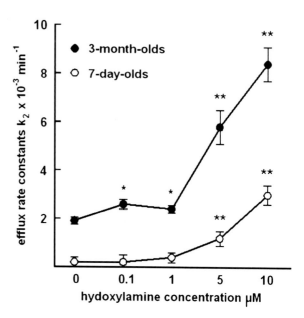

increase in taurine release. The enhanced adenosine release further potentiates the effect of taurine release in that it can attenuate the release of excitatory transmitters.

References

Allan RD, Dickenson HW, Duke RK, Johnston GAR (1991) ZAPA, a substrate for the neuronal high affinity GABA uptake system in rat brain slices. Neurochem Int 18:63–67

Ben-Ari Y (2002) Excitatory actions of GABA during development: the nature of nurture. Nat Rev Neurosci 3:728–739

Bredt SD, Ferris CD, Snyder SG (1992) Nitric oxide synthase regulatory sites. Phosphorylation by cyclic AMP-dependent protein kinase, protein kinase C, and calcium/calmodulin protein kinase, identification of flavin and calmodulin binding sites. J Biol Chem 267:10976–10981

Bruhwyler J, Chleide E, Liègeois J, Carrer F (1993) Nitric oxide: a new messenger in the brain. Neurosci Biobehav Rev 17:373–384

Burke SP, Nadler JV (1988) Regulation of glutamate and aspartate release from slices of the hippocampal CA1 area: effects of adenosine and baclofen. J Neurochem 51:1541–1551

Chebib M, Johnston GAR (1997) Stimulation of [³H]GABA and β-[³H]alanine release from rat brain slices by cis-4-aminocrotonic acid. J Neurochem 68:786–794

Cunha RA (2001) Adenosine as a neuromodulator and as a homeostatic regulator in the nervous system: different roles, different sources and different receptors. Neurochem Int 38:107–125

DeMaster EG, Raij L, Archer SL, Weir EK (1989) Hydroxylamine is a vasorelaxant and a possible intermediate in the oxidative conversion of L-arginine to nitric oxide. Biochem Biophys Res Commun 163:527–533

East SJ, Garthwaite J (1991) NMDA receptor activation in rat hippocampus induces cyclic GMP formation through the L-arginine-nitric oxide pathway. Neurosci Lett 123:17–19

Fredholm BB, Dunwiddie TV (1988) How does adenosine inhibit transmitter release? Trends Pharmacol Sci 9:130–134

Freund TF, Buzsáki G (1988) Alterations in excitatory and GABAergic inhibitory connections in hippocampal transplants. Neuroscience 27:373–385

Frotscher M, Cs L, Lübbers K, Oertel WH (1984) Commissural afferents innervate glutamate decarboxylase immunoreactive non-pyramidal neurons in the guinea-pig hippocampus. Neurosci Lett 46:137–143

Garthwaite J (1991) Glutamate, nitric oxide and cell-cell signalling in the nervous system. Trends Neurosci 14:60–67

Kontro P, Oja SS (1987) Taurine and GABA release from mouse cerebral cortex slices: potassium stimulation releases more taurine than GABA from developing brain. Brain Res 465:277–291

Magnusson KR, Koerner JF, Larson AA, Smullin DH, Skilling SR, Beitz AJ (1991) NMDA-, kainate- and quisqualate-stimulated release of taurine from electrophysiologically monitored rat hippocampal slices. Brain Res 549:1–8

Oja SS, Kontro P (1983) Taurine. In: Lajtha A (ed) Handbook of neurochemistry, 2nd edn, vol. 3, Plenum Press, New York, pp 501–533

Ribeiro JA (1999) Adenosine A_{2A} receptor interactions with receptors for other neurotransmitters and neuromodulators. Eur J Pharmacol 375:101–113

Saransaari P, Oja SS (1991) Excitatory amino acids evoke taurine release from cerebral cortex slices from adult and developing mice. Neuroscience 45:509–523

Saransaari P, Oja SS (1997) Taurine release from the developing and ageing hippocampus: stimulation by agonists of ionotropic glutamate receptors. Mech Ageing Dev 99:219–232

Saransaari P, Oja SS (1998) Release of endogenous glutamate, aspartate, GABA and taurine from hippocampal slices from adult and developing mice in cell-damaging conditions. Neurochem Res 23:567–574

Saransaari P, Oja SS (1999a) Involvement of metabotropic glutamate receptors in taurine release in the adult and developing mouse hippocampus. Amino Acids 16:165–179

Saransaari P, Oja SS (1999b) Taurine release modified by nitric oxide-generating compounds in the developing and adult mouse hippocampus. Neuroscience 89:1103–1111

Saransaari P, Oja SS (2000a) Taurine release modified by GABAergic agents in hippocampal slices from adult and developing mice. Amino Acids 18:17–30

Saransaari P, Oja SS (2000b) Modulation of the ischemia-induced taurine release by adenosine receptors in the developing and adult mouse hippocampus. Neuroscience 97:426–430

Saransaari P, Oja SS (2002) Taurine release in the developing and adult mouse hippocampus: involvement of cyclic guanosine monophosphate. Neurochem Res 27:15–20

Saransaari P, Oja SS (2008) Taurine in neurotransmission. In: Vizi ES (ed) Handbook of neurochemistry and molecular neuroscience, 3rd edn, Neurotransmitter systems. Springer, New York, pp 325–342

Saransaari P, Oja SS (2010) Modulation of taurine release in ischemia by glutamate receptors in mouse brain stem slices. Amino Acids 38:739–746

Schuman ER, Madison DV (1994) Nitric oxide and synaptic function. Annu Rev Neurosci 17:153–183

Sturman JA (1993) Taurine in development. Physiol Rev 73:119–147

Taber KH, Lin C-T, Liu J-W, Thalmann R, Wu J-Y (1986) Taurine in hippocampus: localization and postsynaptic action. Brain Res 386:113–121

Trenkner E (1990) The role of taurine and glutamate during early postnatal cerebellar development of normal and weaver mutant mice. Adv Exp Med Biol 268:239–244

Yoon KW, Rothman SM (1991) Adenosine inhibits excitatory but not inhibitory synaptic transmission in the hippocampus. J Neurosci 11:1375–1380

Chapter 12
Evaluation of the Taurine Concentrations in Dog Plasma and Aqueous Humour: A Pilot Study

Serge-George Rosolen, Nathalie Neveux, José-Alain Sahel, Serge Picaud, and Nicolas Froger

Abstract In the 70s, the amino acid taurine was found essential for photoreceptor survival. Recently, we found that taurine depletion can also trigger retinal ganglion cell degeneration both *in vitro* and *in vivo*. Therefore, evaluation of taurine levels could be a crucial biomarker for different pathologies of retinal ganglion cells such as glaucoma. Because different breeds of dog can develop glaucoma, we performed taurine measurements on plasma and aqueous humour samples from pet dogs. Here, we exposed results from a pilot study on *normal* selected breed of pet dogs, without any ocular pathology. Samples were collected by veterinarians who belong to the *Réseau Européen d'Ophtalmologie Vétérinaire et de Vision Animale*. Following measurements by high-performance liquid chromatography (HPLC), the averaged taurine concentration was 162.3 μM in the plasma and 51.8 μM in the aqueous humour. No correlation was observed between these two taurine concentrations, which exhibited a ratio close to 3. Further studies will determine if these taurine concentrations are changed in glaucomatous dogs.

S.-G. Rosolen (✉)
Réseau Européen d'Ophtalmologie Vétérinaire et de Vision Animale (REOVVA),
Asnières, France

Clinique Vétérinaire Voltaire, Asnières, France

INSERM, U968, Institut de la Vision, Paris, France

UPMC Université Paris 06, UMR_S 968, Institut de la Vision, Paris, France

CNRS, UMR 7210, Institut de la Vision, Paris, France
e-mail: sg.rosolen@orange.fr

N. Neveux
Stress cellulaire: Physiopathologie, stratégies nutritionnelles et thérapeutiques innovantes, EA 4466, Faculté de Pharmacie, Université Descartes Paris V, Paris, France

Service Inter-hospitalier de biochimie, CHU Cochin-St-Vincent de Paul, AP-HP, Paris, France

A. El Idrissi and W.J. L'Amoreaux (eds.), *Taurine 8*, Advances in Experimental Medicine and Biology 775, DOI 10.1007/978-1-4614-6130-2_12,
© Springer Science+Business Media New York 2013

Abbreviations

RGC   Retinal ganglion cells
HPLC   High-performance liquid chromatography

12.1   Introduction

Taurine is an enigmatic free sulphur beta amino acid in very high concentration in the retina. In the 70s, this depletion was found to trigger photoreceptor degeneration in cats (Hayes et al. 1975), rats (Pasantes-Morales et al. 1983) and mice (Rascher et al. 2004). This discovery was achieved on cats because this species is unable to synthesize taurine and it can only obtain taurine from its diet. Taurine concentration was therefore examined in dog models of hereditary degenerative diseases (Schmidt and Aguirre 1985) and this supplementation was reported to prevent visual loss in human patients (Pasantes-Morales et al. 2002). Recently, we have correlated the retinal toxicity of the antiepileptic drug, vigabatrin, to a taurine depletion detected in the plasma of vigabatrin-treated rats and patients (Jammoul et al. 2009). However, if some

J.-A. Sahel
INSERM, U968, Institut de la Vision, Paris, France

UPMC Université Paris 06, UMR_S 968, Institut de la Vision,
Paris, France

CNRS, UMR 7210, Institut de la Vision, Paris, France

Centre Hospitalier National d'Ophtalmologie des Quinze-Vingts,
Paris, France

Institute of Ophthalmology, University College of London, London, UK

Fondation Ophtalmologique Adolphe de Rothschild, Paris, France

French Academy of Sciences, Paris, France

S. Picaud
INSERM, U968, Institut de la Vision, Paris, France

UPMC Université Paris 06, UMR_S 968, Institut de la Vision, Paris, France

CNRS, UMR 7210, Institut de la Vision, Paris, France

Centre Hospitalier National d'Ophtalmologie des Quinze-Vingts, Paris, France

Fondation Ophtalmologique Adolphe de Rothschild, Paris, France

N. Froger
INSERM, U968, Institut de la Vision, Paris, France

UPMC Université Paris 06, UMR_S 968, Institut de la Vision, Paris, France

CNRS, UMR 7210, Institut de la Vision, Paris, France

Centre Hospitalier National d'Ophtalmologie des Quinze-Vingts, Paris, France hotoreceptor degeneration is part of this retinal toxicity (Ravindran et al. 2001; Duboc et al. 2004), primary affected cells in patients appeared to be retinal ganglion cells (RGCs) (Ravindran et al. 2001; Frisen and Malmgren 2003; Buncic et al. 2004; Wild et al. 2006). Animal models also presented this vigabatrin-induced retinal ganglion cell degeneration, which was suppressed by taurine supplementation (Jammoul et al. 2010). Furthermore, we found that pharmacological blockade of taurine transporter also leads to RGC degeneration (Gaucher et al. 2012). Consistent with these results, some cell loss was reported in the RGC layer in taurine-transporter knockout mice during the massive degenerative process affecting photoreceptors although it was not clear which cells were degenerating in the RGC layer (Heller-Stilb et al. 2002). Finally, we demonstrated that taurine is acting directly on RGCs to induce their survival and that taurine supplementation could rescue RGCs in different animal models of pathologies such as glaucoma or *retinitis pigmentosa* (Froger et al. 2012). In agreement with these results, a decrease in the retinal taurine concentration was reported in glaucomatous dogs (Madl et al. 2005).

Considering these results on the critical role of taurine in RGC survival, taurine appears as an important biomarker to follow during pathological conditions such as glaucoma. While rodents are commonly used as animal models, *large* animal models (i.e. dog, cat, pig) have become increasingly attractive to assess efficacy and safety of a variety of treatment modalities that are being considered for clinical trials in human patients. A significant advantage in using the dog is that it enables intraocular drug delivery studies (Le Meur et al. 2007), surgical interventions and *in vivo* imaging procedures (Rosolen et al. 2001; Rosolen et al. 2012) that cannot always be done in the much smaller rodent eye. Considering dogs, breeds offer a unique possibility to find families with high consanguinity. As a consequence, progressive retinal atrophy (PRA) has been described in more than 100 breeds. These canine forms of retinal diseases share great genetic and phenotypic similarities with their human counterparts. Therefore, dogs provide unique animal models to (i) improve our understanding of the corresponding pathogenic mechanisms and (ii) demonstrate the *proof of principle* for novel therapeutic strategies. Yet, the dimensions of the canine eye enable intravitreal implantation of devices or injection of compounds in solution that have, respectively, a size or volume targeted to the human eye (Tao et al. 2002).

To prepare the future use of taurine as a biomarker and potential treatment in RGC degenerative diseases, we here report taurine measurements in plasma and aqueous humour samples from different breeds of dogs. This pilot study included 71 healthy pet dogs, without any detected ocular diseases.

12.2 Methods

All procedures were in accordance with the Association for Research in Vision and Ophthalmology statement for the use of animals in ophthalmic and vision research. All animals included in the study were examined by 11 different veterinarians who belong to the REOVVA network (*Réseau Européen d'Ophtalmologie Vétérinaire et de Vision Animale*) after obtaining informed consent from the owners.

12.2.1   Animals

Results were obtained from a cohort of 69 adult pet dogs (age: min = 18 months, max 7 years; 33 males and 36 females), among 36 different breeds, without detected ocular pathologies. All animals were in good physical condition and without any visual defect following a complete clinical examination. The complete ophthalmologic examination included slit-lamp and indirect ophthalmoscopy. For aqueous humour collections, animals were sedated with a single intramuscular injection of medetomidine (0.1 mg/kg). The cornea of the right eye was topically anesthetized with two drops of chlorhydrate of oxybuprocaine.

12.2.2   Sample Collections

Blood samples (5 ml) were taken from radial vein in awake dogs and placed in tubes containing EDTA. After centrifugation, plasma was removed and frozen until use for amino acid dosages. Aqueous humour samples were taken on sedated dogs, with a needle (21 G; syringe volume: 0.2 ml) and then directly frozen at −20°C until use for amino acid dosages.

12.2.3   Concentration of Taurine and Its Precursors of Biosynthesis in Plasma and Aqueous Humour

Plasma and aqueous humour were deproteinized with a 30% (w/v) sulfosalicylic acid solution, and the supernatants were stored at −20°C until analysis. Amino acids, including taurine, methionine and cystine, the two latter being precursors in taurine biosynthesis, were measured by ion exchange chromatography.

12.2.4   Statistical Analysis

Data exposed are mean ± SEM. One-way ANOVA, followed by a Bonferroni post-hoc, was used to compare mean between groups of dogs. To compare the means of two groups, a Student's t-test was applied. Differences were considered significant at $*p < 0.05$, $**p < 0.01$ and $***p < 0.001$.

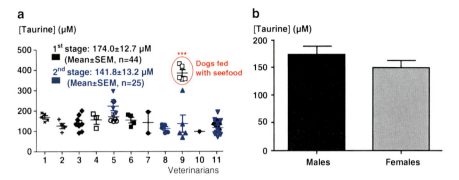

Fig. 12.1 Taurine plasmatic concentrations in healthy dogs and influence of diet and sex. (**a**) Graphic representing the plasma taurine concentrations in normal dogs ($n=69$). The *black* symbols represent values from the first stage while the *blue* symbols represent values from the second stage of samples. The *ellipse* represents the values of dogs fed with seafood ($n=5$, $p<0.01$, one-way ANOVA followed by Bonferroni post-hoc test). Data are expressed in micromolar in plasma. (**b**) Histogram representing plasma taurine concentrations according to sex, showing no difference between males and females. Data expressed as micromolar are mean $\pm$ SEM from $n=33$ males and $n=36$ females

12.3   Results

12.3.1   Plasmatic Taurine Concentration in Healthy Selected Breed of Dogs

Figure 12.1 illustrates the taurine plasma measurements in the different dogs according to the different veterinarians. In the first stage, plasma samples were collected from 44 dogs (black symbols; Fig. 12.1a) by ten veterinarians, while in the second stage, samples were collected from 25 dogs (blue symbols, Fig. 12.1b) by four veterinarians. When mixing all the different measurements, the averaged taurine plasma concentration was found at 162.3 ± 9.5 μM (mean $\pm$ SEM, $n=69$). However, the figure shows a surprising group with animals ($n=5$) exhibiting a plasmatic taurine level greater than the average value. For these five animals the difference with the averaged value was statistically significant. Indeed, in this group of dogs, the plasma taurine level was 2.3-fold higher (385.2 ± 16.6 μM, $n=5$; $p<0.001$, red points; Fig. 12.1a) than the remaining dogs 144.9 ± 6.1 μM (mean $\pm$ SEM, $n=64$). To understand the origin of this major difference, we compared the nature of the dog breeds in this sample with respect to the other cohorts. When all animals were classified according to their sex, males and females were distributed in all groups. In fact, no significant differences were observed in the taurine plasmatic level between sexes (Fig. 12.1b). Similarly, the difference in the group could not be attributed to a bias in the recruitment of particular breed in this group (not shown). We then questioned the food supply in this animal group because taurine can be taken up from our diet. Surprisingly, we found out that it received an unusual seafood-enhanced diet unlike all the other dogs. This result suggests that nutrition can impact the plasmatic taurine level also in dogs.

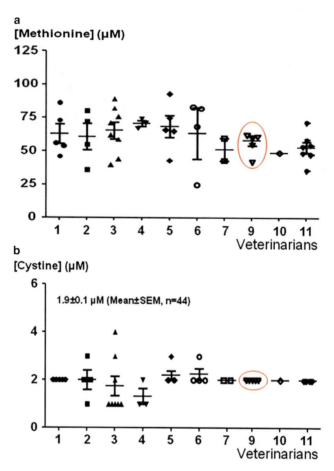

Fig. 12.2 Plasmatic concentration of methionine and cystine, the precursor for taurine biosynthesis in healthy dogs. (**a**) Values of methionine concentration in plasma collected from 44 dogs. The group included into the *red* ellipse corresponds to dogs fed with seafood (*n* = 5). Data are expressed in micromolar in plasma. (**b**) Values of cystine concentration in plasma collected from 44 dogs. The group included into the *red* ellipse corresponds to dogs fed with seafood (*n* = 5). Data are expressed in micromolar in plasma

12.3.2 Plasmatic Concentration of Precursors for Taurine Biosynthesis: Methionine and Cystine in Healthy Selected Breed of Dogs

The concentrations of precursors of taurine biosynthesis were measured on plasma samples collected during the first stage, from the 44 dogs by ten veterinarians. The methionine concentration was found at 61.3 ± 2.3 μM (*n* = 44; Fig. 12.2a), while the cystine concentration reached 1.9 ± 0.1 μM (*n* = 44; Fig. 12.2b).

Fig. 12.3 Plasmatic concentration of taurine in the aqueous humour from healthy dogs. Values of taurine levels in aqueous humour collected from two groups of five dogs. Data are expressed in micromolar in aqueous humour

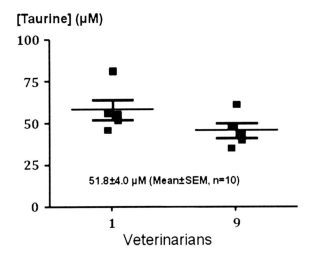

Interestingly, in the group of dogs which exhibited a significant increase in plasma taurine (red ellipses, Fig. 12.1a), there are no differences in either methionine concentration (53.3 ± 3.8 μM, mean ± SEM, $n=5$, red ellipse; Fig. 12.2a) or cystine concentration (2.0 ± 0.0 μM, mean ± SEM, $n=5$ red ellipse; Fig. 12.2b) as compared to the other groups ($p > 0.05$, one-way ANOVA).

12.3.3   Taurine Concentration in Aqueous Humour from Healthy Selected Breeds of Dogs

In a subset of 10 dogs from the second stage, samples of aqueous humour were collected and we were able to measure the taurine levels in this liquid. This taurine concentration was measured at 51.8 ± 4.0 μM (mean ± SEM; $n=10$, Fig. 12.3).

12.4   Discussion

This pilot study assessed the concentration of taurine and precursors of taurine biosynthesis in the plasma of healthy dogs. We found that the plasmatic taurine concentration is ~150 μM in *normal* pet dogs, without any differences between males and females. This value appears in the same range as described in another study on water dogs, in which taurine plasma concentration reached 115 μM (Alroy et al. 2005). In other species, taurine plasma concentration was found lower in cats (52 μM; Earle and Smith 1992), whereas it was higher in rodents like rats (Jammoul et al. 2009) or mice (Gaucher et al. 2012). In humans, the taurine concentration was reported at 39 and 53 μM, while the normal values were considered to range from

40 to 90 μM (Durelli and Mutani 1983; Brons et al. 2004). The normal range appears fairly similar in infants (Jammoul et al. 2009).

The ability of mammals to synthesize taurine from sulphur precursors differs according to species (Huxtable 1989). Dietary sources of taurine are therefore required for species unable to synthesize sufficient taurine such as cats. Previous studies had shown that taurine supplementation can directly increase the taurine plasma level in rodents (Jammoul et al. 2009; Jammoul et al. 2010), as well as in humans (Brons et al. 2004; Shao and Hathcock 2008). Interestingly, our results suggest that the plasma taurine concentration can be manipulated in dogs by introducing a peculiar taurine-rich diet. Indeed, one group of dogs fed with seafood, known to present a high taurine content (Yamori et al. 2009), showed a significant higher taurine plasma concentration. Additional data will be needed to confirm that taurine-enriched diets can increase the plasma taurine concentration in patients.

The discovery of taurine depletion as a cause of retinal degeneration was reported in cats 37 years ago (Hayes et al. 1975). However, in this case, the degenerative process appeared to include only photoreceptors. More recently, the taurine depletion was found to trigger the retinal toxicity of vigabatrin in patients and animals (Jammoul et al. 2009). However, in this retinal toxicity, the primary site of degeneration seems to occur in patients at the level of RGCs even if photoreceptors are also damaged (Ravindran et al. 2001; Wild et al. 2006; Jammoul et al. 2010). We have then demonstrated that taurine directly enhances RGC survival and that a local retinal depletion could thus lead *in vivo* to RGC degeneration as in glaucoma animal models (Froger et al. 2012). These results are consistent with retinal decrease in taurine concentration in dogs affected with primary glaucoma (Madl et al. 2005). The taurine levels may represent a crucial marker for vulnerability to retinal disease. To start assessing the use of taurine as a marker of glaucoma, we also detected the taurine level in the aqueous humour from healthy dogs with an average value of ~50 μM. No correlation was found between the taurine levels in plasma and aqueous humour (not shown), which is consistent with the notion that a local retinal deficit can occur despite a normal plasma taurine concentration. This absence of correlation could rely on regulation of the high taurine uptake system located in the retinal pigment epithelium and the endothelial cells (Hillenkamp et al. 2004). Hence the aqueous humour taurine concentration could provide a closer measure to the retinal taurine concentration and thus a more accurate marker for retinal diseases. To answer this question, future studies will evaluate the taurine levels in both the aqueous humour and plasma of glaucomatous dogs.

12.5 Conclusion

Future taurine measurements will be performed in glaucomatous dogs to determine if taurine concentrations could provide a new diagnostic marker for retinal diseases with RGC degeneration leading subsequently to a therapeutic decision for taurine supplementation.

Acknowledgments Authors acknowledge Dr. Azoulay, T.; Dr. Boderiou, Y.; Boillot, T.; Cantaloube, B.; Durieux, P.; Goulle, F.; Isard, P-F.; Maller, D.; Michaud, Y.; Philippe, R.; and Rosolen, SG. from the REOVVA group (www.reovva.com) for their contribution to this work. This work was supported by INSERM, Université Pierre et Marie Curie (Paris VI), Fondation Ophtalmologique A. de Rothschild (Paris), Fédération de la Recherche Médicale and Fondation Bailly, the city of Paris and the Regional Council of Ile-de-France.

References

Alroy J, Rush JE, Sarkar S (2005) Infantile dilated cardiomyopathy in Portuguese water dogs: correlation of the autosomal recessive trait with low plasma taurine at infancy. Amino Acids 28:51–56

Brons C, Spohr C, Storgaard H, Dyerberg J, Vaag A (2004) Effect of taurine treatment on insulin secretion and action, and on serum lipid levels in overweight men with a genetic predisposition for type II diabetes mellitus. Eur J Clin Nutr 58:1239–1247

Buncic JR, Westall CA, Panton CM, Munn JR, MacKeen LD, Logan WJ (2004) Characteristic retinal atrophy with secondary "inverse" optic atrophy identifies vigabatrin toxicity in children. Ophthalmology 111:1935–1942

Duboc A, Hanoteau N, Simonutti M, Rudolf G, Nehlig A, Sahel JA, Picaud S (2004) Vigabatrin, the GABA-transaminase inhibitor, damages cone photoreceptors in rats. Ann Neurol 55:695–705

Durelli L, Mutani R (1983) The current status of taurine in epilepsy. Clin Neuropharmacol 6:37–48

Earle KE, Smith PM (1992) The effect of dietary supplementation with cysteic acid on the plasma taurine concentration of cats maintained on a taurine-restricted diet. Adv Exp Med Biol 315:23–32

Frisen L, Malmgren K (2003) Characterization of vigabatrin-associated optic atrophy. Acta Ophthalmol Scand 81:466–473

Froger N, Cadetti L, Lorach H, Martins J, Bemelmans AP, Dubus E et al (2012) Taurine Provides Neuroprotection against Retinal Ganglion Cell Degeneration. PLoS One 7(10):e42017

Gaucher D, Arnault E, Husson Z, Froger N, Dubus E, Gondouin P, Dherbecourt D, Degardin J, Simonutti M, Fouquet S, Benahmed MA, Elbayed K, Namer IJ, Massin P, Sahel JA, Picaud S (2012) Taurine deficiency damages retinal neurones: cone photoreceptors and retinal ganglion cells. Amino Acids 43:1979–1993

Hayes KC, Carey RE, Schmidt SY (1975) Retinal degeneration associated with taurine deficiency in the cat. Science 188:949–951

Heller-Stilb B, van Roeyen C, Rascher K, Hartwig HG, Huth A, Seeliger MW, Warskulat U, Haussinger D (2002) Disruption of the taurine transporter gene (taut) leads to retinal degeneration in mice. FASEB J 16:231–233

Hillenkamp J, Hussain AA, Jackson TL, Constable PA, Cunningham JR, Marshall J (2004) Compartmental analysis of taurine transport to the outer retina in the bovine eye. Invest Ophthalmol Vis Sci 45:4099–4105

Huxtable RJ (1989) Taurine in the central nervous system and the mammalian actions of taurine. Prog Neurobiol 32:471–533

Jammoul F, Degardin J, Pain D, Gondouin P, Simonutti M, Dubus E, Caplette R, Fouquet S, Craft CM, Sahel JA, Picaud S (2010) Taurine deficiency damages photoreceptors and retinal ganglion cells in vigabatrin-treated neonatal rats. Mol Cell Neurosci 43:414–421

Jammoul F, Wang Q, Nabbout R, Coriat C, Duboc A, Simonutti M, Dubus E, Craft CM, Ye W, Collins SD, Dulac O, Chiron C, Sahel JA, Picaud S (2009) Taurine deficiency is a cause of vigabatrin-induced retinal phototoxicity. Ann Neurol 65:98–107

Le Meur G, Stieger K, Smith AJ, Weber M, Deschamps JY, Nivard D, Mendes-Madeira A, Provost N, Pereon Y, Cherel Y, Ali RR, Hamel C, Moullier P, Rolling F (2007) Restoration of vision in

RPE65-deficient Briard dogs using an AAV serotype 4 vector that specifically targets the retinal pigmented epithelium. Gene Ther 14:292–303

Madl JE, McIlnay TR, Powell CC, Gionfriddo JR (2005) Depletion of taurine and glutamate from damaged photoreceptors in the retinas of dogs with primary glaucoma. Am J Vet Res 66:791–799

Pasantes-Morales H, Quiroz H, Quesada O (2002) Treatment with taurine, diltiazem, and vitamin E retards the progressive visual field reduction in retinitis pigmentosa: a 3-year follow-up study. Metab Brain Dis 17:183–197

Pasantes-Morales H, Quesada O, Carabez A, Huxtable RJ (1983) Effects of the taurine transport antagonist, guanidinoethane sulfonate, and beta-alanine on the morphology of rat retina. J Neurosci Res 9:135–143

Rascher K, Servos G, Berthold G, Hartwig HG, Warskulat U, Heller-Stilb B, Haussinger D (2004) Light deprivation slows but does not prevent the loss of photoreceptors in taurine transporter knockout mice. Vision Res 44:2091–2100

Ravindran J, Blumbergs P, Crompton J, Pietris G, Waddy H (2001) Visual field loss associated with vigabatrin: pathological correlations. J Neurol Neurosurg Psychiatry 70:787–789

Rosolen SG, Saint-MacAry G, Gautier V, Legargasson JF (2001) Ocular fundus images with confocal scanning laser ophthalmoscopy in the dog, monkey and minipig. Vet Ophthalmol 4:41–45

Rosolen SG, Riviere ML, Lavillegrand S, Gautier B, Picaud S, Legargasson JF (2012) Use of a combined slit-lamp SD-OCT to obtain anterior and posterior segment images in selected animal species. Vet Ophthalmol 15 (suppl.2):105–115

Schmidt SY, Aguirre GD (1985) Reductions in taurine secondary to photoreceptor loss in Irish setters with rod-cone dysplasia. Invest Ophthalmol Vis Sci 26:679–683

Shao A, Hathcock JN (2008) Risk assessment for the amino acids taurine, L-glutamine and L-arginine. Regul Toxicol Pharmacol 50:376–399

Tao W, Wen R, Goddard MB, Sherman SD, O'Rourke PJ, Stabila PF, Bell WJ, Dean BJ, Kauper KA, Budz VA, Tsiaras WG, Acland GM, Pearce-Kelling S, Laties AM, Aguirre GD (2002) Encapsulated cell-based delivery of CNTF reduces photoreceptor degeneration in animal models of retinitis pigmentosa. Invest Ophthalmol Vis Sci 43:3292–3298

Wild JM, Robson CR, Jones AL, Cunliffe IA, Smith PE (2006) Detecting vigabatrin toxicity by imaging of the retinal nerve fiber layer. Invest Ophthalmol Vis Sci 47:917–924

Yamori Y, Liu L, Mori M, Sagara M, Murakami S, Nara Y, Mizushima S (2009) Taurine as the nutritional factor for the longevity of the Japanese revealed by a world-wide epidemiological survey. Adv Exp Med Biol 643:13–25

Chapter 13
Protective Effect of Taurine on Down-Regulated Expression of Thyroid Hormone Receptor Genes in Brains of Mice Exposed to Arsenic

Yachen Wang, Fengyuan Piao, Yachen Li, Xianghu Wang, and Huai Guang

Abstract This study aimed at evaluating protective effect of taurine on the down-regulated expressions of thyroid hormone receptor (TR) genes in brains of mice exposed to arsenic (As). The SPF mice were randomly divided into As exposure group, protective group, and control group. The As exposure group was administered with 4 ppm As_2O_3 through drinking water for 60 days. The protective group was treated with both 4 ppm As_2O_3 and 150 mg/kg taurine. The control group was given with drinking water alone. The gene expressions of TR in the mouse brains of the three groups were analyzed by real-time PCR. Their protein expressions were examined by Western blot and immunohistochemistry. Our results showed that the gene expression of TRβ, a very important regulator of Camk4 transcription, was down-regulated in cerebral and cerebellar tissues of mice exposed to As. The expression of TRβ1 protein in the cerebral or cerebellar tissue significantly decreased in the group exposed to As compared to the control group. However, the expressions of TRβ gene and TRβ1 protein were significantly rescued in the group coadministered with taurine as antioxidant. These results indicated that taurine may have the protective effect on the down-regulated expressions of TR in brains of mice exposed to As.

Abbreviations

TR Thyroid hormone receptor
As Arsenic
LTP Long-term potentiation

Y. Wang • F. Piao (✉) • Y. Li • X. Wang • H. Guang
Department of Occupational and Environmental Health, Dalian Medical University,
No 9 Western Section of Lushun South Road, Dalian, Liaoning 116044, P.R. China
e-mail: piaofy_dy@yahoo.com.cn

A. El Idrissi and W.J. L'Amoreaux (eds.), *Taurine 8*, Advances in Experimental
Medicine and Biology 775, DOI 10.1007/978-1-4614-6130-2_13,
© Springer Science+Business Media New York 2013

LTD   Long-term depression
Creb   cAMP response element-binding protein
TRE   Thyroid hormone-response element

13.1 Introduction

Arsenic (As) is one of the most common heavy metal contaminants found in the environment, particularly in water. Its toxicity is a global health problem affecting many millions of people. Epidemiological studies revealed that chronic exposure to inorganic As via drinking water resulted in a dose-dependent reduction in intellectual functions in children (Wasserman et al. 2004). Animal studies have shown that As crosses the blood-brain barrier and invades the brain parenchyma, and there is a noticeable correlation between dose of As exposure and brain concentration in guinea pigs and rats (Kannan et al. 2001). Deficits in learning tasks as well as behavioral alterations have been also observed in rats following sodium arsenite treatment (Zhang et al. 1999; Rodríguez et al. 2001). These literatures indicated that the brain could be a major target organ for As-induced neurotoxicity. However, the molecular mechanism by which As adversely affects intelligence is poorly understood.

It is commonly thought that learning and memory are induced by the modulation of the strength of synaptic connections between these neurons in the brain, a process known as synaptic plasticity (Carew and Sahley 1986; Mozzachiodi and Byrne 2009; Owen and Brenner 2012). Long-term potentiation (LTP) and long-term depression (LTD) are the key models of the synaptic plasticity. Moreover, The LTP and LTD are believed to underlie aspects of learning and memory in mammals and other vertebrates (Malenka and Bear 2004). The cAMP response element-binding protein (Creb) activation is essential to the maintenance of long-term memory (LTP/LTD). The activation of the Creb needs to be phosphorylated by Ca^{2+}/calmodulin-dependent protein kinase IV (Camk4) (Ahn et al. 1999; Shaywitz and Greenberg 1999). Our previous study (Wang et al. 2009) showed that exposure to As down-regulated expressions of the Camk4 gene and protein. It indicated that As intake may adversely affect learning and memory by repressing the Camk4 expression. Thyroid hormone receptor (TR), a ligand-mediated transcription factor, binds to DNA sequence known as a thyroid hormone-response element (TRE) to activate or repress transcription of target genes. Recently, studies have shown that the TR complexes may be involved in regulating gene transcription of the CaMk4 (Murata et al. 2000). Therefore, we are interested in whether As disrupts Camk4 pathway through down-regulating expression of TR in the brain of mice.

It has been reported that As exposure resulted in marked elevation in ROS, causing oxidative DNA damage, severe pathological changes, and even apoptosis in neural cells (Chattopadhyay et al. 2002a, b). It implies that ROS is involved in mechanism of As-induced neurotoxicity. Taurine (2-aminoethanesulfonic acid), a conditionally essential amino acid, has been considered as an antioxidant (Huxtable 1992; Son et al. 2007). It is a derivative of the sulfur-containing amino acid, cysteine, and is present in many tissues of mammals with high concentrations. A number of investigators reported that taurine protects many of the body's organs against toxicity and oxidative stress due to heavy metals and other toxin as well as drugs (Dogru-Abbasoglu et al. 2001; Tabassum et al. 2006). Although biochemical and physiologic function of taurine is still undefined, considerable evidence shows that it can act as a direct antioxidant by scavenging ROS (Timbrell et al. 1995; Wright et al. 1986).

Therefore, in the present study, we examined the gene expressions of TRα1 and TRβ in cerebrum and cerebellum of mice administered As alone or both of As and taurine by real-time PCR. Their protein expressions were also determined in these tissues of mice in the same groups by Western blot. Moreover, location of TR protein expression was also observed in cerebrum and cerebellum of mice by immunohistochemistry. This study aimed at investigating the protection of taurine against the toxic effect of As exposure on the expression of TR in brain of mice.

13.2 Methods

13.2.1 Chemicals

As_2O_3, HNO_3, H_2O_2, and taurine were purchased from Sigma Chemical Company (St. Louis, USA). When used, As_2O_3 was weighed and dissolved in dilute NaOH solution, and then the pH of the 40 ppm As_2O_3 stock solution was adjusted to 7.2. Affymetrix GeneChip Arrays (Mouse Genome 430 2.0 Array) and related kits were from Affymetrix (Santa Clara, USA).

13.2.2 Animals and Treatment

SPF mice (age 9 weeks) weighing 26.3–30.9 g were purchased from Experimental Animal Center, Dalian Medical University. The animals were maintained on a standard diet and water ad libitum. They were caged under a 12 h dark-light cycle in standard conditions of temperature (18–22°C) and humidity (50%). These mice were randomly divided into five groups. Group 1 received drinking water alone as control. Group 2 received 4 ppm As_2O_3. Group 3 received both of 4 ppm As_2O_3 and 150 mg/kg taurine. As_2O_3 was given through drinking water for 60 days. Taurine was administered by gavation twice a week. The animal experiment was performed in accordance with the Animal Guideline of Dalian Medical University and in agreement with the Ethical Committee of Dalian Medical University.

13.2.3   Quantitative Real-Time PCR

Total RNA was extracted from mouse cerebrum and cerebellum tissues by using RNAiso Plus according to the manufacturer's instructions (Takara, Japan). The RNA was quantified by using a spectrophotometer. Only RNA samples with an A_{260}/A_{280} of 1.9 or higher were used for reverse transcription. One µg of total RNA was reverse-transcribed using Reverse Transcription Kit (Takara, Japan). Quantitative real-time PCR was carried out with SYBR Green PCR kit (Takara, Japan) using a TP800 Real-Time PCR Detection System (Takara, Japan). The primers for TRα, TRβ, and β-actin are shown in Table (designed by Takara, Dalian). The reaction conditions were as follows: an initial denaturation at 95°C for 5 min, followed by 40 cycles of 95°C for 30 s, 55°C for 30 s, and 72°C for 30 s. The β-actin mRNAs were used as internal control probe.

Table Specific primer sequences used in real-time PCR

Gene	Primer	Sequences
TRα	Forward	5'-GACAAGGCCACCGGTTATCACTAC-3'
	Reward	5'-CAGCAGCTGTCATACTTGCAGGA-3'
TRβ	Forward	5'-GACAAGGCCACCGGTTATCACTAC-3'
	Reward	5'-CAGCAGCTGTCATACTTGCAGGA-3'
β-actin	Forward	5'-CTGTCGAGTCGCGTCCACCCG-3'
	Reward	5'-ATATGCCGGAGCCGTTGTCGAC-3'

13.2.4   Western Blot

Mouse cerebrum and cerebellum tissue were homogenized in ice-cold RIPA Tissue Protein Extraction Reagent (Biyuntian, China) supplemented with 1% proteinase inhibitor mix and incubated at 4°C for 30 min. After incubation, debris was removed by centrifugation at $13,000 \times g$ for 20 min at 4°C and the lysates were stored at −80°C until used. The total protein concentration in the lysates was determined using the BCA protein assay kit (Biyuntian, China). The proteins (50 µg/ lane) were mixed with an equal volume of SDS-PAGE loading buffer and separated by SDS-PAGE under nonreducing conditions using 10% SDS-PAGE gels and then electrotransferred to Hybond-P PVDF membrane (Millipore, France). The membrane was blocked with blocking buffer containing defatted milk power for 1 h and incubated overnight at 4°C with 1 µg/ml rat anti-mouse TRα1 monoclone antibody (1:500) (Santa Cruz, Biotechnology, sc-740), rat anti-mouse TRβ1 mono-clone antibody (1:50) (Santa Cruz, Biotechnology, sc-738), and rat anti-mouse TRβ2 monoclone antibody (1:50) (Santa Cruz, Biotechnology, sc-67124), respectively. The membrane was washed three times with Tris-buffered saline containing 0.05% Tween-20 (TBST) for 15 min and then incubated at room temperature for 1 h with horseradish peroxidase-conjugated goat anti-mouse IgG (Sigma).

The signals were visualized using an enhanced ECL chemoluminescence kit and quantified densitometrically using UVP BioSpectrum Multispectral Imaging System (Ultra-Violet Products Ltd. Upland, CA, USA).

13.2.5 Immunohistochemistry

The fixed cerebrum and cerebellum were trimmed, washed, dehydrated, and embedded in paraffin according to standard protocols. Paraffin blocks of the cerebrum and cerebellum were cut at 5 μm thickness from the midportion of tissue and mounted onto poly-L-lysine-coated glass slides. The sections were deparaffinized using xylene (3×10 min) at 23°C and a series of decreasing ethanol concentrations according to standard protocols. Endogenous peroxidase activity was blocked by submerging the slides in methanol containing 3% hydroperoxide for 10 min. Samples were then heated in boiling water bath for antigen retrieval (10 mmol/L citrate buffer, pH 6, 20 min). The sections were allowed to cool in citrate buffer, washed thrice (deionized water, PBS, pH 7.4, 3 min each), and incubated in blocking solution for 30 min. Next, they were washed with PBS (3×10 min) and incubated overnight with monoclone antibody for TRβ1(J52) at 4°C (Santa Cruz, Biotechnology, sc-738). After being washed with PBS (3×10 min), the sections were incubated with horseradish peroxidase-conjugated goat anti-mouse IgG (1:400 PBS) for 1 h at room temperature. Finally, peroxidase was visualized using 0.05% diaminobenzidine (DAB, Sigma), in 0.05 mol/L Tris buffer, pH 7.6, containing 0.01% hydrogen peroxide.

13.2.6 Statistical Analysis

Data were presented as mean ± standard deviation (SD). All data were analyzed with SPSS 11.0 for windows. Difference in mean values between groups was tested with the one-way ANOVA and LSD test. p values less than 0.05 were considered significant.

13.3 Results

13.3.1 Protective Effect of Taurine on Expression of TRβ Gene in Brain of Mice Received As

Because there were no significant changes in mRNA expression of TRα between groups, the mRNA expression of TRβ in cerebrum and cerebellum of mice is shown in Figs. 13.1 and 13.2. The mRNA expression of TRβ in cerebrum and cerebellum

Y. Wang et al.

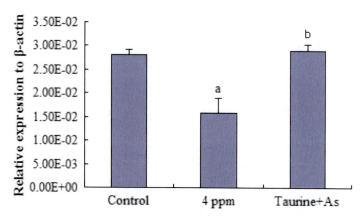

Fig. 13.1 The mRNA expression of TRβ in cerebrum of mice. Control: mice received drinking water alone; 4 ppm: mice exposed to 4 ppm As_2O_3; taurine + As: mice exposed to 4 ppm As_2O_3 with 150 mg/kg taurine. TRβ protein and control β-actin were detected by quantitative real-time PCR. Values represent means ± SD ($n = 6$). a: $p < 0.05$ vs. control; b: $p < 0.05$ vs. 4 ppm

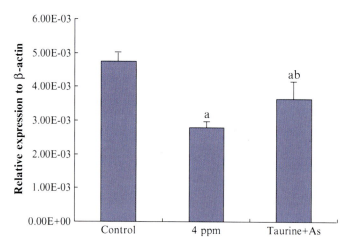

Fig. 13.2 The mRNA expression of TRβ in cerebellum of mice. Control: mice received drinking water alone; 4 ppm: mice exposed to 4 ppm As_2O_3; taurine + As: mice exposed to 4 ppm As_2O_3 with 150 mg/kg taurine. TRβ protein and control β-actin were detected by quantitative real-time PCR. Values represent means ± SD ($n = 6$). a: $p < 0.05$ vs. control; b: $p < 0.05$ vs. 4 ppm

of mice was significantly lower in the group which received As than that in controls ($p < 0.05$). Moreover, the mRNA expression of TRβ in cerebrum and cerebellum of mice was also significantly lower in the group which received As than that in the group exposed to As with taurine ($p < 0.05$).

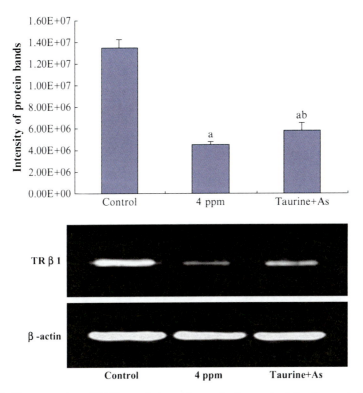

Fig. 13.3 The expression of TRβ1 protein in cerebrum of mice received As alone or both of As and taurine. Control: mice received drinking water alone; 4 ppm: mice exposed to 4 ppm As_2O_3; taurine + As: mice exposed to 4 ppm As_2O_3 with 150 mg/kg taurine. TRβ1 protein and control β-actin were detected by Western blot. Values represent means ± SD ($n = 6$). a: $p < 0.05$ vs. control; b: $p < 0.05$ vs. 4 ppm

13.3.2   Protective Effect of Taurine on Expressions of TRβ1 Protein in the Brain of Arsenic-Treated Mice

Because there were no significant changes in protein expressions of TRα and TRβ2 between groups, the expression of TRβ1 protein in cerebrum and cerebellum of mice is shown in Figs. 13.3 and 13.4. The expression of TRβ1 protein in cerebrum and cerebellum of mice was significantly lower in the groups received As alone or both of As and taurine than that in controls ($p < 0.05$). Moreover, the mRNA expression of TRβ1 in cerebrum and cerebellum of mice was also significantly lower in the group received As than that in the group exposed to As with taurine ($p < 0.05$).

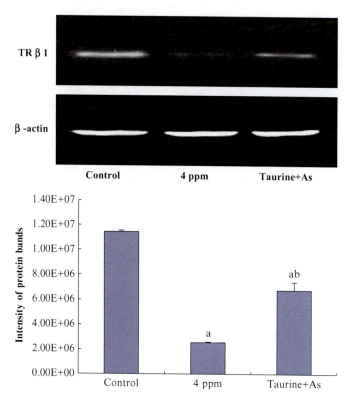

Fig. 13.4 The expression of TRβ1 protein in cerebellum of mice received As alone or both of As and taurine. Control: mice received drinking water alone; 4 ppm: mice exposed to 4 ppm As_2O_3; taurine + As: mice exposed to 4 ppm As_2O_3 with 150 mg/kg taurine. TRβ1 protein and control β-actin were detected by Western blot. Values represent means ± SD ($n = 6$). a: $p < 0.05$ vs. control; b: $p < 0.05$ vs. 4 ppm

13.3.3 Protective Effect of Taurine on the Distribution of TRβ1 Protein Expression in Brain of Arsenic-Treated Mice

The distribution of TRβ1 protein expression in cerebrum and cerebellum of mice is shown in Figs. 13.5 and 13.6. Its expression was mainly localized in nucleus. The result showed that the expression of TRβ1 protein in the cerebrum decreased obviously in the mice exposed to As (Fig. 13.5b) compared with the control or the group which received both of As and taurine (Fig. 13.5a or c). The TRβ1 protein expression in the cerebrum also decreased in the mice exposed to As (Fig. 13.6b) compared with the control or the group received both of As and taurine (Fig. 13.6a or c).

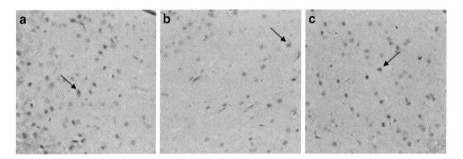

Fig. 13.5 Effect of As on TRβ1 protein expression in cerebral section of mice. (**a**) Control group; (**b**) the group which received 4 mg/L As$_2$O$_3$; (**c**) the group which received 4 ppm As$_2$O$_3$ with 150 mg/kg taurine. After the treatment, localization of TRβ1 protein in cerebral section was observed by immunohistochemical analyses (×200). The positive cells of TRβ1 protein expression appear brown (*arrow*), with stained nucleus

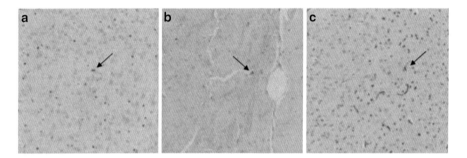

Fig. 13.6 Effect of As on TRβ1 protein expression in cerebellar section of mice. (**a**) Control group; (**b**) the group which received 4 mg/L As$_2$O$_3$; (**c**) the group which received 4 ppm As$_2$O$_3$ with 150 mg/kg taurine. After the treatment, localization of TRβ1 protein in cerebellar section was observed by immunohistochemical analyses (×200). The positive cells of TRβ1 protein expression appear brown (*arrow*), with stained nucleus

13.4   Discussion

In the present study, we observed the expressions of TR genes and their proteins in cerebrum and cerebellum of mice administered As alone or both of As and taurine. Our results showed that the expression of TRβ1 gene and their protein in cerebrum and cerebellum of mice was significantly lower in the group which received As than that in controls ($p<0.05$). However, the expression of TRβ gene and TRβ1 protein in cerebrum and cerebellum of mice was significantly higher in the group exposed to As with taurine than that in the group which received As ($p<0.05$). These results indicate that As exposure can down-regulate the expression of TRβ gene and TRβ1 protein in cerebrum and cerebellum of mice, and this impairment can be mitigated by coadministration of taurine.

Various studies reported that As could participate in the cellular oxidation-reduction reactions and resulted in the formation of excess ROS such as superoxide anion (O-2) and hydroxyl radical (OH.) via a chain reaction (Iwama et al. 2001; Garcia-Chavez et al. 2003). As can cross the blood-brain barrier and accumulates in the tissues. It has been reported that As exposure induces oxidative stress in the rat brain (Flora 1999; Samuel et al. 2005) and cultured cells from the brain of the human fetus and newborn rats (Chattopadhyay et al. 2002a). Because neurons contain a comparatively low content of protective enzymes and free radical scavengers, e.g., catalase, glutathione peroxidase, glutathione (GSH), and vitamin E (Rodríguez et al. 2003), they are very sensitive to oxidative stress. It was reported that As exposure resulted in marked elevation in ROS, causing oxidative DNA damage, severe pathological changes, and even apoptosis in neural cells (Chattopadhyay et al. 2002a, b). It implies that ROS may be involved in mechanism of As-induced neurotoxicity. Taurine is the most abundant free amino acid in many tissues. It protects many of the body's organs against toxicity and oxidative stress caused by various toxic substances (Dogru-Abbasoglu et al. 2001). Some investigation results showed that taurine acts as an antioxidant (Mahalakshmi et al. 2003) and could scavenge ROS (Niittynen et al. 1999). Balkan et al. (2001) reported that taurine treatment ameliorated the hepatic oxidative stress due to thioacetamide as well as chronic ethanol (Balkan et al. 2002) induced toxicity individually. Our previous study also showed that coadministration of taurine protected against pathological changes and nucleic acid damage due to ROS in brain tissue of mice exposed to arsenic (Piao et al. 2005). These literatures imply that taurine has the ability to scavenge the As-induced ROS in brain tissue. In the present study, As exposure down-regulated the expression of TRβ gene in the brain of mice and the coadministration of taurine mitigated the down-regulated the expression of TRβ gene induced by As. Our results indicate that taurine can protect against toxic effect of As exposure on the TR expression and the protection may be associated with anti-oxidation of taurine. Future studies should focus on the dose-effect relationship of taurine to prevent the As-induced down-regulation of TRβ gene expression, as well as determine the exact molecular mechanisms of these protective effects.

13.5 Conclusion

Taurine can protect against toxic effect of As exposure on the TR expression, and the protection may be associated with the anti-oxidative properties of taurine.

Acknowledgements This work was supported by National Nature Science Foundation of China [No. 30571584].

References

Ahn S, Ginty DD, Linden D (1999) A late phase of cerebellar long-term depression requires activation of CaMKIV and CREB. Neuron 23:559–568

Balkan J, Dogru-Abbasoglu S, Kanbagli O, Cevikbas U, Aykac-Toker G, Uysal M (2001) Taurine has a protective role against thioacetamide induced liver cirrhosis by decreasing oxidative stress. Hum Exp Toxicol 20:251–254

Balkan J, Kanbagli O, Aykac-Toker G, Uysal M (2002) Taurine treatment reduces hepatic lipids and oxidative stress in chronically bethanol treated rats. Biol Pharm Bull 25:1231–1233

Carew TJ, Sahley CL (1986) Invertebrate learning and memory: from behavior to molecules. Annu Rev Neurosci 9:435–487

Chattopadhyay S, Bhaumik S, Nag Chaudhury A, Das Gupta S (2002a) Arsenic induced changes in growth development and apoptosis in neonatal and adult brain cells in vivo and in tissue culture. Toxicol Lett 128:73–84

Chattopadhyay S, Bhaumik S, Purkayastha M, Basu S, Nag Chaudhuri A, Das Gupta S (2002b) Apoptosis and necrosis in developing brain cells due to arsenic toxicity and protection with antioxidants. Toxicol Lett 136:65–76

Dogru-Abbasoglu S, Kanbagli O, Balkan J, Cevikbas U, Aykac-Toke G, Uysal M (2001) The protective effect of taurine against thioacetamide hepatotoxicity of rats. Hum Exp Toxicol 20:23–27

Flora SJ (1999) Arsenic-induced oxidative stress and its reversibility following combined administration of N-acetylcysteine and meso 2,3-dimercaptosuccinic acid in rats. Clin Exp Pharmacol Physiol 26:865–869

Garcia-Chavez E, Santamaria A, Diaz-Barriga F, Mandeville P, Juarez BI, Jimenez-Capdeville ME (2003) Arsenite-induced formation of hydroxyl radical in the striatum of awaking rats. Brain Res 976:82–89

Huxtable RJ (1992) Physiological action of taurine. Physiol Rev 72:101–163

Iwama K, Nakajo S, Aluchi T, Nakaya K (2001) Apoptosis induced by arsenic trioxide in leukemia U937cells is dependent on activation of p38, inactivation of ERK and the calcium dependent production of superoxide. Int J Cancer 92:518–526

Kannan GM, Tripathi N, Dube SN (2001) Toxic effects of arsenic (III) on some hematopoietic and central nervous system variables in rats and guinea pigs. J Toxicol Clin Toxicol 39(7):675–682

Mahalakshmi K, Pushpakiran G, Anuradha CV (2003) Taurine prevents acrylonitrile-induced oxidative stress in rat brain. Pol J Pharmacol 55:1037–1043

Malenka RC, Bear MF (2004) LTP and LTD: an embarrassment of riches. Neuron 44:5–21

Mozzachiodi R, Byrne JH (2009) More than synaptic plasticity: role of nonsynaptic plasticity in learning and memory. Trends Neurosci 33:17–26

Murata M, Koibuchi N, Fukuda H, Murata M, Chin WW (2000) Augmentation of thyroid hormone receptor-mediated transcription by Ca2+/calmodulin-dependent protein kinase type. Endocrinology 141(6):2275–2278

Niittynen L, Nurminen ML, Korpela R, Vapaatalo H (1999) Role of arginine, taurine and homocysteine in cardiovascular diseases. Ann Med 31:318–326

Owen GR, Brenner EA (2012) Mapping molecular memory: navigating the cellular pathways of learning. Cell Mol Neurobiol 32:919–941

Piao F, Ma N, Hiraku Y, Murata M, Oikawa S, Cheng F, Zhong L, Yamauchi T, Kawanishi S, Yokoyama K (2005) Oxidative DNA damage in the brain of mice exposed to arsenic at environmental-relevant levels in relation to neurotoxicity. J Occup Health 47:445–449

Rodríguez VM, Carrizales L, Jiménez-Capdeville ME, Dufour L, Giordano M (2001) The effects of sodium arsenite exposure on behavioral parameters in the rat. Brain Res Bull 55:301–308

Rodríguez VM, Jiménez-Capdeville ME, Giordano M (2003) The effects of arsenic exposure on the nervous system. Toxicol Lett 145:1–18

Samuel S, Kathirvel R, Jayavelu T, Chinnakkannu P (2005) Protein oxidative damage in arsenic induced rat brain: influence of DL-α-lipoic acid. Toxicol Lett 155:27–34

Shaywitz AJ, Greenberg ME (1999) CREB: a stimulus-induced transcription factor activated by a diverse array of extracellular signals. Annu Rev Biochem 68:821–861

Son HY, Kim H, Kwon YH (2007) Taurine prevents oxidative damage of high glucose-induced cataractogenesis in isolated rat lenses. J Nutr Sci Vitaminol 53:324–330

Tabassum H, Rehman H, Banerjee BD, Raisuddin S, Parvez S (2006) Attenuation of tamoxifen-induced hepatotoxicity by taurine in mice. Clin Chim Acta 370:129–136

Timbrell JA, Seabra V, Watereld CJ (1995) The in vivo and in vitro protective properties of taurine. Gen Pharmacol 26:453–462

Wang Y, Li S, Piao F, Hong Y, Liu P, Zhao Y (2009) Subchronic exposure to arsenic downregulated the expression of Camk4 as an important gene related to the late phase of cerebellar LTD of mice. Neurotoxicol Teratol 31(5):318–322

Wasserman GA, Liu X, Parvez F, Ahsan H, Factor-Litvak P, van Geen A, Slavkovich V, LoIacono NJ, Cheng Z, Hussain I, Momotaj H, Graziano JH (2004) Water arsenic exposure and children's intellectual function in Araihazar, Bangladesh. Environ Health Perspect 112:1329–1333

Wright CE, Tallan HH, Linn YY (1986) Taurine: biological update. Annu Rev Biochem 55:427–453

Zhang C, Ling B, Liu J, Wang G (1999) Effect of fluoride-arsenic exposure on the neurobehavioral development of rats offspring. Wei Sheng Yan Jiu 28:337–338

Chapter 14
Taurine Exerts Robust Protection Against Hypoxia and Oxygen/Glucose Deprivation in Human Neuroblastoma Cell Culture

Po-Chih Chen, Chunliu Pan, Payam M. Gharibani, Howard Prentice, and Jang-Yen Wu

Abstract Stroke is one of the leading causes of mortality and disability worldwide. There is no effective treatment for stroke despite extensive research. Taurine is a free amino acid which is present at high concentrations in a range of organs including the brain, heart, and retina in mammalian systems. It had been shown that taurine can significantly increase cell survival under stroke conditions using both in vivo and in vitro models. Recently, we have found that several agents including granulocyte colony-stimulating factor (G-CSF), a stem cell enhancer and facilitator; S-methyl-N-diethylthiolcarbamate sulfoxide (DETC-MeSO), an NMDA receptor partial antagonist; sulindac, a potent antioxidant; and taurine, a neuroprotectant and calcium regulator, are effective in protecting against stroke-induced neuronal injury when used alone or in combination in both animal and tissue/cell culture models. In this chapter, we demonstrate that taurine can protect human neuroblastoma cells measured by ATP assay under conditions of hypoxia or oxygen/glucose deprivation (OGD). In addition, we found that taurine exerts its protective function by suppressing the OGD-induced upregulation of endoplasmic reticulum (ER) stress markers and proapoptotic proteins. A model depicting the mode of action of taurine in protecting neuroblastoma cells under OGD conditions is presented.

Po-Chih Chen and Chunliu Pan contributed equally to this work.

P.-C. Chen
Department of Biomedical Science, Charles E. Schmidt College of Medicine,
Florida Atlantic University, 777 Glades Road, Boca Raton, FL 33431, USA

Department of Neurology and Sleep Center, Shung Ho Hospital,
Taipei Medical University, Taipei, Taiwan
e-mail: chenpochih1979@gmail.com

C. Pan • P.M. Gharibani • H. Prentice(✉) • J.-Y. Wu(✉)
Department of Biomedical Science, Charles E. Schmidt College of Medicine,
Florida Atlantic University, 777 Glades Road, Boca Raton, FL 33431, USA
e-mail: chunliupan@gmail.com; pgharibani@gmail.com; hprentic@fau.edu; jwu@fau.edu

A. El Idrissi and W.J. L'Amoreaux (eds.), *Taurine 8*, Advances in Experimental
Medicine and Biology 775, DOI 10.1007/978-1-4614-6130-2_14,
© Springer Science+Business Media New York 2013

14.1   Introduction

Taurine, 2-aminoethanesulfonic acid, is one of the most abundant amino acids in mammals and is found at high concentrations in various tissues, including brain, heart, and kidney (for review, Huxtable 1992). Taurine plays important physiological functions in the brain including serving as a neurotransmitter/modulator and trophic factor, and in neuronal migration in the cerebellum and visual cortex (for review, Wu and Prentice 2010).

In addition, taurine has been shown to have protective functions in various systems including the nervous system, heart, lung, and kidney. presumably through its regulation of calcium homeostasis and its antiapoptotic property (Takatani et al. 2004; Wu and Prentice 2010). In animal studies, taurine had been shown to be effective in reducing the infarct size in a rat stroke model (Ghandforoush-Sattari et al. 2011). Taurine has also been applied clinically in several disorders, including cardiovascular diseases, metabolic disease, alcoholism, retinal degeneration, and hepatic and renal diseases (Birdsall 1998; Bidri and Choay 2003).

In our previous study, we have demonstrated that taurine exerts a protective effect on PC 12 cells under oxidative stress (Pan et al. 2010). Here we report that taurine also has protective effects on a neuroblastoma cell line under hypoxia or oxygen/glucose deprivation conditions.

14.2   Methods

The neuroblastoma SH-SY5Y cell line, F-12 media, EMEM media with L-glutamine, and trypsin-EDTA solution were purchased from ATCC (Manassas, VA, USA). Fetal bovine serum and penicillin–streptomycin were purchased from Sigma (St. Louis, MO, USA).

Rabbit anti-ATF4, rabbit anti-XBP-1, rabbit, rabbit anti-PUMA, and rabbit anti-IRE1 antibodies were purchased from Abcam (Cambridge, MA, USA). RIPA buffer was purchased from Thermo Scientific (Rockford, IL, USA). Rabbit anti-p-eIF2α antibody, was purchased from Cell Signaling Technology (Boston, MA, USA). Rabbit anti-GADD34 antibody and secondary mouse and rabbit antibodies were purchased from Santa Cruz Biotechnology (Santa Cruz, CA, USA). Adenosine 5'-triphosphate (ATP) Bioluminescent Assay Kit was purchased from Promega (Madison, WI, USA).

14.2.1   Cell Culture

SH-SY-5Y human neuroblastoma cells were maintained at 37°C 5% CO_2 in complete medium (Eagle's Minimum Essential Medium (EMEM) 44.5% and F12 medium 44.5%, fetal bovine serum to a final concentration of 10%, penicillin–streptomycin 1%). Cultured cells were plated in 6 (cell density 2×10^6) or 96-well dishes (5×10^5 cell/ml). Dishes contained complete medium at 2 ml/well for 6-well dishes and 100 μl/well for 96-well dishes. After plating 1–2 days, the medium was replaced with incomplete medium (50% EMEM, 50%F12 medium plus 10 μM retinoic acid) to induce cell differentiation.

14.2.2   Glucose Deprivation

After 5–7 days in complete medium, cells were changed to medium without glucose (154 mM NaCl, 5.6 mM KCl, 2.3 mM $CaCl_2$, 1.0 mM $MgCl_2$, 3.6 mM $NaHCO_3$, 5 mM Hepes, pH 7.2). Taurine was added to each well to a final concentration of 10 mM, and 1 h later, cells were subjected to 20 h of hypoxia.

14.2.3   Hypoxia/Reoxygenation

To generate hypoxic conditions, neuroblastoma SH-SY5Y cells in 6- or 96-well plates were placed in the hypoxia chamber with oxygen levels maintained at 0.3–0.4%. The level of oxygen was continuously monitored using an oxygen electrode. Neuroblastoma cells with or without taurine treatment were subjected to 20 h of hypoxia. Reoxygenation was performed by removing cultured plates from the hypoxic chamber and transferring them into normal culture incubator for 20 h.

14.2.4   Measurement of Cell Viability: ATP Assay

Neuroblastoma cells in 96-well plates were incubated with and without 10 mM taurine and then exposed to hypoxic conditions for 24 h to induce cell death. ATP solution (Promega) was added to each well, and cells were incubated for 10 min, after which the amount of ATP was quantified by a luciferase reaction. The luminescence intensity was determined using a luminometer (SpectraMax, Molecular Devices) after transferring the lysate to a standard opaque wall multi-well plate. The ATP content was determined by running an internal standard and expressed as a percentage of untreated cells (control).

14.2.5   Western Blot Analysis

Neuroblastoma cells were lysed in RIPA buffer (25 mM Tris–HCl pH 7.6, 150 mM NaCl, 1% NP-40, 1% sodium deoxycholate, 0.1% SDS) containing 1% (v/v) mammalian protease inhibitor cocktail from Sigma. Cellular proteins were separated on SDS-PAGE and then transferred to a nitrocellulose membrane. The membrane was blocked in blocking buffer (20 mM Tris–HCl, 150 mM NaCl,

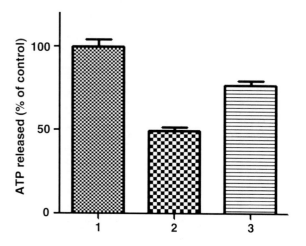

Fig. 14.1 Effect of taurine on the viability of neuroblastoma cells under hypoxic conditions. The percentage of cell viability in the neuroblastoma SH-SY5Y cell line under hypoxic (0.3% O$_2$, 20 h) conditions is shown (*1* normal control, *2* hypoxia, *3* 10 mM taurine + hypoxia). 10 mM taurine was administered by preincubation for 1 h followed by hypoxic exposure for 20 h

0.1% Tween-20, 5% milk) for 1.5 h at room temperature. After blocking, the primary antibody was incubated for 1 h, followed by 1 h incubation with the appropriate HRP-conjugated secondary antibody at room temperature. Extensive washes with a blocking buffer were performed between each step. The protein immunocomplex was visualized by ECL detection reagents.

14.3 Results

14.3.1 Taurine Protects Neuroblastoma SH-SY5Y Cells Against Hypoxic Stress

Using the ATP assay it was found that hypoxia conditions elicited approximately a 50% decrease in viability compared to the control group. This decrease was substantially reversed in the 10 mM taurine-treated group which showed approximately 75% cell survival (Fig. 14.1).

14.3.2 Taurine Restored the Expression of PUMA Under Oxygen/Glucose Deprivation Conditions

In the oxygen glucose deprivation (OGD) study, we analyzed the expression of PUMA (p53 upregulated modulator of apoptosis), a Bcl-2 family member originally identified in differential gene expression studies as a p53-inducible gene (Yu et al. 2001). Two BH3-containing proteins are encoded from the puma gene, Puma-a and Puma-b, both of which are induced by p53, bind Bcl-2, and Bcl-xL; localize to the mitochondria; and promote cytochrome c release and apoptosis. Western blot analysis shows the expression of PUMA was markedly increased in the OGD group and decreased in the OGD plus taurine treatment group (Fig. 14.2).

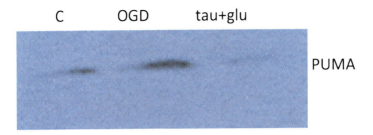

Fig. 14.2 Effect of taurine on the expression of PUMA (p53 upregulated modulator of apoptosis) in neuroblastoma cells under oxygen/glucose deprivation (OGD) conditions. Neuroblastoma cells were treated with or without taurine under OGD conditions. Western blot analysis was conducted with an anti-PUMA antibody. *C* control, *OGD* oxygen/glucose deprivation, *tau + OGD* oxygen/glucose deprivation plus treatment with 10 mM taurine

14.3.3   Taurine Treatment Had No Significant Effect on the Levels of p-eIF2α ATF4 and GADD34 Expression Under Hypoxic Conditions

In hypoxic conditions, we analyzed levels of expression of phosphorylated-eukaryotic initiation factor 2α (p-eIF2α), a downstream component of the PERK pathway which plays a role in inhibition of protein synthesis. The results showed markedly increased levels of p-eIF2-alpha both with and without taurine treatment compared with the control group (Fig. 14.3).

Activating transcription factor 4 (ATF4), which is translated as a compensatory response during a block of expression of eIF-2-alpha, showed no significant change in either the hypoxia condition or hypoxia plus taurine condition. In a previous study, the autoregulatory loop in PKR-like endoplasmic reticulum kinase (PERK) phosphorylates eIF-2α and in turn inhibits protein synthesis and allows ATF4 translation, subsequently leading to growth arrest and DNA damage-inducible protein-34 (GADD34) increases under cellular stress (Ma and Hendershot 2003). Our data showed p-eIF 2α was increased markedly both in the hypoxia group, compatible with conditions of cell stress, whereas GADD 34 expression was increased by 30% in the hypoxia group but not in the taurine plus hypoxia group compared with control.

14.3.4   Taurine Reversed the Increased Expression of XBP-1 and pIRE-1 Under Hypoxic Conditions

We further analyzed inositol-requiring kinase-1 (IRE-1), a ser/thr protein kinase that possesses endonuclease activity. IRE-1 is important for altering gene expression as a response to ER stress signals. IRE-1 senses and responds to unfolded proteins in the lumen of the endoplasmic reticulum via its N-terminal domain, leading to enzyme

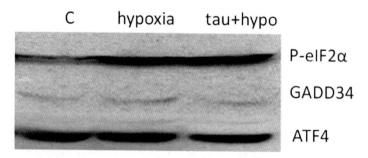

Fig. 14.3 Effect of taurine on the expression of p-eIF2α, GADD 34, and ATF4 in neuroblastoma cells under hypoxic conditions. Neuroblastoma cells were preincubated with 10 mM taurine before with or without 20 h hypoxia condition followed by Western blot analysis. Phosphorylated-eukaryotic initiation factor 2α (p-eIF2α), growth arrest and DNA damage-inducible protein 34 (GADD 34), and activating transcription factor 4 (ATF4) revealed no difference between any of the groups. *C* control, *tau + hypo* hypoxia with taurine treatment

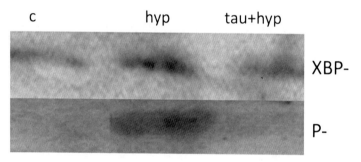

Fig. 14.4 Effect of taurine on the expression of p-IRE1 and XBP-1 in neuroblastoma cells under hypoxic conditions. Neuroblastoma cells were exposed to normoxic conditions or subjected to 20 h of hypoxia with or without a preexposure to 10 mM taurine. Cells were harvested for western blot analysis using antibodies to phosphorylated inositol-requiring kinase-1 (pIRE-1) and X-box binding protein 1 (XBP-1). *C* control, *tau + hyp* hypoxia with taurine treatment

autoactivation. The active endoribonuclease domain induces splicing of X-box binding protein 1 (XBP-1) mRNA. XBP-1 is a transcription factor that has been shown to be the target of the endonuclease activity of IRE-1 in mammals (Yoshida et al. 2003). Spliced XBP-1 then generates a new C-terminus, converting it into a potent unfolded-protein response (UPR) transcriptional activator and subsequently triggering growth arrest and apoptosis. The current results also demonstrated an increase in XBP-1 and p-IRE1 under hypoxic conditions. The increase in p-IRE1 and XBP-1 expression was reversed in the hypoxia plus taurine treatment condition (Fig. 14.4).

14.4   Discussion

Taurine has been shown to exert protective effects against neuronal damage by inhibiting the reverse mode Na^+/Ca^{2+} exchanger (Buddhala et al. 2012) and by decreasing calcium influx through L-, P/Q-, and N-type voltage-gated calcium channels as well as N-methyl-D-aspartic acid (NMDA) receptors (Wu et al. 2005).

Taurine treatment also decreases expression of caspase-3 and calpains and increases the ratio of levels of the antiapoptic protein Bcl-2 and proapoptotic protein Bax. In the ischemic hypothalamic nucleus of mice, taurine also attenuated the expression of caspase-8 and caspase-9 (Taranukhin et al. 2008; Leon et al. 2009).

Taurine has also been found to prevent mitochondrial dysfunction and subsequent apoptosis in hypoxic retinal ganglion cells in culture (Chen et al. 2009). The effects of neuroprotection by taurine have also been seen in in vivo studies including models of epilepsy and of stroke (Sun et al. 2011). In addition, taurine has been shown to reduce cell swelling under conditions of oxygen–glucose deprivation and reoxygenation-induced damage in rat brain cortical slices (Ricci et al. 2009).

ER stress occurs when misfolded or unfolded proteins accumulate in the ER, and the cell is capable of triggering caspase-12 or CHOP-mediated apoptosis if it is unable to repair these misfolded or unfolded proteins. ER stress is known to be activated in various neurodegenerative diseases, including Alzheimer's disease, Huntington's chorea, Parkinson's disease, and amyotrophic lateral sclerosis (Lindholm et al. 2006; Reijonen et al. 2008).

There are at least three known signaling pathways of ER stress identified by the components double-stranded RNA-activated protein kinase 1 (PKR)-like endoplasmic reticulum kinase (PERK), activating transcription factor 6 (ATF6), and IRE-1, respectively. In our previous study, we demonstrated that taurine decreases the expression of ATF6 and IRE-1 while exerting no effect on the PERK pathway in primary neuronal cultures (Pan et al. 2011).

In the current study, we employed the SH-SY5Y neuroblastoma cell line under OGD and hypoxic conditions, and our data revealed a robust pro-survival effect of taurine that was similar to that of our previous cell culture studies. After hypoxia and reoxygenation, neuronal viability without taurine treatment dropped to about 50% (percentage of control). The presence of 10 mM taurine improved the cell viability to greater than 70% (percentage of control neurons). This finding is compatible with our previous studies indicating that taurine decreases cell apoptosis. We also demonstrate that taurine attenuates the ER stress produced by hypoxia, but this protection did not involve the PERK-eIF2-ATF 4 pathway. In our ongoing studies, we will further characterize the role of the p-IRE1 pathway in mediating protection by taurine against hypoxia OGD-induced cell damage as well as examine the potential involvement of the ATF6 pathway in mediating protection in the neuroblastoma cell line.

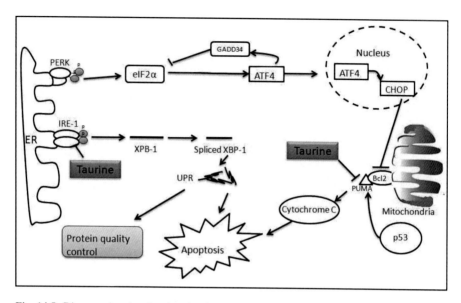

Fig. 14.5 Diagram showing the site of action of taurine in improving survival of neuroblastoma SH-SY5Y cells under conditions of oxygen/glucose deprivation or hypoxia. Taurine decreases the expression of IRE-1 and XPB-1 under hypoxic conditions (*left*). Taurine also decreases the expression of PUMA under oxygen/glucose deprivation conditions (*right*)

14.5   Conclusion

In this study, we found taurine exerts a protective effect on the SH-SY5Y neuroblastoma cell line under OGD and hypoxic conditions. Taurine attenuates OGD and hypoxia-induced apoptosis and ER stress. Our understanding of the mechanisms is depicted schematically in Fig. 14.5. The full mechanism of neuroprotective function of taurine is still not fully understood, and further studies will characterize in more detail the components of the apoptotic and ER stress pathways that are regulated by taurine under conditions of OGD and hypoxia.

References

Birdsall TC (1998) Therapeutic applications of taurine. Altern Med Rev 3(2):128–136
Bidri M, Choay P (2003) Taurine: a particular aminoacid with multiple functions. Ann Pharm Fr 61(6):385–391
Buddhala C, Prentice H, Jang-Yen W (2012) Modes of action of taurine and granulocyte colony-stimulating factor in neuroprotection. J Exp Clin Med 4(1):1–7
Chen K, Zhang Q, Wang J, Liu F, Mi M, Xu H, Chen F, Zeng K (2009) Taurine protects transformed rat retinal ganglion cells from hypoxia-induced apoptosis by preventing mitochondrial dysfunction. Brain Res 1279:131–138

Ghandforoush-Sattari M, Mashayekhi SO, Nemati M, Ayromlou H (2011) Changes in plasma concentration of taurine in stroke. Neurosci Lett 496(3):172–175

Huxtable RJ (1992) Physiological actions of taurine. Physiol Rev 72(1):101–163

Leon R, Wu H, Jin Y, Wei J, Buddhala C, Prentice H, Wu JY (2009) Protective function of taurine in glutamate-induced apoptosis in cultured neurons. J Neurosci Res 87(5):1185–1194

Lindholm D, Wootz H, Korhonen L (2006) ER stress and neurodegenerative diseases. Cell Death Differ 13(3):385–392

Ma Y, Hendershot LM (2003) Delineation of a negative feedback regulatory loop that controls protein translation during endoplasmic reticulum stress. J Biol Chem 278(37):34864–34873

Pan C, Giraldo GS, Prentice H, Wu JY (2010) Taurine protection of PC12 cells against endoplasmic reticulum stress induced by oxidative stress. J Biomed Sci 17(Suppl 1):S17

Pan C, Prentice H, Price AL, Wu JY (2011) Beneficial effect of taurine on hypoxia- and glutamate-induced endoplasmic reticulum stress pathways in primary neuronal culture. Amino Acids 43:845–855

Reijonen S, Putkonen N, Nørremølle A, Lindholm D, Korhonen L (2008) Inhibition of endoplasmic reticulum stress counteracts neuronal cell death and protein aggregation caused by N-terminal mutant huntingtin proteins. Exp Cell Res 314(5):950–960

Ricci L, Valoti M, Sgaragli G, Frosini M (2009) Protection by taurine of rat brain cortical slices against oxygen glucose deprivation- and reoxygenation-induced damage. Eur J Pharmacol 621(1–3):26–32

Sun M, Gu Y, Zhao Y, Xu C (2011) Protective functions of taurine against experimental stroke through depressing mitochondria-mediated cell death in rats. Amino Acids 40(5):1419–1429

Takatani T, Takahashi K, Uozumi Y, Shikata E, Yamamoto Y, Ito T, Matsuda T, Schaffer SW, Fujio Y, Azuma J (2004) Taurine inhibits apoptosis by preventing formation of the Apaf-1/caspase-9 apoptosome. Am J Physiol Cell Physiol 287(4):C949–C953

Taranukhin AG, Taranukhina EY, Saransaari P, Djatchkova IM, Pelto-Huikko M, Oja SS (2008) Taurine reduces caspase-8 and caspase-9 expression induced by ischemia in the mouse hypothalamic nuclei. Amino Acids 34(1):169–174

Wu H, Jin Y, Wei J, Jin H, Sha D, Wu JY (2005) Mode of action of taurine as a neuroprotector. Brain Res 1038(2):123–131

Wu JY, Prentice H (2010) Role of taurine in the central nervous system. J Biomed Sci 17(Suppl 1):S1

Yoshida H, Matsui T, Hosokawa N, Kaufman RJ, Nagata K, Mori K (2003) A time-dependent phase shift in the mammalian unfolded protein response. Dev Cell 4(2):265–271

Yu J, Zhang L, Hwang PM, Kinzler KW, Vogelstein B (2001) PUMA induces the rapid apoptosis of colorectal cancer cells. Mol Cell 7(3):673–682

Chapter 15
The Effects of Chronic Taurine Supplementation on Motor Learning

Allison Santora, Lorenz S. Neuwirth, William J. L'Amoreaux, and Abdeslem El Idrissi

Abstract Taurine is one of the most abundant nonprotein amino acids shown to be essential for the development, survival, and growth of vertebrate neurons. We previously demonstrated that chronic taurine supplementation during neonatal development results in changes in the GABAergic system (El Idrissi, Neurosci Lett 436:19–22, 2008). The brains of mice chronically treated with taurine have decreased levels of $GABA_A$ β subunits and increased expression of GAD and GABA, which contributes to hyperexcitability. This down regulation of $GABA_A$ receptor subunit expression and function may be due to a sustained interaction of taurine with $GABA_A$ receptors. This desensitization decreases the efficacy of the inhibitory synapses at the postsynaptic membrane. If changes occur in the GABAergic system as a possible compensatory mechanism due to taurine administration, then it is important to study all aspects by which taurine induces hyperexcitability and affects motor behavior. We therefore hypothesized that modification of the GABAergic system in response to taurine supplementation influences motor learning capacity in mice. To test this hypothesis, the rotarod task was employed after chronic taurine supplementation in drinking water (0.05% for 4 weeks). Control animals receiving no taurine supplementation were also tested in order to determine the difference in motor learning ability between groups. Each animal was trained on the rotarod apparatus for 7 days at an intermediate speed of 24 rpm in order to establish baseline performance. On the testing day, each animal was subjected to eight different predefined speeds (5, 8, 15, 20, 24, 31, 33, and 44 rpm). From our observations, the animals

A. Santora (✉)
Department of Biology, 6S-143, College of Staten Island/CUNY,
2800 Victory Blvd, Staten Island, NY 10314, USA
e-mail: allison.santora@gmail.com

L.S. Neuwirth • W.J. L'Amoreaux • A. El Idrissi
Department of Biology, 6S-143, College of Staten Island/CUNY,
2800 Victory Blvd, Staten Island, NY 10314, USA

City University of New York Graduate School, New York, NY, USA

A. El Idrissi and W.J. L'Amoreaux (eds.), *Taurine 8*, Advances in Experimental
Medicine and Biology 775, DOI 10.1007/978-1-4614-6130-2_15,
© Springer Science+Business Media New York 2013

that underwent chronic taurine supplementation appeared to have a diminished motor learning capacity in comparison to control animals. The taurine-fed mice displayed minor improvements after repeated training when compared to controls. During the testing session the taurine-fed mice also exhibited a shorter latency to fall, as the task requirements became more demanding.

15.1   Introduction

Taurine, 2-aminoethanesulfonic acid, is the second most abundant nonprotein amino acid in the central nervous system of mammals (Huxtable and Lleu 1992). Taurine is crucial for the development, survival, and growth of vertebrate neurons (Hayes et al. 1975). High concentrations of taurine are incorporated into fetal and early postnatal rodents via their mothers (Sturman et al. 1977). Within the brain, taurine concentrations increase until weaning, and subsequently decline reaching stable concentrations in adulthood that are comparable, but second to those of glutamate which is the main excitatory neurotransmitter. Activation of glutamate receptors leads to a depolarization of the postsynaptic membrane causing extracellular calcium influx as well as mobilization of calcium from intracellular stores (Jaffe and Brown 1994). Many physiological processes rely on calcium as a vital second messenger (Kater et al. 1988), but despite that, excessive elevation of intracellular calcium levels results in structural damage to neurons (El Idrissi and Trenkner 2004). Hyperexcitability of the brain is prevented by γ-aminobutyric acid (GABA), the predominant inhibitory neurotransmitter. When GABA is released from presynaptic neurons, it acts by binding to the ionotropic $GABA_A$ receptor located on the postsynaptic neuron. The outcome of this activity permits chloride influx, and subsequent hyperpolarization of the postsynaptic membrane.

Taurine is structurally related to GABA and acts itself as an inhibitory amino acid during development. Taurine is a partial agonist of the $GABA_A$ receptor (Frosini et al. 2003), and activates chloride influx into postsynaptic neurons through this receptor (El Idrissi and Trenkner 2004). Increases in chloride concentrations within the cell results in hyperpolarization of the postsynaptic membrane, and therefore reduces excitability. In addition to acting as a partial agonist of the $GABA_A$ receptor, taurine has also been shown to activate the corticostriatal pathway by behaving as an endogenous ligand for glycine receptors (Chepkova et al. 2002). Moreover, taurine has been shown to activate Cl^- influx through $GABA_A$ receptors in cerebellar granule cells in vitro (El Idrissi and Trenkner 2004). In the same study, cultures were pretreated with taurine (1 mM) for 24 h prior to the addition of glutamate, and afterwards Ca^{2+} uptake was shown to be significantly lower than in control cultures. Taurine also prevents neuronal damage associated with excitotoxicity. This is achieved through the regulation of cytoplasmic and intramitochondrial calcium homeostasis after glutamate receptor activation (El Idrissi and Trenkner 1999). The mechanisms by which taurine accomplishes modulation of neuronal excitability are direct enhancement of GABAergic function and indirect depression of glutamatergic neurotransmission (El Idrissi and Trenkner 2004). During development of the nervous system, when GABA is excitatory, taurine might play a critical role as a regulator of neuronal excitability through calcium modulation, and thereby, compensate for the lack of receptor-mediated neuronal inhibition (El Idrissi and Trenkner 2004). Furthermore, taurine may also play a vital role in neuroprotection since levels in the brain significantly increase during stressful conditions (Wu et al. 1998).

Taurine deficiency has been established in a number of neuropathological disorders, such as epilepsy (Barbeau et al. 1975; Joseph and Emson 1976), mental depression (Perry 1976), and alcohol withdrawal syndrome (Ikeda 1977). To date, taurine supplementation has been demonstrated to affect behavior in rodents. For example, acute taurine injections have been shown to increase the threshold of pharmacologically induced convulsions, and also significantly improved survivability when compared to controls (El Idrissi et al. 2003). On the other hand, chronic supplementation of taurine in drinking water increases brain excitability in mice, which occurs mainly through alteration in the inhibitory GABAergic system (L'Amoreaux et al. 2010). Chronic taurine supplementation induces down regulation of $GABA_A$ receptor expression due to a sustained interaction of taurine with $GABA_A$ receptors. This process decreases the efficacy of the inhibitory synapses at the postsynaptic membrane. The β subunit of the $GABA_A$ receptor is a key subunit that is present in virtually all of these receptors (Barnard et al. 1998) and is considered to be required for receptor assembly and function (Connolly et al. 1996). Within the hippocampus of taurine-fed mice, a reduction of $GABA_A$ β subunit expression was observed (El Idrissi 2006). In conjunction with this, the taurine-fed mice had reduced expression of $GABA_A$ receptors in the hippocampus.

Motor coordination is an acquired skill that manifests through a process of adaptation. Learning to walk, swim, ride a bike, or excel at a physical sport are examples of motor learning (Crawley 2007). The learning of skilled movement is controlled by interactions between the supplementary motor area, prefrontal parietal cortex, basal ganglia, and cerebellum (Rustay et al. 2003). The basal ganglia are involved with the automatic execution of learned motor plans, and in the preparation of movement (Afifi 2003). When cortical signals are received and processed by basal ganglia systems, the suppression of competing motor programs occurs when the neurotransmitter GABA inhibits thalamic nuclei. Thalamic nuclei provide the link between the basal ganglia and the motor, supplementary motor, premotor, prefrontal, and limbic cortices (Afifi 2003). On the other hand, the cerebellum influences movements primarily by modifying the activity patterns of the upper motor neurons located in the cerebral cortex. The primary function of the cerebellum is evidently to detect the difference, or "motor error," between an intended movement and the actual movement, and, through its influence over upper motor neurons, to reduce the error (Purves et al. 2008). Therefore, the cerebellum is the fundamental processing center required for the learning of compound movements. Furthermore, the cerebellar circuitry is highly reliant on synaptic integration derived from GABA-mediated inhibition.

GABAergic modification plays an important part in motor learning and plasticity. If changes occur in the GABAergic system as a possible compensatory mechanism due to taurine administration, then it is important to study all aspects by which chronic taurine supplementation induces hyperexcitability, and affects behavior. We therefore hypothesized that modification of the GABAergic system in response to taurine supplementation influences motor learning capacity in mice. To test this hypothesis, the rotarod task was employed after chronic taurine supplementation. Control animals receiving no taurine supplementation were also tested in order to determine the difference in motor learning ability between groups. From our observations, the animals that underwent chronic taurine supplementation appeared to have a diminished motor learning capacity in comparison to control animals.

15.2 Methods

15.2.1 Rotarod Task

Experiments were carried out on FVB/NJ adult male mice. Animals were given either distilled water and served as controls or a solution of taurine dissolved in distilled water at 0.05%. After 4 weeks of chronic taurine supplementation rotarod task was employed. The rotarod (IITC Life Science Inc., Woodland Hills, CA) is an apparatus which is used to gauge the ability of an animal to maintain balance on a rotating rod as well as motor learning ability in rodents (Lalonde et al. 1995). Three measures were obtained during each trial from the apparatus: (1) latency to fall measured in seconds, (2) distance traveled prior to fall measured in meters, and (3) animal velocity.

During the training phase, the rotation was set to an intermediate speed (24 rpm) in order to establish baseline performance. Each mouse was placed on the rotating rod for a maximum of 60 s. Latency to fall off the rotarod was recorded within this time period. Each animal underwent five trials per day for 7 days. After each trial, the animal was returned to its home cage for an intertrial interval (ITI) of 5 min. On the testing day, each animal was subjected to eight different predefined speeds (5, 8, 15, 20, 24, 31, 33, and 44 rpm). The mice were given two trials at each speed level with an ITI of 5 min.

15.2.2 Statistical Analysis

Analyses were performed using Statistica V 6.1 (Statsoft, Inc. Tulsa, OK). During the training session, the independent variables were treatment (control or taurine-fed animals) and training day, whereas the dependent variables were latency to fall, distance traveled, and animal velocity. For analysis of the testing day, the independent variables were treatment (control or taurine-fed animals) and rpm speed, while the dependent variables were latency to fall, distance traveled, and animal velocity. Therefore, a multifactorial analysis of variance was used in order to study the interaction effects among treatments. Significance was set at a confidence level of 95%, and data are presented as mean ± SD.

15.3   Results

15.3.1   The Effects of Taurine on Latency to Fall

Control mice displayed longer latencies to fall and had substantial day-to-day improvements in performance when compared to taurine-fed animals (Fig. 15.1a). Two-way ANOVA (training day × treatment) on latency to fall showed a significant effect of training day ($F_{1,6} = 12.15$, $p < 0.01$), suggesting a learning component to rotarod performance. A main effect of treatment was also observed ($F_{1,6} = 12.86$, $p < 0.01$) showing that taurine significantly decreased latency to fall compared to control group performance. Therefore, the taurine-supplemented mice were not improving their performance over training days.

After training, mice were tested at eight different rpm speeds (Fig. 15.1b). Two-way ANOVA (rpm × treatment) on latency to fall revealed a main effect of rpm ($F_{1,7} = 18.45$, $p < 0.01$) with faster rotational speeds leading to shorter latency to fall. Also, an effect of treatment was observed ($F_{1,7} = 17.89$, $p < 0.01$) showing that taurine significantly decreased the latency to fall compared to controls. An rpm × treatment interaction was also found ($F_{1,7} = 3.03$, $p < 0.01$), indicating that the taurine-fed mice fell off earlier at higher rpm.

15.3.2   The Effects of Taurine on Distance Traveled

The taurine-fed mice displayed minor improvements in distance traveled after repeated training (Fig. 15.2a). A main effect of treatment was observed ($F_{1,6} = 100.09$, $p < 0.01$), indicating that the taurine-supplemented mice traveled shorter distances when compared to control animals. An effect of training day was also observed ($F_{1,6} = 17.85$, $p < 0.01$), and a training day × treatment interaction ($F_{1,6} = 9.14$, $p < 0.01$). This interaction suggests that control group distance traveled improved over training days, while taurine-fed mice displayed stable performance.

In Fig. 15.2b, the distance traveled is represented for both groups at different rpm speeds. Two-way ANOVA (rpm × treatment) on distance traveled revealed a main effect of treatment ($F_{1,7} = 36.32$, $p < 0.01$), implying that taurine-fed mice traveled shorter distances. A significant effect was also observed for rpm speed ($F_{1,7} = 2.38$, $p < 0.05$), with an rpm × treatment interaction ($F_{1,7} = 3.41$, $p < 0.01$). This interaction insinuates that taurine supplementation caused mice to travel shorter distances at faster rpm speeds.

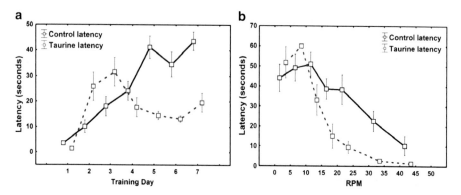

Fig. 15.1 The effects of taurine on rotarod performance when evaluating latency to fall. Taurine (0.05%) was supplemented in the drinking water for 4 weeks. Data represent mean ± SD. Control, $n = 3$; taurine-fed, $n = 3$. Mice were trained for 7 days (**a**). Control mice displayed longer latencies to fall and had substantial day-to-day improvements in performance when compared to taurine-fed animals. After training, mice were tested at eight different rpm speeds (**b**). Taurine-fed mice exhibited shorter latencies to fall at faster speeds

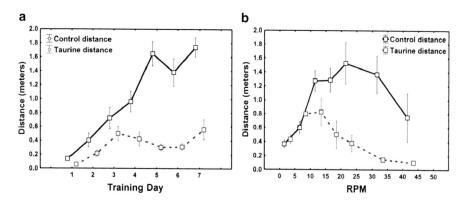

Fig. 15.2 Effects of taurine when assessing distance traveled on the rotarod apparatus. Taurine (0.05%) was supplemented in the drinking water for 4 weeks. Data represent mean ± SD. Control, $n = 3$; taurine-fed, $n = 3$. Mice were trained for 7 days (**a**). The taurine-fed mice displayed minor improvements after repeated training. During the testing phase (**b**) the taurine-fed mice demonstrated poorer performance as the ramp speed increased

15.3.3 The Effects of Taurine on Animal Velocity

Two-way ANOVA (training day × treatment) on animal velocity showed a main effect of treatment ($F_{1,6} = 737.77$, $p < 0.01$). The control animals exhibited stable performance over the course of training, while the taurine-fed mice showed minor improvements (Fig. 15.3a). A significant effect was also seen for training day ($F_{1,6} = 30.45$, $p < 0.01$), with a training day × treatment interaction effect ($F_{1,6} = 38.65$, $p < 0.01$).

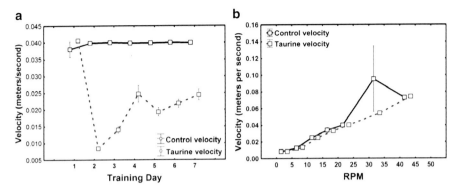

Fig. 15.3 The effects of taurine supplementation on movement velocity. The control animals exhibited stable performance over the course of training (**a**), while the taurine-fed mice showed minor improvements. During the testing phase (**b**), all animals were achieving similar velocities during the task

This interaction implies that taurine affected animal velocity with respect to training day.

In Fig. 15.3b, the animal velocity is depicted for both groups at each rpm speed during the testing phase. A two-way ANOVA (rpm×treatment) on animal velocity revealed an effect of rpm ($F_{1,7} = 4.34$, $p < 0.05$) with faster rotational speeds leading to increased velocities. However, the effect of treatment on animal velocity was nonsignificant ($F_{1,7} = 0.943$, $p > 0.05$), and the interaction effect was also nonsignificant ($F_{1,7} = 1.00$, $p > 0.05$). Therefore, taurine treatment did not significantly affect animal velocity. Taurine-fed and control animals did not differ significantly at each rpm speed.

15.4   Discussion

Motor coordination is a complex behavioral domain that can reflect balance, muscle strength, and patterned gait, as well as sensory competence. Difficulties in motor performance can confound behavioral assays of learning and memory, exploration, and motivation (Rustay et al. 2003). On the other hand, the learning of skilled movement is a finely tuned process that is dependent on interactions between the supplementary motor area, prefrontal parietal cortex, basal ganglia, and cerebellum. Given the neurocircuitry involved in motor learning and performance, one might expect to see differences in learning and performance on the rotarod among a set of inbred strains (Rustay et al. 2003), or after drug treatment. Results from this study suggest that there may be differences in the underlying structure and/or function of the brain regions involved in rotarod performance after chronic taurine supplementation.

Acquisition of a motor skill was examined after chronic taurine supplementation during the training phase of this study. The control animals continued to show improvement in their ability to maintain balance on the rotarod apparatus (Fig. 15.1a). This was evident by their capacity to achieve longer latencies prior to falling in comparison to the taurine-fed mice. The results of this study suggest that chronic taurine supplementation affects the ability of the animal to coordinate and maintain balance since refinement was not observed over repeated training. Furthermore, the control animals displayed progressive improvement in the distance traveled from day to day, which is in contrast to taurine-fed animals.

During the testing phase as illustrated in Fig. 15.1b, an interaction effect can be seen with regard to chronic taurine supplementation and increased task difficultly. The control group exhibited subtle declines in the latency to fall when the rotarod speed was increased from 15 to 24 rpm. Beyond this speed, control animals were able to coordinate and maintain balance on the rotarod to a greater degree than taurine-fed animals. Additionally, the distance traveled by the control animals steadily inclined from 5 to 24 rpm (Fig. 15.2b). Unarguably, this makes sense since faster speeds at longer durations correlate nicely with distance traveled. Conversely, animals that received chronic taurine supplementation displayed poorer performances at speeds beyond 15 rpm.

Motor coordination is the ability of the organism to perform compound movements smoothly, whereas motor learning is the ability to adapt motor coordination to changes in task demands. Both processes are generally thought of as functions of the cerebellum and a deficit in either one of them could result in impaired performance on rotarod task parameters.

15.5 Conclusion

In summary, this study shows that taurine regulates motor learning behavior in mice. Our data show that chronic taurine supplementation may have contributed to motor learning deficits. The taurine-fed animals displayed minor improvements after repeated training when compared to controls. During the testing session the taurine-fed animals also exhibited a shorter latency to fall, as the task requirements became more demanding.

References

Afifi AK (2003) The basal ganglia: a neural network with more than motor function. Semin Pediatr Neurol 10(1):3–10

Barbeau A, Inoue N, Tsukada Y, Butterworth RF (1975) The neuropharmacology of taurine. Life Sci 17:669–678

Barnard EA, Skolnick P, Olson RW, Mohler H, Seighart W, Biggio G, Braestrup C, Bateson AN, Langer SZ (1998) International union of pharmacology. XV. Subtypes of gamma-aminobutyric acid A receptors: classification on the basis of subunit structure and receptor function. Pharmacol Rev 50:291–313

Chepkova AN, Doreulee N, Yanovsky Y, Mukhopadhyah D, Haas HL, Sergeeva OA (2002) Long-lasting enhancement of corticostriatal neurotransmission by taurine. Eur J Neurosci 16:1523–1530

Connolly CN, Wooltorton JRA, Smart TG, Moss SJ (1996) Subcellular localization of γ-aminobutyric acid type A receptors is determined by receptor b subunits (polatityion-channel). Proc Natl Acad Sci USA 93:9899–9904

Crawley JN (2007) What's wrong with my mouse?: Behavioral phenotyping of transgenic and knockout mice, 2nd edn. Wiley, New York

El Idrissi A, Trenkner E (1999) Growth factors and taurine protect against excitotoxicity by stabilizing calcium homeostasis and energy metabolism. J Neurosci 19:9459–9468

El Idrissi A, Messing J, Scalia J, Trenkner E (2003) Prevention of epileptic seizures through taurine. Adv Exp Med Biol 526:515–525

El Idrissi A, Trenkner E (2004) Taurine as a modulator of excitatory and inhibitory neurotransmission. Neurochem Res 29(1):189–197

El Idrissi A (2006) Taurine 6. Taurine and brain excitability. Adv Exp Med Biol 583(5):315–322

El Idrissi A (2008) Taurine improves learning and retention in aged mice. Neurosci Lett 436:19–22

Frosini M, Sesti C, Dragoni S, Valoti M, Palmi M, Dixon HB, Machetti F, Sgaragli G (2003) Interaction of taurine and structurally related analogues with the GABAergic system and taurine binding sites of rabbit brain. Br J Pharmacol 138:1163–1171

Hayes KC, Carey RE, Schmidt SY (1975) Retinal degeneration associated with taurine deficiency in the cat. Science 188:949–951

Huxtable RJ, Lleu PL (1992) A possible relationship between taurine and synaptogenesis in the developing rat brain. Pharmacol Res 26(Suppl 1):146

Ikeda HC (1977) Effects of taurine on alcohol withdrawal. Lancet 2(8036):509

Jaffe DB, Brown TH (1994) Metabotropic glutamate receptor activation induces calcium waves within hippocampal dendrites. J Neurophysiol 72:471–474

Joseph MH, Emson PC (1976) Taurine and cobalt induced epilepsy in the rat: a biochemical and electrocorticographic study. J Neurochem 27:1495–1501

Kater SB, Mattson MP, Cohan C, Conner J (1988) Calcium regulation of the neuronal growth cone. Trends Neurosci 11:315–321

Lalonde R, Bensoula AN, Fiali M (1995) Rotarod sensorimotor learning in cerebellar mutant mice. Neurosci Res 22:423–426

L'Amoreaux WJ, Marsillo A, El Idrissi A (2010) Pharmacological characterization of GABAA receptors in taurine-fed mice. J Biomed Sci 17(Suppl 1):S14

Perry TL (1976) Hereditary mental depression with taurine deficiency: further studies, including a therapeutic trial of taurine administration. In: Huxtable R, Barbeau A (eds) Taurine. Raven, New York, pp 365–374

Purves D et al (2008) Neuroscience, 4th edn, Modulation of movement by the cerebellum. Sinauer Associates Inc., Sunderland, MA, pp 475–494

Rustay NR, Wahlsten D, Crabbe JC (2003) Influence of task parameters on rotarod performance and sensitivity to ethanol in mice. Behav Brain Res 141(2):237–249

Sturman JA, Rassin DK, Gaull GE (1977) Taurine in developing rat brain: transfer of 35S taurine to pups via the milk. Pediatr Res 11:28–33

Wu JY, Tang XW, Schloss JV, Faiman MD (1998) Regulation of taurine biosynthesis and its physiological in the brain. Adv Exp Med Biol 442:339–345

Chapter 16
Changes in Gene Expression at Inhibitory Synapses in Response to Taurine Treatment

Chang Hui Shen, Eugene Lempert, Isma Butt, Lorenz S. Neuwirth, Xin Yan, and Abdeslem El Idrissi

Abstract We have previously shown that chronic supplementation of taurine to mice significantly ameliorated the age-dependent decline in memory acquisition and retention. We also showed that concomitant with the amelioration in cognitive function, taurine caused significant alterations in the GABAergic and somatonergic system. These changes include increased levels of the neurotransmitters GABA and glutamate, increased expression of both isoforms of GAD and the neuropeptide somatostatin, decreased hippocampal expression of the beta (β) 2/3 subunits of the $GABA_A$ receptor, an increase in the number of somatostatin-positive neurons, and an increase in the amplitude and duration of population spikes recorded from CA1 in response to Schaefer collateral stimulation and enhanced paired pulse facilitation in the hippocampus. These specific alterations of the inhibitory system caused by taurine treatment oppose those naturally induced by aging, suggesting a protective role of taurine in this process. In this study, we further investigated the effects of taurine on gene expression of relevant proteins of the inhibitory synapses using qRT-PCR method and found that taurine affects gene

C.H. Shen (✉)
Department of Biology, College of Staten Island, Building 6S, Room 130,
2800 Victory Boulevard, Staten Island, NY 10314, USA

Institute for Macromolecular Assemblies,
City University of New York, Staten Island, NY 10314, USA

CUNY Graduate School, New York, NY, USA
e-mail: ChangHui.Shen@csi.cuny.edu

E. Lempert • I. Butt
Department of Biology, College of Staten Island, Building 6S, Room 130,
2800 Victory Boulevard, Staten Island, NY 10314, USA

L.S. Neuwirth • X. Yan • A. El Idrissi
Department of Biology, College of Staten Island, Building 6S, Room 130,
2800 Victory Boulevard, Staten Island, NY 10314, USA

CUNY Graduate School, New York, NY, USA

A. El Idrissi and W.J. L'Amoreaux (eds.), *Taurine 8*, Advances in Experimental
Medicine and Biology 775, DOI 10.1007/978-1-4614-6130-2_16,
© Springer Science+Business Media New York 2013

expression of various subunits of the $GABA_A$ receptors and GAD. Increased understanding the effects of taurine on gene expression will increase our understanding of age-related taurine-mediated neurochemical changes in the GABAergic system and will be important in elucidating the underpinnings of the functional changes of aging. Taurine might help forestall the age-related decline in cognitive functions through interaction with the GABAergic system.

Abbreviations

qRT-PC   Quantitative reverse transcriptase polymerase chain reaction
GAD   Glutamic acid decarboxylase
GABA   γ-Aminobutyric acid

16.1 Introduction

Taurine is stored at millimolar concentration in all mammalian tissue and has several cytoprotective properties, such as calcium handling, osmoregulation, antioxidation, and detoxication (Huxtable 1992; Satoh and Sperelakis 1998; Schaffer et al. 2000). During early critical periods of developmental maturation, the brain is very sensitive to environmental factors. In this study, we supplemented mice with taurine in drinking water for 4 weeks and examined changes in the inhibitory system. After 4 weeks of treatment with taurine, mice showed increased levels of the inhibitory neurotransmitter GABA and its synthesizing enzyme, glutamic acid decarboxylase (GAD), indicating that chronic taurine treatment induces biochemical changes to the inhibitory GABAergic system. Previously, we showed an increase in both isoforms of GAD using Western blots (El Idrissi and Trenkner 2004). Here we investigated the role of taurine on gene expression of proteins that are relevant to the inhibitory synapses and found that taurine alters gene expression of various subunits of the $GABA_A$ receptors and of GAD, the enzyme responsible for GABA synthesis.

16.2 Methods

16.2.1 Sample Preparation

Brains of controls and taurine-fed mice (4 months old) were dissected into cortex, hippocampus, cerebellum, diencephalons and brainstem within 3 min of the sacrifice and frozen on dry ice.

Table 16.1 Oligonucleotides used in the real-time qRT-PCR reaction

GAPDH *ORF*	
Forward primer	5′-ACAGGGTGGTGGACCTCATG-3′
Reverse primer	5′-GTTGGGATAGGGCCTCTCTTG-3′
GABA$_A$ β1 *ORF*	
Forward primer	5′-CTGCATCCTGATGGAACTGTTC-3′
Reverse primer	5′-CTCATCCAGAGGGTATCTTCGAA-3′
GABA$_A$ β2 *ORF*	
Forward primer	5′-GTGGGCACGAGGGTTAGAAC-3′
Reverse primer	5′-GATCCACCACAGCAGCCATT-3′
GABA$_A$ β3 *ORF*	
Forward primer	5′-CCACGGAGTGACAGTGAAAA-3′
Reverse primer	5′-CACGCTGCTGTCGTAGTGAT-3′
GAD65 *ORF*	
Forward primer	5′-GGTCAACTTCTTCCGCATGGT-3′
Reverse primer	5′-TGTCCGAGGCGTTCGATT-3′

16.2.2   RNA Preparation

RNA was prepared from tissue samples as described in the manufacturer's manual (TRIzol Reagent; Invitrogen 15596-026). Briefly, tissue samples were homogenized in 1 ml of TRIzol Reagent per 100 mg of tissue using a Teflon glass for 1 h. After centrifugation, RNA was extracted with chloroform and precipitated with isopropyl alcohol. Finally, samples were resuspended in 100 µl of DEPC treated H$_2$O.

16.2.3   Preparation of cDNA and Real-Time PCR Analysis

Equal amounts (10 µg) of total RNA were treated with RNase-free DNase (Qiagen cat. #79254) at 37°C for 1 h, and purified by phenol/chloroform (3:1) extraction and ethanol precipitation. One microgram of pure RNA was used in SYBR GreenER Two-Step qRT-PCR kit (Invitrogen cat. #11765-100) for the first-strand cDNA synthesis and real-time PCR reaction preparation as described in the manufacturer's manual.

The real-time PCR primers are described in Table 16.1. All experiments were repeated twice, and in each experiment, PCR reactions were done in triplicate in a 7,500 sequence detection system (Applied Biosystems). Target DNA sequence quantities were estimated as described previously (Ford et al. 2007; Zhang et al. 2009; Wimalarathna et al. 2011; Andrew et al. 2012). Briefly, target DNA sequence quantities were estimated from the threshold amplification cycle number (C_T) using Sequence Detection System software (Applied Biosystems). A ΔC_T value was calculated for each sample by subtracting their C_T value from the C_T value for the corresponding *GAPDH* to normalize the differences in cDNA aliquots. Each relative mRNA level was then expressed as $2^{(-\Delta C_T)} \times 100\%$ of *GAPDH*.

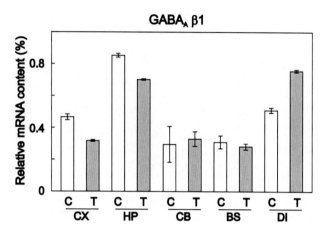

Fig. 16.1 Relative GABA$_A$β1 mRNA expression in various brain regions from controls and taurine-fed mice

16.2.4   Statistic Analysis

Statistical significance was determined by Student's *t*-test. Each value was expressed as the mean ± S.D. Differences were considered statistically significant when the calculated *P* value was less than 0.05.

16.3   Results

16.3.1   Taurine Supplementation Affects Gene Expression of Various GABA$_A$ Receptor Subunits

For GABA$_A$ β1, the relative expression levels were $0.93 \pm 0.04\%$, $1.71 \pm 0.02\%$, $0.59 \pm 0.23\%$, $0.62 \pm 0.08\%$, and $1.02 \pm 0.03\%$ under control conditions for cortex (CX), hippocampus (HP), cerebellum (CB), brain stem (BS), and diencephalon (DI), respectively (Fig. 16.1). The relative expression levels were $0.64 \pm 0.01\%$, $1.40 \pm 0.01\%$, $0.66 \pm 0.09\%$, $0.56 \pm 0.04\%$, and $1.51 \pm 0.02\%$ under taurine conditions for CX, HP, CB, BS, and DI, respectively. Thus, treatment with taurine results in the transcriptional down-regulation of GABA$_A$ β1 in the CX and HP regions, an up-regulation in the DI with no significant effects on CB and BS.

For GABA$_A$ β2, the relative expression levels were $2.90 \pm 0.02\%$, $3.37 \pm 0.07\%$, $6.68 \pm 2.10\%$, $1.73 \pm 0.06\%$, and $4.87 \pm 0.19\%$ under control conditions for CX, HP, CB, BS, and DI, respectively (Fig. 16.2). The relative expression levels were $2.60 \pm 0.09\%$, $2.68 \pm 0.06\%$, $6.86 \pm 1.47\%$, $3.27 \pm 1.67\%$, and $4.09 \pm 0.37\%$ under

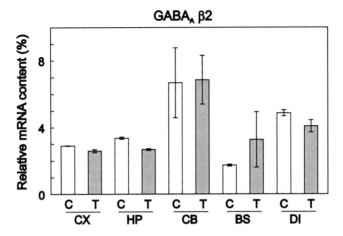

Fig. 16.2 Relative GABA$_A$β2 mRNA expression in various brain regions from controls and taurine-fed mice

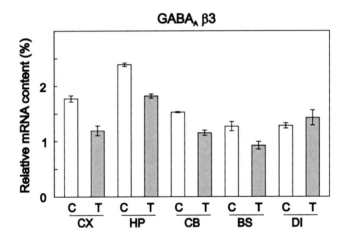

Fig. 16.3 Relative GABA$_A$β3 mRNA expression in various brain regions from controls and taurine-fed mice

taurine conditions for CX, HP, CB, BS, and DI, respectively. As such, the treatment of taurine results in the transcriptional down-regulation of GABA$_A$ β2 at CX, HP, and DI regions, but not at CB and BS regions.

For GABA$_A$ β3, the relative expression levels were $1.77 \pm 0.05\%$, $2.39 \pm 0.03\%$, $1.53 \pm 0.01\%$, $1.27 \pm 0.08\%$, and $1.29 \pm 0.05\%$ under control conditions for CX, HP, CB, BS, and DI, respectively (Fig. 16.3). The relative expression levels were $1.19 \pm 0.09\%$, $1.82 \pm 0.04\%$, $1.15 \pm 0.05\%$, $0.92 \pm 0.07\%$, and $1.43 \pm 0.14\%$ under taurine conditions for CX, HP, CB, BS, and DI, respectively. The treatment of taurine results in the transcriptional down-regulation of GABA$_A$ β3 at CX, HP, CB, and BS regions, but not at DI region.

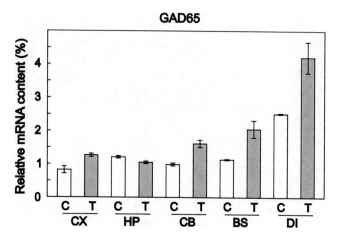

Fig. 16.4 Relative GAD mRNA expression in various brain regions from controls and taurine-fed mice

16.3.2 Taurine Supplementation Up-Regulates GAD Gene Expression

We have shown previously that taurine supplementation increases the level of GAD65 expression at the protein level. Here we show that taurine affects the gene expression level of this enzyme. The relative expression levels of GAD were $0.82 \pm 0.11\%$, $1.21 \pm 0.04\%$, $0.99 \pm 0.04\%$, $1.13 \pm 0.02\%$, and $2.52 \pm 0.02\%$ under control conditions for CX, HP, CB, BS, and DI, respectively (Fig. 16.4). The relative expression levels were $1.26 \pm 0.05\%$, $1.05 \pm 0.04\%$, $1.61 \pm 0.11\%$, $2.05 \pm 0.26\%$, and $4.21 \pm 0.46\%$ under taurine conditions for CX, HP, CB, BS and DI, respectively. Clearly, the treatment of taurine results in the transcriptional up-regulation of GAD65 at CX, CB, BS, and DI regions, but not at HP region.

16.4 Discussion

In the adult brain, the excitability of neuronal circuits is controlled by inhibitory GABAergic interneurons. In this study, we supplemented taurine in drinking water (0.05%) for four continuous weeks and tested the effects of taurine on gene expression of proteins that control the function of GABAergic inhibitory synapses. We found that taurine-fed mice showed biochemical changes in the GABAergic system. Chronic treatment with taurine in drinking water caused an increase in the levels of glutamate and GABA as well as the enzyme responsible for GABA synthesis, glutamic acid decarboxylase (GAD). We also found a reduced hippocampal expression of the β subunit of $GABA_A$ receptors.

We have shown previously that the protein expression level of GAD65 is up-regulated and of GABAergic is down-regulated in response to the treatment of taurine. It is instructive to conclude that the regulation of $GABA_A$ $\beta 1$, $GABA_A$ $\beta 2$, $GABA_A$ $\beta 3$, and GAD65 expression is at either the transcription level or translation level dependent on the brain region.

In CX region, the regulation of $GABA_A$ $\beta 1$, $GABA_A$ $\beta 2$, $GABA_A$ $\beta 3$, and GAD65 expression is at the transcription level. In HP region, the regulation of $GABA_A$ $\beta 1$, $GABA_A$ $\beta 2$, and $GABA_A$ $\beta 3$ expression is at the transcription level, but the regulation of GAD65 expression is at the translation level. In both CB and BS regions, the regulation of $GABA_A$ $\beta 3$ and GAD65 expression is at the transcription level, but the regulation of $GABA_A$ $\beta 1$ and $GABA_A$ $\beta 2$ expression is at the translation level. In DI region, the regulation of $GABA_A$ $\beta 2$ and GAD65 expression is at the transcription level, but the regulation of $GABA_A$ $\beta 1$ and $GABA_A$ $\beta 3$ expression is at the translation level.

GAD, which is responsible for GABA synthesis in GABAergic neurons, has two isoforms, 65 and 67 kDa (GAD65 and GAD67), encoded by different genes (Erlander et al. 1991). The expression of both isoforms has been shown to be activity dependent (Ramirez and Gutierrez 2001; Nishimura et al. 2001) and to be influenced by the effectiveness of GABAergic inhibition (Riback et al. 1988; 1993). Since reduced $GABA_A$ receptor expression would increase excitability, the increased GAD expression could be a compensatory mechanism for reduced efficacy of the inhibitory system. This is particularly interesting because increased GAD can be a compensatory response to the increased excitability (Ramirez and Gutierrez 2001; El Idrissi, and Trenkner 2004) that would be the net result of decreased GABAergic inhibition.

We suggest that taurine-fed mice have elevated extracellular taurine levels, which would lead to sustained activation or at least binding to $GABA_A$ receptors. Such a chronic interaction of taurine with $GABA_A$ receptors may lead to down-regulation of $GABA_A$ receptor function or expression. In response to these changes, there is increased synthesis of GABA by GABAergic neurons, as compensatory mechanism to reduced postsynaptic inhibition. Furthermore, we found an increase in the number of GAD-positive neurons. This suggests that neuronal plasticity in this system is not limited to the actual inhibitory synapses where there is decrease receptor expression on postsynaptic membranes and increase neurotransmitter synthesis on the presynaptic side but rather the entire system compensates for this state of excitability by up-regulating the number of inhibitory interneurons.

It has been shown that the degradation of cerebral cortical function during old age is due to the significant age-related loss of the GABAergic function (Leventhal, et al. 2003). Leventhal and colleagues (2003) were able to show pharmacologic reversal of the age-related loss of orientation and directional selectivity by iontophoretic application of GABA or muscimol in aged primate visual cortical neurons in vivo. Gleich et al. (2003) improved behavioral measures of temporal coding in young adult gerbils who displayed slowed temporal processing by pharmacologically increasing GABA levels. The findings of the current study reinforce the role of GABA inhibition in the maintenance of functional neuronal circuits characterized by a critical balance between excitatory and inhibitory inputs and may have important implications for the treatment of the sensory, motor, and cognitive declines that accompany old age.

16.5   Conclusion

In summary, this study shows that supplementation to mice in drinking water affects gene expression of the $GABA_A$ receptor subunits and GAD65. We suggest that these alterations in gene expression of important regulators of the function of inhibitory synapses occur as compensation to elevated brain excitability after chronic taurine supplementation.

Acknowledgments This work was supported by an NSF Grant (MCB-0919218) and a PSC-CUNY award (64243-0042) to C. H. S. the College of Staten Island/CUNY and CDN-IBR.

References

Andrew H, Zhang A, Ke Y, El Idrissi A, Shen C-H (2012) Decreased expression of $GABA_A\beta$ Subunits in the brains of mice lacking the fragile X mental retardation protein. J Mol Neurosci 46:272–275

El Idrissi A, Trenkner E (2004) Taurine as a modulator of excitatory and inhibitory neurotransmission. Neurochem Res 29:189–197

Erlander MG, Tillakaratne NJK, Feldblum S, Patel N, Tobin AJ (1991) Two genes encode distinct glutamate decarboxylases. Neuron 7:91–100

Ford J, Odeyale O, Eskandar A, Kouba N, Shen C-H (2007) A SWI/SNF- and INO80-dependent nucleosome movement at the INO1promoter. Biochem Biophys Res Commun 361:974–979

Gleich O, Hamann I, Klump GM, Kittel M, Strutz J (2003) Boosting GABA improves impaired auditory temporal resolution in the gerbil. Neuroreport 14:1877–1880

Huxtable RJ (1992) Physiological actions of taurine. Physiol Rev 72:101–163

Leventhal AG, Wang Y, Pu M, Zhou Y, Ma Y (2003) GABA and its agonists improved visual cortical function in senescent monkeys. Science 300:812–815

Nishimura T, Schwarzer C, Furtinger S, Imai H, Kato N, Sperk G (2001) Changes in the GABA-ergic system induced by trimethyltin application in the rat. Mol Brain Res 97:1–6

Ramirez M, Gutierrez R (2001) Activity-dependent expression of GAD67 in the granule cells of the rat hippocampus. Brain Res 917:139–146

Riback CE, Byun MY, Ruiz GT, Reiffenstein RJ (1988) Increased levels of amino acid neurotransmitters in the inferior colliculus of the genetically epilepsy-prone rat. Epilepsy Res 2:9–13

Riback CE, Lauterborn JC, Navetta MS, Gall CM (1993) The inferior colliculus of GEPRs contains greater numbers of cells that express glutamate decarboxylase (GAD67) mRNA. Epilepsy Res 14:105–113

Satoh H, Sperelakis N (1998) Review of some actions of taurine on ion channels of cardiac muscle cells and others. Gen Pharmacol 30:451–463

Schaffer S, Takahashi K, Azuma J (2000) Role of osmoregulation in the actions of taurine. Amino Acids 19:527–546

Wimalarathna RN, Tsai C-H, Shen C-H (2011) Transcriptional regulation of genes involved in yeast phospholipid biosynthesis. J Microbiol 49:265–273

Zhang A, Shen C-H, Ma S-Y, Ke Y, El Idrissi A (2009) Altered expression of Autism-associated genes in the brain of Fragile X mouse model. Biochem Biophys Res Commun 379:920–923

Chapter 17
Taurine Effects on Emotional Learning and Memory in Aged Mice: Neurochemical Alterations and Differentiation in Auditory Cued Fear and Context Conditioning

Lorenz S. Neuwirth, Nicholas P. Volpe, and Abdeslem El Idrissi

Abstract Previously we have shown FVB/NJ mice given taurine acutely (i.e. 43 mg/kg/s.c. [aTau]) is anxiolytic, whereas chronically (0.05% w/v for >4 weeks [cTau]) produces anxiogenic phenotypes under select aversive behavioral experiments, but negated emotional contributions to acquisition learning and retention. Hyperexcitability induced in c-Tau mice is further exacerbated under stressful conditions compromising discrimination between cognitive vs. emotional learning. In the present study, we investigated differences between a-Tau and c-Tau mice using the auditory cued tone (ACTC) and context conditioning (CC) tests. Consistent with previous results, a-Tau mice exhibit less fear and increased inhibition, whereas c-Tau mice exhibit increased fear and decreased inhibition to ACTC and CC. Once fear conditioned, taurine mice become hypersensitive to novel environments and ACTC. Taurine brain levels are noted to increase in response to stressors as a neuroprotective mechanism against hyperexcitability. We suggest that c-Tau mice have increased accumulation of cysteamine (Cyst) and depleted somatostatin (SS) expression resulting in fear disregulation through GABAergic projection neurons in the limbic system, which are not seen in a-Tau mice. Our findings suggest that taurine causes not only varied phenotypic profiles of emotional fear learning, but are further complicated by the inability to associate cues with aversive stimuli due to potential auditory sensory overloading.

L.S. Neuwirth (✉) • A. El Idrissi
Department of Biology, 6S-320, Neuroscience College of Staten Island,
2800 Victory Blvd, Staten Island, NY 10314, USA

The Graduate Center, Staten Island, NY, USA

The Center for Developmental Neuroscience and Developmental Disabilities,
The City University of New York, Staten Island, NY, USA
e-mail: lorenz989@hotmail.com

N.P. Volpe
Department of Biology, 6S-320, Neuroscience College of Staten Island,
2800 Victory Blvd, Staten Island, NY 10314, USA

A. El Idrissi and W.J. L'Amoreaux (eds.), *Taurine 8*, Advances in Experimental
Medicine and Biology 775, DOI 10.1007/978-1-4614-6130-2_17,
© Springer Science+Business Media New York 2013

Abbreviations

aTau   Acute taurine
cTau   Chronic taurine
ACTC   Auditory cued tone conditioning
CC   Context conditioning
Cyst   Cysteamine
SS   Somatostatin

17.1 Introduction

Taurine, 2-aminoethane-sulfonic acid, is a sulfur-containing amino acid found in relatively high concentrations in the mammalian central nervous system second only to glutamate (Huxtable and Peterson 1989). Taurine has been shown to be crucial for the early development, survival, and growth of vertebrate neurons (Hayes et al. 1975), whereas deficiency of taurine can result in neuropathological conditions such as epilepsy (Barbeau et al. 1975; Joseph and Emson 1976), mental depression (Perry 1976), and the alcohol withdrawal syndrome (Ikeda 1977). However, much less is known of mature neurons and taurine physiology. With this understanding, taurine research has been directed towards investigating psychophysiological conditions, including but not limited to, learning and memory brain functions in an effort to determine its role in age-related neuroprotection. However, taurine physiology and emotional learning remain to be elucidated.

Studies have shown that taurine supplementation given perinatally and continued throughout postnatal development may retard learning later in life due to disruptions in neurodevelopmental sensitive/critical periods (Suge et al. 2007). In contrast, despite natural age-related declines in GABAergic neurotransmission, supplemental taurine has been shown to facilitate inhibitory effects on behavior in aged mice (Hruska et al. 1975) and rats (Barbeau et al. 1975; Baskin et al. 1974) in addition to forestalling cognitive age-dependent declines mediated by neurochemical recovery of GABAergic function (El Idrissi 2008). Considering these findings, taurine not only has the potential to be a pharmacotherapy for mediating age-dependent consequences of cognitive decline (El Idrissi 2008), but can also ameliorate neurodevelopmental perturbations such as neuronal excitability and reducing hyperarousal via similar GABAergic modulation in autistic, epileptic, and anxiety-like disorders (Chen et al. 2004; El Idrissi et al. 2011; El Idrissi et al. 2009; El Idrissi and L'Amoreaux 2008; El Idrissi and Trenkner 2004; El Idrissi et al. 2003; El Idrissi and Trenkner 1999; Kong et al. 2006); thus, requiring additional research on how taurine may also regulate emotionality in a similar manner.

Anxiety disorders have been reported to affect between 10 and 30% of the general population resulting from neuroendocrine, neurotransmitter, and neuroanatomical disruptions (Martin et al. 2009) which are further complicated by comorbid psychiatric disorders (Wittchen 2002). Excess anxiety can hinder one's quality of life. Therapeutics for anxiety focus extensively on benzodiazepines to treat the spectrum of anxiety disorders despite their well-known side effects such as sedation, muscle relaxation, amnesia, and dependence when used acutely (Rickels and Schweizer 1997) in addition to producing dependency and withdrawal symptoms when used chronically (Lydiard et al. 1997). The development of and search for new anxiolytic drugs remains as an area of considerable public interest.

The GABAergic system plays a crucial role in the regulation of anxiety. Taurine has been shown to be an agonist that interacts with $GABA_A$ receptors and mimics the actions of GABA (El Idrissi and Trenkner 2004). Previously, El Idrissi et al. (2011, 2009) investigated the effects of chronic versus acute exposure to taurine and its effects on anxiety and have shown that FVB/NJ mice given acute taurine (43 mg/kg/s.c. [a-Tau]) is anxiolytic, whereas chronic taurine (0.05% w/v for >4 weeks [c-Tau]) produces an anxiogenic phenotype; thereby demonstrating neurochemical alterations induced by differences in physiological concentrations of taurine which can be assessed at the behavioral level. Chronic supplementation of taurine has been shown to increase glutamate and GABA neurotransmitter levels, the GABA synthesizing enzyme *glutamic acid decarboxylase* (GAD) and the expression of somatostatin immunoreactivity in the brains of c-Tau mice (El Idrissi and Trenkner 2004; Levinskaya et al. 2006).

Our previous studies (Levinskaya et al. 2006; El Idrissi et al. 2005; El Idrissi et al. 2003) have shown c-Tau mice exhibited biochemical changes in the GABAergic system similar to those observed in the Fragile X Syndrome mouse, a well-established model of genetically induced hyperexcitability, evidenced by the reduction of $GABA_A$ receptors, increased GAD expression and a lower threshold for seizure induction. Hence, $GABA_A$ receptors play a major role not only in regulating inhibition but also in explaining the increased seizure susceptibility/hyperexcitability in the brain due to a reduction in the expression of $GABA_A$ receptors that mediate excitation–inhibition balancing. Increased GAD expression responsible for the synthesis of GABA (i.e., neurotransmitter agonist for $GABA_A$ receptors) was also noted. Riback et al. (1993) proposed that these biochemical changes in GAD synthesis and $GABA_A$ receptor expression in the brain may be compensatory in regulating the reduced inhibition observed in other models of elevated excitability.

Neuronal excitability is a tightly developmentally regulated process (Ben-Ari 2002). Synchronized brain oscillations and rhythms are kept within a normal, yet narrow range, through feed-forward and -backward inhibitions which are mediated by inhibitory interneurons that organize signaling frequencies from the hippocampus to select brain regions (Khalilov et al. 2005). These hippocampal interneurons continuously adjust their inhibitory output to match the levels of excitatory input from impinging prefrontal cortical neurons that interface with a given circuit and across to other circuits such as the amygdala that regulate emotional learning and memory (Barad et al. 2006; LeDoux 2000; LeDoux 1994; Orsini and Maren 2012; Sotres-Bayon et al. 2006). Thus, when there is reduced postsynaptic inhibition, feedback from these interneurons causes the presynaptic neurons to increase their inhibitory output by releasing more GABA into the synaptic cleft. In the example of c-Tau mice, reduced $GABA_A$ receptor expression on postsynaptic membranes would induce an increase in GAD synthesis, which in turn, triggers the presynaptic release of GABA increasing its bioavailability in presynaptic terminals (El Idrissi et al. 2010; El Idrissi and Trenkner 2004). Therefore suggesting, increased GAD may represent a secondary response to the direct effects of chronic taurine exposure inducing hyperexcitability. In addition, these alterations in neurochemical signaling that occur within GABAergic circuits may also disrupt circuitry in the prefrontal cortex, hippocampus, and amygdala responsible for emotional learning.

In addition to the observed neurochemical changes in the GABAergic system, we have also observed compensation of somatostatin (SS) expression in the brains of c-Tau mice with evidenced amelioration of some Fragile X Syndrome features in the mouse model (El Idrissi et al. 2011; El Idrissi et al. 2009). The modifications in SS, an important neuropeptide encoded by a single gene, also known as *somatotroph release inhibiting factor* comprises few peptides originating from different posttranslational processing of the prepro-somatostatin precursor containing 116 amino acids. Somatostatin can act as a neurotransmitter, as well as a neuromodulator with widespread distribution in the CNS (Engine and Treit 2009) with the highest expression levels in the amygdala (Hayashi and Oshima 1986), thus affecting anxiety, depressive, and other emotional psychological states. To date only two biologically active somatostatin isoforms have been identified: the tetradecapeptide (SS-14) with its sequence comprising the entire C-terminus of the amino-terminally extended octacosapeptide (SS-28).

Interestingly, SS-14 and SS-28 are found in the periphery and central nervous system, which is predominated by the SS-14 isoform. The relative proportions of SS-14 and SS-28 vary among the many SS-expressing tissues. Notably, SS-14 and SS-28 display overlapping physiological functions (Krantic et al. 2004). Somatostatin, in particular SS-28, has been shown to play a role in mediating emotional regulation of anxiety, depression, and stress through inhibiting the release of various regulatory hormones critical for maintaining stable emotional responses (Faron-Górecka et al. 2011). In addition, specific effects on serum levels of SS-22 binding in rats exposed to chronic stress were observed in the medial hebenular nucleus, a critical diencephalic structure which bridges the limbic system with the fore and midbrain (Faron-Górecka et al. 2011). Moreover, alterations in SS expression levels in response to stress have been reported in rats to induce neuromodulatory effects in multiple brain structures that can cause variable behavioral responses as a consequence of stress exposure (Kusmider et al. 2011), thus attesting to the importance of SS neuromodulatory influences on emotional learning and memory.

We previously reported the role of brain SS and its relationship in influencing cognitive function is reversed when mice are treated with injections of cysteamine (100 mg/kg/s.c.), a drug which appears to deplete SS selectively and reversibly (El Idrissi et al. 2011). The mechanism by which cysteamine reduces SS expression levels remains to be elucidated. We suggest that it most likely acts through interactions with the disulfide bonds of SS, thus rendering the molecule both immunologically and biologically inactive (El Idrissi et al. 2011).

However, administration of supplemental taurine in animal research may vary from one experiment to another resulting in different threshold effects on inhibitory behavior in conjunction with age-related factors and model organisms. An exemplar of such opposing reports used a postnatal oral supplementation of taurine (0.9% ~1.1 g/kg/day for a month as a chronic exposure) which resulted in impaired learning and memory retention on the passive avoidance and psychomotor behaviors in the open field test dose dependently (Sanberg and Fibiger 1979; Sanberg and Ossenkopp 1977). It is important to note that the concentration of taurine in the previously mentioned study exceeded the 0.05% w/v postnatal concentration used in our previous reports (El Idrissi et al. 2011; El Idrissi 2008; El Idrissi et al. 2009) and yielded opposing learning and memory results in the passive avoidance test. Therefore, it is possible that the concentration of taurine used by Sanberg and Fibiger (1979) and Sanberg and Ossenberg (1977) may have induced reversal effects on inhibitory learning given the levels in which their animals were treated chronically in contrast to our reports.

Interestingly, we observed in our passive avoidance tests that taurine-treated mice exhibited a freezing response that produced ceiling effects in passive avoidance learning (i.e., mice froze in the light chamber without attempting to move horizontally or cross over to the dark chamber where there was an aversive foot shock). These results evidence that the mice learned and taurine further improved memory and retention when comparing young versus old age mice (El Idrissi 2008); however, freezing was not considered to obstruct learning but rather enhanced it. Thus, the passive avoidance test was insensitive to separate out any emotional behavioral features (i.e., freezing) from the cognitive features (i.e., avoiding crossing over) in which learning and memory was previously investigated by our group (El Idrissi et al. 2011; El Idrissi 2008). Moreover, reports have shown that taurine can also induce sedation, analgesia (Baskin et al. 1974), and anti-anxiety effects (Chen et al. 2004; El Idrissi et al. 2011; El Idrissi et al. 2009; Kong et al. 2006) which creates contradictions in assessing utility of avoidance testing given the nature of an electric foot shock and taurine dosage (Sanberg and Fibiger 1979; Suge et al. 2007). It has been well established that taurine acts as an agonist for $GABA_A$ receptors and chronic supplementation of taurine in mice induces neurochemical alterations in the inhibitory system (El Idrissi and Trenkner 2004). When evaluating exposure differences, chronic taurine supplementation results in increased neuronal excitability which elicits effects opposite of acute injections of taurine (El Idrissi et al. 2011; El Idrissi et al. 2009). In an attempt to elucidate the effects of taurine on age, emotional memory, aversive foot shock, and exposure, we investigated the neurobehavioral effects of a-Tau and c-Tau on aged mice (i.e., 6–7 months) in the auditory cued tone (ACTC) and context conditioning (CC) paradigm.

17.2 Methods

17.2.1 Subjects and Treatment

Experimentally naïve FVB/NJ mice were used in these experiments and were tested under The College of Staten Island (CUNY) IACUC approval procedures. Only males were used in these experiments. The animals were maintained under controlled temperature (24 ± 1 °C) and humidity ($55 \pm 5\%$), on a 12-h light (7:00–17:00 h):12-h dark (17:00–07:00 h) cycle. Food and water were available ad libitium. Taurine was either injected acutely (a-Tau: 43 mg/kg/s.c.) 15 min prior to testing or supplemented chronically in the drinking water (c-Tau: 0.05% w/v) when the mice were 4 weeks of age and continued for 4 months until mice were tested between 6 and 7 months old. Subject sample sizes were as follows: Cont $N=4$, a-Tau $N=4$, and c-Tau $N=4$. An additional Cont $N=4$ and c-Tau $N=4$ were used for foot shock dose–response testing to assess sensitivity thresholds as an effect of treatment in these experiments. Acutely treated mice were not evaluated due to required injections which would otherwise positively influence freezing behavior, as would be the case in using a sham when assessing absolute sensitivity threshold for pain.

17.2.2 Animal Handling and Testing Acclimation Procedures

Mice were handled consistently 1 h per day for 1 week prior to experimentation to diminish experimenter effects that could interfere with testing results. At 6–7 months of age, mice were brought to a peripheral testing room and remained in the dark within their home cage for 1 h under red light (100 W/30 lx) in order to acclimate. Post-acclimation mice were individually placed into a transfer cage and brought into the adjacent test room in the dark with green ambient light (100 W/23 lx) for 5 min. After the acclimation to the testing room mice were gently picked up and placed into the context fear-conditioning chamber (Med Associates, VT) and testing ensued within the green light. Mice were handled, habituated, and tested at the same times each day to ensure stability in acquisition and retention performances based on circadian rhythms (Chaudhury and Colwell 2002). Animals were also weighed daily to determine injection dosage and body weight comparisons between treatment groups. However, no significant differences in body weight were noted as a function of treatment condition.

17.2.3 Foot Shock Dose–Response Pain Threshold Sensitization

Mice (Cont $N=4$ and c-Tau $N=4$) were housed together by treatment condition. Prior to testing they were separated into single cages to diminish any observer-related transfer of emotional behavior confounds from the potential demonstrator under study (Knapska et al. 2006). Once separated mice were habituated for 10 min prior to testing. Mice were exposed to a foot shock dose–response sensitization test to determine pain threshold for selecting the most appropriate criteria for foot shock amperage in the subsequent ACTC and fear CC test. Such evaluation is critical in determining whether or not a drug treatment, genotype or mouse strain may have variations in hypersensitivity to foot shock which may facilitate associative memory as a false-positive effect, whereas in contrast hyposensitivity may result in a false-negative evaluation of associative memory impairment. In this test we scored behavioral observations of increasing levels of pain threshold beginning at 0.1–1.0 mA in increments of 0.1 mA. Mice were placed into the test chamber for 60 s and allowed to freely explore. Once this time elapsed, immediately a foot shock was presented for 5 s and the frequency of mice exhibited behaviors were recorded. A 60 s intertrial interval (ITI) was provided as a break between aversive test trials. The sequence repeated with an increase of 0.1 mA foot shocks up till 1.0 mA. After testing, the mice were then housed separately for 24 h to best prevent aggressive behavior towards their cage mates and repaired one-by-one the following day. These mice were only used for the foot shock pain threshold testing and spared from the subsequent ACTC and fear CC testing to eliminate carry over preconditioning effects.

17.2.4 Context Fear and Auditory Cued Conditioning

A separate group of experimentally naïve mice (Cont $N=4$, a-Tau $N=4$, and c-Tau $N=4$) were used for context fear conditioning. Fear conditioning is considered a valid one-trial test for assessing emotional learning and memory in animal models using *Pavlovian* conditioned stimuli immediately preceding an aversive stimulus (i.e., electrical shock) (LeDoux 2000; Fanselow 1990). Fear responding is observed through animal freezing behavior as a measurement of learning associative cues with aversive stimuli (Blanchard and Blanchard 1969). Freezing is defined in this context as the absence of all movement with the exception of respiration (De Oca et al. 1998). The ACTC fear-conditioned stimuli in the testing paradigm were adapted from Wehner and Radcliffe (2004) and are as follows: (a) *Day 1 Acquisition Phase*: 120 s acclimation, followed by a tone emitted for 30 s duration, after 10 s of the tone onset the mice were presented with a light that illuminated for 10 s then shut off and finally during the last 2 s of the tone presentation a 0.5 mA floor grid shock was given for 5 s in duration as the unconditioned aversive stimulus. Prior to and following the delivery of the shock mice latency to break three infrared beams were measured every 10 s for 120 s followed by a 70 s ITI. Four trials were presented during day 1 which was considered the learning acquisition phase. (b) *Day 2 Retention Phase*: Identical testing procedures were administered as in day 1 except that there was only one trial which was presented without shock and the latency to break the infrared beams were measured every 10 s for 300 s. (c) *Day 3 Altered Context Phase*: In order to assess the cued specific learning in separation of the contextual environment in which training occurred alterations were experimentally manipulated to evaluate fear generalization between CC environment and altered (i.e., novel) environment without tone and then with the conditioned tone presentation. The floor grids were covered with a smooth black rubber matt and a black plexi-glass diagonal divider was inserted into the chamber to separate the chamber into two equal triangular compartments. Opposite of the side in which the mice were tested an inaccessible petri dish was placed containing vanilla extract to increase exploratory locomotor behavior in the mice as an internal positive test control in the altered environment.

17.2.5 Data Analysis

In the foot shock dose–response pain threshold sensitization test two observers scored the four categories of mice behavior in progressive order which were the following: flinching, running, jumping, and squealing. Data were taken as manual tally counts for each observation in serial order of increasing mA foot shock during observations lasting 5 s. The independent scores from both observers were analyzed using a Pearson's product–moment correlation coefficient to assess the inter-score agreements between behaviors and shock intensities. The context fear chamber apparatus has three infrared sensors on each side that permit freezing records by measuring the latency to break the infrared beams every 10 s. The data were transduced on line during testing using Freeze Monitoring Software MED-PC-IV® software. An excel spreadsheet was generated containing all the parameters specified.

The mean latency to freeze value of the control group during the first 10 s of experimentation was used as a baseline measure. Data from all treatment groups, within and between, were normalized against this value using the following formula: (*animal latency to freeze raw data following exposure of auditory cue with foot shock* × *100/control mean value of first 10 s exposure to context* = *% time freezing difference from baseline*). This formula was computed for all comparisons of baseline verses all experimental manipulations across the 3 day testing paradigm for relative comparisons of auditory cued fear conditioning and taurine treatment effects on emotional learning and memory.

Day 2 retention data were compared against the trial 4 data from day 1. Retention data comparisons were made by normalizing day 1 trial 4 data against day 2 retention data using the following formula: (*animal latency to freeze raw data from day 2 trial 1 following exposure of auditory cue without foot shock* × *100/control mean value of first 10 s following exposure of auditory cue with foot shock from day 1 trial 4* = *% time freezing difference of auditory cued conditioned retention from acquisition*).

Day 3 altered context data were compared against itself in the presence or absence of a fear-conditioned auditory cue. Altered context data comparisons were made by normalizing day 3 altered context data against day 3 altered context with fear-conditioned auditory cue data using the following formula: (*animal latency to freeze raw data from day 3 trial 1 following exposure of an auditory cue in altered context* × *100/control mean value of first 10 s exposure to altered context* = *% time freezing difference from baseline*).

17.2.6 Statistical Analyses

All data were post-analyzed off line and statistics were computed in *Statistica* V. 6.1 (Statsoft, Inc. Tulsa, OK). The foot shock dose–response pain threshold sensitization data were conducted using a Pearson's product–moment correlation coefficient to assess the inter-score agreements between observers for behavior and shock intensity relationships and multifactorial repeated measures ANOVA to identify treatment, condition, and treatment×condition interaction effects. Significant differences were determined by Duncan's post hoc comparisons test. The ATCT and fear CC data were analyzed using a repeated measure ANOVA to identify treatment, condition, and treatment×condition interaction effects. Significant differences were determined by Tukey's HSD post hoc comparisons test. Significance levels were set at alpha 0.05 with a confidence level of 95%. Data are presented as mean ± SEM.

17.3 Results

17.3.1 Chronic Taurine Treatment Increases Sensitivity to Pain

Results taken from the two independent observers revealed a positive correlation between behavioral observations and shock intensity ($r=0.98$). In the foot shock dose–response pain threshold sensitization test control mice exhibited a steady increase (0.1–0.3 mA) and declination of flinching behavior (0.4–1.0 mA) as a function of increasing foot shock intensity (Fig. 17.1a). In comparison, control exhibited a linear increase in running (0.3–0.5 mA), jumping (0.3–0.5 mA), and squealing (i.e., 0.3–0.6 mA) followed by variable responses at intensities above 0.5 mA (Fig. 17.1b–d). In contrast, c-Tau mice were less sensitive to exhibit flinching between 0.1 and 0.5 mA (Fig. 17.1a), whereas they were more sensitive to running and squealing between 0.3 and 0.5 mA (Fig. 17.1b, d) and were not different with respect to jumping behaviors (Fig. 17.1c). At a shock intensity of 0.3 mA separation of c-Tau mice being more sensitive to pain was observed evidenced by a leftward shift in pain sensitivity from controls (Fig. 17.1b, c) were noted to be statistically significant for running ($*p<0.02$) and squealing behaviors ($**p<0.01$). Controls were noted to be more sensitive to flinching at 0.3 mA than c-Tau mice ($**p<0.01$) (Fig. 17.1a), but would require higher shock intensities to elicit running, jumping, and squealing behaviors, which evidences a higher threshold to pain when compared to c-Tau mice (Fig. 17.1b–d).

Our observations indicated that using a foot shock stimulus intensity between 0.2 and 0.4 mA would create bias towards a greater sensitization of fear when subjecting c-Tau mice to the context fear test more than controls; thus, causing a false-positive detection in enhancement of fear learning not related to the context, but

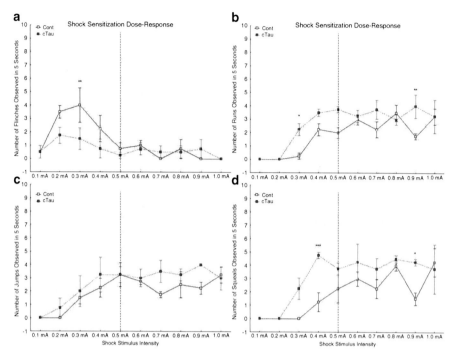

Fig. 17.1 c-Tau have increased sensitivity pain threshold to aversive foot shock. Comparison of control vs. c-Tau mice pain sensitivity as a function of increased shock intensity (i.e., *dotted vertical line* indicates determination point of optimum test parameters reducing treatment effect bias). (**a**) Shows control mice having a higher sensitivity to flinching behavior and (**b–d**) requires greater foot shock intensities to elicit running, jumping, and squealing behaviors. In contrast, c-Tau mice are less sensitive to flinching (**a**) and more sensitive to running and squealing behavior (**b–d**), whereas both groups show no differences in jumping in response to increased foot shock (**c**). Statistical values are representative of a MANOVA with Duncan's post hoc test alpha level 0.05% ($p < 0.05$ *****, $p < 0.01$ ******, and $p < 0.001$ *******)

rather induced by the treatment effects to foot shock alone. In order to avoid such errors when discriminating treatment versus experimental effects and fear learning outcomes we decided to use the 0.5 mA foot shock intensity since behavioral differences at this level were negligible, but yet equally responsive to pain sensitivity (i.e., represented as the dotted vertical line in all graphs in Fig. 17.1a–d). In addition, foot shock intensities above 0.5 mA were more variable which would also hinder appropriate detection of experimental outcomes when inferring emotional learning and memory. Therefore, in determining the most appropriate parameters for the context fear test we selected the 0.5 mA aversive shock intensity since it allows for clear discrimination between treatment and experimental conditions to be observed without any pain hypersensitivity, or ceiling effect bias (Wehner and Radcliffe 2004; De Oca et al. 1998).

17.3.2   Context Fear-Conditioned Anxiolytic and Anxiogenic Effects in Taurine Exposed Mice in Response to Aversive Stimuli and Not Its Paired Associations

Mice were subjected to context fear conditioning as described above. On day 1 after 4 trials of pairing the sound and light (i.e., conditioning stimuli—CS) with aversive foot shock (i.e., the unconditioned stimulus—UCS) mice learned to exhibit an unconditioned response (UCR) to the aversive foot shock and a conditioned response (CR) to the paired associations of sound and light which preceded the presentation of the foot shock. On trial 1 of learning acquisition trials both a-Tau and c-Tau mice froze significantly more than controls (***$p < 0.001$). No differences were noted in trial 2 and 3, but at trial 3 all treatment groups freezing responses were significantly different from trial 1 evidencing fear acquisition learning occurring as a function of repeated exposures (Cont ***, a-Tau †††, c-Tau ###, $p < 0.001$). Notably, at trial 4 a-Tau mice exhibited an anxiolytic phenotype with a reduction in freezing behavior (†††$p < 0.001$), whereas c-Tau mice exhibited an anxiogenic phenotype with an increase in freezing behavior (###$p < 0.001$) when compared to opposing taurine treatment respectively (Fig. 17.2). On trial 4, a-Tau mice were noted to have a near onefold reduction in freezing behavior during fear acquisition learning, whereas c-Tau mice had approximately a onefold increase in freezing behavior (Fig. 17.3). This suggests that c-Tau mice are hypersensitive to aversive stimuli which corroborates with previous reports of hyperarousal when treated with taurine chronically (Chen et al. 2004; El Idrissi et al. 2011; El Idrissi et al. 2009; El Idrissi and L'Amoreaux 2008; El Idrissi and Trenkner 2004; El Idrissi et al. 2003; El Idrissi and Trenkner 1999; Kong et al. 2006).

17.3.3   Taurine Treated Mice Have Selective Responsiveness to Aversive Stimuli and Not Their Paired Associations

Following day 1 test procedures mice were reexposed to the same experimental parameters 24 h later with the exception of the aversive foot shock to assess their retention of fear acquisition learning in response to the learned paired stimuli (i.e., sound and light-CS). The results of day 1 fear acquisition learning evidenced a taurine effect on the mice UCR freezing behavior based on number of exposures when presented with an aversive stimulus. Retention assessment 24 h later revealed that as a function of age all groups were observed to have a reduction in CR freezing behavior in the absence of the aversive foot shock when compared to baseline freezing behaviors from the control group (i.e., data in the CR condition shown below the horizontal dotted line as a baseline marker). In addition, the fear retention mice behavior (i.e., CR) from control, a-Tau, and c-Tau mice were significantly different from control mice fear acquisition learning (i.e., UCR) (***$p < 0.001$). Interestingly, the c-Tau mice fear acquisition behavior (i.e., UCR) was also significantly different

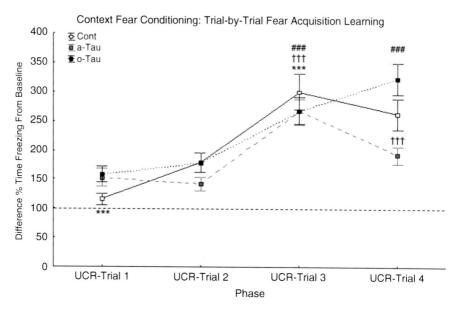

Fig. 17.2 Fear acquisition learning curves in response to taurine treatment. Mice treated with taurine after initial exposure show an increased freezing response to the aversive stimulus when compared to control mice (***$p < 0.001$). *Horizontal dotted line* represents baseline level of Control mice prior to stimulus presentation. After the second exposure no significant differences in freezing behavior as an effect of treatment are noted; however, there begins to be a separation in the effects of a-Tau vs. c-Tau mice freezing responses. Following the third exposure all treatments are statistically significant from trial one (Cont ***, a-Tau †††, c-Tau ###, $p < 0.001$) evidencing learned fear in all groups. After the fourth trial a statistically significant difference in freezing behaviors were observed between a-Tau and c-Tau mice (†††, ###, $p < 0.001$) but not when compared to controls. This suggests that controls may have difficulty with fear acquisition learning as a decline in cognitive functioning as a consequence of age and that taurine exposure facilitate learning and memory of emotional fear. Statistical values are representative of a repeated measure ANOVA with Tukey's HSD post hoc test alpha level 0.05%

from the retention behavior (i.e., CR) of control, a-Tau, and c-Tau mice, respectively (###$p < 0.001$); however, it is important to note that a-Tau mice exhibited no difference between learning acquisition and retention (Fig. 17.4). These data implicate that in both a-Tau and c-Tau mice taurine facilitated the preservation of the GABAergic system in contrast to control mice forestalling the age-dependent decline in GABAergic circuits that mediate this fear learning and behavioral response which are consistent with previous reports (Barbeau et al. 1975; Baskin et al. 1974; El Idrissi 2008; Hruska et al. 1975). Moreover, these data suggest that taurine-treated mice do not respond to contextually familiar environments (i.e., CC) in which learned conditioned stimuli (i.e., CS) were paired with an aversive stimulus (i.e., UCS); thus, taurine mice present with stimulus selectivity learning profiles in which taurine is most sensitive to aversive consequences and chronicity of exposure further enhances this behavioral effect.

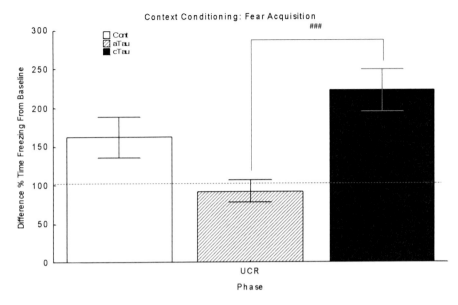

Fig. 17.3 Anxiolytic and anxiogenic effects of taurine in context fear conditioning fear acquisition learning. Mice treated with taurine, dependent upon duration of exposure reveal different phenotypes when tested for fear learning in response to an aversive stimulus. *Horizontal dotted line* represents baseline level of control mice prior to stimulus presentation. We observed that a-Tau mice are anxiolytic, whereas c-Tau mice are anxiogenic. These findings corroborate with our previous reports when assessing anxiety levels as a function of taurine treatment (i.e., acute vs. chronic). This suggests that c-Tau mice are hypersensitive to aversive stimuli which corroborates with previous reports of hyperarousal when treated with taurine chronically. Statistical values are representative of a repeated measure ANOVA with Tukey's HSD post hoc test alpha level 0.05% ($p < 0.001$ ###)

17.3.4 Taurine Increases Fear Responding to Learned Auditory Cue in Novel Rather than Contextually Familiar Environments

The following day (i.e., 36 h from initial fear acquisition training) mice were exposed to an altered version of the test chamber. The floor was covered with a soft rubber black matt to alter the tactile floor stimuli from the metal foot shock bars. The chamber, initially square, was now divided into two equal triangular shapes using a black plexi-glass divider. In the inaccessible triangular zone a petri-dish was filled with vanilla extract to increase locomotor behavior to explore this novel olfactory cue as an experimental control, while freezing behavior data was collected as a new baseline within this novel altered context. Once the new baseline data were established mice were presented with a tone for 180 s to assess the auditory cues influence on freezing behavior in the novel altered context. This type of testing allows for cued trace memory of the previously learned associations (i.e., CS) with the aversive stimulus that would, in turn, elicit freezing behavior (i.e., CR) in the mice if fear acquisition learning was uninterrupted.

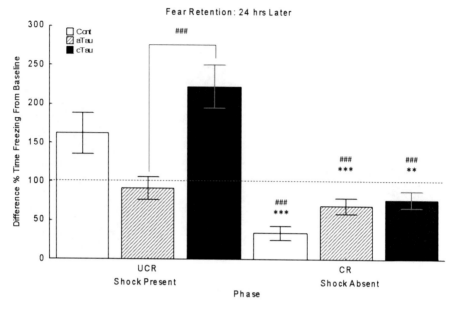

Fig. 17.4 Taurine-treated mice have selective responsiveness to aversive stimuli and not its paired associations. *Horizontal dotted line* represents baseline level of control mice prior to stimulus presentation. The decline in control mice retention of learned fear may be due to a decline in cognitive functions as a consequence of age due to reduced activity of the GABAergic neuronal circuits mediating this behavioral response (Cont *** $p < 0.001$). Interestingly, taurine treatment, whether acute or chronic, preserves the retention of fear learning and memory 24 h later when compared to controls. We observed that control and c-Tau mice have significant differences in fear acquisition learning when compared to all treatment groups with respect to fear retention (Cont *** $p < 0.001$ and c-Tau ### $p < 0.001$). Mice treated with taurine show improved retention in fear acquisition learning 24 h in contrast to control mice (a-Tau *** $p < 0.001$ and c-Tau ** $p < 0.05$). Notably, there are no differences observed in the effect of a-Tau on fear acquisition learning vs. retention. Statistical values are representative of repeated measures ANOVA with Tukey's HSD post hoc test alpha level 0.05%

We observed that in the novel altered context both a-Tau and c-Tau mice exhibit slightly elevated freezing behaviors than control mice, but they were neither statistically significant from control mice nor different from the type of taurine treatment (Fig. 17.5). In contrast, when presented with the auditory tone control mice freezing behavior increased from baseline, as did the taurine-treated mice. In response to the tone both a-Tau and c-Tau mice exhibited a significant increase in freezing behavior when compared to their own baseline rates of freezing (a-Tau †††, c-Tau ### $p < 0.001$) (Fig. 17.5). However, only c-Tau mice exhibited a significant difference in freezing behavior from control mice in response to the tone (**$p < 0.01$) (Fig. 17.5). This suggests that c-Tau mice may be hypersensitive to auditory stimuli consistent with previous reports (Chen et al. 2004; El Idrissi et al. 2011; El Idrissi et al. 2009; El Idrissi and L'Amoreaux 2008; El Idrissi and Trenkner 2004; El Idrissi et al. 2003; El Idrissi and Trenkner 1999; Kong et al. 2006).

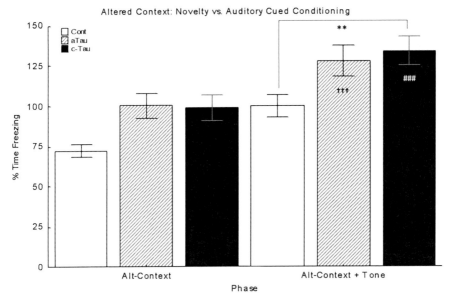

Fig. 17.5 Taurine increases fear responding to learned auditory cue in novel rather than contextually familiar environments. Mice treated with taurine show slightly elevated freezing behavior to the novel altered context when compared to controls, but they are not statistically different. When presented with a tone in the novel environment all treatment conditions increase in freezing behavior. However, only a-Tau and c-Tau mice freezing behavior were significantly different from baseline without tone when compared to with tone (a-Tau †††, c-Tau ###, $p<0.001$). Control mice freezing behavior were different from c-Tau mice (**$p<0.01$) when presented with the tone, but a-Tau mice were not different from neither control nor c-Tau mice. The reduced responsiveness by the control mice to the tone in the novel altered environment may be due to contextually environment specific learning having a greater effect than auditory stimuli alone. In contrast, both taurine-treated mice showed reduced responsiveness to familiar contextual environment (i.e., CC) and enhanced responsiveness to auditory stimuli in novel altered environments. This suggests that c-Tau mice may be hypersensitive to auditory stimuli consistent with previous reports and hyperaroused to novel environments. Statistical values are representative of a repeated measure ANOVA with Tukey's HSD post hoc test alpha level 0.05%

17.4 Discussion

Our study was aimed at elucidating the effects of taurine on age, emotional learning and memory, aversive foot shock sensitivity, and behavioral responses in the ACTC and context conditioning CC paradigm. Taurine has shown to produce two opposing behavioral phenotypes: an anxiolytic behavioral signature in response to a-Tau and an anxiogenic behavioral signature in response to c-Tau. It is noteworthy to mention that the stress of the aversive foot shock in these studies may enhance the distinct behavioral phenotypes when compared to non-aversive anxiety assessments (El Idrissi et al. 2009) while also corroborating with parallel reports on aversive learning tests

(El Idrissi et al. 2011; El Idrissi 2008). Peripheral SS signaling in the A-δ myelinated and C unmyelinated nerve fibers, along with many other neuromodulators in response to painful aversive stimuli, may also play a critical role in transmitting nociceptive signals from the PNS to CNS under such learning and memory tests (Besson 1999).

In addition, to these bottom up influences on neuronal signaling, the brain can also influence both animal and human responsiveness to pain through many descending pathways with indirect effects on perception of pain through more complex top-down processing (McLay et al. 2001). Such neuromodulatory influence of SS on passive and active avoidance learning have shown hippocampal impairments (Matsuoka et al. 1994; Vecsei and Widerlov 1990; Vecsei et al. 1989) and further prevents learning and retention when pretreated with cysteamine (cyst) a selective somatostatin-depleting substance (El Idrissi et al. 2011; Schettini et al. 1998; DeNoble et al. 1989; Haroutunian et al. 1987; Baskhit and Swerdlow 1986; Szabo and Reichlin 1981). Moreover, somatostatin has been shown to play a critical role in impairing acquisition of context fear memory and reduces LTP in hippocampal CA1 evidenced by SST$^{-/-}$ mice and wild-type mice treated with cyst dose-dependently (i.e., 50 and 150 mg/kg) prior to training (Kluge et al. 2008). Interestingly, such elimination of SS does not impair auditory tone learning and post-training administration of SS induces nonspecific enhancements of fear response (Kluge et al. 2008). Thus, somatostatin appears to shape critical activity patterns between the frontal cortex, amygdala, and hippocampus modulating the formation, consolidation, and retention of fear with emphasis for contextual parameters.

In our model of c-Tau mice, we have shown increased SS expression which may enhance cognitive abilities in some learning and memory tests (El Idrissi et al. 2009). However, when considering the aversive test parameters (i.e., foot shock) one must be careful in interpreting such effects of fear induced learning. Taurine increased fear acquisition learning in aged mice when treated chronically, whereas acute exposure had limited negligible effects on performance. In addition, c-Tau and a-Tau mice exhibited similar freezing responses when tested for retention learning in the absence of the foot shock evidencing contextual disruption 24 h prior. Notably, a-Tau and c-Tau mice had intact tone fear learning and c-Tau mice freezing behavior was enhanced when compared to controls. Consistent with our previous reports on the alterations of the GABA$_{AR}$ expression levels when treated with taurine chronically (El Idrissi et al. 2005), Van Nobelen and Kokkinidis (2006) reported that aversive foot shock associated arousal and learning are mediated through GABA amygdaloid neurotransmission and not glutamate; thus if GABA$_{AR}$ expression and GAD synthesis are increased in response to chronic taurine treatment, this will enhance GABA neurotransmission resulting in enhanced fear learning and arousal sensitivity to foot shocks. In addition, we suggest that the increased levels of cyst and SS in c-Tau mice may have drastic effects on learning and memory based on bioavailability and turnover of these two compounds. These findings parallel those of Kluge et al. (2008) showing similar results in the context fear paradigm with cysteamine treatment and SST$^{-/-}$ mice.

17.5 Conclusion

Our data suggests that stress-induced taurine differences, either acute or chronic, have opposing aversive conditioning profiles. Consistent with our previous findings (El Idrissi et al. 2011; El Idrissi et al. 2009; El Idrissi and L'Amoreaux 2008; El Idrissi and Trenkner 2004; El Idrissi et al. 2003; El Idrissi and Trenkner 1999), we suggest that acute taurine exposure produces less fear and increased inhibition, whereas chronic taurine exposure produces increased fear and decreased inhibition to aversive stimuli. Interestingly, once aversive learning is evoked mice become hypersensitive to novel environments. Taurine levels in the brain have been suggested to increase in response to stressors as a neuroprotective mechanism to prevent hyperexcitability (Riback et al. 1993; Suge et al. 2007). However, we propose that chronic taurine supplementation increases the accumulation of cysteamine, which in turn, depletes somatostatin expression resulting in disregulation of fear inhibition through GABAergic projection neurons in the amygdala and periaqueductal gray areas, thus resulting in the exaggerated freezing response observed in our study. Interestingly, these effects are not seen with acute taurine exposure and may be different in younger mice with more optimal cognitive functions. These findings suggest that taurine causes not only varied phenotypic profiles of emotional fear induced learning, but are further complicated by the inability to associate cues with aversive stimuli due to potential sensory overloading consistent with our parallel mouse model of hyperexcitability/hyperarousal within the Fragile X Syndrome.

Acknowledgments This work was supported by PSC-CUNY and CSI. We would like to thank the Louis Stokes Alliance for Minority Participation (LSAMP-NSF) and the CSI-CSTEP program for supporting author L.S. Neuwirth. We would also like to acknowledge Michael Johnson Jr. for assistance with collecting and analyzing the behavior data.

References

Baskhit C, Swerdlow N (1986) Behavioral changes following central injection of cysteamine in rats. Brain Res 365:159–163

Barad M, Gean PW, Lutz B (2006) The role of the amygdala in the extinction of conditioned fear. Biol Psychiatry 60:322–328

Barbeau A, Inoue N, Tsukada Y, Butterworth RF (1975) The neuropharmacology of taurine. Life Sci 17(5):669–677

Baskin SI, Hinkamp DL, Marquis WJ, Tilson HA (1974) Effects of taurine on psychomotor activity in the rat. Neuropharmacology 13(7):591–594

Ben-Ari Y (2002) Excitatory actions of GABA during development: the nature of the nurture. Nat Rev Neurosci 3(9):728–739

Besson JM (1999) The neurobiology of pain. Lancet 353:1610–1615

Blanchard RJ, Blanchard DC (1969) Crouching as an index of fear. J Comp Physiol Psychol 67:370–375

Chaudhury D, Colwell CS (2002) Circadian modulation of learning and memory in fear-conditioned mice. Behav Brain Res 133:95–108

Chen SW, Kong WX, Zhang YJ, Li YL, Mi XJ, Mu XS (2004) Possible anxiolytic effects of taurine in the mouse elevated plus maze. Life Sci 75(12):1503–1511

DeNoble VJ, Helper DJ, Barto RA (1989) Cysteamine-induced depletion of somatostatin produces differential cognitive deficits in rats. Brain Res 482:42–48

De Oca BM, DeCola JP, Maren S, Fanselow MS (1998) Distinct regions of the periaqueductal gray are involved in the acquisition and expression of defensive responses. J Neurosci 18:3426–3432

El Idrissi A, Iskra BS, Neuwirth LS (2011) Neurobehavioral effects of taurine in Fragile X syndrome. In: El Idrissi A, L'Amoreaux WJ (eds) Taurine in health and disease, vol 644. Plenum Press, New York, pp 306–345

El Idrissi A, Neuwirth LS, L'Amoreaux WL (2010) Taurine regulation of short term synaptic plasticity in Fragile X mice. J Biomed Sci 17(Suppl 1):S15

El Idrissi A, Boukarrou L, Heany W, Malliaros G, Sangdee C, Neuwirth LS (2009) Effects of taurine on anxiety-like and locomotor behavior of mice. In: Azuma J, Schaffer SW, Takashi I (eds) Taurine 7: taurine for the future healthcare, vol 643. Springer, New York, pp 207–215

El Idrissi A (2008) Taurine improves learning and retention in aged mice. Neurosci Lett 436:19–22

El Idrissi A, L'Amoreaux WJ (2008) Selective resistance of taurine-fed mice to isoniazid-potentiated seizures: in vivo functional test for the activity of glutamic acid decarboxylase. Neuroscience 156(3):693–699

El Idrissi A, Ding X-H, Scalia J, Trenkner E, Brown WT, Dobkin C (2005) Decreased GABAA receptor expression in the seizure-prone fragile X mouse. Neurosci Lett 377:141–146

El Idrissi A, Trenkner E (2004) Taurine as a modulator of excitatory and inhibitory neurotransmission. Neurochem Res 29:189–197

El Idrissi A, Messing J, Scalia J, Trenkner E (2003) Prevention of epileptic seizures through taurine. In: Lombardini JB, Schaffer SW, Azuma J (eds) Taurine 5 beginning the 21st century, Adv Exp Med Biol, Kluwer Press, New York, 526, pp 515–525

El Idrissi A, Trenkner E (1999) Growth factors and taurine protect against excitotoxicity by stabilizing calcium homeostasis and energy metabolism. J Neurosci 19:9459–9468

Engine E, Treit D (2009) Anxiolytic and antidepressant actions of somatostatin: the role of sst2 and sst3 receptors. Psychopharmacology (Berl) 206:281–289

Fanselow MS (1990) Factors governing one-trial contextual conditioning. Anim Learn Behav 18:264–270

Faron-Górecka A, Kusmider M, Zurawek D, Gaska M, Gruca P, Papp M, Dziedzicka-Wasylewska M (2011) P.1.028 Serum levels of somatostatin-28 and its binding sites in medial habenular nucleus differentiate rats responding and non responding to chronic mild stress. Eur Neuropsychopharmacol 21:S131–S132. doi:10.1016/S0924-977X(11)70151-1

Haroutunian V, Mantin R, Campbell GA, Tsuboyama GK, Davis KL (1987) Cysteamine-induced depletion of central somatostatin-like immunoactivity: effects on behavior, learning, memory, and brain neurochemistry. Brain Res 403:234–242

Hayashi M, Oshima K (1986) Neuropeptides in cerebral cortex of macaque monkey (Macaca fuscata fuscata): regional distribution and ontogeny. Brain Res 364:360–368

Hayes KC, Carey SY, Schmidt SY (1975) Retinal degeneration associated with taurine deficiency in the cat. Science 188(4191):949–951

Hruska RE, Thut PD, Huxtable RJ, Bressler R (1975) Suppression of conditioned drinking by taurine and related compounds. Pharmacol Biochem Behav 3(4):593–599

Huxtable RJ, Peterson A (1989) Sodium-dependent and sodium-independent binding of taurine to rat brain synaptosomes. Neurochem Int 14(1):79–84

Ikeda HC (1977) Effects of taurine on alcohol withdrawal. Lancet 2(8036):509

Joseph MH, Emson PC (1976) Taurine and cobalt induced epilepsy in the rat: a biochemical and electrocorticographic study. J Neurochem 27:1495–1501

Khalilov I, Le Van Quyen M, Gozlan H, Ben-Ari Y (2005) Epileptogenic actions of GABA and fast oscillations in the developing hippocampus. Neuron 48:787–796

Kluge C, Stoppel C, Szinyei C, Stork O, Pape HC (2008) Role of the somatostatin system in contextual fear memory and hippocampal synaptic plasticity. Learn Mem 4:252–260

Knapska E, Nikolaev E, Boguszewski P, Walasek G, Blaszczyk J, Kaczmarek L, Werka T (2006) Between-subject transfer of emotional information evokes specific pattern of amygdala activation. Proc Natl Acad Sci U S A 103(10):3858–3862

Kong WX, Chen SW, Li YL, Zhang YJ, Wang R, Min L, Mi X (2006) Effects of taurine on rat behaviors in three anxiety models. Pharmacol Biochem Behav 83(2):271–276

Krantic S, Goddard I, Saveanu A, Giannetti N, Fombonne J, Cardoso A, Jaquet P, Enjalbert A (2004) Novel modalities of somatostatin actions. Eur J Endocrinol 151:643–655

Kusmider M, Faron-Górecka A, Zurawek D, Gaska M, Gruca P, Papp M, Dziedzicka-Wasylewska M (2011) P.1.029 Alterations in somatostatin binding sites in brains of rats subjected to chronic mild stress. Eur Neuropsychopharmacol 21:S132. doi:10.1016/S0924-977X(11)70152-3

LeDoux JE (2000) Emotion circuits in the brain. Annu Rev Neurosci 23:155–184

LeDoux JE (1994) Emotion, memory and the brain. Sci Am 270:50–57

Levinskaya N, Trenkner E, El Idrissi A (2006) A Increased GAD-positive neurons in the cortex of taurine-fed mice. Adv Exp Med Biol 583:411–417

Lydiard RB, Ballenger JC, Rickles K, for the Abecarmil Work Group. (1997) A double-blind evaluation of the safety and efficacy of abecarmil, alprazolam and placebo in outpatients with generalized anxiety disorder. J. Clin. Psychiatry 58:11–18

Martin EI, Ressler KJ, Binder E, Nemeroff CB (2009) The neurobiology of anxiety disorders: brain imaging, genetics, and psychoneuroendocrinology. Psychiatr Clin North Am 32:549–575

Matsuoka N, Maeda N, Yamaguchi I, Satoh M (1994) Possible involvement of brain somatostatin in the memory formation of rats and cognitive enhancing action of FR121196 in passive avoidance task. Brain Res 642:11–19

McLay RN, Pan W, Kastin AJ (2001) Effects of peptides on animal and human behavior: a review of studies published in the first twenty years of the journal Peptides. Peptides 22:2181–2255

Orsini CA, Maren S (2012) Neural and cellular mechanisms of fear and extinction memory formation. Neursci Biobehav Rev 2012(36):1773–1802

Perry TL (1976) Hereditary mental depression with taurine deficiency: further studies, including a therapeutic trial of taurine administration. In: Huxtable R, Barbeau A (eds) Taurine, Adv Exp Med Biol, Raven Press, New York, 526, pp 365–374

Riback CE, Lauterborn JC, Navetta MS, Gall CM (1993) The inferior colliculus of GEPRs contains greater numbers of cells that express glutamate decarboxylase (GAD67) mRNA. Epilepsy Res 14:105–113

Rickels K, Schweizer E (1997) The clinical presentation of generalized anxiety in primary-care setting: practical concepts of classification and management. J Clin Psychiatry 58:4–9

Sanberg PR, Fibiger HC (1979) Impaired acquisition and retention of a passive avoidance response after chronic ingestion of taurine. Psychopharmacology 29(62)1:97–99

Sanberg PR, Ossenkopp KP (1977) Dose-response effects on some open-field behaviors in the rat. Psychopharmacology (Berl) 53(2):207–209

Schettini G, Florio T, Magri G, Grimaldi M, Meucci O, Landolfi E, Marino A (1998) Somatostatin and SMS 201-995 reverse the impairment of cognitive functions induced by cysteamine depletion of brain somatostatin. Eur J Pharmacol 151:399–407

Sotres-Bayon F, Cain CK, LeDoux JE (2006) Brain mechanisms of fear extinction: historical perspectives on the contribution of prefrontal cortex. Biol Psychiatry 60:329–336

Suge R, Nobuo H, Furube M, Yamamoto T, Hirayama A, Hirano S, Nomura M (2007) Specific timing of taurine supplementation affects learning ability in mice. Life Sci 81:1228–1234

Szabo S, Reichlin S (1981) Somatostatin in rat tissues is depleted by cysteamine administration. Endocrinology 109:2255–2257

Van Nobelen M, Kokkinidis L (2006) Amygdaloid gaba, not glutamate neurotransmission mRNA transcription controls foot-shock associated arousal in the acoustic startle paradigm. Neuroscience 137:707–716

Vecsei L, Widerlov E (1990) Preclinical and clinical studies with cysteamine and pantethine related to the central nervous system. Prog Neuropsychopharmacol Biol Psychiatry 14:835–862

Vecsei L, Pavo I, Zsigo J, Penke B, Widerlov E (1989) Comparative studies of somatostatin-14 and some of its fragments on passive avoidance behavior, open field activity and on barrel rotation phenomenon in rats. Peptides 10:1153–1157

Wehner JM, Radcliffe RA (2004) Cued and contextual fear conditioning in mice. Behav Neurosci 27:8.5C.1–8.5C.14

Wittchen HU (2002) Generalized anxiety disorder: prevalence, burden, and cost to society. Depress Anxiety 16:162–171

Chapter 18
Rising Taurine and Ethanol Concentrations in Nucleus Accumbens Interact to Produce the Dopamine-Activating Effects of Alcohol

Mia Ericson, PeiPei Chau, Louise Adermark, and Bo Söderpalm

Abstract Alcohol misuse and addiction is a worldwide problem causing enormous individual suffering as well as financial costs for the society. To develop pharmacological means to reduce suffering, we need to understand the mechanisms underlying the effects of ethanol in the brain. Ethanol is known to increase extracellular levels of both dopamine and taurine in the nucleus accumbens (nAc), a part of the brain reward system, but the two events have not been connected. In previous studies we have demonstrated that glycine receptors in the nAc are involved in modulating both basal- and ethanol-induced dopamine output in the same brain region. By means of in vivo microdialysis in freely moving rats we here demonstrate that the endogenous glycine receptor ligand taurine mimics ethanol in activating the brain reward system. Furthermore, administration of systemic ethanol diluted in an isotonic (0.9% NaCl) or hypertonic (3.6% NaCl) saline solution was investigated with respect to extracellular levels of taurine and dopamine in the nAc. We found that ethanol given in a hypertonic solution, contrary to an isotonic solution, failed to increase concentrations of both taurine and dopamine in the nAc. However, a modest, non-dopamine elevating concentration of taurine in the nAc disclosed a dopamine elevating effect of systemic ethanol also when given in a hypertonic solution. We conclude that the elevations of taurine and dopamine in the nAc are closely related and that in order for ethanol to induce dopamine release, a simultaneous

M. Ericson (✉) • P. Chau • L. Adermark
Addiction Biology Unit, Section of Psychiatry and Neurochemistry,
Institute of Neuroscience and Physiology, The Sahlgrenska Academy at University
of Gothenburg, Gothenburg, Sweden
e-mail: Mia.Ericson@neuro.gu.se

B. Söderpalm
Addiction Biology Unit, Section of Psychiatry and Neurochemistry,
Institute of Neuroscience and Physiology, The Sahlgrenska Academy at University
of Gothenburg, Gothenburg, Sweden

Beroendekliniken, Sahlgrenska University Hospital, Gothenburg, Sweden

A. El Idrissi and W.J. L'Amoreaux (eds.), *Taurine 8*, Advances in Experimental
Medicine and Biology 775, DOI 10.1007/978-1-4614-6130-2_18,
© Springer Science+Business Media New York 2013

215

increase of extracellular taurine levels in the nAc is required. These data also provide support for the notion that the nAc is the primary target for ethanol in its dopamine-activating effect after systemic administration and that taurine is a prominent participant in activating the brain reward system.

Abbreviations

nAc	Nucleus accumbens
GlyR	Glycine receptor
VTA	Ventral tegmental area
nAChR	Nicotinic acetylcholine receptor
DA	Dopamine

18.1   Introduction

Alcoholism is a worldwide chronic disease causing enormous individual suffering as well as socioeconomic costs. The underlying mechanism for development of this disabling disease remains unknown, which is why it is of great importance to study the actions of alcohol in the brain in order to find potential new targets for treating alcohol addiction.

Alcohol, as well as other drugs of abuse, activates the mesolimbic dopamine (DA) system, a central part of the brain reward system, resulting in increased DA release in the nucleus accumbens (nAc) (DiChiara and Imperato 1988; Wise and Rompre 1989; Drevets et al. 1999; Boileau et al. 2003), which has been associated with the reinforcing properties of the drugs. In a series of studies we have demonstrated that glycine receptors (GlyR) located in the nAc are involved in modulating both basal- and ethanol-induced dopamine output in the same brain region (Molander and Söderpalm 2005a, b). We found that ethanol as well as glycine increased DA in the nAc and that this is executed in a GlyR-dependent manner. These effects are not local but involve activation of nicotinic acetylcholine receptors (nAChRs) in the ventral tegmental area (VTA), possibly due to inhibition of GABAergic projection neurons modulating acetylcholine release in the VTA (Ericson et al. 2003; Larsson et al. 2005; for review see Söderpalm et al. 2009).

Besides glycine there are several amino acids with affinity for the GlyR (Pan and Slaughter 1995). Taurine has been demonstrated to act as an agonist or a partial agonist at the GlyR and has been attributed inhibitory, neuromodulatory, neuroprotectant and osmoregulatory properties, to mention a few (Huxtable 1989; 1992; Saransaari and Oja 2000; Olive 2002). Interestingly, taurine levels in the nAc increase after systemic and local ethanol administration (De Witte et al. 1994; Dahchour et al. 1996; Adermark et al. 2011), an increase that is reduced and abolished by increased osmolarity of the ethanol solution (Quertemont et al. 2003). Taurine administration also decreases ethanol intake and alters ethanol aversion (Quertemont et al. 1998).

In addition, chronic ethanol administration decreases taurine levels in the whole brain, an effect that returns to normal upon ethanol withdrawal (Iwata et al. 1980). Whether ethanol-induced taurine and dopamine release are connected events or separate from each other has not been determined.

In the present study we aimed to explore whether taurine per se can influence DA output and if manipulation of the ethanol-induced elevation of extracellular taurine levels influences the DA output after ethanol administration. To this end we used in vivo microdialysis in freely moving Wistar rats while monitoring both DA and taurine.

18.2   Methods

18.2.1   In Vivo Microdialysis

Male Wistar rats were implanted with a custom made I-shaped dialysis probe in the nAc alone or in combination with a probe in the VTA, as previously described (Lidö et al. 2009). Two days after surgery the sealed inlet and outlet of the probes were cut open and connected to a microperfusion pump via a swivel allowing the animal to move around freely. The probes were perfused with Ringer solution at a rate of 2 µl/ min and dialysate samples (40 µl) were collected every 20 min. The rats were perfused with Ringer solution for 1 h in order to obtain a balanced fluid exchange before baseline sampling began. Dopamine and taurine were analyzed in the dialysis samples by means of HPLC as previously described (Lidö et al. 2009). Animals were sacrificed directly after the experiment, brains were removed, and probe placements were verified using a vibroslicer. Only rats with correctly placed dialysis probes were included in statistical analysis.

18.2.2   Experimental Design

In the first set of experiments rats were perfused with vehicle (Ringer) or taurine (1, 10, or 100 mM) via reversed dialysis in the nAc. Following this the rats received pretreatment with either the GlyR antagonist strychnine (2 µM perfused in the nAc) or the nAChR antagonist mecamylamine (100 µM perfused in the VTA) before administration of 10 mM taurine via the nAc dialysis probe. Extracellular levels of DA were monitored in the nAc for 3 h after drug administration.

In the second set of experiments four groups of drug naïve rats received an acute injection of 0.9% NaCl (i.p.), 3.6% NaCl (i.p.), ethanol 2.5 g/kg diluted in 0.9% NaCl (i.p.), or ethanol 2.5 g/kg diluted in 3.6% NaCl (i.p.). Half of the animals in each group received the addition of 50 µM taurine in the perfusate (nAc) at the time of injection. Extracellular levels of DA and taurine were monitored in the nAc for 3 h after the systemic injection.

18.2.3   Statistical Analysis

Data were statistically evaluated using Student's *t*-test or a one-way ANOVA followed by Fishers PLSD. A probability value (*P*) less than 0.05 were considered statistically significant. All values are expressed as means ± SEM.

18.3   Results

18.3.1   Taurine Mimics the Dopamine Elevating Properties of Ethanol

In the first study, where different concentrations of taurine were administered into the nAc by reversed microdialysis, the two higher concentrations (10 or 100 mM in the perfusate) increased nAc DA output while the low dose (1 mM) had no effect. The medium concentration, 10 mM, elevated DA in a pattern similar to ethanol over the 3 h of measuring, which is why this concentration was selected for further studies (Fig. 18.1).

Administration of the GlyR antagonist strychnine (2 μM locally in the nAc) alone did not influence the DA levels. However, strychnine perfusion 40 min prior to co-perfusion with taurine (10 mM in the nAc) completely abolished the DA elevating effects of taurine, just as previously demonstrated with ethanol (Ericson et al. 2003). Also in line with studies on the DA elevating effects of ethanol, administration of the nAChR antagonist mecamylamine (100 μM locally in the VTA) prevented taurine (10 mM in the nAc) from increasing accumbal DA levels (Fig. 18.1).

18.3.2   Ethanol, Dopamine, and the Necessity of Taurine

In the second set of experiments, we explored the extracellular DA as well as taurine response to ethanol when administered in a normal (0.9%) or hypertonic (3.6%) saline solution. In line with the findings from Quertemont et al. (2003), we found that ethanol diluted in a hypertonic saline solution (3.6%) completely prevented the ethanol-induced increase of taurine 40 min after the administration, whereas systemic ethanol diluted in an isotonic saline solution (0.9%) elevated the extracellular levels of taurine by approximately 50%. None of the saline solutions influenced the taurine levels per se (Fig. 18.2a). Furthermore, concomitant measurement of DA in the same samples revealed an ethanol-induced DA response similar to that of taurine. Ethanol diluted in a hypertonic saline solution was unable to increase DA output in the nAc, whereas when administered in an isotonic saline solution ethanol produced the expected increase (Fig. 18.2b).

In a final set of rats we added a small amount of taurine (50 μM in the nAc perfusate), unable to influence DA per se, at the time of the systemic injection (saline

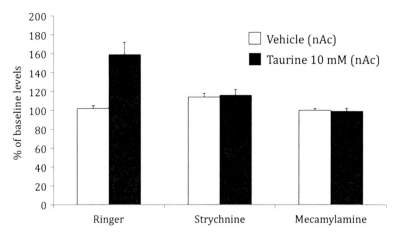

Fig. 18.1 Effect of taurine (10 mM) or vehicle (Ringer) administration on extracellular dopamine levels in the nucleus accumbens 40 min after initiation of taurine/vehicle administration as measured by in vivo microdialysis. The rats received no pretreatment (Ringer), 40 min pretreatment with the glycine receptor antagonist strychnine (2 μM perfused locally in the nAc) or 40 min pretreatment with the nAChR antagonist mecamylamine (100 μM perfused locally in the VTA). Taurine increased dopamine levels, an effect that was prevented by both accumbal strychnine treatment and ventral tegmental mecamylamine treatment. Data are presented as means ± SEM, $n = 8-12$

or ethanol). The addition of taurine completely restored ethanol's ability to increase DA when administered in a hypertonic (3.6%) saline solution (Fig. 18.3b). However, the small amount of taurine did not influence DA levels observed after administration of ethanol in normal (0.9%) saline solution (Fig. 18.3a).

18.4   Discussion

In the present study we found that taurine increases DA in the mesolimbic DA system, a part of the brain reward pathway. More specifically, taurine elevated DA levels in the nAc, a phenomenon that has been linked to positive reinforcement and perhaps also to the development of addiction (Koob 1992; Spanagel 2009). Since ethanol is known to produce enhanced extracellular levels of both taurine and DA it is interesting to note that taurine on its own can raise DA levels. Further studies demonstrated that taurine appears to use the same mechanisms as ethanol to influence DA, since, as with ethanol, pretreatment with either strychnine in the nAc or mecamylamine in the VTA completely abolished both ethanol- and taurine-induced elevations of DA.

In a series of studies we have previously demonstrated the importance of both accumbal GlyRs and ventral tegmental nAChRs for the reinforcing and DA elevating effects of ethanol. Based on these studies we have suggested that ethanol influences DA via a neuronal nAc-VTA-nAc circuitry (Söderpalm et al. 2009) and the data presented here suggest that taurine exerts is effect on DA via the same mechanism.

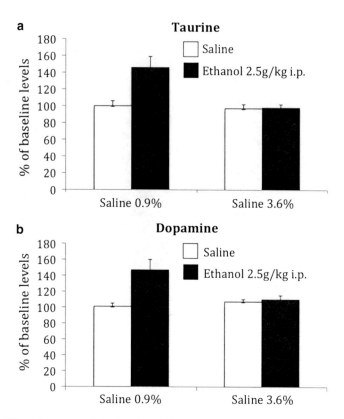

Fig. 18.2 Effect of a systemic injection (i.p.) of 0.9% NaCl, 3.6% NaCl, ethanol (2.5 g/kg) diluted in 0.9% NaCl or ethanol (2.5 g/kg) diluted in 3.6% NaCl on extracellular (**a**) taurine levels or (**b**) dopamine levels in the nucleus accumbens 40 min after the systemic injection. Administration of ethanol in a hypertonic saline solution prevented the drug from increasing both taurine and dopamine. Data are presented as means ± SEM, $n = 6$–12 and were detected by using in vivo microdialysis in freely moving Wistar rats

An interesting study by Quertemont et al. (2003) demonstrated that the ethanol-induced increase in taurine levels could be modified by altering the osmolarity of the saline solution that ethanol was diluted in. Here we repeated this finding and found a similar phenomenon for taurine and DA when concomitantly measured in the same animal. Several studies have demonstrated the osmoregulatory properties of taurine, where, for example, a change in the sodium milieu surrounding the cells greatly influences taurine release (Korpi and Oja 1983). Ethanol has also been shown to induce cell swelling in astrocytes, which leads to an increased release of taurine into the extracellular space (Kimelberg et al. 1993; Allansson et al. 2001). In fact, a recent study found that inhibition of ethanol-induced astrocyte cell swelling also prevents the increase in microdialysate concentration of taurine and DA induced by local administration of ethanol in the nAc (Adermark et al. 2011). It is thus possible that taurine is released in response to ethanol-mediated cell swelling, and that this swelling is counterbalanced when ethanol is administered in a hyperosmotic solution. However, it could also be speculated that the administration of high

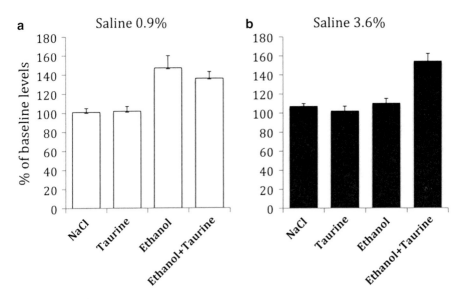

Fig. 18.3 Administration of vehicle (NaCl i.p.), taurine (50 µM perfused locally in the nAc), ethanol (2.5 g/kg i.p.), or the combination of taurine and ethanol using (**a**) normal saline solution (0.9% NaCl) or (**b**) hypertonic saline solution (3.6% NaCl) measuring nucleus accumbens dopamine levels by means of in vivo microdialysis in male Wistar rats. Data are presented as means ± SEM, $n = 6$–8, 40 min after the systemic injection

amounts of sodium disrupts neurotransmission in general, leaving, e.g., the cells unable to release taurine and/or DA or compromising the GlyR previously shown to be involved in the DA releasing effect of ethanol. To address this issue we mimicked the taurine elevation normally induced by ethanol by perfusing a relatively low concentration of taurine in the nAc concomitantly with systemic administration of the hypertonic ethanol solution. The addition of the inert amount of taurine appeared to be the missing component for ethanol to produce the elevation of DA. Overall, these results demonstrate that the hypertonic solution does not compromise mechanisms involved in DA release and indicate that GlyR function is intact under these conditions. Since the concentration of taurine used did not influence DA release per se the results moreover suggest that a concomitant rise in ethanol and taurine concentrations is required in order to obtain DA release after ethanol and that the two substances act in synergy at GlyRs in the nAc. It should be noted that taurine also can act as a ligand at $GABA_A$ receptors, and thus modulate DA output by influencing GABAergic neurotransmission in the nAc. However, taurine-induced currents in medium spiny neurons are only partially depressed by the $GABA_A$ receptor antagonist gabazine indicating that taurine primarily affect neurotransmission in the nAc by interacting with GlyR (Sergeeva and Haas 2001).

These results also may have implications for the debate concerning the site of action of ethanol in its DA activating and reinforcing effects. Based on pharmacological in vivo studies using microdialysis with or without concomitant monitoring of systemic ethanol intake, we have suggested that the primary site of action is in the nAc (cf. Söderpalm et al. 2009), whereas other investigators, based mainly on in vitro studies and on studies of intracerebral self-administration of ethanol argue that the VTA is the important site in this respect (Brodie et al. 1999; McBride et al. 1999). The present finding that local perfusion of an inert concentration of taurine in the nAc completely rescues the DA activating effect of systemic ethanol shows that an ethanol-induced event, i.e., taurine elevation, in this particular area in combination with ethanol itself is extremely important, and, again, points to GlyR in the nAc as the primary site of action for ethanol in this context. The alternative, more far-fetched interpretation would be that a concomitant elevation of taurine in the nAc is a prerequisite for obtaining DA release after some ethanol interaction in the VTA, but this alternative, if true, would still require ethanol-induced taurine release in the nAc.

18.5   Conclusion

Taurine has the ability to increase DA output in the nAc on its own, via mechanisms similar to ethanol. In addition it appears that in order for systemic ethanol to increase DA in the nAc, a phenomenon that has been related to the reinforcing properties of the drug, a concomitant extracellular increase of accumbal taurine is required. The studies presented here suggest that taurine could be a target for development of new pharmacotherapies against alcoholism.

Acknowledgments   The authors are thankful for the technical assistance from Mrs Rosita Stomberg. This work was supported by Swedish Medical Research Council (Grants No: 2009-2289, 2009-4477, 2010-3100), governmental support under the LUA/ALF agreement, Wilhelm and Martina Lundgrens Scientific Foundation, the Swedish Brain foundation.

References

Adermark L, Clarke RBC, Olsson T, Hansson E, Söderpalm B, Ericson M (2011) Implications for glycine receptors and astrocytes in ethanol-induced elevation of dopamine levels in the nucleus accumbens. Addict Biol 16(1):43–54

Allansson L, Khatibi S, Olsson T, Hansson E (2001) Acute ethanol exposure induces $[Ca^{2+}]_i$ transients, cell swelling and transformation of actin cytoskeleton in astroglial primary cultures. J Neurochem 76:472–479

Boileau I, Assaad JM, Pihl RO, Benkelfat C, Leyton M, Diksic M, Tremblay RE, Dagher A (2003) Alcohol promotes dopamine release in the human nucleus accumbens. Synapse 49:226–231

Brodie MS, Pesold C, Appel SB (1999) Ethanol directly excites dopaminergic ventral tegmental area reward neurons. Alcohol Clin Exp Res 23(11):1848–1852

Dahchour A, Quertemont E, De Witte P (1996) Taurine increases in the nucleus accumbens microdialysate after acute ethanol administration to naive and chronically alcoholised rats. Brain Res 735:9–19

De Witte P, Dahchour A, Quertemont E (1994) Acute and chronic alcohol injections increase taurine in the nucleus accumbens. Alcohol Alcohol Suppl 2:229–233

DiChiara G, Imperato A (1988) Drugs abused by humans preferentially increase synaptic dopamine concentrations in the mesolimbic system of freely moving rats. Proc Natl Acad Sci USA 85:5274–5278

Drevets WC, Price JC, Kupfer DJ, Kinahan PE, Lopresti B, Holt D, Mathis C (1999) PET measures of amphetamine-induced dopamine release in ventral versus dorsal striatum. Neuropsychopharmacology 21(6):694–709

Ericson M, Molander A, Löf E, Engel JA, Söderpalm B (2003) Ethanol elevates accumbal dopamine levels via indirect activation of ventral tegmental nicotinic acetylcholine receptors. Eur J Pharmacol 467:85–93

Huxtable RJ (1992) Physiological actions of taurine. Physiol Rev 72:101–163

Huxtable RJ (1989) Taurine in the central nervous system and the mammalian actions of taurine. Prog Neurobiol 32:471–533

Iwata H, Matsuda T, Lee E, Yamagami S, Baba A (1980) Effect of ethanol on taurine concentration in the brain. Experientia 36:332–333

Kimelberg HK, Cheema M, O'Connor ER, Tong H, Goderie SK, Rossman PA (1993) Ethanol-induced aspartate and taurine release from primary astrocyte cultures. J Neurochem 60:1682–1689

Koob GF (1992) Neural mechanisms of drug reinforcement. In: Kalivas PW, Samson HH (eds) The neurobiology of drug and alcohol addiction, vol 654. The New York Academy of Sciences, New York, pp 171–191

Korpi ER, Oja SS (1983) Characteristics of taurine release from cerebral cortex slices induced by sodium-deficient media. Brain Res 289:197–204

Larsson A, Edström L, Svensson L, Söderpalm B, Engel JA (2005) Voluntary ethanol intake increases extracellular acetylcholine levels in the ventral tegmental area in the rat. Alcohol Alcohol 40:349–358

Lidö HH, Stomberg R, Fagerberg A, Ericson M, Söderpalm B (2009) Glycine reuptake inhibition: a novel principle for prevention of ethanol-induced dopamine release. Alcohol Clin Exp Res 33(7):1–7

McBride WJ, Murphy JM, Ikemoto S (1999) Localization of brain reinforcement mechanisms: intracranial self-administration and intracranial place-conditioning studies. Behav Brain Res 101(2):129–152

Molander A, Söderpalm B (2005a) Glycine receptors regulate dopamine release in the rat nucleus accumbens. Alcohol Clin Exp Res 29:17–26

Molander A, Söderpalm B (2005b) Accumbal strychnine-sensitive glycine receptors: an access point for ethanol to the brain reward system. Alcohol Clin Exp Res 29:27–37

Olive MF (2002) Interactions between taurine and ethanol in the central nervous system. Amino Acids 23:345–357

Pan ZH, Slaughter MM (1995) Comparison of the actions of glycine and related amino acida on isolated third order neurons from the tiger salamander retina. Neuroscience 64:153–164

Quertemont E, Devitgh A, De Witte P (2003) Systemic osmotic manipulations modulate ethanol induced taurine release: a brain microdialysis study. Alcohol 29:11–19

Quertemont E, Goffaux V, Vlaminck AM, Wolf C, De Witte P (1998) Oral taurine supplementation modulates ethanol-conditioned stimulus preference. Alcohol 16:201–206

Saransaari P, Oja SS (2000) Taurine and neural cell damage. Amino Acids 19:509–526

Sergeeva OA, Haas HL (2001) Expression and function of glycine receptors in striatal cholinergic interneurons from rat and mouse. Neuroscience 104(4):1043–1055

Söderpalm B, Löf E, Ericson M (2009) Mechanistic studies of ethanol´s interaction with the mesolimbic dopamine reward system. Pharmacopsychiatry 42(Suppl 1):S87–S94

Spanagel R (2009) Alcoholism: a systems approach from molecular physiology to addictive behavior. Physiol Rev 89(2):649–705

Wise RA, Rompre PP (1989) Brain dopamine and reward. Annu Rev Psychol 40:191–225

Part II
Taurine and the Immune System

Chapter 19
Thiotaurine Prevents Apoptosis of Human Neutrophils: A Putative Role in Inflammation

Elisabetta Capuozzo, Laura Pecci, Alessia Baseggio Conrado, and Mario Fontana

Abstract Thiotaurine, a metabolic product of cystine, contains a sulfane sulfur atom that can be released as H_2S, a gaseous molecule with a regulatory activity on inflammatory responses. The influence of thiotaurine on human leukocyte spontaneous apoptosis has been evaluated by measuring caspase-3 activity in human neutrophils. Addition of 100 µM thiotaurine induced a 55% inhibition of caspase-3 activity similar to that exerted by 100 µM H_2S. Interestingly, in the presence of 1 mM GSH, an increase of the inhibition of apoptosis by thiotaurine has been observed. These results indicate that the bioactivity of thiotaurine can be modulated by GSH, which promotes the reductive breakdown of the thiosulfonate generating H_2S and hypotaurine. As thiotaurine is able to incorporate reversibly reduced sulfur, it is suggested that the biosynthesis of this thiosulfonate could be a means to transport and store H_2S.

Abbreviations

GSH	Glutathione
HTAU	Hypotaurine
TTAU	Thiotaurine

E. Capuozzo • L. Pecci • A. Baseggio Conrado • M. Fontana (✉)
Dipartimento di Scienze Biochimiche, Sapienza Università di Roma,
Piazzale Aldo Moro 5, Rome 00185, Italy
e-mail: mario.fontana@uniroma1.it

A. El Idrissi and W.J. L'Amoreaux (eds.), *Taurine 8*, Advances in Experimental
Medicine and Biology 775, DOI 10.1007/978-1-4614-6130-2_19,
© Springer Science+Business Media New York 2013

19.1 Introduction

Thiotaurine (2-aminoethane thiosulfonate) is a biomolecule structurally related to hypotaurine and taurine. Thiosulfonates (RSO_2SH), including thiotaurine, have been occasionally detected among the products of biochemical reactions involving sulfur compounds. Thiotaurine is a metabolic product of cystine in vivo (Cavallini et al. 1959, 1960) and is produced by a spontaneous transulfuration reaction involving thiocysteine (RSSH) and hypotaurine (RSO_2H) (De Marco et al. 1961). Moreover, an enzyme capable of oxidizing thiols to sulfinates and thiosulfonates, in the presence of inorganic forms of sulfur has been detected in a number of animal tissues (Cavallini et al. 1961). A sulfurtransferase, which catalyzes the transfer of sulfur from mercaptopyruvate to hypotaurine with production of thiotaurine has been also reported (Sörbo 1957; Chauncey and Westley 1983).

Recently, it has been shown that hydrogen sulfide (H_2S), an endogenously generated gaseous molecule, plays relevant signal roles, modulating several pathophysiological functions (Predmore et al. 2012). Though desulfuration of cysteine constitutes the main source of H_2S in mammals, thiotaurine contains a sulfane sulfur atom that can be released as H_2S (Westley and Heyse 1971). It is widely recognized that hypotaurine, taurine, and H_2S exert a regulatory activity on inflammatory responses (Green et al. 1991; Whiteman and Winyard 2011). However, thiotaurine has never been investigated for a bioactivity in inflammation.

In the present study, the influence of thiotaurine on human leukocyte spontaneous apoptosis has been evaluated. Neutrophil apoptosis is an important process because it provides a signal for neutrophil removal promoting resolution of inflammation, and because it results in the loss of functional neutrophil responsiveness (Savill and Fadok 2000; Simon 2003). On the other hand, increased survival in the inflamed tissue permits neutrophils to fulfill their effector functions most efficiently (Lee et al. 1993; Savill et al. 2002). Thus, modulation of apoptosis may have a major effect on the inflammatory process.

As several studies suggest a critical role of caspase-3 in both spontaneous and Fas receptor-mediated apoptosis in neutrophils (Weinmann et al. 1999; Ottonello et al. 2002), we tested the effect of thiotaurine on neutrophil apoptosis by measuring the caspase-3 activity in cell lysates of human neutrophils.

19.2 Materials and Methods

19.2.1 Chemicals

Thiotaurine was prepared from hypotaurine and elemental sulfur (Cavallini et al. 1959). L-Glutathione reduced, hypotaurine, sodium hydrosulfide (NaHS), sulfur, N,N-dimethyl-p-phenylenediamine   sulfate,   acetyl-Asp-Glu-Val-Asp-7-amido-4-

methylcoumarin (Ac-DEVD-AMC, caspase-3 substrate) were obtained from Sigma-Aldrich, Inc (St. Louis, MO, USA). All other chemicals were analytical grade.

19.2.2   Isolation of Neutrophils

Leukocytes were purified from heparinized human blood freshly drawn from healthy donors. Leukocyte preparations containing 90–98% neutrophils were obtained by one-step procedure involving centrifugation of blood samples layered on Ficoll-Hypaque medium (Polymorphprep, Axis-Shield, Oslo, Norway) (Ferrante and Thong 1980). The cells were suspended in isotonic phosphate-buffered saline (PBS), pH 7.4, with 5 mM glucose and stored on ice. Each preparation produced cells with a viability higher than 90% up to 6 h after purification. The incubations were carried out at 37°C.

19.2.3   Measurement of H_2S

Aliquots of the sample were mixed with distilled water to a final volume of 0.5 mL. Then 0.25 mL zinc acetate (1% w/v), 0.25 mL N,N-dimethyl-p-phenylenediamine sulfate (20 mM in 7.2 M HCl) and 0.2 mL $FeCl_3$ (30 mM in 1.2 M HCl) were added. After 15 min at room temperature, the absorbance of the resulting solution was measured at 670 nm (Siegel 1965). All samples were assayed in duplicate and H_2S was calculated against a calibration curve of sodium hydrosulfide (NaHS, 2–100 μM).

19.2.4   Detection of Neutrophil Apoptosis by Caspase-3 Activity Assay

Caspase-3 activity was tested in neutrophil lysates by measuring the release of the fluorescent 7-amino-4-methylcoumarin (AMC) moiety from the synthetic substrate   acetyl-Asp-Glu-Val-Asp-7-amido-4-methyl-coumarin   (Ac-DEVD-AMC) (Nicholson et al. 1995). Neutrophils (5×10^6 cells), preincubated in PBS with 5 mM glucose at 37°C for 3.5 h, were collected by centrifugation and lysed in 0.5 mL of 50 mM HEPES buffer, pH 7.4, containing 5 mM 3-[3-(cholamido-propyl) dimethylammonio]-1-propanesulfonate (CHAPS), 5 mM dithiothreitol (DTT), 10 μM 4-amidinophenylmethanesulfonyl fluoride (APMSF), 10 μg/mL pepstatin, and 10 μg/mL aprotinin. The reaction was started by adding 100 μL aliquots of the lysates in 2 mL solutions containing 16 μM AcDEVD-AMC, 20 mM HEPES, 0.1% CHAPS, 5 mM DTT, and 2 mM EDTA, pH 7.4. The assay mixture was incubated at 20°C in the dark for 1 h. The fluorescence (excitation wavelength 360 nm, emission wavelength 460 nm) increase was compared with an appropriate blank control containing 10 µM acetyl-Asp-Glu-Val-Asp-al, a specific caspase-3 inhibitor (Nicholson et al. 1995) or standard preparations of recombinant caspase-3 (Sigma). A calibration curve obtained with standard AMC solutions was employed for quantitative analysis.

19.2.5 HPLC Analysis

Hypotaurine and thiotaurine were determined by HPLC using the *o*-phthaldialdehyde reagent (Hirschberger et al. 1985). Analyses were performed as previously described (Fontana et al. 2005), using a Waters 474 scanning fluorescence detector ($\lambda_{ex} = 340$ nm, $\lambda_{em} = 450$ nm). The elution times of hypotaurine and thiotaurine were 22 min and 27 min, respectively.

19.2.6 Statistics

Results are expressed as means ± SEM for at least three separate experiments performed in duplicate. Graphics and data analysis were performed using GraphPad Prism 4 software.

19.3 Results

19.3.1 Influence of Thiotaurine on Human Neutrophil Spontaneous Apoptosis

Spontaneous apoptosis was evaluated by measuring caspase-3 activity in lysates of neutrophils (5×10^6 cells/mL) that were preincubated at 37°C for 3.5 h. When the preincubation step was performed in the presence of thiotaurine (TTAU), a concentration-dependent decrease of caspase-3 activity was observed (Fig. 19.1). As thiotaurine contains a sulfane sulfur atom that can be released as H_2S, the influence of NaHS on caspase-3 activity has been also evaluated. With 100 µM thiotaurine the reduction of caspase-3 activity was 55 ± 3%, similar to that exhibited by 100 µM NaHS (57 ± 3%). Control experiments (not shown) indicated that neither TTAU, nor NaHS, at concentrations ranging from 0.01 to 0.2 mM, affected the activity of recombinant caspase-3.

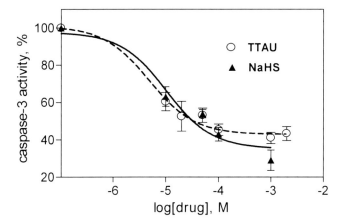

Fig. 19.1 Effect of thiotaurine and NaHS on caspase-3 activity. Neutrophils (5×10^6 cells/mL) were incubated at 37°C for 3.5 h with different concentrations of thiotaurine (TTAU) or NaHS. Caspase-3 activity was determined as described in Sect. 2

19.3.2   Effect of Glutathione on Thiotaurine-Induced Inhibition of Caspase-3 Activity

It is reported that glutathione (GSH) regulates neutrophil apoptosis by affecting caspase-3 activity (O'Neill et al. 2000). This effect has been attributed to its antioxidant activity (Wedi et al. 1999). To gain insights into the mechanism of inhibition by TTAU, the inhibitory effect of this thiosulfonate on caspase-3 activity has been compared with that of GSH (Fig. 19.2).

Under our experimental conditions, the inhibitory effect of 1 mM GSH ($58 \pm 3\%$) on caspase-3 activity is similar to that of 0.1 mM TTAU. Interestingly, the inhibition of spontaneous apoptosis by 0.1 mM TTAU increases to $76 \pm 4\%$ when GSH is present in the preincubation step.

19.3.3   Reductive Breakdown of Thiotaurine by Glutathione: Generation of H_2S and Hypotaurine

It is well known that thiol compounds such as GSH promote reductive breakdown of thiosulfonates generating H_2S and sulfinates (Chauncey and Westley 1983). Spontaneous and GSH-catalyzed H_2S release has been analyzed in the presence or in the absence of human neutrophils. Figure 19.3 shows that the release of H_2S is stimulated by GSH and it increases with the incubation time. Furthermore, it can be seen that the amount of H_2S results lower in the presence of cells. This result may depend on different factors, such as H_2S binding to proteins (Cavallini et al. 1970) or H_2S uptake by cells (Mathai et al. 2009).

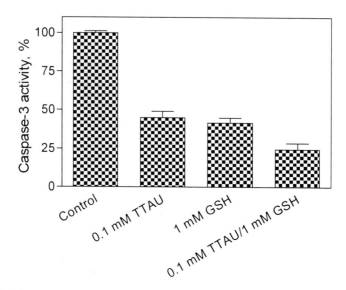

Fig. 19.2 Influence of glutathione on thiotaurine inhibition of caspase-3 activity. Neutrophils (5×10^6 cells/mL) were incubated at 37°C for 3.5 h in the absence (control) and in the presence of 0.1 mM thiotaurine (TTAU) or 1 mM glutathione (GSH) or both compounds. Caspase-3 activity was determined as described in Sect. 2

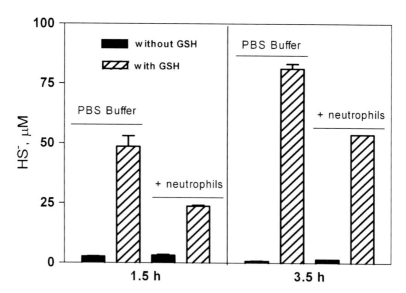

Fig. 19.3 Generation of H_2S by thiotaurine: effect of glutathione. 1 mM thiotaurine (TTAU) was added to neutrophils (5×10^6 cells/mL) and incubated at 37°C for 1.5 and 3.5 h. When present, glutathione (GSH) was 1 mM. The controls (PBS Buffer) were performed in the same conditions without neutrophils. H_2S was determined spectrophotometrically as described in Sect. 2

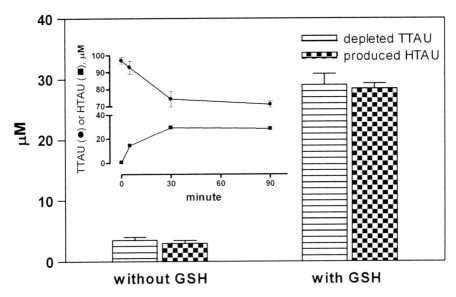

Fig. 19.4 Generation of hypotaurine by thiotaurine: effect of glutathione. 0.1 mM thiotaurine (TTAU) was added to neutrophils (5×10^6 cells/mL) and incubated at 37°C. When present, glutathione (GSH) was 1 mM. Hypotaurine (HTAU) and TTAU concentrations were determined by HPLC as described in Sect. 2. The amounts of depleted TTAU and produced HTAU, after 90 min incubation in the absence or in the presence of GSH, are compared. *Inset* time-course of the reaction of TTAU with GSH

Figure 19.4 shows that, in human leukocytes, GSH promotes the generation of hypotaurine (HTAU) as the main metabolite of TTAU. The production of HTAU increases with time (up to 30 min) and a stoichiometry of approximately 1 mol of HTAU produced/mol of TTAU depleted is observed (inset). The fate of TTAU has been evaluated also in the absence of cells; the results were similar to those observed with human leukocytes. Furthermore, GSH-mediated breakdown of TTAU in human neutrophils activated by phorbol 12-myristate 13-acetate (PMA) produces also taurine, the oxidative product of HTAU (not shown).

19.4 Discussion

These results indicate that the thiosulfonate, thiotaurine, may exert regulatory effects on inflammation influencing lifespan of human neutrophils. Mature circulating neutrophils are constitutively committed to apoptosis. During inflammatory response, survival of neutrophils recruited into the inflamed area is significantly prolonged. Increased survival in the inflamed tissue permits neutrophils to fulfill their effector functions most efficiently. On the other hand, macrophage-mediated elimination of apoptotic neutrophils from the inflamed area has been recognized as a crucial mechanism for promoting resolution of inflammation (Savill and Fadok 2000; Simon 2003).

It is recognized that the production of reactive oxygen species by activated cells accelerate the apoptosis and that superoxide release is required for spontaneous apoptosis (Ottonello et al. 2002; Scheel-Toellner et al. 2004). Moreover, the spontaneous and FAS-mediated apoptosis are prevented by antioxidants, such as GSH (Wedi et al. 1999). This effect has been ascribed to the ability of GSH to scavenge reactive oxygen species (Watson et al. 1997). It has been also shown that thiotaurine is highly effective in counteracting the damaging effect of oxidants (Acharya and Lau-Cam 2012). Thus, it is possible that the delay of spontaneous apoptosis of human neutrophils by thiotaurine may be related to its antioxidant activity. On the other hand, our results show that the inhibitory effect of thiotaurine on caspase-3 activity was higher than that of GSH. Moreover thiotaurine, in the presence of GSH, is more effective in influencing neutrophil apoptosis. These findings suggest that alternative or additional mechanisms of inhibition can be involved. It is well-known that GSH can act as a catalyst of the reductive breakdown of thiotaurine with generation of hypotaurine and H_2S (Chauncey and Westley 1983). Accordingly, we found that human neutrophils generate H_2S from thiotaurine with GSH as a necessary reductant in the reaction. It has been previously reported that H_2S promotes the short-term survival of neutrophils by inhibition of caspase-3 cleavage (Rinaldi et al. 2006). Our results confirm the effect of H_2S on prolonging the survival of neutrophils. Hence, it is likely that the sulfane sulfur of thiotaurine released as H_2S in the presence of GSH, may contribute to the observed effect on neutrophil survival.

The biological relevance of thiotaurine in mammalian is still a challenge to biochemical research. Biological roles have been sporadically reported (Costa et al. 1990; Baskin et al. 2000). On the contrary, in some marine organisms a key role for thiotaurine in the transport of sulfur has been strongly demonstrated (Pruski et al. 2001; Pruski and Fiala-Médioni 2003). Morevover, the metabolic origin of thiotaurine in mammalians is subject to debate, as is its fate. One pathway for thiotaurine metabolism is via transulfuration reactions with hypotaurine being the main intermediate (Cavallini et al. 1961; De Marco and Tentori 1961). These reactions can be spontaneous or catalyzed by sulfur transferases (De Marco et al. 1961; Chauncey and Westley 1983). Our experiments show that hypotaurine is the main metabolite of thiotaurine with a 1:1 stoichiometry, suggesting a role of thiotaurine as a biochemical intermediate in the transport, storage, and release of sulfide also in mammalians. This hypothesis is further supported by the fact that hypotaurine, present in leukocytes at millimolar concentration (Learn et al. 1990), can readily incorporate H_2S formed during inflammation with production of thiotaurine (De Marco and Tentori 1961).

Since thiotaurine as well as hypotaurine, taurine, and H_2S can modulate leukocyte functional responses, it would be worthy to investigate the metabolic and functional interplay between these sulfur compounds at inflammatory sites.

References

Acharya M, Lau-Cam CA (2012) Comparative evaluation of the effects of hypotaurine, taurine and thiotaurine on alterations of the cellular redox status and activities of antioxidant and glutathione-related enzymes by acetaminophen in the rat. Amino Acids 42:1512

Baskin SI, Prabhaharan V, Bowman JD, Novak MJ (2000) In vitro effects of anionic sulfur compounds on the spectrophotometric properties of native DNA. J Appl Toxicol 20:S3–S5

Cavallini D, De Marco C, Mondovì B (1959) Chromatographic evidence of the occurence of thiotaurine in the urine of rats fed with cystine. J Biol Chem 234:854–857

Cavallini D, De Marco C, Mondovì B, Tentori L (1960) Radioautographic detection of metabolites of ^{35}S-DL-cystine. J Chromatogr 3:20–24

Cavallini D, De Marco C, Mondovì B (1961) Detection and distribution of enzymes for oxidizing thiocysteamine. Nature 192:557–558

Cavallini D, Federici G, Barboni E (1970) Interaction of proteins with sulfide. Eur J Biochem 14:169–174

Chauncey TR, Westley J (1983) The catalytic mechanism of yeast thiosulfate reductase. J Biol Chem 258:15037–15045

Costa M, Vesci L, Fontana M, Solinas SP, Duprè S, Cavallini D (1990) Displacement of [^{3}H] GABA binding to bovine brain receptors by sulfur-containing analogues. Neurochem Int 17:547–551

De Marco C, Tentori L (1961) Sulfur exchange between thiotaurine and hypotaurine. Experientia 17:345–346

De Marco C, Coletta M, Cavallini D (1961) Spontaneous transulfuration of sulfinates by organic persulfides. Arch Biochem Biophys 93:179–180

Ferrante A, Thong YH (1980) Optimal conditions for simultaneous purification of mononuclear and polymorphonuclear leukocytes from human peripheral blood by the Hypaque-Ficoll method. J Immunol Methods 36:109–117

Fontana M, Amendola D, Orsini E, Boffi A, Pecci L (2005) Oxidation of hypotaurine and cysteine sulphinic acid by peroxynitrite. Biochem J 389:233–240

Green TR, Fellman JH, Eicher AL, Pratt KL (1991) Antioxidant role and subcellular location of hypotaurine and taurine in human neutrophils. Biochim Biophys Acta 1073:91–97

Hirschberger LL, De La Rosa J, Stipanuk M (1985) Determination of cysteinesulfinate, hypotaurine and taurine in physiological samples by reversed-phase high-performance liquid chromatography. J Chromatogr B 343:303–313

Learn DB, Fried VA, Thomas EL (1990) Taurine and hypotaurine content of human leukocytes. J Leukoc Biol 48:174–182

Lee A, Whyte MKB, Haslett C (1993) Inhibition of apoptosis and prolongation of neutrophil functional longevity by inflammatory mediators. J Leukoc Biol 54:283–288

Mathai JC, Missner A, Kügler P, Saparov SM, Zeidel ML, Lee JK, Pohl P (2009) No facilator required for membrane transport of hydrogen sulfide. Proc Natl Acad Sci USA 106:16633–16638

Nicholson DW, Ali A, Thornberry NA, Vaillancourt JP, Ding CK, Gallant M, Gareau Y, Griffin PR, Labelle M, Lazebnik YA, Munday NA, Raju SM, Smulson ME, Yamin T-T, Yu VL, Miller DK (1995) Identification and inhibition of the ICE/CED-3 protease necessary for mammalian apoptosis. Nature 376:37–43

O'Neill AJ, O'Neill S, Hegarty NJ, Coffey RN, Gibbons N, Brady H, Fitzpatrick JM, Watson RW (2000) Glutathione depletion-induced neutrophil apoptosis is caspase 3 dependent. Shock 14:605–609

Ottonello L, Frumento G, Arduino N, Bertolotto M, Dapino P, Mancini M, Dallegri F (2002) Differential regulation of spontaneous and immune complex-induced neutrophil apoptosis by proinflammatory cytokines. Role of oxidants, Bax and caspase-3. J Leukoc Biol 72:125–132

Predmore BL, Lefer DJ, Gojon G (2012) Hydrogen sulfide in biochemistry and medicine. Antioxid Redox Signal 17:119–140

Pruski AM, Fiala-Médioni A (2003) Stimulatory effect of sulphide on thiotaurine synthesis in three hydrothermal-vent species from the East Pacific Rise. J Exp Biol 206:2923–2930

Pruski AM, De Wit R, Fiala-Médoni A (2001) Carrier of reduced sulfur is a possible role for thiotaurine in symbiotic species from hydrotermal vents with thiotrophic symbionts. Hydrobiologia 461:9–13

Rinaldi L, Gobbi G, Pambianco M, Micheloni C, Mirandola P, Vitale M (2006) Hydrogen sulfide prevents apoptosis of human PMN via inhibition of p38 and caspase 3. Lab Invest 86:391–397

Savill J, Fadok V (2000) Corpse clearance defines the meaning of cell death. Nature 407:770–776

Savill J, Dransfield I, Gregory C, Haslett C (2002) A blast from the past: clearance of apoptotic cells regulates immune responses. Nat Rev Immunol 2:965–975

Scheel-Toellner D, Wang K, Craddock R, Webb PR, McGettrick HM, Assi LK, Parkes N, Clough LE, Gulbins E, Salmon M, Lord JM (2004) Reactive oxygen species limit neutrophil life span by activating death receptor signaling. Blood 104:2557–2564

Siegel LM (1965) A direct microdetermination for sulfide. Anal Biochem 11:126–132

Simon HU (2003) Neutrophil apoptosis pathways and their modifications in inflammation. Immunol Rev 193:101–110

Sörbo B (1957) Enzymic transfer of sulfur from mercaptopyruvate to sulfite or sulfinates. Biochim Biophys Acta 24:324–329

Watson RW, Rotstein OD, Jimenez M, Parodo J, Marshall JC (1997) Augmented intracellular glutathione inhibits Fas-triggered apoptosis of activated human neutrophils. Blood 89:4175–4181

Wedi B, Straede J, Wieland B, Kapp A (1999) Eosinophil apoptosis is mediated by stimulators of cellular metabolisms and inhibited by antioxidants: involvement of a thiol-sensitive redox regulation in eosinophil cell death. Blood 94:2365–2373

Weinmann P, Gaehtgens P, Walzog B (1999) Bcl-X$_l$- and Bax-α-mediated regulation of apoptosis of human neutrophils via caspase-3. Blood 93:3106–3115

Westley J, Heyse D (1971) Mechanisms of sulfur transfer catalysis. Sulfhydryl-catalyzed transfer of thiosulfonate sulfur. J Biol Chem 246:1468–1474

Whiteman M, Winyard PG (2011) Hydrogen sulphide and inflammation: the good, the bad, the ugly and the promising. Expert Rev Clin Pharmacol 4:13–32

Chapter 20
Protection by Taurine Against INOS-Dependent DNA Damage in Heavily Exercised Skeletal Muscle by Inhibition of the NF-κB Signaling Pathway

Hiromichi Sugiura, Shinya Okita, Toshihiro Kato, Toru Naka, Shosuke Kawanishi, Shiho Ohnishi, Yoshiharu Oshida, and Ning Ma

Abstract Taurine protects against tissue damage in a variety of models involving inflammation, especially the muscle. We set up a heavy exercise bout protocol for rats consisting of climbing ran on a treadmill to examine the effect of an intraabdominal dose of taurine (300 mg/kg/day) administered 1 h before heavy exercise for ten consecutive days. Each group ran on the treadmill at 20 m/min, 25% grade, for 20 min or until exhaustion within 20 min once each 10 days. Exhaustion was the point when an animal was unable to right itself when placed on its side. The muscle damage was associated with an increased accumulation of 8-nitroguanine and 8-OHdG in the nuclei of skeletal muscle cells. The immunoreactivities for NF-κB and iNOS were also increased in the exercise group. Taurine ameliorated heavy exercise-induced muscle DNA damage to a significant extent since it reduced the accumulation of 8-nitroguanine and 8-OHdG, possibly by down-regulating the expression of iNOS through a modulatory action on NF-κB signaling pathway. This study demonstrates for the first time that taurine can protect against intense exercise-induced nitrosative inflammation and ensuing DNA damage in the skeletal muscle of rats by preventing iNOS expression and the nitrosative stress generated by heavy exercise.

H. Sugiura
Faculty of Health Science and Pharmaceutical Sciences,
Suzuka University of Medical Science, Mie 510-0293, Japan

Department of Sports Medicine, Graduate School of Medicine,
Nagoya University, Nagoya, Japan

S. Okita • T. Kato • T. Naka • S. Kawanishi • S. Ohnishi • N. Ma(✉)
Faculty of Health Science and Pharmaceutical Sciences,
Suzuka University of Medical Science, Mie 510-0293, Japan
e-mail: maning@suzuka-u.ac.jp

Y. Oshida
Department of Sports Medicine, Graduate School of Medicine,
Nagoya University, Nagoya, Japan

A. El Idrissi and W.J. L'Amoreaux (eds.), *Taurine 8*, Advances in Experimental
Medicine and Biology 775, DOI 10.1007/978-1-4614-6130-2_20,
© Springer Science+Business Media New York 2013

Abbreviations

ROS	Reactive oxygen species
O_2^-	Superoxide anion
NF-κB	Nuclear factor kappa-B
iNOS	Inducible NO synthase
8-OHdG	8-Hydroxydeoxyguanosine

20.1 Introduction

Heavy exercise is thought to increase oxidative stress and to damage muscle tissue. Taurine protects against tissue damage in a variety of experimental models involving oxidative stress, especially during exercise. The mechanism of taurine protection is not well understood, but the ability of taurine to attenuate the toxic effects of HOCl/OCl via formation of taurine chloramine (TauCl) and its subsequent effects are thought to be important. Nitrosative stress-mediated activation of inflammatory mediators is currently being emphasized as an important factor mediating inflammation-related disorders. Reactive oxygen species (ROS) and reactive nitrogen species (RNS) are capable of causing damage to various cellular constituents, such as nucleic acids, proteins and lipids. ROS can induce the formation of oxidative DNA lesion products, including 8-hydroxyldeoxyguanosine (8-OHdG) which is considered to be mutagenic. On the other hand, nitric oxide (NO) is generated specifically during inflammation via inducible nitric oxide synthase (iNOS) in inflammatory cells. Excess NO production plays a crucial role in an enormous variety of pathological processes, including apoptosis. NO can react with superoxide anion (O_2^-) to form peroxynitrite (ONOO-), a highly reactive nitrogen species capable of causing nitrosative and oxidative DNA damage. In turn, ONOO- can mediate the formation of 8-OHdG and 8-nitroguanine, a marker of nitrosative DNA damage (Yermilov et al. 1995). 8-Nitroguanine formed in DNA is chemically unstable and, thus, can be spontaneously released, resulting in the formation of an apurinic site (Yermilov et al. 1995). The apurinic site can form a pair with adenine during DNA synthesis, leading to G → T transversions (Loeb and Preston 1986). It has been demonstrated that 8-nitroguanine is formed via NO production associated with inflammation in *Helicobacter pylori* infected patients (Ma et al. 2004). 8-Nitroguanine is considered to be not only a marker of inflammation, but also a potential mutagenic DNA lesion product-capable of mediating apoptosis and carcinogenesis. Nitrosative stress arises mainly from the large accumulation of NO following the overexpression of iNOS in damaged tissue to form ONOO- when NO reacts with O_2^- to form 8-nitroguanine.

The aim of this study was to determine the cytoprotective role of taurine against nitrosative stress in intense exercise-induced damage of the skeletal muscle.

20.2 Methods

20.2.1 Animals and Experimental Design

All experimental protocols were approved by the Animal Ethics Committee of Suzuka University of Medical Science, Japan. In total, 18 male rats (24-months-old; 250–260 g bodyweight) were housed in cages (max. 6 per cage) with water and food ad libitum. The animals were randomly divided into the following groups: exercise plus saline ($n=6$); exercise plus taurine ($n=6$); and control ($n=6$). All the animals were maintained under a 12-h light and 12-h dark cycle at 24°C.

20.2.2 Taurine Supplementation and Exercise Protocol

We set up a heavy exercise protocol which consisted of an inclined treadmill and examined the effect of intraabdominally administered taurine 1 h before heavy exercise at a dose of 300 mg/kg/day for ten consecutive days. All animals were initially acclimated to running on a motor-driven treadmill designed for rats (model MK-680, Muromachi Kikai, Tokyo, Japan), beginning at 8 m/min for 10 min during the week preceding the exercise experiments. This protocol accustomed the rats to the locomotion intended for the final exercise experiments without stimulating development of skeletal muscle as an adaptation to training. Each group ran on the treadmill at 20 m/min, 25% grade, for 20 min or until exhaustion once in 10 days. Exhaustion was determined as the point when an animal was unable to right itself when placed on its side. The workload was ~75% of the rats' maximal aerobic capacity (VO_2 max) (Brooks and White 1978). To return to basal physiological conditions and to prevent the influence of acute exercise, the animals were killed by decapitation 48 h after the exercise session. Rats were deeply anesthetized with an IP injection of sodium pentobarbital and were perfused transcardially with a fixative that contained 4% paraformaldehyde in 0.01 M phosphate buffer, pH 7.4. After the perfusion, the soleus was dissected out and allowed to stand in the same fixative for 4 h. Then the tissue was rinsed several times with phosphate buffer, dehydrated with a graded alcohol series and acetone, and embedded in paraffin. Sections 6-μm in thickness were mounted on albumin-coated slides.

20.2.3 Immunofluorescence Study

Anti-8-nitroguanine rabbit polyclonal antibody was prepared as described earlier (Pinlaor et al. 2004). The immunoreactivity of 8-nitroguanine and of other biomarkers in skeletal muscle fiber was assessed by single or double immunofluorescence labeling studies, as previously described (Pinlaor et al. 2004). Briefly, deparaffinized and dehydrated sections (6 μm thickness) were incubated with 5% skim milk followed by incubation with rabbit polyclonal anti-8-nitroguanine antibody (2 μg/mL) or other antibody mouse monoclonal anti-8-OHdG (5 μg/mL; Japan Institute for Control of Aging, Fukoroi, Japan), mouse monoclonal anti-iNOS (1:400, Sigma, St. Louis, MO), rabbit polyclonal anti-iNOS (1:500, CalbiochemNovabiochem Corporation, Darmstadt, Germany), or mouse monoclonal anti-NF-κB p65 (2 μg/mL, Santa Cruz Biotechnology) antibody overnight at room temperature. Then, the sections were incubated for 3 h with Alexa 594-labeled goat antibody against rabbit IgG or Alexa 488-labeled goat antibody against mouse IgG (1:400; Molecular Probes, Eugene, OR). The stained sections were examined using a confocal laser scanning microscope (FV-1000, Olympus, Tokyo, Japan) or by fluorescence microscopy (BX53, Olympus, Tokyo, Japan).

20.2.4 Statistical Analysis

Data are presented as the mean ± S.E.M. values. The two-tailed Student's t-test was performed. Differences were considered statistically significant at $p < 0.05$.

20.3 Results

20.3.1 Analyses for 8-Nitroguanine and 8-OHdG Immunoreactivities in the Soleus Muscle

In each experimental group, the number of muscle cells positive for 8-nitroguanine and 8-OHdG were compared with the total number of muscle cells. The exercise group had the highest percentage of 8-nitroguanine positive cells ($35.1 \pm 0.1\%$) followed by the exercise plus taurine group ($30.1 \pm 0.1\%$) and control ($23.1 \pm 0.0\%$) groups. The highest 8-OHdG immunoreactivity was also observed with the exercise group ($34.8 \pm 0.1\%$) followed by the exercise plus taurine group ($26.5 \pm 0.1\%$) and the control ($16.1 \pm 0.0\%$) groups (Fig. 20.1).

The mean diameter of muscle fibers positively and negatively immunoreactive for 8-nitroguanine was 26.2 ± 0.8 μm and 39.6 ± 1.3 μm in controls, 54.9 ± 4.8 μm and 76.9 ± 8.3 μm in the exercise group, and 28.3 ± 1.1 μm and 39.7 ± 2.1 μm in the exercise plus taurine group, respectively, showing that 8-nitroguanine-positive fibers were significantly smaller than negative ones. The mean diameter of muscle fibers showing positive and negative immunoreactivities for 8-OHdG was 27.3 ± 0.4 μm and 39.3 ± 2.5 μm in controls, 51.9 ± 2.8 μm and 76.5 ± 5.5 μm in the exercise group, and 29.1 ± 1.5 μm and 38.9 ± 1.0 μm in the exercise plus taurine group, respectively. These results demonstrate that the immunoreactive fibers were smaller than the non-immunoreactive ones (Fig. 20.2).

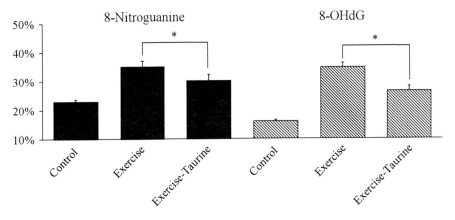

Fig. 20.1 Number of soleus muscle cells showing positive and negative immunoreactivity for 8-nitroguanine or 8-OHdG. Data are presented as the mean ± S.E.M., and were analyzed by the two-tail Student's t-test. Exercise versus Exercise-Taurine: $*p < 0.05$

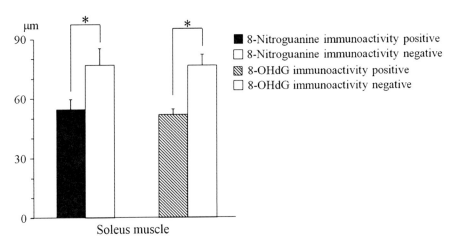

Fig. 20.2 The mean diameter size of soleus muscle cells in the exercise group showing positive or negative immunoreactivity for 8-OHdG or 8-nitroguanine. Comparison of positively immunoreactive cells versus negatively immunoreactive cells at $*p < 0.05$

20.3.2 8-Nitroguanine and 8-OHdG Formation in the Soleus Muscle

The formation of 8-nitroguanine and 8-OHdG in fibers of soleus muscle is shown in Fig. 20.3. Strong immunoreactivities for 8-nitroguanine (red) and 8-OHdG (green) were clearly observed in the nuclei of muscle fibers of all rats after exercise; but they were decreased in the taurine-treated group, and became negative in the control (normal) group. While 8-nitroguanine and 8-OHdG were colocalized in almost all fibers, 8-nitroguanine was predominated than over 8-OHdG in the exercise group (Fig. 20.3).

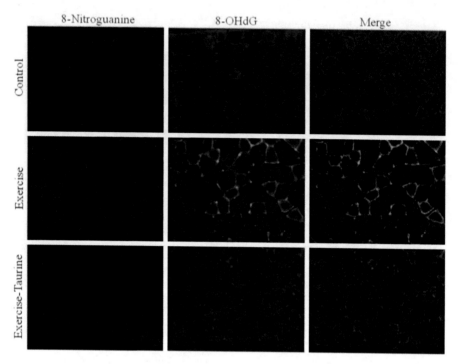

Fig. 20.3 8-Nitroguanine and 8-OHdG formation in heavily exercised rats with and without a taurine reatment. Double immunofluorescence staining of paraffin sections shows the localization of 8-OHdG and 8-nitroguanine in the nuclei of fibers of the soleus muscle

20.3.3   8-Nitroguanine and NF-κB Formation in the Soleus Muscle

8-Nitroguanine formation and NF-κB expression in fibers of soleus muscle are shown in Fig. 20.4. 8-Nitroguanine immunoreactivity (red) was observed in the nuclei of the fibers whereas NF-κB expression (green) was strongly observed in both the cytoplasm and nucleus. The immunoreactivities of 8-nitroguanine and NF-κB were decreased significantly in the taurine-treated group. Little to no 8-nitroguanine and NF-κB expression were observed in control muscle fibers (Fig. 20.4).

20.3.4   iNOS and NF-κB Formation in the Soleus Muscle

Strong immunoreactivities for iNOS and NF-κB were observed in soleus muscle fibers of all rats after heavy exercise. Intense staining of muscle cells was noted in both the nucleus and cytoplasm of the heavily exercised group. However, the taurine-

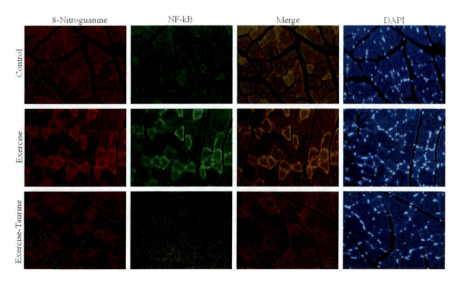

Fig. 20.4 8-Nitroguanine formation and NF-κB immunoreactivities in fibers of the soleus muscle. In heavily exercised rats, double immunofluorescence staining of paraffin sections shows the localization of NF-κB and 8-nitroguanine and the immunoreactivities of 8-nitroguanine and NF-κB colocalized primarily in the nuclei of the muscle fibers

treated group showed a marked reduction in the number of taurine-containing cells with stained nucleus and cytoplasm (Fig. 20.5). The immunoreactivity for iNOS and NF-κB decreased significantly in the muscle cells of rats receiving taurine, and was either drastically decreased or absent in the muscle fibers of control rats.

20.4   Discussion

In muscle cells, the levels of reactive oxygen species (ROS), constantly generated by cellular metabolic processes, are regulated by intracellular oxidative defense systems driven by antioxidant enzymes such as catalase, superoxide dismutase (SOD) (Silva et al. 2011), and glutathione peroxidase (Sen et al. 1997). Heavy exercise activates cellular metabolism and, thus, gives rise to a large number of ROS in the form of free radicals. An imbalance between the levels of ROS and antioxidant defenses results in oxidative and nitrosative stresses in the muscle fibers. Acute heavy exercise has been reported to activate NF-κB in skeletal muscle cells (Ji et al. 2004). A study of the activation mechanism of NF-κB revealed that acute exercise promotes cellular oxidative stress and activation of NF-κB through oxidation of glutathione in muscle cells (Sen et al. 1997).

In this study, oxidative and nitrosative stresses were induced in the skeletal muscle through heavy-load exercise (75% of $\dot{V}O_2$ max) performed for 20 min or

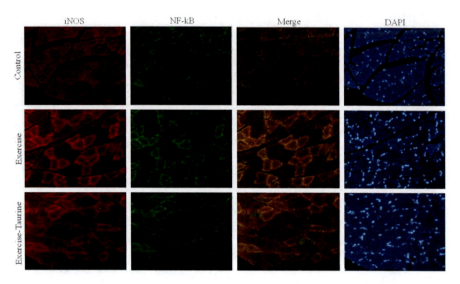

Fig. 20.5 iNOS and NF-κB immunoreacitivies in the soleus muscle. In heavily exercised rats, double immunofluorescence staining of paraffin sections shows the localization of iNOS and NF-κB and the immunoreactivity of 8-nitroguanine and NF-κB colocalizes primarily in the fibers of the extensor digitorum longus muscle

until exhaustion. Exercise activated inflammatory mediators such as iNOS and NF-κB in skeletal muscle cells and significantly increased the expression of the oxidative/nitrosative stress indicators 8-OHdG and 8-nitroguanine. However, administration of taurine prevented heavy exercise from activating NF-κB and reduced oxidative and nitrosative stresses.

The antioxidant action of taurine is related to its ability to scavenge free radicals generated by various enzymatic processes by directly causing reduction reactions (Aruoma et al. 1988; Shi et al. 1997). Taurine is also found to reduce protein carbonylation by effectively scavenging hypochlorous acid (HOCl) generated by myeloperoxidase (Prutz 1996). Taurine functions like the NF-κB blocker sulphasalazine (Gurujeyalashmi et al. 2000), and taurine chloramine (TauCl), a product of the condensation of taurine with HOCl, is reported to inhibit the activation of NF-κB (Barua et al. 2001). In turn, TauCl can interact with hydrogen peroxide (H_2O_2), which is produced as an inflammatory response during exercise, to buffer the effects of H_2O_2 before the appearance of oxidative stress. Consistent with the results of previous studies, those generated here suggest that taurine is cytoprotective by reducing both oxidative and nitrosative stresses.

ATP production and oxygen supply to the mitochondria need to be increased to activate skeletal muscles during exercise. An increase in oxygen consumption is also likely to increase the generation of superoxide anion radicals (O_2^-). Taurine appears to reduce the production of O_2^- via redox reaction, particularly in intracellular places prone to produce high levels of O_2^- particularly the mitochondria (Hansen et al. 2006). In this study, immunofluorescence analysis demonstrated that the diameter of 8-nitroguanine in immunoreactive muscle fibers was significantly smaller than that of nonimmunoreactive muscle fibers, suggesting that the immunoreactive muscle fibers were type I. Type I muscle fibers contain a large number of mitochondria and their aerobic metabolism consumes a large number of oxygen molecules to produce ATP during exercise. Therefore, it is likely that type I muscle fibers generate a high level of free radicals from oxygen and are vulnerable to the noxious effects of oxidative and nitrosative stress. Compared with type II muscle fibers, type I muscle fibers contain a higher level of polyunsaturated fatty acid that make them more susceptible to lipid peroxidation (Nikolaidis and Mougios 2004; Nikolaidis et al. 2006) as well as to other types of oxidative damage. Because type I muscle fibers also contain a higher number of mitochondria than type II muscle fibers, they may produce a higher level of free radical production during exercise and even at rest (Moyes 2003). In fact, heavy exercise-induced muscle inflammation is accompanied by an increase formation of 8-nitrogauanine in the muscle fibers. In this study, taurine produced a decrease in NF-κB as well as in iNOS immunoreactivity in the muscle fibers of rats. This finding indicates that taurine is inhibiting proinflammatory factors by suppressing iNOS activity through an inhibitory effect on a signaling pathway activating NF-κB.

20.5 Conclusion

In conclusion, this study demonstrates for the first time that taurine offers a strong protective effect against intense exercise-induced nitrosative DNA damage in the skeletal muscle of rats. Since this sulfur-containing compound prevented iNOS expression, it may act as a modulator of nitrosative stress in the muscle during periods of heavy exercise. Hence, upregulated expression of iNOS in skeletal muscles could be responsible for the nitrosative muscle damage taking place during heavy exercise, probably through inflammatory damage mediated by a NF-κB-activating signaling pathway.

Acknowledgments This work was supported by Grants-in-Aid for Scientific Research from the Ministry of Education, Science, Sports, and Culture of Japan.

References

Aruoma OI, Halliwell B, Hoey BM, Butler J (1988) The antioxidant action of taurine, hypotaurine and their metabolic precursors. Biochem J 256:251–255

Barua M, Liu Y, Quinn MR (2001) Taurine chloramine inhibits inducible nitric oxide synthase and TNF-alpha gene expression in activated alveolar macrophages: decreased NF-κB activation and IκB kinase activity. J Immunol 167:2275–2281

Brooks GA, White TP (1978) Determination of metabolic and heart rate responses of rats to treadmill exercise. J Appl Physiol 45:1009–1015

Gurujeyalashmi G, Wang Y, Giri SN (2000) Taurine and niacin block lung and fibrosis by down-regulating bleomycin-induced activation of transcription nuclear factor-κB in mice. J Pharmacol Exp Ther 293:82–90

Hansen SH, Andersen ML, Birkedal H, Cornett C, Wibrand F (2006) The important role of taurine in oxidative metabolism. Adv Exp Med Biol 583:129–135

Ji LL, Gomez-Cabrera MC, Steinhafel N, Vina J (2004) Acute exercise activates nuclear factor (NF)-κB signaling pathway in rat skeletal muscle. FASEB J 18:1499–1506

Loeb LA, Preston BD (1986) Mutagenesis by a apurinic/apyrimidinic sites. Annu Rev Genet 20:201–230

Ma N, Adachi Y, Hiraku Y, Horiki N, Horike S, Imoto I, Pinlaor S, Murata M, Semba R, Kawanishi S (2004) Accumulation of 8-nitroguanine in human gastric epithelium induced by *Helicobacter pylori* infection. Biochem Biophys Res Commun 319:506–510

Moyes CD (2003) Controlling muscle mitochondrial content. J Exp Biol 206:4385–4391

Nikolaidis MG, Mougios V (2004) Effects of exercise on the fatty-acid composition of blood and tissue lipids. Sports Med 34:1051–1076

Nikolaidis MG, Petridou A, Mougios V (2006) Comparison of the phospholipid and triacylglycerol fatty acid profile of rat serum, skeletal muscle and heart. Physiol Res 55:259–265

Pinlaor S, Hiraku Y, Ma N, Yongvanit P, Semba R, Oikawa S, Murata M, Sripa B, Sithithaworn P, Kawanishi S (2004) Mechanism of NO-mediated oxidative and nitrative DNA damage in hamsters infected with Opisthorchis viverrini: a model of inflammation-mediated carcinogenesis. Nitric Oxide 11:175–183

Prutz WA (1996) Hypochlorous acid interactions with thiols, nucleotides, DNA, and other biological substrates. Arch Biochem Biophys 332:110–120

Sen CK, Khanna S, Resznick AZ, Roy S, Packer L (1997) Glutathione regulation of tumor necrosis factor-α-induced NF-κB activation in skeletal muscle-derived L6 cells. Biochem Biophys Res Commun 237:645–649

Shi X, Flynn DC, Porter DW, Leonard SS, Vallyathan V, Castranova V (1997) Efficacy of taurine based compounds as hydroxyl radical scavengers in silica induced peroxidation. Ann Clin Lab Sci 27:365–374

Silva LA, Silveira PC, Ronsani MM, Souza PS, Scheffer D, Vieira LC, Benetti M, De Souza CT, Pinho RA (2011) Taurine supplementation decreases oxidative stress in skeletal muscle after eccentric exercise. Cell Biochem Funct 29:43–49

Yermilov V, Rubio J, Becchi M, Friesen MD, Pignatelli B, Ohshima H (1995) Formation of 8-nitroguanine by the reaction of guanine with peroxynitrite *in vitro*. Carcinogenesis 16: 2045–2050

Chapter 21
Effect of Taurine Chloramine on Differentiation of Human Preadipocytes into Adipocytes

Kyoung Soo Kim, Hyun-Mi Choi, Hye-In Ji, Chaekyun Kim, Jung Yeon Kim, Ran Song, So-Mi Kim, Yeon-Ah Lee, Sang-Hoon Lee, Hyung-In Yang, Myung Chul Yoo, and Seung-Jae Hong

Abstract We investigated whether taurine chloramine (TauCl), which is endogenously produced by immune cells such as macrophages that infiltrate adipose tissue, affects the differentiation of preadipocytes into adipocytes or modulates the expression of adipokines in adipocytes. To study the physiological effects of TauCl on human adipocyte differentiation and adipokine expression, preadipocytes were cultured under differentiation conditions for 14 days in the presence or the absence of TauCl. Differentiated adipocytes were also treated with TauCl in the presence or the absence of IL-1β (1 ng/ml) for 7 days. The culture supernatants were analyzed for adipokines such as adiponectin, leptin, IL-6, and IL-8. At concentrations of 400–600 μM, TauCl significantly inhibited the differentiation of human preadipocytes into adipocytes in a dose-dependent manner. It did not induce the dedifferentiation of adipocytes or inhibit fat accumulation in adipocytes. Expression of major transcription factors of adipogenesis and adipocyte marker genes was decreased after treatment with TauCl, in agreement with its inhibition of differentiation. These results suggest that TauCl may inhibit the differentiation of

K.S. Kim (✉) • H.-M. Choi • H.-I. Ji • M.C. Yoo
East-West Bone & Joint Disease Research Institute, Kyung Hee
University Hospital at Gangdong, 149 Sangil-dong, Gangdong-gu, Seoul 134-727, Korea
e-mail: labrea46@yahoo.co.kr

C. Kim
Laboratory for Leukocyte Signaling Research, Department of Pharmacology
and BK21 Program, Inha University School of Medicine, Incheon, Korea

J.Y. Kim
Department of Pathology, Inje University Sanggye Paik Hospital
Seoul, Korea

R. Song • S.-M. Kim • Y.-Ah. Lee • S.-H. Lee • H.-In. Yang • S.-J. Hong (✉)
Division of Rheumatology, Kyung Hee University Medical School, Seoul, Korea
e-mail: hsj718@paran.com

A. El Idrissi and W.J. L'Amoreaux (eds.), *Taurine 8*, Advances in Experimental
Medicine and Biology 775, DOI 10.1007/978-1-4614-6130-2_21,
© Springer Science+Business Media New York 2013

preadipocytes into adipocytes. Thus, TauCl or more stable derivatives of TauCl could potentially be a safe drug therapy for obesity-related diseases.

Abbreviations

MSCs	Mesenchymal stem cells
C/EBP	CCAAT/enhancer-binding protein
PPARγ	Peroxisome proliferator-activated receptor γ
FABP	Fatty acid-binding protein
LPL	Lipoprotein lipase
GLUT4	Glucose transporter 4
TauCl	Taurine chloramine
HOCl	Hypochlorous acid

21.1 Introduction

The inhibition of adipogenesis is one of the most important mechanisms involved in reducing body fat, which also includes apoptosis, lipolysis, and fatty acid oxidation. Recent studies have proposed that obesity is a state of systemic, chronic low-grade inflammation (Itoh et al. 2011). During the course of obesity, adipose tissue is characterized by the infiltration of immune cells such as macrophages.

The formation of adipose tissue involves the commitment of mesenchymal stem cells (MSCs) (which have the potential to differentiate into various cell lineages) to the preadipocyte lineage and the differentiation of preadipocytes into mature adipocytes (Gesta et al. 2007; Roufosse et al. 2004). In vitro, MSCs proliferate and develop into preadipocytes after reaching confluency. Preadipocytes remain competent for proliferation but, when allowed to become confluent and treated with substances such as insulin and glucocorticoids, they cease cell division and differentiate into adipocytes (MacDougald and Mandrup 2002). During adipogenesis, a number of morphological and physiological changes occur. The cells change from fibroblast-like preadipocytes to spherical adipocytes and accumulate large fat droplets containing triglycerides (Jessen and Stevens 2002). Then, upstream regulators, including CCAAT/enhancer-binding protein β (C/EBP-β), C/EBP-δ, and sterol-regulatory element-binding protein 1c (SREBP1c), regulate the expression of peroxisome proliferator-activated receptor γ (PPARγ) and C/EBP-α, which are major transcription factors for adipogenesis. The concerted action of these adipogenic transcription factors ultimately drives the expression of adipocyte marker genes, such as fatty acid-binding protein (FABP) 4, lipoprotein lipase (LPL), fatty acid synthetase, adiponectin, glucose transporter 4 (GLUT4), leptin, and others, which are responsible for the synthesis and storage of triglycerides in lipid droplets (Rosen et al. 2002; Rosen and MacDougald 2006).

Taurine (2-aminoethanesulfonic acid) is a simple sulfur-containing amino acid; it is one of the most abundant intracellular free amino acids in mammalian tissues and blood cells. It modulates a variety of cellular functions, including antioxidation, modulation of ion movement, osmoregulation, modulation of neurotransmitters, and conjugation of bile acids (Ito et al. 2011). Decreased tissue taurine concentrations are characteristic of many pathological states (Szymanski and Winiarska 2008). Furthermore, it was reported that taurine deficiency creates a vicious circle that may promote obesity (Tsuboyama-Kasaoka et al. 2006). Both neutrophils and mono-cytes contain high levels of taurine. Taurine acts as a scavenger of hypochlorous acid (HOCl), which is produced by the myeloperoxidase/hydrogen peroxide/chlo-ride system of activated neutrophils and monocytes (Thomas et al. 1985). It reacts with HOCl to form taurine chloramine (TauCl). TauCl has been shown to play a major role in downregulating the expression of inflammatory mediators such as chemokines, cytokines, cyclooxygenase-2, and inducible nitric oxide synthase in various types of cells (Schuller-Levis and Park 2004).

In this study, we investigated whether TauCl, which is endogenously produced by immune cells such as macrophages that have infiltrated adipose tissue, affects the differentiation of preadipocytes into adipocytes.

21.2   Methods

21.2.1   Preadipocyte Cell Culture and Differentiation into Adipocytes

Human preadipocytes were purchased from Cell Applications (San Diego, CA, USA) and maintained in Preadipocyte Growth Medium Kit (Cell Applications). Preadipocytes were seeded into six-well plates (1.5×10^5 cell/well in 2 ml of medium) or 60-mM dishes (2.5×10^5 cell/60 mM dish in 2 ml of medium) and cul-tured until confluent. For differentiation, the culture medium was changed to Adipocyte Differentiation Medium (Cell Applications) and cultured for 2 weeks by changing the medium every 2 days in the presence or the absence of TauCl at differ-ent concentrations.

21.2.2   Preparation of TauCl

TauCl was synthesized by mixing equimolar amounts of sodium hypochlorite (Aldrich Chemical, Milwaukee, MI, USA) and taurine (Sigma, St. Louis, MO, USA) as described somewhere (Kim et al. 2007). TauCl formation was verified by UV absorption (200–400 nM). Endotoxin-free or low-endotoxin-grade water and buffers were used. Stock solutions of TauCl were kept at 4°C and used within 3 days.

21.2.3   Oil Red O Staining

Lipid accumulation was examined with oil red O staining (Ramirez-Zacarias et al. 1992). Cultured cells were rinsed twice with phosphate-buffered saline (PBS) and fixed in 10% (v/v) formaldehyde for 1 h. After the formaldehyde was removed, cells were rinsed three times with deionized water and stained with a saturated solution of oil red O in 60% isopropanol solution for 2 h at room temperature. Microscopic images (Olympus, Tokyo, Japan) of the stained cells were obtained after removing the staining solution. Finally, the dye retained in the cells was eluted with isopropanol and quantified by measuring the optical absorbance at 500 nm.

21.2.4   Semiquantitative and Real-Time Reverse Transcription Polymerase Chain Reaction

Complementary DNA was synthesized from 1 μg total RNA in 20 μl reverse transcription reaction mixture containing 5 mmol/l MgCl2, 1× RT buffer, 1 mmol/l dNTP, 1 U/μl RNase inhibitor, 0.25 U/μl AMV reverse transcriptase, and 2.5 μmol/l random 9-mers as described previously (Kim et al. 2007). For semiquantitative PCR, aliquots of cDNA were amplified with the primers in a 25 μl PCR mixture containing 1× PCR buffer, 0.625 units of TaKaRa Ex Taq™ HS, and 0.2 μmol/l of specific upstream primers, in accordance with the manufacturer's protocol (TaKaRa Bio, Kyoto, Japan). The PCR conditions for the Leptin, GLUT-4, and β-actin were as follows: 25–30 cycles at 95°C for 45 s, 55–60°C for 45 s, and 72°C for 45 s. PCR products were subjected to electrophoresis in 1.5% agarose gels containing ethidium bromide, and the bands were visualized under UV light. The primers were synthesized by Bioneer Co. Ltd (Seoul, Republic of Korea), and their sequences are as follows: leptin (5'-CGCAGTCAGTCTCCTCCAAA-3', 5'-GGTTCTCCAGGTCGTTGGAT-3'), GLUT4 (5'-TGGCTGAGCTGAAGGATGAG -3', 5'-CCAACAACACCGAGACCAAG-3'), and β-actin (5'-TCATGAGGTAGTCAG TCAGG-3', 5'-CTTCTACAATGAGCTGCGTG-3').

21.2.5   Western Blot Analysis

The cells are prepared for Western blot analysis and the samples were separated using 12 or 15% SDS-PAGE, and were then transferred to Hybond-ECL membranes (Amersham, Arlington Heights, IL, USA) as described previously (Choi et al. 2009). The membranes were first blocked with 6% nonfat milk dissolved in TBST buffer (10 mM Tris–Cl [pH 8.0], 150 mM NaCl, 0.05% Tween 20). The blots were then probed with various rabbit polyclonal antibodies for C/EBP-α (Cell Applications, Inc), PPAR-γ (Cell Applications, Inc), SREBP1 (Santa Cruz BioTechnology), and β-actin (Santa Cruz BioTechnology) diluted according to the manufacturer's protocol in TBS at 4°C overnight, and incubated with 1:1,000 dilutions of goat anti-rabbit IgG secondary antibody coupled with horseradish peroxidase. The blots were developed using the ECL method (Amersham). For re-probing, the blots were incubated in the stripping buffer (100 mM 2-mercaptoethanol, 2% SDS, 62.5 mM Tris–HCl [pH 6.7]) at 50°C for 30 min with occasional agitation.

21.2.6   Statistical Analysis

The in vitro experimental data are expressed as the mean ± standard error of the mean (SEM) of quadruplicate samples. Differences between groups were compared with the Mann–Whitney test. Prism software 4 (Graphpad Software, San Diego, CA) was used for statistical analysis and graphing. Differences were considered significant at $P < 0.05$.

21.3   Results

21.3.1   Effect of TauCl on Differentiation of Preadipocytes into Adipocytes

To study the physiological effects of TauCl on human adipocyte differentiation, preadipocytes were cultured under differentiation conditions for 14 days in the presence or the absence of TauCl. As shown in Fig. 21.1a, the preadipocytes differentiated into adipocytes in the absence of TauCl, and differentiation was inhibited by TauCl in a dose-dependent manner. In addition, the differentiated adipocytes showed intracellular lipid accumulation. The accumulated lipid droplets were examined by oil red O staining. In accordance with the degree of differentiation, the degree of oil red O staining was dose dependently decreased by TauCl (Fig. 21.1b, c). Fat droplet formation was almost completely blocked by treatment with 600 μM TauCl. To test the cytotoxic effects of TauCl on preadipocytes, MTT assays were conducted after cells were treated with 1 mM TauCl for 14 days. Cell viability was not affected by TauCl at a concentration of 1 mM, suggesting that the inhibitory effect of TauCl on cell differentiation and lipid accumulation is not due to cytotoxicity (data not shown).

21.3.2   Effect of TauCl on Expression of Adipocyte-Specific Proteins

To determine whether the inhibition of cell differentiation and fat accumulation resulted from TauCl-mediated alterations in the differentiation program, we examined the expression of a number of adipogenic proteins by Western blot analysis.

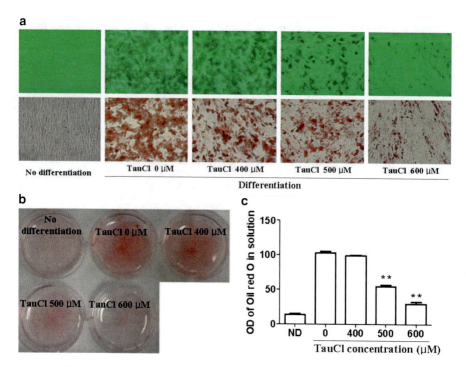

Fig. 21.1 Effect of TauCl on differentiation of preadipocytes into adipocytes. Human preadipocytes were seeded into six-well plates and cultured until confluent. For differentiation, the culture medium was changed to adipocyte differentiation medium and cells were cultured for 2 weeks while changing the medium every 2 days with or without TauCl at different concentrations. (**a**) Microscopic image of differentiated adipocytes before (*top row*) and after (*bottom row*) oil red O staining. (**b**) Image of differentiated adipocytes in six-well plate in the absence or the presence of different concentrations of TauCl. (**c**) Optical absorbance at 500 nm of the dye retained in adipocytes. Three independent experiments were performed in triplicate. The data shown are representative of three independent experiments, and similar results were obtained from all three. Values are expressed as mean ± standard error of the mean (SEM). **$P < 0.01$ versus no treatment with TauCl

As shown in Fig. 21.2a, TauCl treatment reduced the protein levels of the major adipogenic transcription factors PPAR-γ and C/EBP-α in preadipocytes. TauCl treatment at a concentration of 600 μM significantly reduced the protein levels of the major adipogenic transcription factors PPAR-γ and C/EBP-α. In addition, SREBP1, the upstream regulator of these transcription factors, was also downregulated by TauCl treatment in a dose-dependent manner. The effect of TauCl on these factors was specific, because the levels of β-actin were unaffected. Furthermore, the expression of GLUT4, a protein specific to adipocytes, was significantly reduced in the presence of TauCl in a dose-dependent manner (Fig. 21.2b). These data suggest that TauCl downregulates the expression of SREBP1 and the subsequent expression of PPAR-γ and C/EBP-α, resulting in inhibition of adipocyte differentiation.

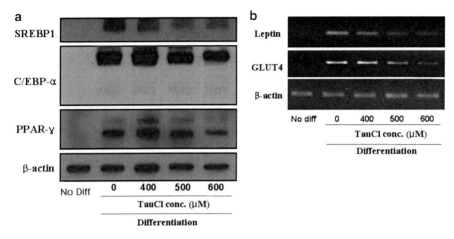

Fig. 21.2 Inhibition of adipogenic gene expression by TauCl. Human preadipocytes were cultured for 14 days in the absence or the presence of TauCl as described in Fig. 21.1. The cells were prepared for (**a**) Western blot analysis and (**b**) semiquantitative RT-PCR. The data shown are representative of three independent experiments, and similar results were obtained from all three

21.3.3   Effect of TauCl on Fat Accumulation and Adipocyte Dedifferentiation

To check if TauCl inhibits fat accumulation in differentiated adipocytes and induces the dedifferentiation of adipocytes into preadipocytes, TauCl was added to fully differentiated adipocytes and cultured for 14 days. As shown in Fig. 21.3, the degree of adipogenesis did not change; furthermore, oil red O staining showed that TauCl did not inhibit intracellular fat accumulation (Fig. 21.3a). Expression of the adipocyte-specific proteins PPAR-γ, C/EBP-α, SREBP1, and GLUT4 was also not affected, even at a concentration of 600 μM TauCl (Fig. 21.3b, c). These results suggest that TauCl does not lead to the dedifferentiation of adipocytes and does not affect fat accumulation in adipocytes.

21.4   Discussion

Adipose tissue, once viewed as simply a storage and release depot for lipids, is now considered to be an endocrine tissue (Halberg et al. 2008; Ronti et al. 2006). Adipose tissue secretes pro- and anti-inflammatory adipokines such as adiponectin, leptin, visfatin, IL-6, and IL-8, which play critical roles in many aspects of the metabolic syndrome (Itoh et al. 2011). Thus, obesity is an important topic in the realm of public health and preventive medicine, since it is considered to be a risk factor associated with the genesis or development of various diseases, including coronary heart

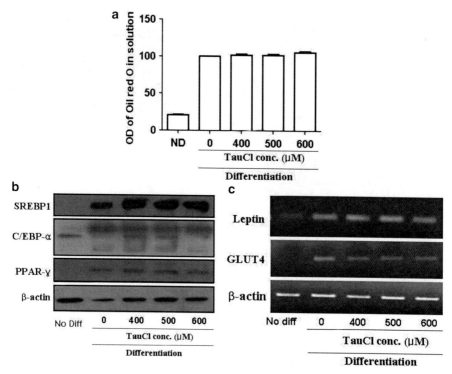

Fig. 21.3 Effect of TauCl on fat accumulation and adipocyte dedifferentiation. TauCl was added to differentiated adipocytes, which were then further incubated in starvation medium for 14 days (medium was changed after 7 days of culture in the absence or the presence of different concentrations of TauCl). (**a**) Dye retained in adipocytes was measured by optical absorbance at 500 nm. (**b**) Western blot analysis and (**c**) semiquantitative RT-PCR of adipogenic gene expression. The data shown are representative of three independent experiments, and similar results were obtained from all three

disease, hypertension, type 2 diabetes mellitus, cancer, respiratory complications, and osteoarthritis (Shehzad et al. 2011; Sowers and Karvonen-Gutierrez 2010). Increases in fat tissue result in part from an imbalance between energy intake and energy expenditure. It is also a result of increased adipogenesis, which includes adipocyte differentiation. Adipogenesis can occur even in adults, as observed in severe human obesity and rodents fed a high-carbohydrate or a high-fat diet (Kirkland et al. 1990). Conversely, when energy intake is less than output, mobilization of triglycerides leads to a decrease in adipose tissue mass.

There is increasing interest in finding safe and effective antiobesity agents for long-term use. Thus, many research groups have focused on developing antiobesity agents from natural products such as plants (Cho et al. 2008). However, the in vitro antiobesity effects of biomaterials extracted from plants and natural products are sometimes clinically negligible in humans (Gades and Stern 2005; Ho et al. 2001).

Obesity is now viewed as a state of systemic, chronic low-grade inflammation, which directly promotes systemic insulin resistance (Apovian et al. 2008; Itoh et al. 2011), which has led to increased interest in anti-inflammatory agents as a means of treating obesity. Some molecules, such as resolvins and protectins, are endogenously generated during the process of inflammation, and are then involved in terminating acute inflammation so that the local tissues can return to homeostasis (Ariel and Serhan 2007).

TauCl is also endogenously produced during the inflammation process, in which it plays an important role (Schuller-Levis and Park 2004). We thought that some of the endogenous by-products produced by the inflammation process might also be involved in the homeostasis of adipose tissue. Thus, we investigated whether TauCl affects the differentiation of adipocytes and the expression of adipokines, which may play important roles in energy metabolism and inflammation. TauCl inhibited human preadipocyte differentiation into adipocytes and intracellular lipid accumulation in a dose-dependent manner without cytotoxicity. However, taurine alone at concentrations of 10–100 mM did not affect adipogenesis (data not shown), suggesting that TauCl has a specific inhibitory effect on the differentiation of preadipocytes into adipocytes. Future studies seeking to develop TauCl as a therapeutic agent to treat obesity-related diseases or disorders should employ animal models with obesity-related diseases. In addition, it will be necessary to develop taurine derivatives that are more stable and effective than TauCl, which is unstable and easily degraded at room temperature. TauCl derivatives should be safer for long-term use than extracts from natural materials, because TauCl is an endogenously generated molecule.

Some concerns should be raised regarding the anti-adipogenic effects of TauCl when used as an antiobesity drug. Adipose tissue comprises approximately 50% adipocytes and 50% other cell types, including preadipocytes; vascular, neural, and immune cells; and leucocytes (Compher and Badellino 2008). The number of adipoctyes is determined during early adulthood, and changes in fat mass are attributed to changes in adipocyte cell size (Spalding et al. 2008). Thus, most adult-onset obesity appears to be related to the hypertrophy of adipocytes; in addition to insulin resistance, enlarged adipocytes appear to be the factor most closely correlated with obesity (Bjorntorp et al. 1971; Weyer et al. 2000). Therefore, TauCl, which had in vitro anti-adipogenic effects in the present study, may not have any therapeutic effects on obesity-related diseases in adult humans. However, the adipocyte turnover rate in humans was recently established to be about 10% per year (Spalding et al. 2008). If endogenous anti-adipogenic molecules such as TauCl could be taken for long periods of time without side effects, they might reduce the number of adipocytes, thereby leading to loss of fat mass.

Adipose tissue also serves as a source of adipokines and cytokines, which have both local and systemic actions in health and disease. The adipocytes, preadipocytes, and macrophages within adipose tissue secrete a wide range of hormones and cytokines, including IL-6, IL-8, IL-1β, and monocyte chemoattractant protein (MCP-1). It is now evident that many of these adipokines have the ability to influence other tissues, such as liver and muscle. For example, IL-6 promotes inflammation not only in adipose tissue but also in endothelial and liver cells

(Klover et al. 2005). IL-6 also promotes insulin resistance by interfering with insulin signaling in adipose tissue (Rotter et al. 2003). In general, pro-inflammatory cytokines have been demonstrated to prevent in vitro adipocyte differentiation from preadipocytes and to enhance lipolytic activity in adipocytes, leading to the so-called dedifferentiation (Coppack 2001; Gregoire et al. 1998).

21.5 Conclusion

TauCl, which is endogenously produced by immune cells such as macrophages that have infiltrated adipose tissue, may inhibit the differentiation of preadipocytes into adipocytes and modulate the expression of adipokines in adipocytes. Thus, TauCl could potentially be developed into a safe drug therapy for obesity-related diseases.

Acknowledgments This research was supported by Basic Science Research Program through the National Research Foundation of Korea (NRF) funded by the Ministry of Education, Science and Technology (2010-0024089 and 2011-0009061).

References

Apovian CM, Bigornia S, Mott M, Meyers MR, Ulloor J, Gagua M, McDonnell M, Hess D, Joseph L, Gokce N (2008) Adipose macrophage infiltration is associated with insulin resistance and vascular endothelial dysfunction in obese subjects. Arterioscler Thromb Vasc Biol 28: 1654–1659

Ariel A, Serhan CN (2007) Resolvins and protectins in the termination program of acute inflammation. Trends Immunol 28:176–183

Bjorntorp P, Berchtold P, Tibblin G (1971) Insulin secretion in relation to adipose tissue in men. Diabetes 20:65–70

Cho EJ, Rahman MA, Kim SW, Baek YM, Hwang HJ, Oh JY, Hwang HS, Lee SH, Yun JW (2008) Chitosan oligosaccharides inhibit adipogenesis in 3T3-L1 adipocytes. J Microbiol Biotechnol 18:80–87

Choi HM, Lee YA, Lee SH, Hong SJ, Hahm DH, Choi SY, Yang HI, Yoo MC, Kim KS (2009) Adiponectin may contribute to synovitis and joint destruction in rheumatoid arthritis by stimulating vascular endothelial growth factor, matrix metalloproteinase-1, and matrix metalloproteinase-13 expression in fibroblast-like synoviocytes more than proinflammatory mediators. Arthritis Res Ther 11:R161

Compher C, Badellino KO (2008) Obesity and inflammation: lessons from bariatric surgery. JPEN J Parenter Enteral Nutr 32:645–647

Coppack SW (2001) Pro-inflammatory cytokines and adipose tissue. Proc Nutr Soc 60:349–356

Gades MD, Stern JS (2005) Chitosan supplementation and fat absorption in men and women. J Am Diet Assoc 105:72–77

Gesta S, Tseng YH, Kahn CR (2007) Developmental origin of fat: tracking obesity to its source. Cell 131:242–256

Gregoire FM, Smas CM, Sul HS (1998) Understanding adipocyte differentiation. Physiol Rev 78:783–809

Halberg N, Wernstedt-Asterholm I, Scherer PE (2008) The adipocyte as an endocrine cell. Endocrinol Metab Clin North Am 37:753–768, x–xi

Ho SC, Tai ES, Eng PH, Tan CE, Fok AC (2001) In the absence of dietary surveillance, chitosan does not reduce plasma lipids or obesity in hypercholesterolaemic obese Asian subjects. Singapore Med J 42:6–10

Ito T, Schaffer SW, Azuma J (2011) The potential usefulness of taurine on diabetes mellitus and its complications. Amino Acids 42(5):1529–1539

Itoh M, Suganami T, Hachiya R, Ogawa Y (2011) Adipose tissue remodeling as homeostatic inflammation. Int J Inflam 2011:720926

Jessen BA, Stevens GJ (2002) Expression profiling during adipocyte differentiation of 3T3-L1 fibroblasts. Gene 299:95–100

Kim KS, Park EK, Ju SM, Jung HS, Bang JS, Kim C, Lee YA, Hong SJ, Lee SH, Yang HI, Yoo MC (2007) Taurine chloramine differentially inhibits matrix metalloproteinase 1 and 13 synthesis in interleukin-1beta stimulated fibroblast-like synoviocytes. Arthritis Res Ther 9:R80

Kirkland JL, Hollenberg CH, Gillon WS (1990) Age, anatomic site, and the replication and differentiation of adipocyte precursors. Am J Physiol 258:C206–C210

Klover PJ, Clementi AH, Mooney RA (2005) Interleukin-6 depletion selectively improves hepatic insulin action in obesity. Endocrinology 146:3417–3427

MacDougald OA, Mandrup S (2002) Adipogenesis: forces that tip the scales. Trends Endocrinol Metab 13:5–11

Ramirez-Zacarias JL, Castro-Munozledo F, Kuri-Harcuch W (1992) Quantitation of adipose conversion and triglycerides by staining intracytoplasmic lipids with Oil red O. Histochemistry 97:493–497

Ronti T, Lupattelli G, Mannarino E (2006) The endocrine function of adipose tissue: an update. Clin Endocrinol (Oxf) 64:355–365

Rosen ED, Hsu CH, Wang X, Sakai S, Freeman MW, Gonzalez FJ, Spiegelman BM (2002) C/EBPalpha induces adipogenesis through PPARgamma: a unified pathway. Genes Dev 16:22–26

Rosen ED, MacDougald OA (2006) Adipocyte differentiation from the inside out. Nat Rev Mol Cell Biol 7:885–896

Rotter V, Nagaev I, Smith U (2003) Interleukin-6 (IL-6) induces insulin resistance in 3T3-L1 adipocytes and is, like IL-8 and tumor necrosis factor-alpha, overexpressed in human fat cells from insulin-resistant subjects. J Biol Chem 278:45777–45784

Roufosse CA, Direkze NC, Otto WR, Wright NA (2004) Circulating mesenchymal stem cells. Int J Biochem Cell Biol 36:585–597

Schuller-Levis GB, Park E (2004) Taurine and its chloramine: modulators of immunity. Neurochem Res 29:117–126

Shehzad A, Ha T, Subhan F, Lee YS (2011) New mechanisms and the anti-inflammatory role of curcumin in obesity and obesity-related metabolic diseases. Eur J Nutr 50:151–161

Sowers MR, Karvonen-Gutierrez CA (2010) The evolving role of obesity in knee osteoarthritis. Curr Opin Rheumatol 22:533–537

Spalding KL, Arner E, Westermark PO, Bernard S, Buchholz BA, Bergmann O, Blomqvist L, Hoffstedt J, Naslund E, Britton T, Concha H, Hassan M, Ryden M, Frisen J, Arner P (2008) Dynamics of fat cell turnover in humans. Nature 453:783–787

Szymanski K, Winiarska K (2008) [Taurine and its potential therapeutic application]. Postepy Hig Med Dosw (Online) 62:75–86

Thomas EL, Grisham MB, Melton DF, Jefferson MM (1985) Evidence for a role of taurine in the in vitro oxidative toxicity of neutrophils toward erythrocytes. J Biol Chem 260:3321–3329

Tsuboyama-Kasaoka N, Shozawa C, Sano K, Kamei Y, Kasaoka S, Hosokawa Y, Ezaki O (2006) Taurine (2-aminoethanesulfonic acid) deficiency creates a vicious circle promoting obesity. Endocrinology 147:3276–3284

Weyer C, Foley JE, Bogardus C, Tataranni PA, Pratley RE (2000) Enlarged subcutaneous abdominal adipocyte size, but not obesity itself, predicts type II diabetes independent of insulin resistance. Diabetologia 43:1498–1506

Chapter 22
Taurine Chloramine Administered In Vivo Increases NRF2-Regulated Antioxidant Enzyme Expression in Murine Peritoneal Macrophages

In Soon Kang and Chaekyun Kim

Abstract Taurine chloramine (TauCl) is produced from taurine by the myeloperoxidase-halide system in activated neutrophils via a stoichiometric reaction between taurine and HOCl. TauCl has been shown to provide cytoprotection against inflammatory tissue injury by inhibiting the overproduction of inflammatory mediators and also by increasing the expression of antioxidant enzymes that are regulated by nuclear factor E2-related factor 2 in murine macrophages. In this study, primary murine macrophages were prepared after either by injection of 3% thioglycolate into mouse peritoneal cavity or by differentiation of the isolated bone marrow cells. TauCl increased HO-1, Prx-1, and Trx-1 expression in murine primary macrophages. Also, when TauCl was injected in combination with 3% thioglycolate, HO-1 expression in the peritoneal macrophages was increased. Our results suggest that TauCl plays a protective role against cytotoxicity of oxidative stress in macrophages by increasing the expression of antioxidant enzymes in vivo.

Abbreviations

ARE	Antioxidant response element
BMDM	Bone marrow-derived macrophages
GPx	Glutathione peroxidase
HO	Heme oxygenase
Keap	Kelch-like ECH-associated protein
LPS	Lipopolysaccharide

I.S. Kang • C. Kim (✉)
Laboratory for Leukocyte Signaling Research, Department of Pharmacology and Toxicology,
Inha Research Institute for Medical Science, Inha University School of Medicine,
Incheon, Korea
e-mail: chaekyun@inha.ac.kr

A. El Idrissi and W.J. L'Amoreaux (eds.), *Taurine 8*, Advances in Experimental
Medicine and Biology 775, DOI 10.1007/978-1-4614-6130-2_22,
© Springer Science+Business Media New York 2013

M-CSF   Macrophage colony stimulating factor
MMP   Metalloproteinase
MPO   Myeloperoxidase
Nrf2   Nuclear factor E2-related factor
Prx   Peroxiredoxin
TauCl   Taurine chloramine
TNF-α   Tumor necrosis factor-α
Trx   Thioredoxin

22.1 Introduction

Taurine chloramine (TauCl), a chlorine derivative of taurine (2-aminoethanesulfonic acid), is produced mainly by the myeloperoxidase (MPO)-halide system in activated neutrophils at inflammatory tissues. The production of TauCl upon removal of highly toxic hypochlorite (HOCl) reduces cellular toxicity and the TauCl released from apoptotic neutrophils possesses its own anti-inflammatory and antioxidant activities in the inflamed tissues. TauCl inhibits the production of inflammatory mediators such as superoxide anion (O_2^-), nitric oxide (NO), tumor necrosis factor-α (TNF-α), interleukins, prostaglandins and proteolytic enzymes including metalloproteinase (MMP)-1 and 13 (Park et al. 1995; Kim et al. 1996; Marcinkiewicz et al. 1995, 1999; Kim and Kim 2005; Kim et al. 2007, 2010b). Furthermore, TauCl increases the expression of antioxidant enzymes, including heme oxygenase 1 (HO-1), peroxiredoxin (Prx), thioredoxin (Trx), glutathione peroxidase (GPx), and catalase, and provides the cytoprotective antioxidant effect against the toxicity of inflammatory mediators given in vitro (Olszanecki and Marcinkiewicz 2004; Jang et al. 2009; Kim et al. 2010a). The anti-inflammatory and antioxidant effects of TauCl allow inhibition of cell death and prevention of chronic inflammatory diseases in mice (Jang et al. 2009; Piao et al. 2011; Wang et al. 2011).

HO-1, Prx, Trx, and GPx are antioxidants whose expressions are upregulated by a redox-sensitive transcription factor, nuclear factor E2-related factor (Nrf2). Under normal state, Nrf2 binds to Kelch-like ECH-associated protein 1 (Keap1) in the cytoplasm. Upon oxidative stimulation, Nrf2 dissociates from Keap1 and translocates into the nucleus, binds to antioxidant response element (ARE) in the nucleus and stimulates the transcription of many oxidative stress-inducible antioxidant enzyme genes including, *ho-1*, *prx*, *trx*, and *gpx*. In RAW 264.7 cells treated with TauCl, the intracellular GSH level at early time points decreases, and this GSH depletion changes cellular redox balance toward oxidative stress and subsequently inactivates Keap1 but activates Nrf2. Thus, TauCl promotes the translocation of activated Nrf2 into the nucleus and its binding to ARE (Jang et al. 2009; Kim et al. 2010a).

HO catalyzes the degradation of free-heme to yield ferrous iron, carbon monoxide (CO), and biliverdin, which is converted to a strong antioxidant bilirubin (Tenhunen et al. 1968). Among the three HO isoforms, HO-1 is inducible and HO activity is increased upon exposure to free heme released from heme enzymes by oxidative stress. Among the products of HO activity, CO itself has anti-inflammatory and cytoprotective effects, and bilirubin serves as a potent intracellular antioxidant; thus, increase of HO expression protects cells from oxidative damage and prevents cell death (Otterbein et al. 2000). In this study, in an effort to support the observations that have been made in vitro, we investigated whether TauCl administered in vivo induces the Nrf2-regulated antioxidants in primary macrophages isolated from mouse peritoneum.

22.2 Methods

22.2.1 Antibodies and Reagents

Antibodies against Nrf2 (Santa Cruz Biotechnology, Santa Cruz, CA), Keap1 (Santa Cruz), Prx-1 (AbFrontier, Seoul, Korea), Trx-1 (AbFrontier), and HO-1 (Enzo, Farmingdale, NY) were purchased from commercial sources. Dulbecco's modified Eagle's medium (DMEM), α-minimal essential medium (α-MEM), fetal bovine serum (FBS), phosphate-buffered saline (PBS), penicillin, and streptomycin were purchased from HyClone (Logan, UT). Other routinely used chemicals were purchased from Sigma (St. Louis, MO) unless otherwise stated. TauCl was synthesized freshly on the day of use by adding 400 mM NaOCl (Aldrich, Milwaukee, MI) to equimolar amounts of taurine. The authenticity of TauCl formation was monitored by its UV absorption at 200–400 nm (Thomas et al. 1986).

22.2.2 Preparation of Murine Peritoneal Macrophages

Murine macrophages were obtained from C57BL/6 mice (Jackson Laboratories, Bar Harbor, MA) that have been injected with 1 ml of 3% sterile thioglycolate broth intraperitoneally. After 4 days, the peritoneal exudate cells were harvested by washing the peritoneal cavity with ice-cold PBS. The cells were incubated in DMEM supplemented with 10% FBS, 100 U/ml penicillin, and 100 μg/ml streptomycin in 5% CO_2 at 37°C. On the next day, adherent cells were removed and suspended cells were cultured for 2 more days. Murine peritoneal macrophages were treated in vitro with 1 μg/ml lipopolysaccharide (LPS) or TauCl (0.1, 0.2, 0.5, and 0.7 mM) for 24 h. To see the in vivo effect of TauCl on peritoneal macrophages, various concentrations of TauCl was injected together with thioglycolate broth prior to the isolation of primary macrophages.

22.2.3 Preparation of Bone Marrow-Derived Macrophages

Murine bone marrow-derived macrophages (BMDM) were prepared as described previously (Yamauchi et al. 2004). Briefly, bone marrow cells were isolated from femurs and tibias flushed with α-MEM using a 22-gauge needle (KOVAX, Seoul, Korea). The cells were pelleted by centrifugation at $500 \times g$ for 10 min at 4°C and the RBC was lysed. Following lysis of RBC, bone marrow cells were resuspended in DMEM containing 20% FBS, penicillin, streptomycin, and 5 ng/ml macrophage colony stimulating factor (M-CSF) and cultured in 5% CO_2 at 37°C. After discarding the adherent cells, suspended cells were cultured in DMEM containing 20% FBS and 30 ng/ml M-CSF. Every 3 days, the medium was replaced with fresh medium. At 10–12 days after harvesting the bone marrow cells, the adherent bone marrow-derived macrophages were collected.

22.2.4 Western Blot Analysis

Cell were lysed in ice-cold lysis buffer containing 50 mM Tris–HCl (pH 7.4), 150 mM NaCl, 1% NP-40, 1 mM EDTA, 0.25% Na-deoxycholate, 0.1% sodium dodecyl sulfate (SDS), 10 μM leupeptin, 20 μg/ml chymostatin, and 2 mM phenyl-methylsulfonyl fluoride. Protein concentration in the lysates was quantified using the bicinchoninic acid protein assay kit (Pierce, Rockford, IL). Equal amounts of lysate proteins were mixed with 5× Laemmli sample buffer and subjected to 12% SDS-polyacrylamide gel electrophoresis. Separated proteins were transferred onto polyvinylidene fluoride membrane (Bio-Rad, Hercules, CA) and immersed in TBST buffer (10 mM Tris–HCl, pH 6.8, 150 mM NaCl, and 0.05% Tween 20) containing 6% skim milk to block nonspecific binding. Then, the separated protein bands were detected using the ECL method (Pierce).

22.2.5 Statistical Analysis

Statistical significance of differences was determined by two-tailed Student's t-test. Each value was expressed as the mean $\pm$ SD. Differences were considered statistically significant when the calculated p value was less than 0.05.

22.3   Results

22.3.1   TauCl Treatment Increases Cellular Content of Antioxidant Enzymes in Murine Peritoneal Macrophages and BMDM

Antioxidant proteins like HO-1, Prx, and Trx protect cells from oxidative cytotoxicity caused by reactive oxygen species such as O_2^-, H_2O_2, and HOCl. To examine the effect of TauCl on the expression of HO-1, Prx-1, and Trx-1 in the isolated peritoneal macrophages from C57BL/6 mice, they were treated with various concentrations of TauCl (0.2, 0.5, and 0.7 mM) for 24 h. HO-1, Prx-1, and Trx-1 expressions in peritoneal macrophage were increased by TauCl treatment (Fig. 22.1a, b).

The transcription factor Nrf2 plays a central role in enhancing the expression of many cytoprotective antioxidant enzyme genes including *ho-1*, *prx*, and *trx* that are induced in response to oxidative stress and electrophiles. Nrf2 is associated with Keap1 and undergoes continuous degradation in the cytosol, and thus Keap1 is essential for the regulation of Nrf2 activity. To explore the effect of TauCl on cellular content of Nrf2 and Keap1 in peritoneal macrophages, cells were treated with various concentrations of TauCl (0.1, 0.2, and 0.5 mM) for 24 h. The cellular content of Nrf2 and Keap1 was not altered by the TauCl treatment (Fig. 22.1c). This result was consistent with our previous results obtained with RAW 264.7 cells treated with TauCl (Kim et al. 2010a). Thus, TauCl induces the expression of HO-1, Prx-1, and Trx-1 in murine peritoneal macrophages without affecting cellular content of Nrf2 and Keap1.

To examine the effect of TauCl on bone marrow-derived macrophages, BMDM were treated with TauCl for 12 h and HO-1 expression was determined. HO-1 expression was increased by TauCl treatment in a dose-dependent manner, and the increase was greater than that induced by LPS treatment (Fig. 22.2).

22.3.2   TauCl Injection Increases HO-1 Expression in Peritoneal Macrophages

To explore the in vivo effect of TauCl on peritoneal macrophages, various concentrations of TauCl (2, 5, and 10 mM) were injected into mouse peritoneal cavity in combination with 3% thioglycolate broth. The peritoneal macrophages were collected at 4 days after the TauCl injection and cell lysates were prepared. HO-1 expression in peritoneal cells obtained from TauCl-injected mice (Fig. 22.3) was increased showing that TauCl administration induces the expression of antioxidant enzymes in the peritoneal macrophages.

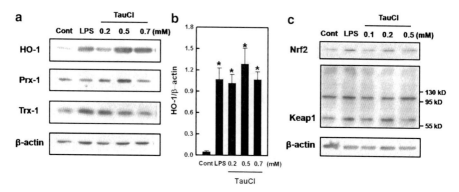

Fig. 22.1 Effect of TauCl on Nrf2-regulated antioxidant enzyme expression in murine peritoneal macrophages. (**a**) Murine peritoneal macrophages were incubated with 1 μg/ml LPS and 0.2, 0.5, and 0.7 mM TauCl for 24 h, and the expression of antioxidant enzyme was analyzed by western blotting ($n=4$). (**b**) Bar graphs represent the relative level of HO-1 expression. Data were expressed as the mean ± SD, *$p<0.05$ ($n=4$ except 0.7 mM TauCl for which $n=2$). (**c**) Expression of Nrf2 and Keap1 were determined at 24 h after TauCl treatment ($n=2$). β-Actin was used as an internal control

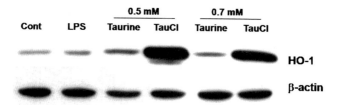

Fig. 22.2 TauCl induces HO-1 expression in murine bone marrow-derived macrophages. BMDM were incubated with 1 μg/ml LPS or 0.5 and 0.7 mM TauCl for 12 h, and HO-1 expression was analyzed ($n=3$)

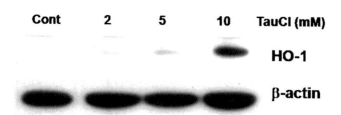

Fig. 22.3 TauCl injection induces HO-1 expression in murine peritoneal macrophages. TauCl was injected into mouse peritoneal cavity in combination with 3% thioglycolate broth. Four days after TauCl injection, peritoneal macrophages were collected and HO-1 expression was analyzed ($n=3$)

22.4 Discussion

Upon oxidative stress, heme is released from heme enzymes and the free heme mediates Fenton reaction producing hydroxyl radical ($\bullet$OH) that cause free radical toxicity. Thus, to remove the released free heme, HO-1 is induced and the products of increased HO activity (CO and bilirubin) bring about anti-inflammatory and cytoprotective effects. In support, CO has been shown to inhibit the production of prostaglandin E2, O_2^- and NO in neutrophils and macrophages (Colville-Nash et al. 1998; Srisook et al. 2005). Thus, induction of HO-1 that occurs in response to oxidative stress is essential to maintain cellular homeostasis against the cytotoxicity of oxidative stress.

It has been reported that TauCl increases HO-1 expression in both the non-activated and LPS-activated J774.2 and RAW 264.7 cells (Olszanecki and Marcinkiewicz 2004; Jang et al. 2009; Kim et al. 2010a). However, it has always been questioned whether the in vitro response shown in cell lines represents the response in vivo. Thus, we assessed whether the antioxidant enzyme expression is increased in response to TauCl using primary macrophages. TauCl induced antioxidant enzyme expression in peritoneal macrophages and BMDM isolated from mouse as similarly with J774 cells and RAW 264.7 cells. Alternatively, in vivo administration of TauCl into the peritoneum of thioglycolate-treated mouse also induced HO-1 expression in peritoneal macrophages. These results show that TauCl increases antioxidant enzyme expression both in vitro and in vivo, and explains the underlying reason by which TauCl ameliorates the collagen-induced arthritis in mice (Wang et al. 2011). The TauCl concentration that was required for in vivo induction of HO-1 expression in peritoneum was as high as 5 mM (Fig. 22.3) while its concentration needed for the macrophages in vitro was as low as 0.2 mM (Fig. 22.1a). Interestingly, there was no detectable cytotoxicity even at 10 mM TauCl injection. We tried to follow the TauCl concentration following the intraperitoneal injection. At this time, there is no available method to measure the in vivo concentration of TauCl.

We have shown that the TauCl-induced HO-1 expression is dependent on the nuclear translocation and binding of Nrf2 to ARE (Kim et al. 2010a; Piao et al. 2011). Although the precise mechanism by which TauCl activates Nrf2 is not clear for the moment, the TauCl-induced initial decrease in cellular GSH level and the resulting oxidative stress may have promoted inactivation of Keap1 that allows Nrf2 activation, its translocation into nucleus and subsequent binding to ARE (Jang et al. 2009; Kim et al. 2010a). In addition, Nrf2 knockdown in macrophages using Nrf2-specific siRNA attenuates the extent of TauCl inducible HO-1 expression (Kim et al. 2010a). These results suggest that TauCl produced in vivo by the neutrophils at the inflammation site can induce the expression of antioxidant enzymes in macrophages and that in turn protect surrounding cells and tissues from the oxidative damage caused by inflammatory mediators produced and secreted by activated macrophages.

Although the precise mechanisms underlying anti-inflammatory and cytoprotective effects of TauCl still need further investigation, TauCl produced in the activated neutrophils by MPO system appears to protect surrounding cells from inflammatory and oxidative injuries by three mechanisms. First, the generation of TauCl removes highly cytotoxic hypochlorite. Second, TauCl inhibits the production of pro-inflammatory mediators such as O_2^-, NO, TNF-α, interleukin-6 and -8, prostaglandins, and MMPs. The inhibition of NO, TNF-α, prostaglandins and MMPs production resulted from the inhibition of nuclear factor-kappa B activation, a main transcription factor for production of inflammatory cytokines and mediators (Barua et al. 2001; Kanayama et al. 2002; Kontny et al. 2000; Kim and Kim 2005). Third, TauCl diminishes oxidative stress by increasing the expression of antioxidant enzymes such as the Nrf2-dependent HO-1, Prx, GPx, and Trx and also the Nrf2-independent catalase (Olszanecki and Marcinkiewicz 2004; Jang et al. 2009; Kim et al. 2010a).

22.5   Conclusion

In summary, TauCl increases the expression of antioxidant enzymes including HO-1 in peritoneal macrophages and BMDM. This shows that TauCl produced at the inflammatory site may protect macrophages and surrounding tissues from cytotoxicity caused by inflammatory mediators by increasing antioxidants as well as by decreasing the production and secretion of pro-inflammatory mediators in vivo.

Acknowledgments We thank Dr. Young-Nam Cha for discussions throughout the study and critical comments on the manuscript, and Mi Ran Cho for the technical support. This work was supported by the NRF of Korea grant funded by the Korea government MEST (2012R1A1A3007097) and Inha University research grant.

References

Barua M, Liu Y, Quinn MR (2001) Taurine chloramine inhibits inducible nitric oxide synthase and TNF-alpha gene expression in activated alveolar macrophages: decreased NF-kappaB activation and IkappaB kinase activity. J Immunol 167:2275–2281
Colville-Nash PR, Qureshi SS, Willis D, Willoughby DA (1998) Inhibition of inducible nitric oxide synthase by peroxisome proliferator-activated receptor agonists: correlation with induction of heme oxygenase 1. J Immunol 161:978–984
Jang JS, Piao SY, Cha Y-N, Kim C (2009) Taurine chloramine activates Nrf2, increases HO-1 expression and protects cells from death caused by hydrogen peroxide. J Clin Biochem Nutr 45:37–43
Kanayama A, Inoue J, Sugita-Konishi Y, Shimizu M, Miyamoto Y (2002) Oxidation of Ikappa Balpha at methionine 45 is one cause of taurine chloramine-induced inhibition of NF-kappa B activation. J Biol Chem 277:24049–24056
Kim C, Jang JS, Cho MR, Agarawal SA, Cha Y-N (2010a) Taurine chloramine induces heme oxygenase-1 expression via Nrf2 activation in murine macrophages. Int Immunopharmacol 10: 440–446

Kim C, Park E, Quinn MR, Schuller-Levis G (1996) The production of superoxide anion and nitric oxide by cultured murine leukocytes and the accumulation of TNF-alpha in the conditioned media is inhibited by taurine chloramine. Immunopharmacology 34:89–95

Kim JW, Kim C (2005) Inhibition of LPS-induced NO production by taurine chloramine in macrophages is mediated though Ras-ERK-NF-kappaB. Biochem Pharmacol 70:1352–1360

Kim KS, Choi H-M, Oh DH, Kim C, Jeong JS, Yoo MC, Yang H-I (2010b) Effect of taurine chloramine on the production of matrix metalloproteinases (MMPs) in adiponectin- or IL-1β-stimulated fibroblast-like synoviocytes. J Biomed Sci 17(S1):S27

Kim KS, Park EK, Ju SM, Jung HS, Bang JS, Kim C, Lee YA, Hong SJ, Lee SH, Yang HI, Yoo MC (2007) Taurine chloramine differentially inhibits matrix metalloproteinase 1 and 13 synthesis in interleukin-1beta stimulated fibroblast-like synoviocytes. Arthritis Res Ther 9:R80

Kontny E, Szczepanska K, Kowalczewski J, Kurowska M, Janicka I, Marcinkiewicz J, Maslinski W (2000) The mechanism of taurine chloramine inhibition of cytokine (interleukin-6, interleukin-8) production by rheumatoid arthritis fibroblast-like synoviocytes. Arthritis Rheum 43: 2169–2177

Marcinkiewicz J, Grabowska A, Bereta J, Stelmaszynska T (1995) Taurine chloramine, a product of activated neutrophils, inhibits in vitro the generation of nitric oxide and other macrophage inflammatory mediators. J Leukoc Biol 58:667–674

Marcinkiewicz J, Nowak B, Grabowska A, Bobek M, Petrovska L, Chain B (1999) Regulation of murine dendritic cell functions in vitro by taurine chloramine, a major product of the neutrophil myeloperoxidase-halide system. Immunology 98:371–378

Olszanecki R, Marcinkiewicz J (2004) Taurine chloramine and taurine bromamine induce heme oxygenase-1 in resting and LPS-stimulated J774.2 macrophages. Amino Acids 27:29–35

Otterbein LE, Bach FH, Alam J, Soares M, Lu HT, Wysk M, Davis RJ, Flavell RA, Choi AMK (2000) Carbon monoxide has anti-inflammatory effects involving the mitogen-activated protein kinase pathway. Nat Med 6:422–428

Park E, Schuller-Levis G, Quinn MR (1995) Taurine chloramine inhibits production of nitric oxide and TNF-alpha in activated RAW 264.7 cells by mechanisms that involve transcriptional and translational events. J Immunol 154:4778–4784

Piao SY, Cha Y-N, Kim C (2011) Taurine chloramine protects RAW 264.7 macrophages from hydrogen peroxide-induced apoptosis by increasing antioxidants. J Clin Biochem Nutr 49:50–56

Srisook K, Kim C, Cha Y-N (2005) Cytotoxic and cytoprotective actions of O_2^- and NO (ONOO$^-$) are determined both by cellular GSH level and HO activity in macrophages. Methods Enzymol 396:414–424

Tenhunen R, Marver HS, Schmid R (1968) The enzymatic conversion of heme to bilirubin by microsomal heme oxygenase. Proc Natl Acad Sci U S A 61:748–755

Thomas EL, Grisham MB, Jefferson MM (1986) Preparation and characterization of chloramines. Methods Enzymol 132:569–585

Wang Y, Cha Y-N, Kim KS, Kim C (2011) Taurine chloramine inhibits osteoclastgenesis and lymphocyte proliferation in mice with collagen-induced arthritis. Eur J Pharmacol 668:325–330

Yamauchi A, Kim C, Li S, Marchal CC, Towe J, Atkinson SJ, Dinauer MC (2004) Rac2-deficient murine macrophages have selective defects in superoxide production and phagocytosis of opsonized particles. J Immunol 173:5971–5979

Chapter 23
Influence of Taurine Haloamines (TauCl and TauBr) on the Development of *Pseudomonas aeruginosa* Biofilm: A Preliminary Study

Janusz Marcinkiewicz, Magdalena Strus, Maria Walczewska, Agnieszka Machul, and Diana Mikołajczyk

Abstract Biofilms are consortia of microorganisms (sessile cells) that form on various surfaces including mucosal membranes or teeth. Bacterial biofilms cause many human infections such as chronic sinusitis, acne vulgaris, periodontal diseases, and chronic wounds. These infections are persistent as they show increased resistance to antibiotics and host defense system. Taurine chloramine (TauCl) and taurine bromamine (TauBr) are the physiological products of activated neutrophils, resulting from the reaction between taurine with hypochlorous acid (HOCl) and hypobromous acid (HOBr), respectively. It has been shown in vitro that taurine haloamines exert antimicrobial properties against various pathogenic bacteria. Moreover, clinical studies have shown that both haloamines are effective in the local treatment of skin and mucose infections, including biofilm-related infections. Nevertheless, it has been not tested yet whether they can kill bacteria hidden in biofilm or disrupt biofilm structure. In this study we have investigated the capacity of TauCl and TauBr to inhibit in vitro the formation of *P. aeruginosa* biofilm. We have also tested their ability to destroy the mature biofilm. Our results suggest that TauBr is able to inhibit in vitro the formation of *P. aeruginosa* biofilm but cannot destroy the mature biofilm and effectively killed hidden bacteria. In further studies, the combined effect of TauBr and DNase, one of suggested biofilm inhibitors, was tested. Together, we conclude that TauBr is a better than TauCl candidate for local therapy of biofilm-related infections. However, a combined therapy, an application of TauBr together with other anti-biofilm agents (e.g., DNase), seems to be more promising.

J. Marcinkiewicz (✉) • M. Walczewska
Department of Immunology, Jagiellonian University Medical College,
Cracow, Poland
e-mail: mmmarcin@cyf-kr.edu.pl

M. Strus • A. Machul • D. Mikołajczyk
Department of Microbiology, Jagiellonian University Medical College,
Cracow, Poland

A. El Idrissi and W.J. L'Amoreaux (eds.), *Taurine 8*, Advances in Experimental
Medicine and Biology 775, DOI 10.1007/978-1-4614-6130-2_23,
© Springer Science+Business Media New York 2013

Abbreviations

Tau	Taurine
TauCl	Taurine chloramine
TauBr	Taurine bromamine
P. aeruginosa	*Pseudomonas aeruginosa*
CFU	Colony forming units

23.1 Introduction

Taurine chloramine (TauCl) and taurine bromamine (TauBr), the main haloamines of neutrophil MPO-halide system, exert anti-inflammatory and microbicidal properties (Gaut et al. 2001; Marcinkiewicz et al. 2005; Nagl et al. 2000a, b). Their bactericidal, fungicidal, antiviral, and antiparasitic activities in vitro have been demonstrated in a number of papers (Nagl et al. 2000a, b, 2003; Marcinkiewicz et al. 2000, 2005; Yazdanbakhsh et al. 1987). Moreover, well-documented outstanding tolerability of TauCl allows the use of TauCl at a high concentration (1% aqueous solution) as a local antiseptic (Gottardi and Nagl 2010). TauBr, in contrast to TauCl, shows its microbicidal properties even at very low physiological concentrations (Marcinkiewicz et al. 2006). However, all these data are related to TauCl/TauBr antimicrobial activity against planktonic form of bacteria. On the other hand, a number of clinical studies have shown that both haloamines are effective in the local treatment of skin and mucosa infections, including biofilm-related infections (Marcinkiewicz et al. 2008; Gstöttner et al. 2003). Nevertheless, it has not been tested yet whether they can kill bacteria hidden in a biofilm or disrupt biofilm structure in such diseases.

Microbial biofilms are the most common mode of growth of bacteria and fungi in nature, especially on epithelial surfaces and represents a microbial survival strategy in an unfriendly environment (O'Toole et al. 2000; Sutherland 2001). Biofilm formation is a dynamic process. Within few hours bacteria adhere to the surface irreversibly, multiply and produce extracellular polymeric substances (EPS), components of the biofilm matrix. The composition of EPS varies depending upon the bacterial strain and the environmental conditions, but, in general EPS consists of exopolysacharides, proteins, and extracellular DNA (Sutherland 2001; Whitchurch et al. 2002). Recently, a tremendous interest has been observed in the role of biofilms in chronic infectious diseases and in the resistance of biofilms to antibiotics, disinfectant chemicals and to phagocytosis (Costerton 2002; Fux et al. 2005). Thus, it is reasonable to put forward a question whether TauCl and TauBr are able to kill bacteria hidden in a biofilm or/and to destroy a protective exopolymeric matrix of forming biofilms.

To address this issue, we tested the capacity of TauCl and TauBr to inhibit in vitro the formation of *Pseudomonas aeruginosa* biofilm. We have also tested their ability to destroy the mature biofilm and to kill biofilm-hidden bacteria (sessile cells). We have chosen *P. aeruginosa* as it has been widely studied as a model organism for biofilm formation (Leid et al. 2005; Ghafoor et al. 2011). Moreover, it is well documented, that *P. aeruginosa* chronic infections (skin wound infections, lung cystic fibrosis) are characterized by a biofilm formation and are resistant to antibiotic/antiseptic treatment and the host immune response (Bjarnsholt et al. 2008; Fazli et al. 2009). Therefore, such type of infections requires a novel therapeutic approach. Promising strategies may include the use of compounds that affect the biofilm structural integrity (e.g., DNase to destroy extracellular DNA, a key matrix component of *P. aeruginosa* biofilm), which will increase therapeutic efficacy of antimicrobial agents (e.g., antibiotics, taurine haloamines).

23.2   Methods

23.2.1   Chemicals

Bovine Pancreatic DNase I from Sigma Aldrich, *N*-chlorotaurine sodium salt (taurine chloramine, TauCl), a kind gift from Prof. Waldemar Gottardi and Prof. Marcus Nagl from the Division of Hygiene and Medical Microbiology, Innsbruck Medical University, Austria.

23.2.2   Preparation of Taurine Chloramine and Taurine Bromamine

TauCl (*N*-chlorotaurine sodium salt) as a crystalline sodium salt (molecular weight 181.57) was prepared as described previously (Gottardi and Nagl 2002). Each preparation of TauCl was monitored by UV absorption spectra ($\lambda = 200$–400 nm) to assure the authenticity of TauCl ($\lambda = 252$ nm) and the absence of dichloramine (TauCl$_2$) ($\lambda = 300$ nm) and unreacted HOCl/OCl$^-$ ($\lambda = 292$ nm). The concentration of synthesized TauCl was determined using the molar extinction coefficient w 429 M^{-1} cm^{-1} at A_{252}.

TauBr was prepared in a two-step procedure. First, NaOBr was synthesized in reaction between equimolar amounts of NaOCl and NaBr (POCH, Poland) in the PBS solution. In such conditions virtually all the OCl$^-$ present reacts with Br$^-$ to form OBr$^-$ and Cl$^-$. The presence and concentration of OBr$^-$ was confirmed by UV spectra ($\lambda = 200$–400 nm). In the second step, 20 mM NaOBr was added dropwise to equal volume of 400 mM taurine. UV absorption spectrum was checked to exclude the formation of taurine dibromamine or chloramines and to estimate the concentration of TauBr (molar extinction coefficient—430 M^{-1}cm^{-1} at A_{288}). Stock solution of TauCl and TauBr was kept at 4°C for a maximum period of 3 days before use.

23.2.3 Bacterial Cultures

All tests were performed in *Pseudomonas aeruginosa* strain KM/1 isolated from a sample taken from a patient with diabetic foot infection. The strain was propagated on 10 ml of Trypticase-Soy Broth (TSB, Difco) at 37°C for 24 h in aerobic conditions. Then the culture was centrifuged (2,000 rpm; 10 min) and washed with 10 ml of saline. Stock suspensions of the strain (1×10^6, 1×10^9 CFU/ml) were prepared by serial diluting of bacteria in saline using McFarland scale.

23.2.4 Growth Conditions and Measurement of Biofilm Formation by *P. aeruginosa*

P. aeruginosa biofilm was set up in sterile plastic 96-well plates with adherent surface (Greiner Bio-One, USA). Twenty microliter quantities of the bacterial suspensions, prepared as described above, were added to each well followed with 180 µl of sterile TSB. The final concentration of the bacteria was 1×10^8 CFU/ml. The plates were centrifuged for 10 min at 2,000 rpm to sediment bacteria on the bottom of each well and then incubated for 72 h (37°C, aerobic conditions). Biofilm quantity (total mass of bacterial polysaccharides) was measured using Congo red dye according to a modified procedure described by Reuter et al. (2010). Briefly, at different time points of the culture (0, 6, 18, 24, 48, and 72 h) the plates were centrifuged, the culture medium was gently removed from wells and immediately 200 µl of 0.1% Congo red solution was added. The plates were left for 30 min at room temperature and washed twice with buffered saline to remove unbound dye. Absorbance was measured at $\lambda = 492$ nm wavelength using spectrophotometer (Awareness Technology Inc.). All measurements were performed in triplicates and mean values ± SD are given.

23.2.5 Bactericidal Activity of the Tested Substances

Serial dilutions of TauBr (10–300 µM) and TauCl (300, 1,000 µM) in 1 ml saline were prepared. Then 100 µl of the bacterial suspensions were added to each tube containing graded concentration of the test substances to obtain final suspensions containing 1×10^5 and 1×10^8 CFU/ml of bacteria. The tubes were thoroughly mixed and incubated at 37°C. Samples of 100 µl were taken immediately after being mixed and after 0.5, 4, 8, and 24 h of the incubation, transferred and distributed on the surfaces of MacConkey agar (Difco) plates. The plates were further incubated for 24 h (37°C, aerobic conditions) and numbers of the grown colonies were counted. The results are shown in CFU/ml.

23.2.6 Effects of the Tested Agents on Biofilm Formation and on the Number of Viable Bacteria

The experiments were performed in two different phases of the biofilm formation by *Pseudomonas aeruginosa*.

Model 1: Tested agents (TauCl, TauBr) were added immediately after setting up of bacteria and early biofilm was measured, starting just after adherence of bacteria to a plastic surface.

Model 2: Tested agents (TauCl, TauBr, and DNase) were added to mature biofilm (24 h after the induction) and the development of late (mature) biofilm was observed in further 48 h.

Experimental study design:

Model 1: Formation of early biofilm under influence of the tested substances was observed by filling wells in 96-well plate with 20 µl of *P. aeruginosa* suspension and 180 µl of TSB, as described above. Then, TauBr and TauCl were added at a final concentration of 300 µM. The plates were gently mixed and the samples were incubated for 48 h at 37°C. Biofilm formation and the number of viable bacteria were checked in the following time points (0, 0.5, 4, 8, 24, and 48 h), as described above.

Model 2A: Influence of 300 µM TauBr and TauCl on the already formed late biofilm of *P. aeruginosa* was measured as in Model-1, but the tested substances were added 24 h after adherence of bacteria to the wells. Then the samples were incubated as above. Biofilm formation and the number of bacteria were measured in the following time points (0, 0.5, 4, 8, 24, and 48 h).

Model 2B: Combined effects of TauBr and DNase on the development of late biofilm and number of viable bacteria were tested. DNase (100, 300 µg/ml) and TauBr (300, 1,000 µM) were added to the 24 h cultures containing standardized suspension of *P. aeruginosa* strain KM/1 (1×10^8 CFU/ml) and TSB medium. First, the enzyme was added and followed by TauBr after 30 min. Such procedure was repeated three times every 12 h. Altogether, the culture was run for 48 h. Biofilm formation and the number of bacteria were measured immediately after addition of the tested substances and then after 8, 24, and 48 h.

23.3 Results and Discussion

23.3.1 Bactericidal Activity of TauCl and TauBr Against Planktonic Form of P. aeruginosa

Previously it has been demonstrated that both haloamines, TauCl and TauBr, show microbicidal activities against various microbes, including major pathogens associ-

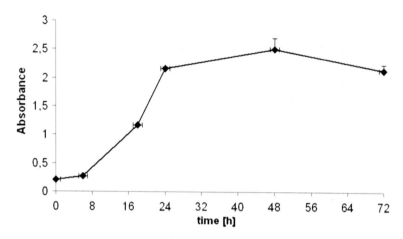

Fig. 23.1 The representative record of kinetics of biofilm formation by *P. aeruginosa* cultured in the Trypticase-Soy Broth (TSB). Congo red staining was used for measuring biofilm formation. Plot depicts mean values ± SD of absorbance [λ=492 nm]. The average was calculated on the basis of three replicates (n=3). A high amount of biofilm was observed after 24 h, followed by plateau and a typical breakdown of biofilm growth after 72 h of the culture

ated with skin and wound infections, such as *Staphylococcus aureus* and *Pseudomonas aeruginosa*. In the majority of tests in vitro bactericidal activity of these haloamines was measured against planktonic form of bacteria used at concentrations ranging from 10^5 to 10^6 CFU/ml (Nagl et al. 2000a, b; Marcinkiewicz et al. 2000, 2006). In such conditions TauBr killed all tested bacteria strains within relatively short time of 10–30 min, with MBC below 100 µM. TauCl, in contrast to TauBr, needs to be used in much higher millimolar concentrations to achieve a significant killing of pathogens within 30 min, which is possible due to its outstanding tolerability. Indeed, TauCl as a local antiseptic is used in 1% solution (55 mM) (Gottardi and Nagl 2010).

In our study we have tested bactericidal properties of TauCl and TauBr against *P. aeruginosa*, one of the most common pathogens of wound infections (Fazli et al. 2009; Tascini et al. 2006; Zhao et al. 2010). To achieve results confirming the ability of TauCl and TauBr to kill *in vivo*, we performed the experiments in conditions related to skin infections. First, planktonic form of *P. aeruginosa* was used at very high concentration (~10^8 bacteria/ml). Second, we have investigated the effect of TauCl and TauBr on biofilm formation by *P. aeruginosa*, the common form of bacteria existing *in vivo* (Fig. 23.1).

TauCl, at a concentration of 1,000 µM, significantly decreased the number of viable bacteria but only in the low density culture (10^5 CFU/ml), as shown in Fig. 23.2. Incubation times of 4 h were needed for a 3 log 10 reduction of planktonic form of *P. aeruginosa* (Fig. 23.2a). At these experimental conditions TauCl was ineffective against the high density bacteria population (10^8 CFU/ml). TauBr, however, confirmed its stronger bactericidal properties than TauCl. TauBr, at a concentration of 300 µM, completely inhibited the growth of the low density population of *P. aeruginosa* within 30 min. To achieve the same bactericidal effect against the high density population of *P. aeruginosa*, TauBr needed 4 h incubation time (Fig. 23.2b).

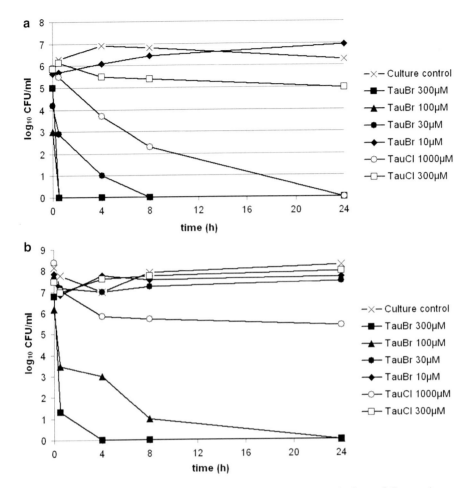

Fig. 23.2 Bactericidal activity of TauCl and TauBr against planktonic form of *P. aeruginosa*. TauBr and TauCl, used at different concentrations, were incubated with *P. aeruginosa* (**a**) 105 CFU/ml or (**b**) 108 CFU/ml. We used *P. aeruginosa* in TSB broth alone, as a control. Results depict CFU/ml measured at different time intervals; detection limit—100 CFU/ml

To demonstrate efficiency of antimicrobial agents against high density bacterial population (>10⁷ CFU/ml) is of great importance as bacteria may achieve such values at a site of inflammation. For example, the number of *Propionibacterium acnes* isolated from acne inflammatory lesions has been estimated in a range of 10⁷–10¹⁰ CFU/ml (Burkhart et al. 1999). In summary, the present data (Fig. 23.2) show that TauBr and TauCl can markedly inhibit a growth of planktonic form of *P. aeruginosa*. It also suggests that taurine haloamines are able to prevent biofilm formation. To confirm this hypothesis *P. aeruginosa* biofilm growth was stimulated in the presence of TauCl and TauBr. As expected, TauBr not only significantly reduced the number of viable bacteria (CFU) but also completely blocked formation of a biofilm as shown by Congo red assay (Fig. 23.3a). Interestingly, in these experimental

276 J. Marcinkiewicz et al.

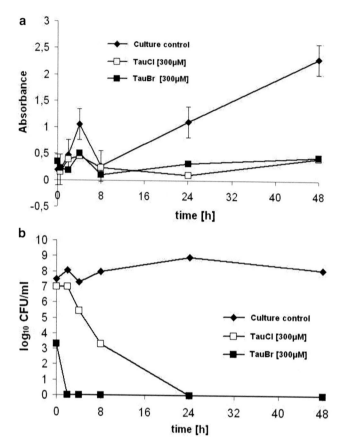

Fig. 23.3 Effects of TauCl and TauBr on the early stages of biofilm formation. Model 1: *P. aeruginosa* suspension in TSB broth was incubated with 300 µM TauBr or TauCl for 48 h at 37°C. We used a culture of *P. aeruginosa* in TSB broth alone, as a control. Biofilm formation (**a**) and the number of viable bacteria (**b**) were measured at the following time points (0, 0.5, 4, 8, 24, and 48 h), as described in Sect. 23.2. Results depict mean values ± SD calculated from three replicates (*n* = 3)

conditions TauCl also affects biofilm formation. However, the bactericidal activity of TauBr was stronger than TauCl and the effect was achieved within a shorter period of time (Fig. 23.3b).

A time factor is crucial in killing bacteria *in vivo* at a gate of infections, as it is necessary to precede bacteria binding to surfaces, the first stage of biofilm formation (Bjarnsholt et al. 2008; Fazli et al. 2009). As microbial biofilms are the most common mode of growth of bacteria in nature (O'Toole et al. 2000), it is reasonable to ask a question whether TauCl and TauBr are able to kill bacteria hidden in a biofilm or to destroy a protective exopolymeric matrix of growing biofilms.

23.3.2   The Effect of TauBr and TauCl on the Development of the Mature P. aeruginosa Biofilm and a Number of P. aeruginosa Sessile Cells

To determine whether taurine haloamines can kill bacteria hidden in biofilms we investigated the activity of TauCl and TauBr against a mature, 24 h old, mature biofilm. In contrast to their high bactericidal efficacy against planktonic form of *P. aeruginosa* in these experimental conditions, neither TauCl nor TauBr could affect biofilm structure (Fig. 23.4a) and kill sessile bacteria (Fig. 23.4b). Neither the absorbance (Congo red assay of biofilm formation) nor the number of bacteria (CFU/ml) changed even after 48 h of culture in the presence of TauCl and TauBr. This lack of effect may be explained by a well known resistance of biofilms to anti-biotics and antiseptics. It has been documented that bacteria in biofilms may be even 50–500 times more resistant to antibiotics than their planktonic counterparts (Mah and O'Toole 2001).

The resistance to TauBr and TauCl may be explained by the fact that biofilm effectively weakens the effect of test substances by limiting their access to the bacteria.

23.3.2.1   The Effect of DNase Alone and in a Combination with TauBr on the Development of the Mature *P. aeruginosa* Biofilm and Survival of Biofilm Hidden Bacteria

To enhance TauBr efficiency in killing of biofilm hidden bacteria, we have decided to destroy the structure of biofilm prior to administration of TauBr. Based on other reports (Allesen-Holm et al. 2006; Fuxman Bass et al. 2010; Montanaro et al. 2011), we have chosen an extracellular DNA, one of the components of *P. aerugi-nosa* matrix, as a target. DNase was used to digest extracellular DNA and to dam-age the structure of the biofilm and facilitate access of TauBr to the sessile cells of *P. aeruginosa* (Martins et al. 2012; Yang et al. 2007). The experiments were performed according to the two protocols, with DNase used at low (Fig. 23.5) and high (Fig. 23.6) concentrations. We found that low concentration of DNase did not affect biofilm formation and further application of TauBr did not reduce the growth of bacteria.

Interestingly, DNase used at a high concentration of 300 μg/ml led to a breach of the three-dimensional structure and reduced total mass of biofilm. However, the number of viable bacteria did not change after application of DNase and TauBr.

Why is bactericidal activity of TauBr against biofilm bacteria suppressed in our experimental conditions? We speculate that the lack of TauBr effectiveness may be caused by the two independent factors. First, the access of TauBr to sessile cells is limited, resulting in a reduction of its concentrations to the noneffective values. Second, maybe still unknown product(s) of *P. aeruginosa* biofilm neutralize(s) TauBr activity. Hydrogen peroxide is one of candidates. H_2O_2, the prod-

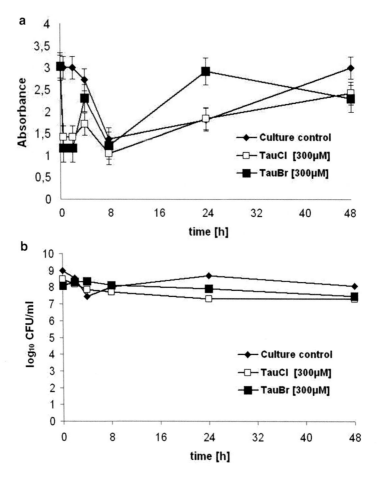

Fig. 23.4 Effects of TauCl and TauBr on the mature *P. aeruginosa* biofilm. Model 2A: TauCl and TauBr, at a concentration of 300 μM, were added to the bacteria culture 24 h after induction of biofilm. The development of biofilm (absorbance of Congo red staining) (**a**) and the number of bacteria (CFU) (**b**) were monitored for the next 48 h. We used a culture of *P. aeruginosa* in TSB broth as a control. Results depict mean values ± SD calculated from three replicates ($n = 3$)

uct of activated phagocytes and some bacteria, interacts with TauBr and being at the same concentration as TauBr, completely inhibits its activity (Marcinkiewicz et al. 2000, 2005). Therefore, in order to improve effectiveness of TauBr, it should be either served in a big excess, or biofilm structure should be destroy. The first strategy is supported by our previous studies. We have shown that *Propionibacterium acnes,* a major pathogen of biofilm-associated acne lesions (Holmberg et al. 2009), was killed by a topical application of 3 mM TauBr (Marcinkiewicz et al. 2008; Marcinkiewicz 2009). On the other hand, the second strategy used in the present studies, which included addition of DNase, was only slightly effective. Maybe DNase was added too late, and like in other reports, DNase did not destroy a structure of mature *P. aeruginosa* biofilm (Whitchurch et al. 2002;

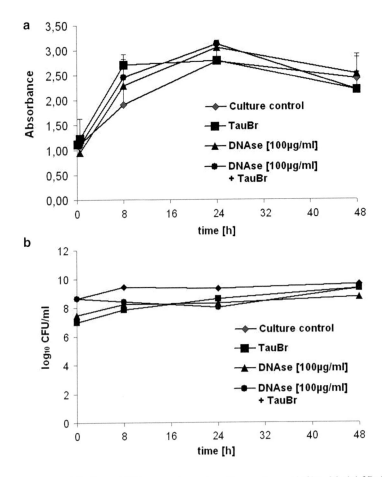

Fig. 23.5 Effects of TauBr and DNase on the mature *P. aeruginosa* biofilm. Model 2B: DNase (100 μg/ml) followed by TauBr (300 μM) was added to the bacteria culture 24 h after induction of biofilm, three times in the 4 h intervals. We used *P. aeruginosa* cultured in TSB broth without tested substances as a control. Both, biofilm growth (**a**) and the number of bacteria CFU/ml (**b**) were monitored as described in Fig. 23.4. Results depict mean values ± SD calculated from three replicates (*n* = 3)

Yang et al. 2007). A resistance of biofilms to the DNase treatment may be explained by the fact that a composition of biofilm extracellular matrix is not only species but also strain specific, and it seems that agents other than extracellular DNA may stabilize some biofilms (Ryder et al. 2007; Mann and Wozniak 2011). For example, various pneumoccocal strains develop biofilms that exhibit extracellular DNA in the biofilm matrix; however, strains with high biofilm forming index positively correlated with greater polysaccharide-associated structural complexity and antibiotic resistance (Hall-Stoodley et al. 2008). Therefore, further studies are necessary to confirm our observations and to test the combined effect of TauBr with other anti-biofilm agents, such as those able to destroy polysaccharides, the second major component of *P. aeruginosa* biofilm matrix.

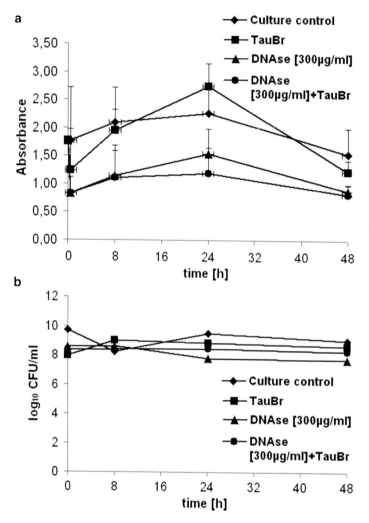

Fig. 23.6 Effects of high concentrations of TauBr (1,000 μM) and DNase (300 μg/ml) on the mature *P. aeruginosa* biofilm (Model 2B). For experimental design see the legend for Fig. 23.4

23.4 Conclusions

In summary, we would like to stress that the present preliminary results hardly touch the problem of unknown behavior and activity of taurine haloamines against bacterial biofilm. It is an important clinical issue as all previous investigations are based on characterizing bactericidal properties of TauCl and TauBr against planktonic form of bacteria. In nature, however, most bacteria, including *P. aeruginosa,* exist as biofilms. Moreover, the present study also allows apprehending why TauCl must be applied as antiseptic in biofilms-associated infections at so extremely high concentration (as 55 mM aqueous solution). In this study we have also shown that even TauBr, despite its strong bactericidal activity, is not sufficiently effective to kill sessile bacteria in biofilm. Combined effect of TauBr and DNase seems to be more promising, but in order to achieve better therapeutic effects, TauBr should be applied with other anti-biofilm agents. Further studies are necessary to prove the therapeutic potential of TauBr in a treatment of biofilm associated infections. Moreover, future experimental models should be extended to other bacteria, as a composition and properties of bacterial biofilms depend on both a microenvironment and bacteria strains.

Acknowledgments We want to thank Prof. Waldemar Gottardi and Prof. Marcus Nagl from the Division of Hygiene and Medical Microbiology, Innsbruck Medical University, Austria, for giving us N-chlorotaurine sodium salt. This paper was supported by Jagiellonian University Medical College grant No K/ZDS/002964, grant No K/ZDS/002861 and grant N N 401 547 040.

References

Allesen-Holm M, Barken KB, Yang L, Klausen M, Webb JS, Kjelleberg S, Molin S, Givskov M, Tolker-Nielsen T (2006) A characterization of DNA release in Pseudomonas aeruginosa cultures and biofilms. Mol Microbiol 59(4):1114–1128

Bjarnsholt T, Kirketerp-Møller K, Jensen PØ, Madsen KG, Phipps R, Krogfelt K, Høiby N, Givskov M (2008) Why chronic wounds will not heal: a novel hypothesis. Wound Repair Regen 16(1):2–10

Burkhart CG, Burkhart CN, Lehmann PF (1999) Acne: a review of immunologic and microbiologic factors. Postgrad Med J 75:328–331

Costerton JW (2002) Anaerobic biofilm infections in cystic fibrosis. Mol Cell 10(4):699–700

Fazli M, Bjarnsholt T, Kirketerp-Møller K (2009) Nonrandom distribution of *Pseudomonas aeruginosa* and *Staphylococcus aureus* in chronic wounds. J Clin Microbiol 47:4084–4089

Fux CA, Costerton JW, Stewart PS, Stoodley P (2005) Survival strategies of infectious biofilms. Trends Microbiol 13(1):34–40

Fuxman Bass JI, Russo DM, Gabelloni ML, Geffner JR, Giordano M, Catalano M, Zorreguieta A, Trevani AS (2010) Extracellular DNA: a major proinflammatory component of Pseudomonas aeruginosa biofilms. J Immunol 1(184(11)):6386–6395

Gaut JP, Yeh GC, Tran DH, Byun J, Henderson JP, Richter GM, Brennan ML, Lusis AJ, Belaaouaj A, Hotchkiss RS, Heinecke JW (2001) Neutrophils employ the myeloperoxidase system to generate antimicrobial brominating and chlorinating oxidants during sepsis. Proc Natl Acad Sci 98:11961–11966

Ghafoor A, Hay ID, Rehm BH (2011) Role of exopolysaccharides in Pseudomonas aeruginosa biofilm formation and architecture. Appl Environ Microbiol 77(15):5238–5246

Gottardi W, Nagl M (2002) Chemical properties of N-chlorotaurine sodium, a key compound in the human defence system. Arch Pharm 335(9):411–421

Gottardi W, Nagl M (2010) N-chlorotaurine, a natural antiseptic with outstanding tolerability. J Antimicrob Chemother 65:399–409

Gstöttner M, Nagl M, Pototschnig C, Neher A (2003) Refractory rhinosinusitis complicating immunosuppression: application of N-chlorotaurine, a novel endogenous antiseptic agent. ORL J Otorhinolaryngol Relat Spec 65(5):303–305

Hall-Stoodley L, Nistico L, Sambanthamoorthy K, Dice B, Nguyen D, Mershon WJ, Johnson C, Hu FZ, Stoodley P, Ehrlich GD, Post JC (2008) Characterization of biofilm matrix, degradation by DNase treatment and evidence of capsule downregulation in Streptococcus pneumoniae clinical isolates. BMC Microbiol 8:173

282   J. Marcinkiewicz et al.

Holmberg A, Lood R, Mörgelin M, Söderquist B, Holst E, Collin M, Christensson B, Rasmussen M (2009) Biofilm formation by Propionibacterium acnes is a characteristic of invasive isolates. Clin Microbiol Infect 15(8):787–795

Leid JG, Willson CJ, Shirtliff ME, Hassett DJ, Parsek MR, Jeffers AK (2005) The exopolysaccharide alginate protects Pseudomonas aeruginosa biofilm bacteria from IFN-gamma-mediated macrophage killing. J Immunol 175(11):7512–7518

Mah TF, O'Toole GA (2001) Mechanisms of biofilm resistance to antimicrobial agents. Trends Microbiol 9:34–39

Mann EE, Wozniak DJ (2011) Pseudomonas biofilm matrix composition and niche biology. FEMS Microbiol Rev. doi:10.1111/j.1574-6976.2011.00322.x

Marcinkiewicz J (2009) Taurine bromamine: a new therapeutic option in inflammatory skin diseases. Pol Arch Med Wewn 119:673–675

Marcinkiewicz J, Mak M, Bobek M, Biedroń R, Białecka A, Koprowski M, Kontny E, Maśliński W (2005) Is there a role of taurine bromamine in inflammation? Interactive effects with nitrite and hydrogen peroxide. Inflamm Res 54(1):42–49

Marcinkiewicz J, Chain B, Nowak B et al (2000) Antimicrobial and cytotoxic activity of hypochlorous acid: interactions with taurine and nitrite. Inflamm Res 49:280–289

Marcinkiewicz J, Biedroń R, Białecka A et al (2006) Susceptibility of Propionibacterium acnes and Staphylococcus epidermidis to killing by MPO-halide system products. Implication for taurine bromamine as a new candidate for topical therapy in treating acne vulgaris. Arch Immunol Ther Exp 54(1):61–68

Marcinkiewicz J, Wojas-Pelc A, Walczewska M, Lipko-Godlewska S, Jachowicz R, Maciejewska A, Białecka A, Kasprowicz A (2008) Topical taurine bromamine, a new candidate in the treatment of moderate inflammatory acne vulgaris. Eur J Dermatol 18:433–439

Martins M, Henriques M, Lopez-Ribot JL, Oliveira R (2012) Addition of DNase improves the in vitro activity of antifungal drugs against Candida albicans biofilms. Mycoses 55(1):80–85

Montanaro L, Poggi A, Visai L, Ravaioli S, Campoccia D, Speziale P, Arciola CR (2011) Extracellular DNA in biofilms. Int J Artif Organs 34(9):824–831. doi:10.5301/ijao.5000051

Nagl M, Nguyen VA, Gottardi W et al (2003) Tolerability and efficacy of N-chlorotaurine in comparison with chloramine T for treatment of chronic leg ulcers with a purulent coating: a randomized phase II study. Br J Dermatol 149:590–597

Nagl M, Teuchner B, Pöttinger E, Ulmer H, Gottardi W (2000a) Tolerance of N-chlorotaurine, a new antimicrobial agent, in infectious conjunctivitis – a phase II pilot study. Ophthalmologica 214(2):111–114

Nagl M, Hess MW, Pfaller K, Hengster P, Gottardi W (2000b) Bactericidal activity of micromolar N-chlorotaurine: evidence for its antimicrobial function in the human defense system. Antimicrob Agents Chemother 44:2507–2513

O'Toole GA, Kaplan HB, Kolter R (2000) Biofilm formation as microbial development. Annu Rev Microbiol 54:49–79

Reuter M, Mallett A, Pearson BM, van Vliet AHM (2010) Biofilm formation by Campylobacter jejuni is increased under aerobic conditions. Appl Environ Microbiol 76(7):2122–2128

Ryder C, Byrd M, Wozniak DJ (2007) Role of polysaccharides in Pseudomonas aeruginosa biofilm development. Curr Opin Microbiol 10(6):644–648

Sutherland IW (2001) The biofilm matrix – an immobilized but dynamic microbial environment. Trends Microbiol 9(5):222–227

Tascini C, Gemignani G, Palumbo F, Leonildi A, Tedeschi A, Lambelet P, Lucarini A, Piaggesi A, Menichetti F (2006) Clinical and microbiological efficacy of colistin therapy alone or in combination as treatment for multidrug resistant Pseudomonas aeruginosa diabetic foot infections with or without osteomyelitis. J Chemother 18:648–651

Whitchurch CB, Tolker-Nielsen T, Ragas PC, Mattick JS (2002) Extracellular DNA required for bacterial biofilm formation. Science 295(5559):1487

Zhao G, Hochwalt PC, Usui ML, Underwood RA, Singh PK, James GA, Stewart PS, Fleckman P, Olerud JE (2010) Delayed wound healing in diabetic (db/db) mice with Pseudomonas aeruginosa biofilm challenge: a model for the study of chronic wounds. Wound Repair Regen 18(5):467–477

Yang L, Barken KB, Skindersoe ME, Christensen AB, Givskov M, Tolker-Nielsen T (2007) Effects of iron on DNA release and biofilm development by Pseudomonas aeruginosa. Microbiology 153(Pt 5):1318–1328

Yazdanbakhsh M, Eckmann CM, Roos D (1987) Killing of schistosomula by taurine chloramine and taurine bromamine. Am J Trop Med Hyg 37:106–110

Part III
Taurine and Diabetes

Chapter 24
Inhibitory Effects of Taurine on STZ-Induced Apoptosis of Pancreatic Islet Cells

Shumei Lin, Jiancheng Yang, Gaofeng Wu, Mei Liu, Qiufeng Lv, Qunhui Yang, and Jianmin Hu

Abstract The present research aims to investigate the inhibition effect of taurine on the apoptosis of pancreatic islet cells induced by Streptozotocin (STZ). One hundred male Wistar rats weighing 180–200 g were randomly divided into two groups, rats in the experimental were intraperitoneally injected with 2% STZ (50 mg/kg bw, dissolved in 0.1 mmol/L pH 4.2 citrate buffer), rats in the control group were injected with the same volume of citrate buffer. Rats with the fasting blood glucose level higher than 16.7 mmol/L were selected and randomly divided into four groups: Rats in M group were STZ-induced diabetes rats, rats in T1, T2, and T3 groups were intragastrically administered with taurine (dissolved in 0.5% sodium carboxymethyl cellulose as thickening agent) once a day for 4 weeks with the contents of 0.6 g/kg, 1.2 g/kg, and 2.4 g/kg bw, respectively, while rats in group C and M were given the same amount of thickening agent as T2 group. Four weeks later, pancreatic tissues were fixed and processed for paraffin section. The results showed that STZ induced a significant increase in apoptotic rate of pancreatic islet cells, up-regulated the expression of bax and Fas and down-regulated the expression of Bcl-2, which were significantly blocked by taurine ($P < 0.05$). The results indicated that taurine could significantly restrain apoptosis of pancreatic islet cells induced by STZ.

Abbreviations

T   Taurine
STZ Streptozotocin

S. Lin • J.Yang • G. Wu • M. Liu • Q. Lv • Q. Yang • J. Hu(✉)
College of Animal Science & Veterinary Medicine, Shenyang Agricultural University,
Shenyang, Liaoning Province, P.R. China
e-mail: hujianmin59@163.com

A. El Idrissi and W.J. L'Amoreaux (eds.), *Taurine 8*, Advances in Experimental
Medicine and Biology 775, DOI 10.1007/978-1-4614-6130-2_24,
© Springer Science+Business Media New York 2013

24.1   Introduction

Apoptosis is the process of programmed cell death that may occur in multicellular organisms. Biochemical events lead to characteristic cell changes and death. These changes include blebbing, cell shrinkage, nuclear fragmentation, chromatin condensation, and chromosomal DNA fragmentation. In type 1 diabetes, apoptosis of pancreatic b-cells is the most critical and final step in the development of autoimmune diabetes (Kurrer et al. 1997; Kim et al. 1999). Perforin, FasL, TNFα, IL-1, IFNγ, and NO have been claimed as effector molecules. They may also be important in the pathogenesis of type 2 diabetes, and recent reports comparing pancreatic tissues from type 2 diabetic patients and nondiabetic subjects showed significantly higher rate of apoptosis in diabetic islets as opposed to the nondiabetic counterparts (Butler et al. 2003).

Taurine, 2-aminoethane sulfonic acid, is present in many tissues and organs, especially in excitable tissues such as heart, retina, brain, and skeletal muscle (Huxtable 1992). It has many important physiological functions such as regulating intracellular Ca^{2+} concentration (Molchanova et al. 2005), modulating inflammatory reactions (Park et al. 2002), acting as a neuromediator and neuromodulator (Kuriyama 1980), as well as anti-oxidization (Tong et al. 2006), osmoregulation (Ozasa and Gould 1982), and cholic acid production (Kase et al. 1986). Many studies have shown that taurine can attenuate apoptosis induced by a number of factors in different cell types. It has been reported that taurine can effectively prevent myocardial ischemia-induced apoptosis by inhibiting the assembly of the Apaf-1/caspase-9 apoptosome (Takatani et al. 2004). Taurine administered prior to stimulation down-regulated FasL protein expression and partially inhibited apoptosis induced by IL-2 (Maher et al. 2005). Taurine can also attenuated hyperglycemia-induced apoptosis in human tubular cells via an inhibition of oxidative stress and could exert a beneficial effect in preventing tubulointerstitial injury in diabetic nephropathy (Verzola et al. 2002). Administration of taurine to St. Thomas' cardioplegic solution improved cardiac function recovery for prolonged hypothermic rat heart preservation by suppressing DNA oxidative stress and cell apoptosis (Oriyanhan et al. 2005). Taurine protected cerebellar neurons of the external granular layer against ethanol-induced apoptosis in 7-day-old mice (Taranukhin et al. 2012). In other investigations, taurine was found to attenuate rat hepatocyte apoptosis and necrosis through inhibition of both nitric oxide and reactive oxygen species (Redmond et al. 1996), and decrease human endothelial cell apoptosis through its antioxidant effect and regulation of intracellular calcium flux (Wang et al. 1996; Wu et al. 1999).

Pancreatic β-cell apoptosis is known to participate in the β-cell destruction process that occurs in diabetes. A better understanding of how it takes place is essential for future development of therapeutic strategies aimed at preventing β-cell loss and diabetes. In this study the possible role that taurine plays as an enhancer streptozotocin (STZ)-mediated rat islet cell apoptosis in vivo and expressions of pro- and anti-apoptotic proteins were investigated.

24.2 Methods

24.2.1 Experimental Design

One hundred male Wistar rats (230–250 g) were provided by Experimental Animal Center of China Medical University. They were maintained under controlled light (14 h-light, 10 h-dark) and temperature ($22 \pm 2°C$), and were given free access to food and water.

One hundred rats were intraperitoneally injected with 2% STZ (50 mg/kg bw, dissolved in 0.1 mmol/L, pH 4.2 citrate buffer), the other 20 rats used as control (C) were injected with the same volume of citrate buffer. The level of blood glucose was measured using blood samples taken from vena caudalis 3 days after STZ injection. The rats with the fasting blood glucose level higher than 16.7 mmol/L were selected and randomly divided into five groups. Rats in M group were STZ-induced diabetes rats, rats in T1, T2, and T3 groups were intragastrically administered with taurine (dissolved in 0.5% sodium carboxymethyl cellulose as thickening agent) once a day for 4 weeks with the contents of 0.6, 1.2, and 2.4 g/kg bw, respectively, rats in C and M groups were given the same amount of thickening agent as T2 group. After 4 weeks of taurine administration, the rats were sacrificed by decapitation, blood glucose levels were determined by Acute-Check Active Blood Glucose Monitoring Meter and Glucose Strip (America Rossi Pharmacy International Group Company).

24.2.2 Tissue Preparation

For Transferase-mediated nick end labeling (TUNEL) and immunohistochemical studies, pancreatic gland (25–15 mm) were sampled and fixed in 10% neutral buffered formalin in phosphate buffered saline (PBS) for 72 h, and then were embedded in paraffin in a routine manner. Sections with 5 μm thickness were floated onto a distilled water bath (45°C), collected on SuperFrost Plus slides, deparaffinized and rehydrated stepwise through an ethanol series, and processed for routine histological analysis using hematoxylin and transferase-mediated nick end labeling (TUNEL) and immunohistochemistry.

24.2.3 Terminal Deoxynucleotidyl Transferase-Mediated Nick End Labeling

Tissue sections were treated with 1 mg/ml proteinase K in PBS at 37°C for 5 min. After being preincubated with terminal deoxynucleotidyl transferase (TdT) buffer

(200 mM potassium cacodylate, 25 mM Tris–HCl, 0.25 mg/ml BSA, pH 6.6) for 10 min, the sections were reacted with TdT buffer containing 100 IU/ml TdT and 2.5 mM biotinylated 16-dUTP (Roche Diagnostics) at 37°C for 2 h. The reaction was terminated by washing with 50 mM Tris–HCl (pH 7.5). Endogenous peroxidase was inactivated by immersing the sections in 0.3% H_2O_2 in methanol for 15 min, and then washed with 0.075% Brij (Sigma) in PBS. To block the nonspecific reaction, the sections were incubated with 500 mg/ml normal rabbit IgG (Sigma) in 5% BSA/PBS for 1 h. The sections were then treated with horseradish peroxidase (HRP) labeled rabbit anti-biotin diluted with 1% BSA/PBS overnight. HRP sites were visualized by 3,3-diaminobenzidine/4HCl (DAB, Boshide, China) and H_2O_2 for 6 min. The frequency of TUNEL-positive cells in the pancreatic gland was evaluated under a 400-fold microscope magnification. The percentage of TUNEL-positive cells in the islet cells was calculated as the number of TUNEL-positive cells out of the number of total cells in the islet cells of diabetic, nondiabetic, and taurine administration rats. For counting the number of TUNEL-positive cells in the pancreatic gland of diabetic, nondiabetic, taurine-added rat islet cells, and so on, five sections were randomly selected per pancreatic gland.

24.2.4   Immunohistochemical Staining

Rabbit polyclonal antibodies to bax (1:200), bcl-2 (1:200), and Fas (1:200) were obtained from Boshide, (Wuhan, China). After being deparaffinized, the sections were autoclaved at 121°C for 10 min in 0.01 M citrate buffer (pH 6.0) for bax, bcl-2, and Fas staining. The sections were then reacted with antibodies to bax, bcl-2, and Fas, respectively, overnight. After being washed with 0.075% Brij in PBS, the sections were reacted with biotinylated labeled rabbit anti-rat antibody (1:200) for 1 h at room temperature, and the biotinylated sites were visualized by DAB and H_2O_2 solution same as the TUNEL method. As negative control, slides were reacted with normal rat IgG instead of the specific antibodies used above. No signal was detected in the negative control slides. All slides were reviewed and scored by two independent observers blind to each other's scoring.

24.2.5   Microscopy

Quantitative measurements were done on a Jiangsu Jetta 801 series of multimedia color morphological analysis system. Measurement of immunoreactive material was based on integral optical density.

24.2.6 Statistics

The data were expressed as the mean number of bcl-2, bax, Fas positive and TUNEL-positive cells per five pancreatic islands in each pancreatic tissue. Data were examined by one-way ANOVA using SPSS16.0. Tukey's multiple-comparisons test was used to determine the significant differences ($P<0.05$).

24.3 Results

24.3.1 Effect of Taurine on the Apoptosis

The apoptosis in islet cells of pancreatic gland was scarce in the control group, but the percentage of apoptosis was more in M group than the C group ($P<0.05$). The rate of apoptosis in all taurine groups was much lower than the M group (Fig. 24.1).

24.3.2 Effect of Taurine on the Expression of Fas

The expression of Fas in islet cells of pancreatic gland was lower in the control group compared with the M group ($P<0.05$). Which was markedly inhibited by intragastrically administered with taurine ($P<0.05$) (Fig. 24.2).

24.3.3 Effect of Taurine on the Expression of bax

The expression of Fas in islet cells of pancreatic gland was much higher in M group compared with the control group ($P<0.05$). As displayed by the image analysis, the integral optical density of bax was decreased by taurine administration ($P<0.05$) (Fig. 24.3).

24.3.4 Effect of Taurine on the Expression of bcl-2

The expression of bcl-2 in islet cells of pancreatic gland was much lower in the M group compared with the control group ($P<0.05$). As displayed by the image analysis, the integral optical density of bcl-2 was decreased by taurine administration ($P<0.05$) (Fig. 24.4).

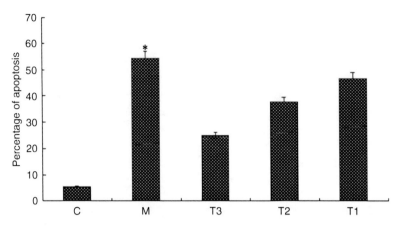

Fig. 24.1 Effect of taurine on the apoptosis of islet cells. The rats were designated as control group, diabetic model group (M), and taurine groups (T) with subscript 1, 2, and 3. That is, taurine groups were intragastrically administered with taurine (dissolved in 0.5% sodium carboxymethyl cellulose as thickening agent) once a day for 4 weeks with the contents of 0.6 g/kg, 1.2 g/kg, and 2.4 g/kg bw, respectively, the rats in group C and M were given the same amount of thickening agent as T2 group. Measurements were made after 4 weeks treatment. Data are the mean $\pm$ SD ($n = 15$). Values with *asterisk* are significantly different ($P < 0.05$)

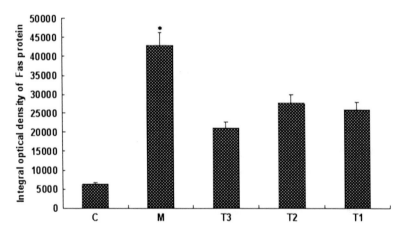

Fig. 24.2 Effect of taurine on the integral optical density of Fas protein in islet cells of pancreatic gland. The rats were designated as control group, diabetic model group (M), and taurine groups (T) with subscript 1, 2, and 3. That is, taurine groups were intragastrically administered with taurine (dissolved in 0.5% sodium carboxymethyl cellulose as thickening agent) once a day for 4 weeks with the contents of 0.6 g/kg, 1.2 g/kg, and 2.4 g/kg bw, respectively, the rats in group C and M were given the same amount of thickening agent as T2 group. Measurements were made after 4 weeks treatment. Data are the mean $\pm$ SD ($n = 15$). Values with *asterisk* are significantly different ($P < 0.05$)

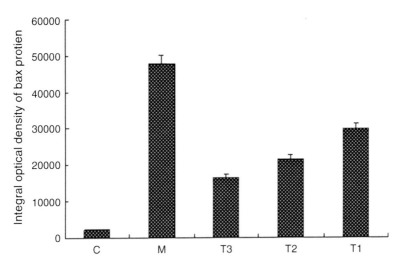

Fig. 24.3 Effect of taurine on the integral optical density of bax protein in islet cells of pancreatic gland. The rats were designated as control group, diabetic model group (M), and taurine groups (T) with subscript 1, 2, and 3. That is, taurine groups were intragastrically administered with taurine (dissolved in 0.5% sodium carboxymethyl cellulose as thickening agent) once a day for 4 weeks with the contents of 0.6 g/kg, 1.2 g/kg, and 2.4 g/kg bw, respectively, the rats in group C and M were given the same amount of thickening agent as T2 group. Measurements were made after 4 weeks treatment. Data are the mean ± SD ($n = 15$). Values with *asterisk* are significantly different ($P < 0.05$)

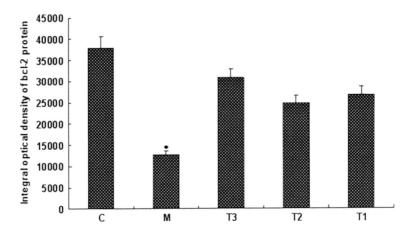

Fig. 24.4 Effect of taurine on the integral optical density of bcl-2 protein in islet cells. The rats were designated as control group, diabetic model group (M), and taurine groups (T) with subscript 1, 2, and 3. That is, taurine groups were intragastrically administered with taurine (dissolved in 0.5% sodium carboxymethyl cellulose as thickening agent) once a day for 4 weeks with the contents of 0.6 g/kg, 1.2 g/kg, and 2.4 g/kg bw, respectively, the rats in group C and M were given the same amount of thickening agent as T2 group. Measurements were made after 4 weeks treatment. Data are the mean ± SD ($n = 15$). Values with *asterisk* are significantly different ($P < 0.05$)

24.4   Discussion

Pancreatic b-cell apoptosis is a pathological feature that is common to both type 1 (T1DM) and type 2 diabetes (T2DM). In T1DM, b-cells are selectively destroyed after lymphoid infiltration of the islet. This autoimmune destruction results in insulin deficiency and hyperglycemia. In T2DM, the reduced secretion of insulin which is in associated with insulin resistance lead to a glucose toxicity effect that, in the presence or absence of hyperlipidemia, contributes to b-cell death by apoptosis.

STZ was originally identified in the late 1950s as an antibiotic (Vavra et al. 1959). The drug was discovered in a strain of soil microbe *Streptomyces achromogenes* by scientists. In the mid-1960s STZ was found to be selectively toxic to the b-cells of the pancreatic islets, which normally regulate blood glucose levels by producing the hormone insulin. This suggested the drug's use as an animal model of diabetes (Mansford and Opie 1968; Rerup 1970). From then on, STZ was widely used to study the mechanism of pancreatic b-cell death and diabetes. STZ enters the b-cell via a glucose transporter (GLUT2) and causes alkylation of DNA. DNA damage induces activation of poly ADP-ribosylation, a process that is more important for the diabetogenicity of STZ than DNA damage itself. Poly ADP-ribosylation leads to depletion of cellular NAD^+ and ATP. Enhanced ATP dephosphorylation after STZ treatment supplies a substrate for xanthine oxidase resulting in the formation of superoxide radicals. Consequently, hydrogen peroxide and hydroxyl radicals are also generated. Furthermore, STZ liberates toxic amounts of nitric oxide that inhibits aconitase activity and participates in DNA damage. As a result of the streptozotocin action, b-cells undergo the destruction (Szkudelski 2001).

The effect of STZ on apoptosis in b-cells has been investigated by many authors and conflicting results have been reported. STZ have been shown to induce apoptosis in b-cells. It was reported that the STZ-induced b-cell apoptosis develops very rapidly, within 8 h after STZ administration (Wada and Yagihashi 2004), while STZ did not induce islet cell apoptosis when incubated in normal glucose (Eizirik et al. 1988; Hoorens and Pipeleers 1999; Liu et al. 2002; Thomas et al. 2002). In this study, the potential role of taurine in STZ-induced apoptosis of pancreatic b-cells was investigated (Fig. 24.1). The results showed that intraperitoneally injected with 2% STZ (50 mg/kg bw) could induce apoptosis of pancreatic b-cells. Some difference results could be due to species, time course, and experimental models used.

The effect of taurine on β-cell apoptosis has already been described in vitro. This study investigated the role of taurine on STZ-induced pancreatic b-cell apoptosis in vivo. The results showed that dose-dependent effects of taurine supplementation compensated in vivo for the STZ-induced pancreatic b-cells apoptosis (Fig. 24.1). In type 1 (autoimmune) diabetes, apoptosis of pancreatic b-cells is the most critical and final step in the development of autoimmune diabetes. It is also important in the pathogenesis of type 2 diabetes. Fas, or CD95, the rodent equivalent for human APO-1 could be the potential initiator for the increased apoptotic rate in islet cells, which belongs to the TNF-receptor family and is expressed in many cells. Fas–FasL interaction led to cleavage of procaspase-8 to caspase-8 which activated caspase-3 and DNA fragmentation (Stennicke and Salvesen 2000; Maedler et al. 2001). The expression of Fas in islet cells has been shown in murine (Lee et al. 1999) and humans (Stassi et al. 1995). In this study, similar to apoptosis, the percentage of Fas positive islet cells was increased by STZ. Taurine supplementation decreased the immunoreactivity of Fas in diabetic rats 4 weeks after STZ administration. These results suggested a possible role of Fas in the induction of apoptosis in islet cells, and also give a possible hypothesis on a mechanism of taurine in the protection of islet cells from apoptosis (Fig. 24.2). A delicate balance normally exists in the body between the anti-apoptotic and pro-apoptotic regulators of apoptosis to ensure the proper survival and turnover of different body cells. Imbalance in the apoptotic pathway occurs in disease scenarios. In the two of apoptosis signal transduction pathways, the ratio of bcl-2 family members is the key factor, especially bcl-2 and bax, which are the most representative members, have similar structure but the opposite function, the ratio of which directly determine the fate of islet cells on accepting the death message. In addition, up-regulation of several anti-apoptotic members of the bcl-2 family proteins, such as bcl-2 and bcl-xL, has been strongly associated with increased resistance to apoptosis and potentially linked with diabetes susceptibility (Garchon et al. 1994; Hanke 2000). On this basis, the effects of taurine and STZ on the expression levels of bcl-2 which is an anti-apoptotic intracellular mediator and bax which is a pro-apoptotic protein were analyzed. In this study, we established rat diabetic model by intraperitoneally injected with 2% STZ (50 mg/kg bw). The results demonstrated a significant increase in the expression of bax, together with a significant decrease of bcl-2 expression in diabetic model rats compared with the controls, which could be partly retrieve by taurine administration (Figs. 24.3 and 24.4).

24.5   Conclusion

The results demonstrated that taurine could inhibit STZ-mediated rat islet cell apoptosis and increase the expression of anti-apoptotic molecules and decrease the expression of apoptosis promoting molecules in rat islet cells.

Acknowledgments   Financial support from the National Research Foundation of China (Project no. 30972154) and Doctoral initiaitng project of liaoning province (Project no. 20111083) is acknowledged.

References

Butler AE, Janson J, Bonner-Weir S, Ritzel R, Rizza RA, Butler PC (2003) Beta-cell deficit and increased beta-cell apoptosis in humans with type 2 diabetes. Diabetes 52:102–110
Eizirik DL, Strandell E, Sandler S (1988) Culture of mouse pancreatic islets in different glucose concentrations modifies B cell sensitivity to streptozotocin. Diabetologia 31:168–174

Garchon HJ, Luan JJ, Eloy L, Bédossa P, Bach JF (1994) Genetic analysis of immune dysfunction in non-obese diabetic (NOD) mice: mapping of a susceptibility locus close to the Bcl-2 gene correlates with increased resistance of NOD T cells to apoptosis induction. Eur J Immunol 24(2):380–384

Hanke J (2000) Apoptosis and occurrence of Bcl-2, Bak, Bax Fas and FasL in the developing and adult rat endocrine pancreas. Anat Embryol 202:303–312

Hoorens A, Pipeleers D (1999) Nicotinamide protects human beta cells against chemically-induced necrosis, but not against cytokine-induced apoptosis. Diabetologia 42:55–59

Huxtable RJ (1992) Physiological actions of taurine. Physiol Rev 72(1):101–163

Kase BF, Prydz K, Björkhem I et al (1986) Conjugation of cholic acid with taurine and glycine by rat liver peroxisomes. Biochem Biophys Res Commun 138(1):167–173

Kim YH, Kim S, Kim KA, Yagita H, Kayagaki N, Kim KW, Lee MS (1999) Apoptosis of pancreatic β-cells detected in accelerated diabetes of NOD mice: no role of Fas-Fas ligand interaction in autoimmune diabetes. Eur J Immunol 29:455–465

Kuriyama K (1980) Taurine as a neuromodulator. Fed Proc 39(9):2680–2684

Kurrer MO, Pakala SV, Hanson HL, Katz JD (1997) Beta cell apoptosis in T cell-mediated autoimmune diabetes. Proc Natl Acad Sci USA 94:213–218

Lee MS, Kim S, Chung JH, Lee MK, Kim KW (1999) Fas is expressed in murine pancreatic islet cells and an insulinoma cell line but does not mediate their apoptosis in vitro. Autoimmunity 29(3):189–199

Liu D, Cardozo AK, Darville MI, Eizirik DL (2002) Double-stranded RNA cooperates with interferon-gamma and IL-1 beta to induce both chemokine expression and nuclear factor-kappa B-dependent apoptosis in pancreatic beta-cells: potential mechanisms for viral-induced insulitis and beta-cell death in type 1 diabetes mellitus. Endocrinology 143:1225–1234

Maedler K, Spinas GA, Dyntar D, Moritz W, Kaiser N, Donath MY (2001) Distinct effects of saturated and monounsaturated fatty acids on beta-cell turnover and function. Diabetes 50:69–76

Maher SG, Condron CEM, Bouchier-Hayes DJ, Toomey DM (2005) Taurine attenuates CD3/interleukin-2-induced T cell apoptosis in an in vitro model of activation-induced cell death (AICD). Clin Exp Immunol 139(2):279–286

Mansford KR, Opie L (1968) Comparison of metabolic abnormalities in diabetes mellitus induced by streptozotocin or by alloxan. Lancet 1(7544):670–671

Molchanova SM, Oja SS, Saransaari P (2005) Mechanisms of enhanced taurine release under Ca^{2+} depletion. Neurochem Int 47(5):343–349

Oriyanhan W, Yamazaki K, Miwa S, Takaba K, Ikeda T, Komeda M (2005) Taurine prevents myocardial ischemia/reperfusion-induced oxidative stress and apoptosis in prolonged hypothermic rat heart preservation. Heart Vessels 20(6):278–285

Ozasa H, Gould KG (1982) Protective effect of taurine from osmotic stress on chimpanzee spermatozoa. Arch Androl 9(2):121–126

Park E, Jia J, Quinn MR et al (2002) Taurine chloramine inhibits lymphocyte proliferation and decreases cytokine production in activated human leukocytes. Clin Immunol 102(2):179–184

Redmond HP, Wang JH, Bouchier-Hayes D (1996) Taurine attenuates nitric oxide- and reactive oxygen intermediate-dependent hepatocyte injury. Arch Surg 131:1280–1287

Rerup CC (1970) Drugs producing diabetes through damage of the insulin secreting cells. Pharmacol Rev 22(4):485–518

Stassi G, Todaro M, Richiusa P, Giordano M, Mattina A, Sbriglia MS, Lo Monte A, Buscemi G, Galluzzo A, Giordano C (1995) Expression of apoptosis-inducing CD95 (Fas/Apo 1) on human beta-cells sorted by flow cytometry and cultured in vitro. Transplant Proc 27:3271–3275

Stennicke HR, Salvesen GS (2000) Caspases – controlling intracellular signals by protease zymogen activation. Biochim Biophys Acta 1477(1–2):299–306

Szkudelski T (2001) The mechanism of alloxan and streptozotocin action in B cells of the rat pancreas. Physiol Res 50:536–546

Takatani T, Takahashi K, Uozumi Y, Shikata E, Yamamoto Y, Ito T, Matsuda T, Schaffer SW, Fujio Y, Azuma J (2004) Taurine inhibits apoptosis by preventing formation of the Apaf-1/caspase-9 apoptosome. Am J Physiol Cell Physiol 287:C949–C953

Taranukhin AG, Taranukhina EY, Saransaari P, Pelto-Huikko M, Podkletnova M, Oja S (2012) Taurine protects cerebellar neurons of the external granular layer against ethanol-induced apoptosis in 7-day-old mice. Amino Acids. doi:10.1007/s00726-012-1254-6

Thomas HE, Darwiche R, Corbett JA, Kay TW (2002) Interleukin-1 plus-interferon-induced pancreatic beta-cell dysfunction is mediated by beta-cell nitric oxide production. Diabetes 51:311–316

Tong L, Li J, Qiao H et al (2006) Taurine protects against ischemia-reperfusion injury in rabbit livers. Transplant Proc 38(5):1575–1579

Vavra JJ, Deboer C, Dietz A, Hanka LJ, Sokolski WT (1959) "Streptozotocin, a new antibacterial antibiotic". Antibiot Annu 7:230–235

Verzola D, Bertolotto MB, Villaggio B, Ottonello L, Dallegri F, Frumento G, Berruti V, Gandolfo MT, Garibotto G, Deferrari G (2002) Taurine prevents apoptosis induced by high ambient glucose in human tubule renal cells. J Investig Med 50(6):443–451

Wada R, Yagihashi S (2004) Nitric oxide generation and poly(ADP ribose) polymerase activation precede beta-cell death in rats with a single high-dose injection of streptozotocin. Virchows Arch 444:375–382

Wang JH, Redmond HP, Watson RW, Condron C, Bouchier-Hayes D (1996) The beneficial effect of taurine on the prevention of human endothelial cell death. Shock 6:331–338

Wu QD, Wang JH, Fennessy F, Redmond HP, Bouchier-Hayes D (1999) Taurine prevents high-glucose-induced human vascular endothelial cell apoptosis. Am J Physiol 277:C1229–C1238

Chapter 25
Taurine's Effects on the Neuroendocrine Functions of Pancreatic β Cells

Christina M. Cuttitta, Sara R. Guariglia, Abdeslem El Idrissi, and William J. L'Amoreaux

Abstract Taurine plays significant physiological roles, including those involved in neurotransmission. Taurine is a potent γ-aminobutyric acid (GABA) agonist and alters cellular events via $GABA_A$ receptors. Alternately, taurine is transported into cells via the high affinity taurine transporter (TauT), where it may also play a regulatory role. We have previously demonstrated that treatment of Hit-T15 cells with

C.M. Cuttitta (✉)
Department of Biology, College of Staten Island, Staten Island, NY 10314, USA

Center for Developmental Neuroscience, College of Staten Island, Staten Island, NY 10314, USA
e-mail: christinacuttitta@gmail.com

S.R. Guariglia
Department of Biology, College of Staten Island, Staten Island, NY 10314, USA

Advanced Imaging Facility, College of Staten Island, Staten Island, NY 10314, USA

Center for Developmental Neuroscience, College of Staten Island, Staten Island, NY 10314, USA

A. El Idrissi
Department of Biology, College of Staten Island, Staten Island, NY 10314, USA

Advanced Imaging Facility, College of Staten Island, Staten Island, NY 10314, USA

Center for Developmental Neuroscience, College of Staten Island, Staten Island, NY 10314, USA

Graduate Program in Biology (Neuroscience), Graduate Center, City University of New York,
365 Fifth Avenue, New York, NY 10016, USA

W.J. L'Amoreaux
Department of Biology, College of Staten Island, Staten Island, NY 10314, USA

Advanced Imaging Facility, College of Staten Island, Staten Island, NY 10314, USA

Center for Developmental Neuroscience, College of Staten Island, Staten Island, NY 10314, USA

Graduate Program in Biology (Neuroscience), Graduate Center, City University of New York,
365 Fifth Avenue, New York, NY 10016, USA

Graduate Program in Biochemistry, Graduate Center, City University of New York,
365 Fifth Avenue, New York, NY 10016, USA

A. El Idrissi and W.J. L'Amoreaux (eds.), *Taurine 8*, Advances in Experimental
Medicine and Biology 775, DOI 10.1007/978-1-4614-6130-2_25,
© Springer Science+Business Media New York 2013

1 mM taurine for 24 h significantly decreases insulin and GABA levels. We have also demonstrated that chronic in vivo administration of taurine results in an up-regulation of glutamic acid decarboxylase (GAD), the key enzyme in GABA synthesis. Here, we wished to test if administration of 1 mM taurine for 24 h may increase release of another β cell neurotransmitter somatostatin (SST) and also directly impact up-regulation of GAD synthesis. Treatment with taurine did not significantly alter levels of SST ($p > 0.05$) or GAD67 ($p > 0.05$). This suggests that taurine does not directly affect SST release, nor does it directly affect GAD synthesis. Taken together with our observation that taurine does promote GABA release via large dense-core vesicles, the data suggest that taurine may alter membrane potential, which in turn would affect calcium flux. We show here that 1 mM taurine does not alter intracellular Ca^{2+} concentrations from 20 to 80 s post treatment ($p > 0.05$), but does increase Ca^{2+} flux between 80 and 200 s post-treatment ($p < 0.005$). This suggests that taurine may induce a biphasic response in β cells. The initial response of taurine via $GABA_A$ receptors hyperpolarizes β cell and sequesters Ca^{2+}. Subsequently, taurine may affect Ca^{2+} flux in long term via interaction with K_{ATP} channels.

Abbreviations

GABA	γ-Aminobutyric acid
Tau	Taurine
GAD	Glutamic acid decarboxylase
TauT	Taurine transporter
LDCV	Large dense-core vesicles
SLMV	Synapse-like microvesicles
SST	Somatostatin

25.1 Introduction

25.1.1 The Neuroendocrine Nature of the Pancreas

The pancreas is a dual function organ with both exocrine and endocrine portions. The endocrine portion contains five cell types, including the α- and β-cells responsible respectively for secreting the peptide hormones glucagon and insulin. These are regarded as two master neuroendocrine cells, responsible for maintaining plasma glucose concentrations. Additional endocrine cells include the PP or F cell that secretes pancreatic polypeptide responsible for actions in the gastrointestinal tract, and G-cells that release gastrin to enhance gastric functions. For glucose homeostasis, it is the relative plasma concentrations of glucose that dictates which of the exocytotic mechanisms predominates release of glucagon or insulin. However, the molecular machinery involved in the process is typically activated through changes in membrane potential facilitated by binding of neurotransmitters to their ionotropic receptors. The final cell type is the δ-cell, which appears to be a master regulator of all islet cells through its release of somatostatin (SST).

Along with the release of crucial glucose catabolic and anabolic hormones, the islet cells also secrete several neurotransmitters. Additional neurotransmitters synthesized or released by islet cells are glutamate (α-cells) and GABA (β-cells). Other neurotransmitters participate in pancreatic regulation, but these are likely from neuronal input. Therefore, the three major neurotransmitters regulating the endocrine pancreas are glutamate, GABA, and somatostatin.

In α-cells, glutamate is stored in LDCV where it is co-released with glucagon (Hayashi et al. 2003). Glutamate may bind to ionotropic glutamate receptors (AMPA-type receptor variant GluR4c-flip) on δ-cells (Muroyama et al. 2004) to stimulate release of somatostatin (SST). Alternately, glutamate may bind to AMPA/KA receptors on α-cells, serving as a positive autocrine mechanism to further glucagon release (Cho et al. 2010; Koh et al. 2012).

In β-cells, insulin is stored in large dense-core vesicles (LDCV) along with some of the γ-aminobutyric acid (GABA) synthesized in the β-cell, while a significant pool of GABA remains in synapse-like microvesicles (SLMV) (Sorenson et al. 1991; Nathan et al. 1995; Braun et al. 2004; Braun et al. 2007). In vivo, plasma glucose levels of ≥ 2.8 mM are sufficient to stimulate the Ca^{2+}-dependent release of insulin from β-cells (Gilon et al. 1991). Any GABA released along with the insulin may be sufficient to bind to $GABA_A$ receptors on α cells, initiating a hyperpolarization of α-cells and reduce glucagon release (Rorsman et al. 1989; Suckale and Solimena 2010).

SST from δ-cells inhibits the release of the insulin and glucagon from α- and β-cells (Ahren 2009). These neurotransmitters allow for communication between islet cells that is complex and sophisticated in allowing for the cell communication in response to a rise or fall in glucose levels. Additionally, regulation of insulin and glucagon release (and thus glucose homeostasis) is also facilitated aided by SST secreted from the pancreatic δ cells in response to high levels insulin and glucagon (Goldsmith et al. 1975; Efendic et al. 1979; Taborsky 1983).

The initiation of insulin exocytosis includes multiple signaling events in the β-cells beginning with the Na^+-dependent, electrogenic glucose uptake. Once glucose has entered the cytoplasm, glycolysis is initiated, resulting in the cytoplasmic accumulation of ATP; this increase in cytoplasmic ATP inhibits ATP-sensitive K^+ (K_{ATP}) channels, resulting in a depolarization of β-cells. This depolarization ultimately opens voltage-sensitive Ca^{2+} channels; ensuing Ca^{2+} flux stimulates exocytosis of the LDCV containing both insulin and GABA. GABA then binds to the $GABA_A$ receptor on α-cells, initiating a depolarization of α-cells and subsequent inhibition of glucagon or to $GABA_A$ receptors on β-cells (Gu et al. 1993; Braun et al. 2010).

25.1.2   Taurine as a Potential Neuromodulator

Taurine is a conditionally essential amino acid whose role as a GABA agonist in the developing brain has been well documented. Taurine concentrations remain high in the neonatal brain for about 6 weeks, and then drops to adult levels and remains as the predominant free amino acid in some tissues (e.g., retina) or second to glutamate (e.g., brain). It is present in high concentrations in excitable cells (neurons, cardiomyocytes and skeletal muscle fibers) and maintains intracellular osmotic balance. Taurine may be synthesized from cysteine (Agrawal et al. 1971) or taken up into cells via the taurine transporter (TauT). In the brain, taurine acts as a GABA agonist, where it binds to $GABA_A$ receptors. Taurine activation of $GABA_A$ chloride channels alters membrane potential via rapid chloride uptake and subsequent hyperpolarization. Alternately, taurine may also exert a neurotransmitter-like effect through activation of glycine receptors. Although there is speculation of a taurine-specific receptor, to date there is no definitive proof of the existence of such a molecule (Wu and Prentice 2010).

In the pancreas, we and others have shown that release of insulin from β cells may be regulated via the GABAergic signaling system (Kawai and Unger 1983; Satin and Kinard 1998; El Idrissi et al. 2009a, 2010, 2012; Braun et al. 2010; L'Amoreaux et al. 2010). Additionally, we have shown that taurine treatment of β-cell lines is sufficient to induce insulin and GABA release (L'Amoreaux et al. 2010). When plasma glucose levels rise, glucose uptake into β-cells leads to increased exocytosis of LDCV containing insulin and GABA. This calcium-dependent exocytosis relies on the activation of voltage-sensitive calcium channels, which are activated upon depolarization of the membrane potential. The depolarization is driven by inhibition of K_{ATP} channels when cytoplasmic ATP levels rise following glycolytic processing of glucose. Taurine also is able to modulate cytoplasmic ATP levels through its interactions with mitochondria (El Idrissi 2008). Additionally, taurine also interacts with the sulfonylurea receptor subunit of K_{ATP} channels to inhibit channel activity (Tricarico et al. 2000). Inhibition of LDCV exocytosis may be derived by an autocrine feedback via activation of β-cell $GABA_A$ receptors (Braun et al. 2010), $GABA_B$ receptors (Gu et al. 1993), or through SST regulation of β-cell activity (McDermott and Sharp 1993; Doyle and Egan 2003). To date, studies have shown taurine's modulatory effects on cell activities through $GABA_A$ receptors, but not $GABA_B$. We have previously demonstrated that taurine can lead to an up-regulation of somatostatin expression in brain (El Idrissi et al. 2009b) and pancreatic islets (El Idrissi et al. 2010). Taurine can stimulate release of SST from neurons (Aguila and McCann 1985), but as yet there are no reports of taurine's efficacy in stimulating SST release in pancreas.

Therefore, we sought to examine a system by which taurine may participate in glucose homeostasis via GABA and SST signaling. Here, we used an isolated β cell line (Hit-T15) to investigate the efficacy of taurine in altering SST and GAD expression. Furthermore, we tested the efficacy of taurine in initiating calcium flux in these cells. Together, our data suggests that taurine may participate in glucose homeostasis.

25.2   Methods

25.2.1   Taurine's Effect on Somatostatin and GAD Levels

Pancreatic β-cell lines Hit T-15 from Syrian hamster were grown in Ham's F12-K medium. For treatments, cells were plated in complete medium on sterile cover glass at 5,000 cells/cm^2 until ~80% confluent. Cells were serum starved for 24 h prior to glucose and taurine treatments. The media were aspirated and replaced with glucose-free Ham's medium supplemented with 1 mM glucose, 1 mM taurine, or 3 mM glucose for 24 h. Following treatments with the supplemented media, cultures were fixed in 4% paraformaldehyde in PBS. The cells were then prepared for immunohistochemical analysis using the appropriate antibodies diluted 1:400. Primary antibodies included mouse anti-GAD and rabbit anti-somatostatin (Life Technologies/Molecular Probes, Carlsbad, CA). Primary antibodies were detected using goat anti-mouse IgG conjugated with Alexa 633, and goat anti-rabbit IgG conjugated with Alexa 488 (Life Technologies/Molecular Probes, Carlsbad, CA). Following incubation, the cover glass were placed on a drop of antifade (Slow Fade Gold Plus with DAPI) and sealed. The data were collected by confocal microscopy (Leica SP2 AOBS Confocal Microscope, Germany). Gain and offset for the acquisitions were identical for these three treatments.

25.2.2   Ca^{2+} Flux

Hit-T15 cells were grown in Hams F12-K medium as described above. For treatments, cells were plated in complete medium on sterile cover glass at 5,000 cells/cm^2 until ~80% confluent. Once confluent cells were incubated for 4 h with a 5 μM/ml solution of the fluorescent Ca^{2+} indicator Fluo-3 (Life Technologies/Molecular Probes, Carlsbad, CA). Following incubation, β-cells were treated with 1 mM glucose, 3 mM glucose, or 1 mM taurine. Live cell imagining (Zeiss Cell Observer; Carl Zeiss, Thornwood, NY), was used to detect Ca^{2+} flux every 15 s for a 5 min period following treatment. Images were obtained using both bright field and FITC filters.

25.2.3   Statistic Analysis

Statistical analyses were performed on intensity values using a one-way ANOVA and Bonferroni post hoc analyses (Prism). Values are expressed as the mean ± SEM. Differences were considered statistically significant when the calculated p value was less than 0.05.

25.3 Results

25.3.1 Taurine Does Not Alter Expression of SST or GAD

We report here for the first time the presence of SST in the Hit-T15 cell line (Fig. 25.1). Previous studies have confirmed the presence of a SST receptor in this cell line (Thermos et al. 1990; Seaquist et al. 1995; Cheng et al. 2002; Yao et al. 2005), but to date we find no evidence in the literature confirming the expression of the neurotransmitter within these cells. The in vivo role of somatostatin is to modulate exocytosis of LDCV in islets, presumably through the inhibition of Ca^{2+} flux (Yao et al. 2005). We speculate that these transformed cells may express somatostatin to serve as an autocrine regulator of cell function.

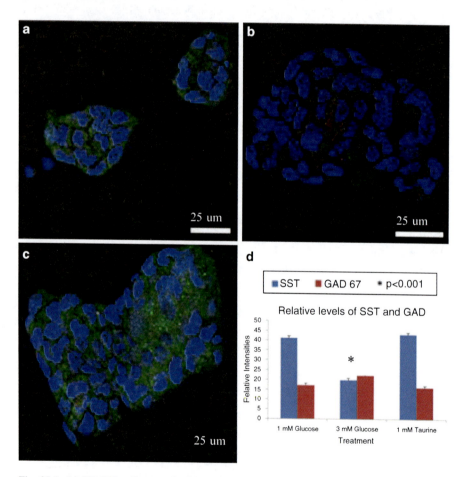

Fig. 25.1 (**a**) Hit-T15 cells treated with 1 mM glucose. Staining for SST (*green*) and GAD67 (*red*). (**b**) Hit-T15 cells treated with 3 mM glucose. SST levels are decreased and GAD67 increased. (**c**) Treatment with 1 mM taurine has no effect on the release of SST or expression of GAD67. (**d**) Only treatment with 3 mM glucose significantly impacted SST levels ($p < 0.001$); neither 3 mM glucose nor 1 mM taurine affected GAD67 expression

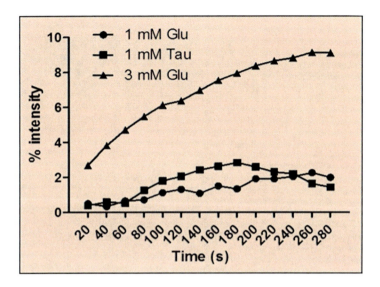

Fig. 25.2 Calcium flux in Hit-T15 β cell line using the fluorescent indicator Fluo-3. Treatment with 3 mM increases Ca^{2+} flux immediately whereas 1 mM taurine exhibited a lag of about 120 s

We also demonstrated here taurine's lack of efficacy in promoting SST release from these cells. Somatostatin levels were significantly decreased ($p < 0.001$) in the presence of 3 mM glucose (Fig. 25.1b, d), but SST levels in the taurine-treated were identical to those cells treated with 1 mM glucose (Fig. 25.1c, d). We also examined the role of taurine in eliciting an increase in GAD67 expression, which we have demonstrated in vivo with taurine treatment (El Idrissi et al. 2009a). We demonstrated that while 3 mM glucose significantly decreased somatostatin and increased GAD levels ($p < 0.05$) compared to those cells treated with 1 mM glucose 1 mM taurine was insufficient in eliciting these responses (Fig. 25.1c, d).

25.3.2   *Taurine May Induce a Latent Ca^{2+} Flux*

We hypothesized that the exocytosis of LDCV in response to taurine was dependent on the flux of Ca^{2+}. To test the efficacy of taurine in eliciting calcium flux, we preloaded Hit-T15 cells with the fluorescent indicator Fluo-3, and then treated cells with subthreshold (1 mM) glucose, suprathreshold (3 mM) glucose, or 1 mM taurine. The 1 mM glucose was insufficient to induce calcium flux, yet 3 mM glucose did cause a significant increase in calcium flux ($p < 0.001$; Fig. 25.2). There were no significant differences in Ca^{2+} flux between 1 mM glucose and 1 mM taurine in the first 60 s following treatment ($p > 0.05$; Fig. 25.2). Following the initial 60 s, there was a significant increase ($p < 0.005$) in Ca^{2+} levels in the taurine-treated cells that persisted for an additional 120 s (Fig. 25.2).

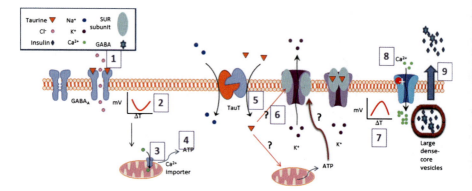

Fig. 25.3 Scheme proposing the dual roles of taurine in eliciting exocytosis of LDCV containing insulin and GABA. In Steps 1–4, taurine elicits sequestration of Ca^{2+}; in Steps 5–7 either interaction with K_{ATP} channels or cytoplasmic ATP increase via the $GABA_A$ receptor or both activates voltage-sensitive Ca^{2+} channels for Ca^{2+} flux required for exocytosis

These data suggests that taurine may initially initiate calcium sequestration, followed by calcium flux via a second mechanism. The data are similar to findings of taurine's role in calcium sequestration in neurons (El Idrissi and Trenkner 2003; El Idrissi 2008). Taurine increases cytoplasmic ATP levels as a consequence of mitochondrial calcium buffering (Han et al. 2004; El Idrissi 2008). Taurine likely increases cytoplasmic ATP levels by increasing mitochondrial Ca^{2+} influx via the Ca^{2+} uniporter, which increases mitochondrial metabolic function (Han et al. 2004). This strongly suggests that in these studies using a β-cell line, taurine is most likely acting via $GABA_A$ receptors initially. Because β-cells in vivo express $GABA_A$ receptors, we hypothesize that taurine's initial action in intact islets is to bind to $GABA_A$ receptors on β cells, causing both a hyperpolarization of the membrane and calcium sequestration. Following prolonged exposure to taurine, taurine may enter the cell via TauT where we hypothesize that intracellular taurine may bind to the K_{ATP} channel, polarizing the membrane and initiating Ca^{2+} flux that subsequently leads to prolonged exocytosis of the LDCV.

25.3.3 Proposed Model for Taurine's Influence on Insulin Release

We propose that taurine may act through two independent mechanisms, one short term and the other long term (Fig. 25.3). Initially, taurine binds to $GABA_A$ receptors (Step 1), initiating rapid Cl^- influx and hyperpolarization of β cells (Step 2). This hyperpolarization triggers calcium sequestration (Step 3), initially buffering the cell and increasing cytoplasmic ATP levels (Step 4). The timing of these initial four step must be further investigated. It is possible that the initial phase (Steps 1–4) will permit exocytosis of insulin by increasing cytoplasmic ATP levels and inhibiting K_{ATP} channels. Alternately, this may be a short-term (2 min) mechanism for β cells to retain intracellular insulin. During the subsequent 2 min interval, Ca^{2+} might be sufficient to premit limited release of insulin (Fig. 25.2).

With long-term (24 h) exposure to taurine, TauT takes up the taurine (Step 5), which may then interact with the K_{ATP} channel's sulfonylurea receptor subunit (Step 6) to inhibit the channel (Tricarico et al. 2000; Schaffer et al. 2010). Through either or both mechanisms, taurine treatment increases cytoplasmic ATP levels, inhibiting K_{ATP} channels and triggering a depolarization of β cells (Step 7), which in turn activates voltage-sensitive calcium channels (Step 8) and thus calcium flux and exocytosis of the LDCV (Step 9). Through our work and the work of others, the hypothesis of the modes of action of taurine is certainly plausible. Our laboratory will continue to examine the roles of taurine in eliciting release of insulin from the β cells.

25.4   Discussion

In the in vivo environment, when interstitial fluid levels reach ~3 mM glucose pancreatic β-cells respond by releasing insulin and inhibiting glucagon release. In our in vitro studies, we used 3 mM glucose as our minimal dose required to initiate insulin release. Conversely, 1 mM glucose served as a control dose in which insulin should not be released. When examining the relative fluorescence intensity of insulin and GABA in the Hit cells, abundant signals for both markers were observed (L'Amoreaux et al. 2010). In those studies, we demonstrated that 1 mM taurine alone is effective in lowering plasma insulin and glucose as the values observed were significantly lower than those cells treated with the subthreshold glucose concentration of 1 mM.

Since taurine can affect release of insulin and GABA from LDCV, we demonstrated here that the mechanism by which taurine promotes exocytosis of these vesicles is different from the mechanism by which SST is released. Further, we provided evidence that the taurine-dependent up-regulation of GAD expression requires a feedback mechanism from α cells. We believe that in chronic administration of taurine during early pancreatic development, that a feedback mechanism between α and β cells provide a mechanism through which taurine alters expression of the β2 subunits of $GABA_A$ receptors, which leads to a requirement for increased GABA and thus increased GAD expression. As in vivo β cells also express $GABA_A$ receptors (Braun et al. 2010), an evaluation of $GABA_A$ receptor expression on this cell line is warranted and is forthcoming.

Taurine likely interacts with pancreatic β cells through two mechanisms. Initially, taurine binds to $GABA_A$ receptors to inhibit exocytosis of LDCV. Taurine is a potent GABA agonist and, as such, is likely playing a role in the feedback mechanism to inhibit further release of insulin via LDCV exocytosis. During a prolonged exposure to taurine, the amino acid is transported into β cells via the TauT transporter.

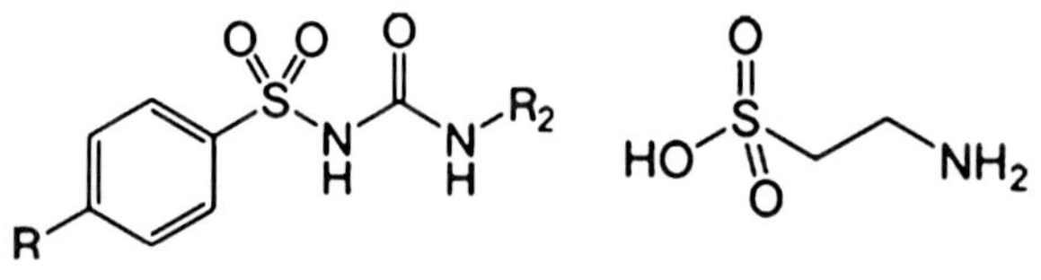

Fig. 25.4 Comparison of
generic sulfonylurea
molecule with taurine

Sulfonylurea   Taurine

Once in the cytoplasm, taurine may bind to the sulfonylurea receptor (SUR) of the ATP-dependent potassium channels. There is strong evidence that taurine can interact with SUR in muscle cells (Tricarico et al. 2000; Schaffer et al. 2010). Use of glibenclamide, a sulfonylurea, in perfused isolated pancreases moderately increases insulin release (Efendic et al. 1979) and also enhances the release of arginine-dependent release of SST (Efendic et al. 1980). Based upon the chemical similarities of the two molecules (Fig. 25.4), it is plausible that taurine may also serve as an antidiabetic agent and work via inhibition of K_{ATP} channels.

25.5   Conclusion

Taurine may serve as a cost-effective treatment for diabetes in that it promotes insulin release from β cells. We present evidence that short-term administration of taurine (such as in the diet) may elicit a response via $GABA_A$ receptors and restrict insulin release. With chronic administration of taurine, binding of taurine to the ATP-dependent potassium channels may elicit insulin release. Further studies are needed to confirm these observations and to determine the role of TauT in regulating insulin release from β cells.

Acknowledgments We thank Jonathan Blaize and Janto Tachjadi for assistance with the confocal microscopy. Support for the confocal microscope comes from the National Science Foundation (DBI 0421046). Support also from the Professional Staff Congress of the City University of New York. We would also like to thank the Organizing Committee of the 18th International Taurine Meeting for the scientific and social programs and the participants that made this a memorable conference.

References

Agrawal HC, Davison AN, Kaczmarek LK (1971) Subcellular distribution of taurine and cysteinesulphinate decarboxylase in developing rat brain. Biochem J 122:759–763
Aguila MC, McCann SM (1985) Stimulation of somatostatin release from median eminence tissue incubated in vitro by taurine and related amino acids. Endocrinology 116:1158–1162
Ahren B (2009) Islet G protein-coupled receptors as potential targets for treatment of type 2 diabetes. Nat Rev Drug Discov 8:369–385

Braun M, Ramracheya R, Bengtsson M, Clark A, Walker JN, Johnson PR, Rorsman P (2010) Gamma-aminobutyric acid (GABA) is an autocrine excitatory transmitter in human pancreatic beta-cells. Diabetes 59:1694–1701

Braun M, Wendt A, Birnir B, Broman J, Eliasson L, Galvanovskis J, Gromada J, Mulder H, Rorsman P (2004) Regulated exocytosis of GABA-containing synaptic-like microvesicles in pancreatic beta-cells. J Gen Physiol 123:191–204

Braun M, Wendt A, Karanauskaite J, Galvanovskis J, Clark A, MacDonald PE, Rorsman P (2007) Corelease and differential exit via the fusion pore of GABA, serotonin, and ATP from LDCV in rat pancreatic beta cells. J Gen Physiol 129:221–231

Cheng H, Yibchok-anun S, Coy DH, Hsu WH (2002) SSTR2 mediates the somatostatin-induced increase in intracellular Ca(2+) concentration and insulin secretion in the presence of arginine vasopressin in clonal beta-cell HIT-T15. Life Sci 71:927–936

Cho JH, Chen L, Kim MH, Chow RH, Hille B, Koh DS (2010) Characteristics and functions of {alpha}-amino-3-hydroxy-5-methyl-4-isoxazolepropionate receptors expressed in mouse pancreatic {alpha}-cells. Endocrinology 151:1541–1550

Doyle ME, Egan JM (2003) Pharmacological agents that directly modulate insulin secretion. Pharmacol Rev 55:105–131

Efendic S, Enzmann F, Nylen A, Uvnas-Wallensten K, Luft R (1979) Effect of glucose/sulfonylurea interaction on release of insulin, glucagon, and somatostatin from isolated perfused rat pancreas. Proc Natl Acad Sci U S A 76:5901–5904

Efendic S, Enzmann F, Nylen A, Uvnas-Wallensten K, Luft R (1980) Sulphonylurea (glubenclamide) enhances somatostatin and inhibits glucagon release induced by arginine. Acta Physiol Scand 108:231–233

El Idrissi A (2008) Taurine increases mitochondrial buffering of calcium: role in neuroprotection. Amino Acids 34:321–328

El Idrissi A, Boukarrou L, L'Amoreaux W (2009a) Taurine supplementation and pancreatic remodeling. Adv Exp Med Biol 643:353–358

El Idrissi A, Boukarrou L, Splavnyk K, Zavyalova E, Meehan EF, L'Amoreaux W (2009b) Functional implication of taurine in aging. Adv Exp Med Biol 643:199–206

El Idrissi A, Trenkner E (2003) Taurine regulates mitochondrial calcium homeostasis. Adv Exp Med Biol 526:527–536

El Idrissi A, Yan X, L'Amoreaux W, Brown WT, Dobkin C (2012) Neuroendocrine alterations in the fragile X mouse. Results Probl Cell Differ 54:201–221

El Idrissi A, Yan X, Sidime F, L'Amoreaux W (2010) Neuro-endocrine basis for altered plasma glucose homeostasis in the Fragile X mouse. J Biomed Sci 17 Suppl 1:S8

Gilon P, Bertrand G, Loubatieres-Mariani MM, Remacle C, Henquin JC (1991) The influence of gamma-aminobutyric acid on hormone release by the mouse and rat endocrine pancreas. Endocrinology 129:2521–2529

Goldsmith PC, Rose JC, Arimura A, Ganong WF (1975) Ultrastructural localization of somatostatin in pancreatic islets of the rat. Endocrinology 97:1061–1064

Gu XH, Kurose T, Kato S, Masuda K, Tsuda K, Ishida H, Seino Y (1993) Suppressive effect of GABA on insulin secretion from the pancreatic beta-cells in the rat. Life Sci 52:687–694

Han J, Bae JH, Kim SY, Lee HY, Jang BC, Lee IK, Cho CH, Lim JG, Suh SI, Kwon TK, Park JW, Ryu SY, Ho WK, Earm YE, Song DK (2004) Taurine increases glucose sensitivity of UCP2-overexpressing beta-cells by ameliorating mitochondrial metabolism. Am J Physiol Endocrinol Metab 287:E1008–E1018

Hayashi M, Yamada H, Uehara S, Morimoto R, Muroyama A, Yatsushiro S, Takeda J, Yamamoto A, Moriyama Y (2003) Secretory granule-mediated co-secretion of L-glutamate and glucagon triggers glutamatergic signal transmission in islets of Langerhans. J Biol Chem 278:1966–1974

Kawai, Unger RH (1983) Effects of gamma-aminobutyric acid on insulin, glucagon, and somatostatin release from isolated perfused dog pancreas. Endocrinology 113:111–113

Koh DS, Cho JH, Chen L (2012) Paracrine interactions within islets of Langerhans. J Mol Neurosci 48(2):429–440

L'Amoreaux WJ, Cuttitta C, Santora A, Blaize JF, Tachjadi J, El Idrissi A (2010) Taurine regulates insulin release from pancreatic beta cell lines. J Biomed Sci 17(Suppl 1):S11

McDermott AM, Sharp GW (1993) Inhibition of insulin secretion: a fail-safe system. Cell Signal 5:229–234

Muroyama A, Uehara S, Yatsushiro S, Echigo N, Morimoto R, Morita M, Hayashi M, Yamamoto A, Koh DS, Moriyama Y (2004) A novel variant of ionotropic glutamate receptor regulates somatostatin secretion from delta-cells of islets of Langerhans. Diabetes 53:1743–1753

Nathan B, Floor E, Kuo CY, Wu JY (1995) Synaptic vesicle-associated glutamate decarboxylase: identification and relationship to insulin-dependent diabetes mellitus. J Neurosci Res 40:134–137

Rorsman P, Berggren PO, Bokvist K, Ericson H, Mohler H, Ostenson CG, Smith PA (1989) Glucose-inhibition of glucagon secretion involves activation of GABAA-receptor chloride channels. Nature 341:233–236

Satin LS, Kinard TA (1998) Neurotransmitters and their receptors in the islets of Langerhans of the pancreas: what messages do acetylcholine, glutamate, and GABA transmit? Endocrine 8:213–223

Schaffer SW, Jong CJ, Ramila KC, Azuma J (2010) Physiological roles of taurine in heart and muscle. J Biomed Sci 17 Suppl 1:S2

Seaquist ER, Armstrong MB, Gettys TW, Walseth TF (1995) Somatostatin selectively couples to G(o) alpha in HIT-T15 cells. Diabetes 44:85–89

Sorenson RL, Garry DG, Brelje TC (1991) Structural and functional considerations of GABA in islets of Langerhans. Beta-cells and nerves. Diabetes 40:1365–1374

Suckale J, Solimena M (2010) The insulin secretory granule as a signaling hub. Trends Endocrinol Metab 21:599–609

Taborsky GJ Jr (1983) Evidence of a paracrine role for pancreatic somatostatin in vivo. Am J Physiol 245:E598–E603

Thermos K, Meglasson MD, Nelson J, Lounsbury KM, Reisine T (1990) Pancreatic beta-cell somatostatin receptors. Am J Physiol 259:E216–E224

Tricarico D, Barbieri M, Camerino DC (2000) Taurine blocks ATP-sensitive potassium channels of rat skeletal muscle fibres interfering with the sulphonylurea receptor. Br J Pharmacol 130:827–834

Wu JY, Prentice H (2010) Role of taurine in the central nervous system. J Biomed Sci 17 Suppl 1:S1

Yao CY, Gill M, Martens CA, Coy DH, Hsu WH (2005) Somatostatin inhibits insulin release via SSTR2 in hamster clonal beta-cells and pancreatic islets. Regul Pept 129:79–84

Chapter 26
Antidiabetic Effect of Taurine in Cultured Rat Skeletal L6 Myotubes

Sun Hee Cheong and Kyung Ja Chang

Abstract Taurine (2-aminoethanesulfonic acid), a sulfur-containing β-amino acid, is found in all animal cells at millimolar concentrations and has been reported to show various health promoting activities including antidiabetic properties. The beneficial effects of taurine in diabetes mellitus have been known. However, the exact mechanism of hypoglycemic action of taurine is not properly defined. In this study, we investigated antidiabetic effect of taurine in the cell culture system using rat skeletal muscle cells. In cultured rat skeletal L6 myotubes, we studied the effect of taurine (0–100 μM) on glucose uptake to plasma membrane from the aspects of AMP-activated protein kinase (AMPK) signaling. Taurine stimulated glucose uptake in a dose-dependent manner by activating AMPK signaling. From these results, it may suggest that taurine show antidiabetic effect by stimulating insulin-independent glucose uptake in rat skeletal muscle.

Abbreviations

AMPK   AMP-activated protein kinase
PI3K   Phosphatidylinositol-3 kinase
GLUT4   Glucose transporter 4

S.H. Cheong
Department of Applied Biological Chemistry, Tokyo University of Agriculture and Technology, Fuchu, Tokyo 183-8509, Japan

K.J. Chang(✉)
Department of Food and Nutrition, Inha University,
Incheon 402-751, Korea
e-mail: kjchang@inha.ac.kr

A. El Idrissi and W.J. L'Amoreaux (eds.), *Taurine 8*, Advances in Experimental
Medicine and Biology 775, DOI 10.1007/978-1-4614-6130-2_26,
© Springer Science+Business Media New York 2013

26.1   Introduction

Taurine, a sulfur containing beta amino acid, is present in most animal tissues and is essential for the normal functioning of several organs (Brosnan and Brosnan 2006). It has been reported that taurine exhibits antioxidative properties, controls blood pressure, membrane stabilizing effect, regulates intracellular Ca^{2+} concentration, inhibits apoptosis, and reduces the levels of pro-inflammatory cytokines in various organs (Aerts and Van Assche 2002; Racasan et al. 2004; Sinha et al. 2007; Das et al. 2008, 2009; Manna et al. 2010). Moreover, taurine is found at high concentrations inside glucagon and somatostatin-containing cells in the pancreatic islets and increases insulin secretion, sensitivity and glucose uptake in different experimental conditions (Cherif et al. 1998; De la Puerta et al. 2010).

Diabetes is the most common and serious metabolic disease. Several trials have been conducted to reduce the hyperglycemia (Moller 2001). Chang (2000) reported that dietary supplementation with taurine was shown to protect pancreatic β-cells in the streptozotocin model of type 1 diabetes. On the other hand, it has been proven that diabetes is associated with a decrease in the levels of endogenous antioxidants, particularly of taurine, so that oxidative damage may be enhanced by the deficiency of taurine, since it frequently becomes depleted in diabetic states (Schaffer et al. 2009). The skeletal muscles which account for the majority of insulin-mediated glucose uptake in the postprandial state play an important role in maintaining glucose homeostasis (Saltiel and Kahn 2001). In skeletal muscle, especially, insulin increases glucose uptake via a signaling that leads to activation of phosphatidylinositol-3 kinase (PI3K) and AKt, resulting in increased translocation of glucose transporter 4 (GLUT4) to the plasma membrane (Saltiel and Kahn 2001). In mammalian cells, the AMP-activated protein kinase (AMPK), which is another GLUT4 translocation promoter, acts as an energy sensor and is activated by an increase in AMP/ATP ratio, by exercise/contraction, and by several compounds including metformin and thiazolidinedione, resulting in stimulation of glucose uptake in skeletal muscles (Zou et al. 2004; Towler and Hardie 2007). Although a number of studies concerning the effect of taurine on diabetes or hyperglycemia have been reported, little has been validated the effect of taurine on glucose uptake using L6 myotubes. Therefore, this study was conducted to investigate the effects of taurine on glucose uptake in vitro using L6 myotubes and to clarify the mechanisms associated with the enhanced glucose uptake.

26.2   Methods

26.2.1   Materials

L6 myoblast cells derived from a rat were purchased from American Type Culture Collection (Rockville, MD, USA; ATCC numbers: CRL-1458). Taurine was obtained from Donga Pharm. (Seoul, Korea). The following items were purchased from the cited commercial sources: Glucose CII Test Kit and compound C from

Wako Pure Chemical Industries Ltd. (Osaka, Japan); Dulbecco's modified Eagle's medium (DMEM), fetal bovine serum (FBS) and Gö6983 from Sigma (St. Louis, MO, USA); anti-phospho-AMPK and anti-AMPK antibodies from Cell Signaling Technology (Beverly, MA, USA); horseradish-peroxidase conjugated anti-mouse and anti-rabbit IgG antibodies from Invitrogen (San Diego, CA, USA); ECL Plus Western blotting detection reagents and Hybond ECL nitrocellulose membrane from GE Healthcare (Buckinghamshire, UK).

26.2.2 Culture of L6 Myoblast Cells

L6 myoblast cells were cultured in DMEM containing 10% (vol/vol) FBS, penicillin G (100 U/ml), and streptomycin (100 μg/ml) in a humidified 5% CO_2 incubator at 37°C. To differentiate into myotubes, the myoblast cells (5×10^4 or 7×10^5) were seeded in Falcon 24-place multiwell plates or 60-mM culture dishes and cultured to 90% confluency in DMEM containing 2% FBS for 1 week.

26.2.3 Determination of Glucose Uptake by Cultured Rat Skeletal L6 Myotubes

Briefly, perfused L6 myoblast cells were subcultured into Falcon 24-place multiwall plates at 5×10^4 cells/well and grown for 11 days in 0.4 ml of 2% FBS/DMEM to allow the formation of myotubes. The medium was renewed every 2 days. Subsequently, the 11-day-old myotubes were incubated in filter-sterilized Krebs–Henseleit buffer (141 mg/l $MgSO_4$, 160 mg/l KH_2PO_4, 350 mg/l KCl, 6,900 mg/l NaCl, 373 mg/l $CaCl_2 \cdot 2H_2O$, 2,100 mg/l $NaHCO_3$, pH 7.4) containing 0.1% bovine serum albumin, 10 mM Hepes, and 2 mM sodium pyruvate (KHH buffer) for 2 h. The myotubes were then cultured for 4 h in KHH buffer containing 11 mM glucose with or without taurine (10–100 μM) and with or without 100 nM insulin or 5 μM compound C, an AMPK inhibitor. The differences in the glucose concentrations in the KHH buffer before and after culture were determined by the absorbance at 505 nM using a microplate reader (Model AD200; Beckman Coulter, Brea, CA, USA) and the Glucose CII Test Kit. The amounts of glucose consumed were calculated.

26.2.4 Western Blot Analysis

Rat skeletal L6 myotubes were solubilized in a lysis buffer [10 mM Tris–HCl pH 7.4, 150 mM NaCl, 1% NP-40, 0.5% sodium deoxycholate, 0.1% sodium dodecyl sulfate (SDS), 0.5 mM dithiothreitol, 0.2 mg/ml Pefabloc SC, 1 mM Na_3VO_4] for 30 min at 4°C. The lysates were then sonicated for 10 s and centrifuged at $12,000 \times g$ for 15 min at 4°C. The protein concentrations of the supernatants were evaluated using a protein assay reagent (Bio-Rad Laboratories). Equal amounts of protein

(20 μg/lane) and prestained molecular weight markers (Wako Pure Chemical Industries Ltd.) were loaded onto 10% premade polyacrylamide gels (Wako Pure Chemical Industries Ltd.), separated by electrophoresis and transferred to nitrocellulose membranes. The membranes were then incubated in a blocking solution comprising 3% BSA in Tris-buffered saline (TBS) for 1 h. After the incubation, the membranes were washed in TBS and incubated with anti-phospho-AMPK and anti-AMPK overnight at 4°C. The membranes were then washed in TBS containing 0.1% (vol/vol) Tween-20 for 30 min and incubated with horseradish-peroxidase-conjugated anti-mouse or anti-rabbit IgG antibodies at a dilution of 1:5,000 for 60 min at room temperature. Immunoreactive bands were detected using ECL Plus Western blotting detection reagents. The intensity of each band was analyzed with a lumino-image analyzer (Model LAS-4000 Mini; Fujifilm, Tokyo, Japan) coupled with image analysis software (Multi Gauge Ver. 3.0; Fujifilm).

26.2.5   Statistical Analysis

All data are presented as the mean ± SEM. The data were evaluated by a one-way analysis of variance. Differences between the mean values were assessed using Tukey-Kramer multiple comparison test. Statistical significance was considered for values of $P < 0.05$.

26.3   Results

26.3.1   Taurine Stimulates Glucose Uptake in Cultured Rat Skeletal L6 Myotubes

We determined the effect of taurine on glucose uptake under normal (11 mM) glucose condition. In this study, taurine dose-dependently and significantly stimulated glucose uptake at concentrations of 25–100 μM (Fig. 26.1). Based on these results, we adopted 100 μM for glucose conditions as the optimal taurine concentrations in the following experiments.

26.3.2   Influence of Insulin on Taurine Induced Glucose Uptake

To determine the regulatory mechanism by which taurine induced the glucose uptake in rat skeletal L6 myotubes, we performed glucose uptake assays using 100 nM insulin condition. In this study, taurine significantly stimulated glucose uptake at concentrations of 100 μM independently of insulin (Fig. 26.2). These results suggest that the stimulatory effect of taurine on glucose uptake is independent on the insulin.

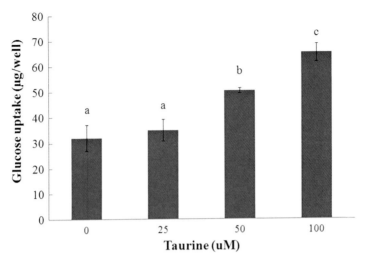

Fig. 26.1 Effect of taurine on glucose uptake in cultured rat skeletal L6 myotubes. L6 myotubes were preincubated in KHH buffer without glucose for 2 h. They were then incubated in KHH buffer containing 11 mM glucose and 0, 25, 50, or 100 µM taurine for 4 h, and the glucose uptake was determined. Each value represents the mean ± SEM for six wells. Values not sharing a common letter are significantly different at $P < 0.05$ by Tukey-Kramer multiple comparison test

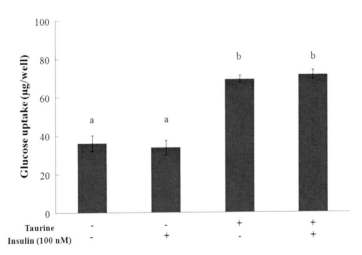

Fig. 26.2 Influence of insulin on taurine induced glucose uptake. L6 myotubes were preincubated in KHH buffer without glucose for 2 h. They were then incubated for 4 h in Krebs–Henseleit buffer with 11 mM glucose in the presence or absence of 0 or 100 µM taurine and 100 nM insulin. Each value represents the mean ± SEM for six wells. Values not sharing a common letter are significantly different at $P < 0.05$ by Tukey-Kramer multiple comparison test

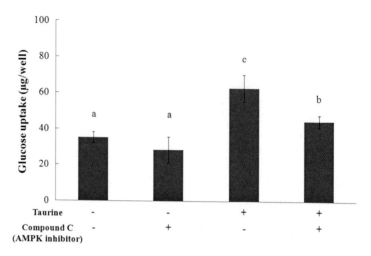

Fig. 26.3 Effects of AMPK inhibitor on taurine-promoted glucose uptake. L6 myotubes were preincubated in KHH buffer without glucose for 2 h. They were then incubated for 4 h in Krebs–Henseleit buffer with 11 mM glucose in the presence or absence of 0 or 100 μM taurine and 5 μM compound C. Each value represents the mean ± SEM for six wells. Values not sharing a common letter are significantly different at $P < 0.05$ by Tukey-Kramer multiple comparison test

26.3.3 Stimulatory Effect of Taurine on Glucose Uptake Is Dependent on the AMPK Pathways

To determine the regulatory mechanism by which taurine induced the glucose uptake in rat skeletal L6 myotubes, we performed glucose uptake assays using kinase inhibitors, namely, compound C, an ATP-competitive inhibitor of AMPK. The promotion of glucose uptake by taurine was completely inhibited by the treatments with compound C (Fig. 26.3). These results suggest that the stimulatory effect of taurine on glucose uptake is dependent on the AMPK pathways.

26.3.4 Taurine Induces the Phosphorylation of AMPK

To examine the activity of AMPK, a well-known main regulator of glucose uptake in skeletal muscle cells, we investigated the temporal expression of phosphorylated AMPK. Taurine time-dependently stimulated the phosphorylation of AMPK under normal glucose conditions (Fig. 26.4). Consequently, these data suggest that the main mechanism of glucose uptake by taurine is mediated by the AMPK pathway.

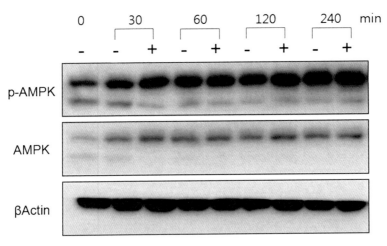

Fig. 26.4 Effect of taurine on the phosphorylation of AMPK. L6 myotubes were preincubated in Krebs–Henseleit buffer without glucose for 2 h. They were then incubated in KHH buffer containing 11 mM glucose in the presence or absence of 0 or 100 μM taurine for 30, 60, 120, or 240 min. Total lysates were analyzed by immunoblotting with anti-phospho-AMPK and anti-AMPK antibodies

26.4 Discussion

Taurine, a β-aminosulfonic acid, is the most abundant amino acid and is essential for sustain several structure and function. Recently, several studies have indicated that taurine exhibits beneficial effects in diabetic patients. In addition, it has been reported that taurine influences various biological functions, including antioxidant, brain and retinal development, cell membrane stabilization, osmoregulation, and hypoglycemic action (Thurston et al. 1980; Pasantes-Morales et al. 1985; El Idrissi and Trenkner 2004). In this study, we confirmed the effect of taurine on glucose uptake to the muscle cell, and clarified the regulatory mechanism of glucose uptake by taurine such as AMPK activation in cultured rat skeletal myotubes under normal glucose condition. Our data showed that taurine dose-dependently and significantly stimulated glucose uptake at concentrations from 25 to 100 μM in cultured rat skeletal L6 myotubes. Especially, rat skeletal L6 myotubes are insulin-insensitive cells on glucose uptake and our data showed that the enhancement of glucose uptake by insulin in the L6 myotubes was significantly lower than that by taurine. Also, the effect of taurine was not significantly different in the absence or presence of insulin. Therefore, it was confirmed that the stimulatory effect of taurine on glucose uptake is independent on the action of insulin, even under insulin-insensitive conditions. Several previous studies indicate that taurine shows hypoglycemic effects by enhancing insulin action, as well as by facilitating the interaction of insulin with its receptor (Lampson et al. 1983; Maturo and Kulakowski 1988). Ribeiro et al. (2009) also reported that plasma taurine level seems be important for β-cell function and insulin action. However, in contrast, it was reported that there was no significant differences after taurine intervention compared to placebo in incremental insulin response, neither during intravenous ted glucose tolerance test (IVGTT) nor in insulin-stimulated glucose disposal during the clamp in type 2 diabetes patients (Brons et al. 2004). Also, Doi et al. (2003) reported that among the branched-chain amino acids, leucine and isoleucine increase glucose uptake in an insulin-independent manner in C_2C_{12} skeletal muscle cells. Several studies have indicated that taurine is involved in glucose homeostasis; however, the specific molecular mechanisms are unknown (Kulakowski and Maturo 1984; Franconi et al. 2004).

Insulin-stimulated glucose uptake by skeletal muscle plays an important role in the maintenance of whole-body glucose homeostasis (Herman and Kahn 2006). AMPK is an important protein to provide energy in mammalian cells (Towler and Hardie 2007). AMPK is activated in the skeletal muscle of mammals by exercise and this activation is associated with an increase in GLUT4-mediated glucose uptake by the tissue (Jessen and Goodyear 2005; Magnoni et al. 2012). Based on these signaling pathways related to glucose uptake, we investigated the signaling pathways for glucose uptake by taurine using kinase inhibitor. In this study, it was indicated that the stimulatory effect of taurine on glucose uptake is stimulated on the AMPK pathways by promoting the phosphorylation of AMPK (AMPK signaling) under normal glucose condition time-dependently. Recently, Solon et al. (2011) reported that taurine acted similarly to insulin, stimulating the activities of the Akt/FOXO1 and JAK2/STAT3 signaling pathways, while inhibiting the AMPK signaling pathway. On the other hand, Carneiro et al. (2009) reported that mice supplemented with taurine had a significant increased tyrosine phosphorylation of the insulin receptor in skeletal muscle, both at basal and insulin-stimulated states. In another animal experiment, it was reported that taurine increased insulin signal transduction through the phosphatidylinositol 3 kinase (PI3K) pathway, resulting in increased glucose uptake (Colivicchi et al. 2004).

These results of this study suggest that taurine has a beneficial effect on glucose uptake in the muscle and that this effect is mediated through a mechanism including the activation of AMPK.

26.5   Conclusion

Our present study shows that taurine improve the glucose uptake by increasing the AMPK phosphorylation in rat skeletal L6 myotubes. These results may suggest that taurine has an antidiabetic effect by stimulating insulin-independent glucose uptake in skeletal muscle and may have hypoglycemic effects in diabetes.

References

Aerts L, Van Assche FA (2002) Taurine and taurine-deficiency in the perinatal period. J Perinat Med 30:281–286. doi:10.1515/JPM.2002.040

Brons C, Spohr C, Storgaard H, Dyerberg J, Vaag A (2004) Effect of taurine treatment on insulin secretion and action on serum lipid levels in overweighted men with genetic predisposition for type II diabetes mellitus. Eur J Clin Nutr 58:1239–1247. doi:10.1038/sj.ejcn.1601955

Brosnan JT, Brosnan ME (2006) The sulfur-containing amino acids: an overview. J Nutr 136:1636S–1640S

Carneiro EM, Latorraca MQ, Araujo E, Beltrá M, Oliveras MJ, Navarro M, Berná G, Bedoya FJ, Velloso LA, Soria B, Martin F (2009) Taurine supplementation modulates glucose homeostasis and islet function. J Nutr Biochem 20:503–511. doi:10.1016/j.jnutbio.2008.05.008

Chang KJ (2000) Effect of taurine and β-alanine on morphological changes of pancreas in streptozotocin-induced rats. Adv Exp Med Biol 483:571–577. doi:10.1007/0-306-46838-7_61

Cherif H, Reusens B, Ahn MT, Hoet JJ, Remacle C (1998) Effects of taurine on the insulin secretion of rat fetal islets from dams fed a low-protein diet. J Endocrinol 159:341–348. doi:10.1677/joe.0.1590341

Colivicchi MA, Raimondi L, Bianchi L, Tipton KF, Pirisino R, Della Corte L (2004) Taurine prevents streptozotocin impairment of hormone-stimulated glucose uptake in rat adipocytes. Eur J Pharmacol 495:209–215. doi:10.1016/j.ejphar.2004.05.004

Das J, Ghosh J, Manna PM, Sil PC (2008) Taurine provides antioxidant defense against NaF-induced cytotoxicity in murine hepatocytes. Pathophysiology 15:181–190. doi:10.1016/j.pathophys.2008.06.002

Das J, Ghosh J, Manna P, Sinha M, Sil PC (2009) Taurine protects rat testes against NaAsO2-induced oxidative stress and apoptosis via mitochondrial dependent and independent pathways. Toxicol Lett 187:201–210. doi:10.1016/j.toxlet.2009.03.001

De la Puerta C, Arrieta FJ, Balsa JA, Botella-Carretero JI, Zamarron I, Vazquez C (2010) Taurine and glucose metabolism: a review. Nutr Hosp 25:910–919. doi:10.3305/nh.2010.25.6.4815

Doi M, Yamaoka I, Fukunaga T, Nakayama M (2003) Isoleucine, a potent plasma glucose-lowering amino acid, stimulates glucose uptake in C_2C_{12} myotubes. Biochem Biophys Res Commun 312:1111–1117. doi:10.1016/j.bbrc.2003.11.039

El Idrissi A, Trenkner E (2004) Taurine as a modulator of excitatory and inhibitory neurotransmission. Neurochem Res 1:189–197. doi:10.1023/B:NERE.0000010448.17740.6e

Franconi F, Di Leo MA, Bennardini F, Ghirlanda G (2004) Is taurine beneficial in reducing risk factors for diabetes mellitus? Neurochem Res 29:143–150

Herman MA, Kahn BB (2006) Glucose transport and sensing in the maintenance of glucose homeostasis and metabolic harmony. J Clin Invest 116:1767–1775. doi:10.1172/JCI29027

Jessen N, Goodyear LJ (2005) Contraction signaling to glucose transport in skeletal muscle. J Appl Physiol 99:330–337. doi:10.1152/japplphysiol.00175.2005

Kulakowski EC, Maturo J (1984) Hypoglycemic properties of taurine: not mediated by enhanced insulin release. Biochem Pharmacol 33:2835–2838. doi:10.1016/0006-2952(84)90204-1

Lampson WG, Kramer JH, Schaffer SW (1983) Potentiation of the actions of insulin by taurine. Can J Physiol Pharmacol 61:457–463. doi:10.1139/y83-070

Magnoni LJ, Vraskou Y, Palstra AP, Planas JV (2012) AMP-activated protein kinase plays an important evolutionary conserved role in the regulation of glucose metabolism in fish skeletal muscle cells. PLoS One 7:e31219. doi:10.1371/journal.pone.0031219

Manna P, Das J, Ghosh J, Sil PC (2010) Contribution of type 1 diabetes to rat liver dysfunction and cellular damage via activation of NOS, PARP, IκBα/NF-κB, MAPKs, and mitochondria-dependent pathways: prophylactic role of arjunolic acid. Free Radic Biol Med 48:1465–1484. doi:10.1016/j.freeradbiomed.2010.02.025

Maturo J, Kulakowski EC (1988) Taurine binding to the purified insulin receptor. Biochem Pharmacol 37:3755–3760. doi:10.1016/0006-2952(88)90411-X

Moller DE (2001) New drug targets for type 2 diabetes and the metabolic syndrome. Nature 414:821–827. doi:10.1038/414821a

Pasantes-Morales H, Wright CE, Gaull GE (1985) Taurine protection of lymphoblastoid cells from iron-ascorbate-induced damage. Biochem Pharmacol 34:2205–2207. doi:10.1016/0006-2952 (85)90419-8

Racasan S, Braam B, van der Giezen DM, Goldschmeding R, Boer P, Koomans HA (2004) Perinatal L-arginine and antioxidant supplements reduce adult blood pressure in spontaneously hypertensive rats. Hypertension 44:83–88. doi:10.1161/01.HYP.0000133251.40322.20

Ribeiro RA, Bonfleur ML, Amaral AG, Vanzela EC, Rocco SA, Boschero AC, Carneiro EM (2009) Taurine supplementation enhances nutrient-induced insulin secretion in pancreatic mice islets. Diabetes Metab Res Rev 25:370–379. doi:10.1002/dmrr.959

Saltiel AR, Kahn CR (2001) Insulin signaling and the regulation of glucose and lipid metabolism. Nature 414:799–806. doi:10.1038/414799a

Schaffer SW, Azuma J, Mozaffari M (2009) Role of antioxidant activity of taurine in diabetes. Can J Physiol Pharmacol 87:91–99. doi:10.1139/Y08-110

Sinha M, Manna P, Sil PC (2007) Taurine, a conditionally essential amino acid, ameliorates arsenic-induced cytotoxicity in murine hepatocytes. Toxicol In Vitro 21:1419–1428. doi:10.1016/j.tiv.2007.05.010

Solon CS, Franci D, Ignacio-Souza LM, Romanatto T, Roman EA, Arruda AP, Morari J, Torsoni AS, Carneiro EM, Velloso LA (2011) Taurine enhances the anorexigenic effects of insulin in the hypothalamus of rats. Amino Acids. doi:10.1007/s00726-011-1045-5

Thurston JH, Hauhart RE, Dirgo JA (1980) Taurine: a role in osmotic regulation of mammalian brain and possible clinical significance. Life Sci 26:1561–1568. doi:10.1016/0024-3205 (80)90358-6

Towler MC, Hardie DG (2007) AMP-activated protein kinase in metabolic control and insulin signaling. Circ Res 100:328–341. doi:10.1161/01.RES.0000256090.42690.05

Zou MH, Kirkpatrick SS, Davis BJ, Nelson JS, Wiles WG IV, Schlattner U, Neumann D, Brownlee M, Freeman MB, Goldman MH (2004) Activation of the AMP-activated protein kinase by the anti-diabetic drug metformin in vivo. Role of mitochondrial reactive nitrogen species. J Biol Chem 279:43940–43951. doi:10.1074/jbc.M404421200

Chapter 27
Protection by Taurine and Thiotaurine Against Biochemical and Cellular Alterations Induced by Diabetes in a Rat Model

Roshil Budhram, Kashyap G. Pandya, and Cesar A. Lau-Cam

Abstract In this study, the actions of taurine (TAU), a sulfonate, and thiotaurine (TTAU), a thiosulfonate, on diabetes-mediated biochemical alterations in red blood cells (RBCs) and plasma and on the RBC membrane, morphology and spectrin distribution were examined in rats. Diabetes was induced in male Sprague–Dawley rats with streptozotocin (60 mg/kg i.p.) and allowed to progress for 14 days. From days to 56, the rats received a daily, 2.4 mmol/kg, oral dose of TAU or TTAU, 2 mL oral dose of physiological saline or 4 U/kg subcutaneous dose of isophane insulin (INS). Naive rats served as the control group. The rats were sacrificed on day 57 and their blood was collected to measure HbA_{1c}, to isolate intact RBCs, and to obtain plasma. A 6-weeks treatment with INS effectively lowered the elevations in plasma glucose, cholesterol, triglycerides, and plasma and RBC malondialdehyde and glutathione disulfide while effectively counteracting the decreases in plasma INS, plasma and RBC glutathione redox status, and plasma and RBC activities of antioxidant enzymes caused by diabetes. Also, INS returned the echynocytic appearance and peripheral location of spectrin seen in RBCs from diabetic rats to the normal discocytic shape and uniform distribution. TAU and TTAU were as effective as INS in inhibiting malondialdehyde formation, changes in redox status and oxidative stress in both the plasma and RBC, but were much less effective in controlling hyperglycemia and hypoinsulinemia. Furthermore TTAU was more effective than INS or TAU in lowering the increase in cholesterol to phospholipids ratio in the RBC membrane and, unlike TAU, it was able to normalize the RBC morphology and spectrin distribution.

R. Budhram • K.G. Pandya • C.A. Lau-Cam (✉)
Department of Pharmaceutical Sciences, College of Pharmacy and Allied Health Professions,
St. John's University, Jamaica, New York, NY, USA
e-mail: claucam@usa.net

A. El Idrissi and W.J. L'Amoreaux (eds.), *Taurine 8*, Advances in Experimental
Medicine and Biology 775, DOI 10.1007/978-1-4614-6130-2_27,
© Springer Science+Business Media New York 2013

Abbreviations

TAU	Taurine
TTAU	Thiotaurine
INS	Insulin
STZ	Streptozotocin
RBCs	Red blood cells
GLC	Glucose
HbA_{1c}	Glycated hemoglobin
LPO	Lipid peroxidation
MDA	Malondialdehyde
CHOL	Cholesterol
TGs	Triglycerides
PLPs	Phospholipids
GSG	Reduced glutathione
GSSG	Glutathione disulfide
CAT	Catalase
GPx	Glutathione peroxidase
SOD	Superoxide dismutase

27.1 Introduction

Type 2 diabetes mellitus is a heterogeneous metabolic disorder characterized by chronic hyperglycemia, resistance of peripheral tissues to the effects of insulin, hypertension, elevated HbA_{1c}, and a common form of dyslipidemia (raised triglycerides, and low high-density lipoprotein cholesterol with or without elevation of low-density lipoprotein cholesterol) (Järvi et al. 1999; Mayfield 1998; Moller 2001).

At present, type 2 diabetes is recognized as a major risk of coronary, cerebral, and peripheral artery disease, and as a determining factor for late-stage complications such as nephropathy, retinopathy, and neuropathy (Moller 2001). From the large number of studies that have been directed at establishing the roles of different factors in the development of diabetic complications, it has become apparent that their etiology is multifactorial and to a large extent dependent on hyperglycemia and on the development of oxidative stress (Monnier et al. 2006). Furthermore, postprandial hyperglycemic spikes have been found to correlate closely with the activation of oxidative stress, an important contributor to the pathogenesis and progression of diabetic tissue damage and to the later development of diabetic complications (Brownlee 2005; Ceriello 2005).

In diabetes, oxidative stress seems to be the result of an increased production of reactive oxygen (ROS) and nitrogen (RNS) species and a sharp decline of antioxidant defenses (Baynes and Thorpe 1999). ROS may arise from a variety of sources, including the autoxidation of monosaccharides, glycated proteins and glycated lipids, from the activities of nitric oxide synthase, NAD(P)H oxidase (Guzik et al. 2002), xanthine dehydrogenase (Alciguzel et al. 2003), xanthine oxidase (Desco et al. 2002), lipoxygenase (Obrosova et al. 2010), and cytochrome P450 monooxygenases (Kalapos et al. 1993).

Additional contributors to the oxidative stress of diabetes are the glucose-induced activation of protein kinase C isoforms and the nuclear factor kappa B (NFκB), increased formation of glucose-derived advanced glycation products (AGEs), increased glucose flux through the polyol pathway (Nishikawa et al. 2000), increased formation of angiotensin II, a promoter of mitochondrial superoxide anion production (Ricci et al. 2008), and the hexosamine pathway (Brownlee 2001).

In recent years, the mitochondrial electron transport chain (ETC) has been implicated as a major source of ROS under the influence of hyperglycemia. According to this postulate, under normoglycemic conditions the oxidative degradation of metabolic products of glucose by the tricarboxylic acid (TCA) cycle will contribute electrons to the electron transport chain (ETC) through the generation of NADH and $FADH_2$. The flow of these electrons down complexes I–IV of the ETC towards diatomic oxygen will generate water and, at the same time, will create a voltage gradient across the inner mitochondrial membrane to drive protons across and the synthesis of ATP (Brownlee 2005). However, under the hyperglycemic conditions of diabetes, there will be more glucose becoming channeled into the TCA cycle and, hence, more electrons will be fed by NADH and $FADH_2$ into the ETC to further energize the mitochondrial membrane to a critical threshold which, when exceeded, will inhibit electron transfer to complex III and divert the electrons, one at a time, to molecular oxygen to yield superoxide anion at the cost of ATP formation (Brownlee 2005; Pieczenik and Neustadt 2007). This problem may be compounded upon injury to mitochondrial component of the ETC by ROS generated in the mitochondrion itself, with complexes I and III appearing particularly susceptible (Pieczenik and Neustadt 2007). Conversely, normalizing mitochondrial superoxide anion production blocks glucose-induced activation of protein kinase C, formation of AGEs, sorbitol accumulation, and NFκB activation of hyperglycemic damage (Nishikawa et al. 2000).

Based on the role played by hyperglycemia-stimulated oxidative stress in the development and progression of diabetes-related complications, numerous antioxidants have been tested on the premise that amelioration of oxidative stress will be of help in the management of diabetes and its complications (Rahimi et al. 2005). One of the compounds that has received extensive evaluation for this purpose has been taurine (TAU), in all likelihood because of its ability to attenuate oxidative stress from different sources by inhibiting lipid peroxidation, maintaining the intracellular glutathione redox status, preserving the activity of antioxidant enzymes (Acharya and Lau-Cam 2010), diminishing hyperlipidemia (Goodman and Shihabi 1990), and preventing the formation of AGEs (Nandhini et al. 2004; Trachtman et al. 1995). More importantly, by forming a TAU–tRNA conjugate with mitochondrial tRNA, TAU of exogenous origin will induce the translation and expression of mitochondrial-encoded protein components of the ETC, which may have become limiting upon a shortage of endogenous

TAU, to restore ATP synthesis at the expense of superoxide anion formation (Schaffer et al. 2009). Two additional benefits that have been reported for TAU in laboratory animals are a hypoglycemic effect (Winiarska et al. 2009) and a stimulatory action on insulin secretion (Tenner et al. 2003).

In a previous study from this laboratory, a short, 5-day, treatment of Goto–Kakizaki rats, genetically predisposed to develop type 2 diabetes, was found to ameliorate oxidative stress and cell membrane injury in red blood cells (RBCs) but failed to normalize spectrin distribution and morphology (Gossai and Lau-Cam 2009). The present study was undertaken in rats made diabetic with streptozotocin to: (a) determine if a chronic treatment with TAU can lead to a further decrease of oxidative stress, to positive effects on membrane lipids, and to normalization of the spectrin distribution and morphology of RBCs; (b) to compare the effects of a sulfonic acid like TAU ($H_2N–CH_2–CH_2–SO_3H$) with those of a thiosulfonate like thiotaurine (TTAU, $H_2N–CH_2–CH_2–SO_2–SH$) on the hyperglycemia, hypoinsulinemia, dyslipidemia, plasma oxidative state and RBC alterations brought about by type 2 diabetes; and (c) to compare the actions of TAU and TTAU with those of an established hypoglycemic agent like insulin.

27.2 Methods

27.2.1 Animals and Treatments

All the experiments were conducted using groups of six male Sprague–Dawley rats, 225–250 g in weight. Diabetes was induced with an intraperitoneal, 60 mg/kg, dose of streptozotocin (STZ) in citrate buffer pH 4.5, and allowed to progress for 14 days. From day 15 until day 56, the diabetic rats received a single, daily, 2.4 mmol/kg/2 mL dose of either TAU or TTAU in physiological saline by intragastric gavage (the treatment groups). Additional diabetic rats were treated by the oral route with citrate buffer pH 4.5 on day 1 and with a daily 2 mL volume of physiological saline (the diabetic group) or subcutaneously with a 4 U/kg dose of isophane insulin (INS) (the reference group) from day 15 onwards. Untreated normal rats served as the control group. All the rats were sacrificed by decapitation on day 57, and their blood was collected in heparinized tubes. One aliquot was used for the analysis of glycated hemoglobin (HbA_{1c}). The rest of the blood was centrifuged at $700 \times g$ and 4 °C for 10 min to obtain the plasma fraction, which was subsequently analyzed for its contents of glucose (GLC), cholesterol (CHOL), triglycerides (TGs), malondialdehyde (MDA), and reduced (GSH) and oxidized (GSSG) glutathione, and for the activities of the antioxidant enzymes catalase (CAT), glutathione peroxidase (GPx) and superoxide dismutase (SOD). After separating the plasma and buffy coat, the pelleted RBCs were suspended in phosphate buffered saline pH 7.4 supplemented with 5 mM glucose (PBSG) to a hematocrit of 20% as described by Sharma and Premachandra (1991), and analyzed for their contents in MDA, GSH, GSSG, and hemoglobin A (Hb), for antioxidant enzymes activities, for morphology, and for spectrin distribution. The lipids in the cell membrane were extracted by the method of Folch et al. (1957), and the extract was analyzed for its contents in CHOL and phospholipids (PLPs). The study received the approval of the Institutional Animal Care and Use Committee of St. John's University, Jamaica, NY, and the animals were cared in accordance with guidelines established by the United States Department of Agriculture.

27.2.2 Biochemical Assays

The following parameters were measured using commercially available assay kits: plasma glucose (colorimetric Procedure No. 510 from Sigma–Aldrich, St. Louis MO); plasma INS (Insulin ELISA kit, Calbiotech Inc., Spring Valley, CA); blood HbA_{1c} (Glycohemoglobin Test, Stanbio Laboratory, Boerne, TX), plasma, and RBCs CHOL (Cholesterol LiquiColor® Procedure No. 1010, Stanbio Laboratory, Boerne, TX); and plasma TGs (Enzymatic Triglycerides Procedure No. 2150, Stanbio Laboratory, Boerne, TX). The concentration of membrane PLPs was measured by the method of Stewart (1979), and those of plasma and RBC GSH and GSSG were derived by the fluorometric method of Hissin and Hilf (1976). The spectrophotometric methods of Aebi (1984), Günzler and Flohé (1985), and Misra and Fridovich (1972) were used to assay the activities of CAT, GPx, and SOD, respectively, in plasma and RBCs.

27.2.3 Microscopic Studies

The changes in RBC morphology induced by diabetes, with and without a pharmacological treatment, were studied by scanning electron microscopy (SEM) using the method of Straface et al. (2002) after fixing the RBCs with 1.5% glutaraldehyde. The distribution of spectrin in the RBCs was determined by the immunohistochemical method of Straface et al. (2002), in which the RBCs are treated with goat serum followed by successive incubations with rabbit anti-chicken spectrin and anti-rabbit IgG (whole molecule) FITC conjugate. Then, the RBCs were examined using a fluorescence confocal microscope.

27.2.4 Statistical Analysis of the Data

The experimental results, reported as mean ± SEM for $n=6$, were analyzed for statistical significance using unpaired Student's t-test followed by one-way analysis of variance (ANOVA) and Tukey's post hoc test. Intergroup differences were considered to be statistically significant when $p \leq 0.05$.

Table 27.1 Effects of INS, TAU, and TTAU on the blood and plasma GLC levels of diabetic rats

Treatment groups	Blood GLC (mg/dL)	Plasma GLC (mg/dL)
Control	122.28 ± 2.53+++	103.59 ± 5.06+++
STZ	477.62 ± 15.45***	428.08 ± 21.74***
STZ-INS	114.10 ± 12.68+++	127.15 ± 10.71*·+++
TAU	110.07 ± 3.57+++	102.05 ± 3.33+++
STZ-TAU	378.68 ± 15.14***·+	333.27 ± 6.60***·+
TTAU	104.20 ± 2.48+++	104.61 ± 2.11+++
STZ-TTAU	330.27 ± 14.24***·++	265.08 ± 8.21***·+++

Values are reported as the mean ± SEM for $n=6$. Differences were significant from Control at *$p<0.05$ and ***$p<0.001$; and from STZ at +$p<0.05$, ++$p<0.01$, and +++$p<0.001$

27.3 Results

27.3.1 Circulating Glucose (GLC) Levels

In comparison to control values, diabetes elevated the blood GLC by 3.9-fold ($p<0.001$) and the plasma glucose by 4-fold ($p<0.001$) (Table 27.1). Following a 6 week treatment with TAU, the blood and plasma GLC of these animals rose by only 3.1-fold and 3.2-fold, respectively (both at $p<0.001$ vs. diabetic group). An identical treatment with an equidose (2.4 mmol/kg) of TTAU was able to lower the diabetic blood and plasma GLC levels further (2.7-fold and 2.6-fold, respectively, $p<0.001$ vs. diabetic group). In contrast, at the end of a 6 week treatment with INS the blood and plasma GLC of diabetic animals had decreased to values comparable to those of control animals. As shown in Table 27.1, an excellent correlation existed between GLC levels measured in blood tail vein and in plasma for all of the experimental groups.

27.3.2 Plasma Insulin (INS) Levels

The plasma INS decreased by ~75% ($p<0.001$) as a result of diabetes (Fig. 27.1). A 6 week treatment with TAU reduced this loss to ~50% below the control value ($p<0.001$), an effect that was further enhanced by a treatment with TTAU (~44% decrease, $p<0.001$ vs. control). In contrast, diabetic rats receiving INS exhibited a plasma INS level that was 20% above the control value ($p<0.05$) at the end of 6 weeks of treatment.

27.3.3 Blood Glycated Hemoglobin (HbA$_{1c}$) Levels

As shown in Fig. 27.2, diabetes elevated the circulating level of HbA$_{1c}$ by >200% above control ($p<0.001$). Both TAU and TTAU were found to drastically lower this increase (to only 60%, $p<0.001$), an effect that was rather similar to that attained

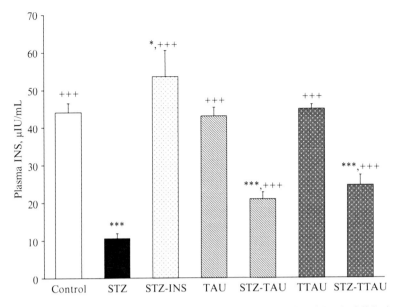

Fig. 27.1 The effects of INS, TAU and TTAU on the plasma and RBC INS level of diabetic rats. Differences were significant from Control at *$p<0.05$ and ***$p<0.001$; and from STZ at $^{+++}p<0.001$. Values are shown as mean ± SEM for $n=6$

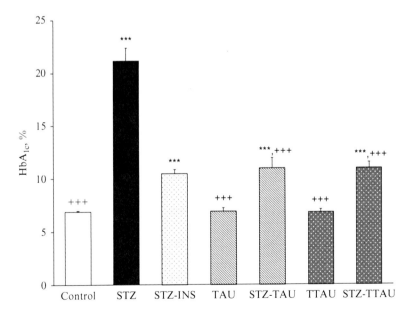

Fig. 27.2 The effects of INS, TAU and TTAU on the plasma and RBC HbA$_{1c}$ content of diabetic rats. Differences were significant from Control at ***$p<0.001$; and from STZ at $^{+++}p<0.001$. Values are shown as mean ± SEM for $n=6$

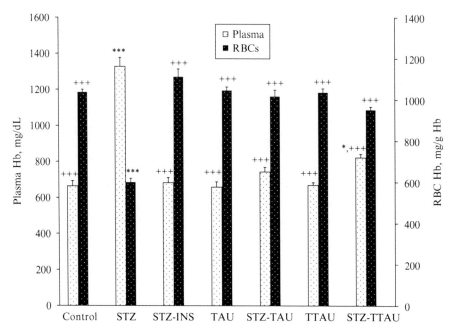

Fig. 27.3 The effects of INS, TAU, and TTAU on the plasma and RBC Hb content of diabetic rats. Differences were significant from Control at *$p<0.05$ and ***$p<0.001$; and from STZ at $^{+++}p<0.001$. Values are shown as mean ± SEM for $n=6$

with an INS treatment (55% increase, $p<0.001$). However all comparisons against STZ were much lower ($p<0.001$), thus indicating significant protection.

27.3.4 Plasma and RBC Hemoglobin (HbA) Levels

Chronic diabetes caused a marked loss of HbA from the RBC (−42%, $p<0.001$) into the circulation (up by 100%, $p<0.001$) (Fig. 27.3). TAU and TTAU were about equally protective in attenuating these effects (~12% loss) and not significantly different from the effect exerted by INS (<5% loss). Conversely, while STZ lowered the RBC HbA content by 42% ($p<0.001$), all three treatment agents kept the intracellular HbA to a level comparable to the control value (only ≤7% loss).

27.3.5 Plasma and RBC LPO

The occurrence of LPO as a result of diabetes was investigated by measuring the levels of MDA in the plasma and RBCs. As shown in Fig. 27.4, diabetes elevated

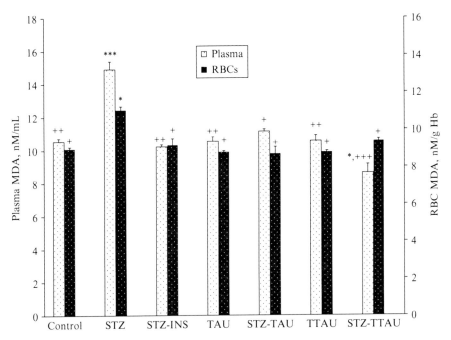

Fig. 27.4 The effects of INS, TAU, and TTAU on the plasma and RBC MDA content of diabetic rats. Differences were significant from Control at $*p < 0.05$ and $***p < 0.001$; and from STZ at $^+p < 0.05$, $^{++}p < 0.01$, and $^{+++}p < 0.001$. Values are shown as mean $\pm$ SEM for $n = 6$

the MDA to a significant extent, with the levels being greater in the plasma (+41%, $p < 0.01$) than in the RBC (+23%). In either case, all the treatment compounds were highly protective, with the values approximating those of control animals.

27.3.6   *Plasma and RBC Reduced (GSH) and Disulfide (GGSG) Glutathione Levels*

As shown in Fig. 27.5, diabetes reduced the plasma and RBC levels of GSH (by 46%, $p < 0.001$, and 18%, $p < 0.05$, respectively, vs. controls), elevated the accompanying levels of GSSG (by 115%, $p < 0.001$, and 15%, $p < 0.05$, respectively) (Fig. 27.6), and lowered the corresponding GSH/GSSG ratios (by 75%, $p < 0.001$, and 27%, $p < 0.01$, respectively) (Fig. 27.7). In terms of the GSH levels, all the treatment compounds effectively counteracted the alterations caused by diabetes. Thus, in the plasma TAU raised the GSH content to almost baseline values (−3%) and TTAU and INS were found to reduce the losses to 17% ($p < 0.05$) and 26% ($p < 0.01$), respectively, below the control values. On the other hand, in RBCs while TAU and TTAU were equipotent in abolishing the loss of GSH, INS was able to limit the decrease in GSH to only about 12%.

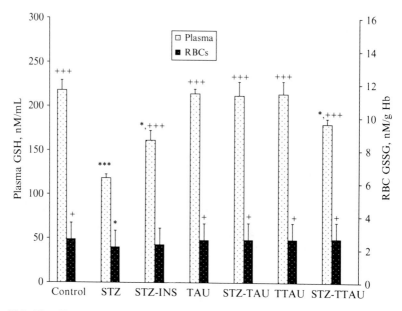

Fig. 27.5 The effects of INS, TAU, and TTAU on the plasma and RBC GSH content of diabetic rats. Differences were significant from Control at *$p < 0.05$ and ***$p < 0.001$; and from STZ at ++$p < 0.01$ and +++$p < 0.001$. Values are shown as mean ± SEM for $n = 6$

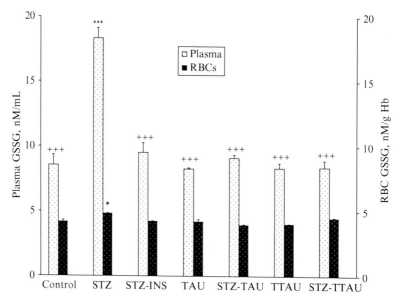

Fig. 27.6 The effects of INS, TAU, and TTAU on the plasma and RBC GSSG content of diabetic rats. Differences were significant from Control at *$p < 0.05$ and ***$p < 0.001$; and from STZ at +++$p < 0.001$. Values are shown as mean ± SEM for $n = 6$

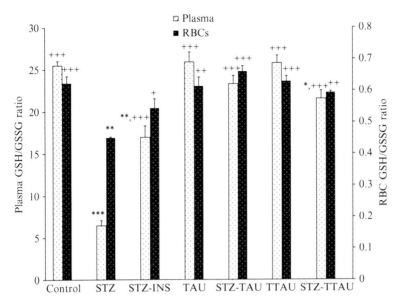

Fig. 27.7 The effects of INS, TAU, and TTAU on the plasma and RBC GHS/GSSG ratio of diabetic rats. Differences were significant from Control at $*p < 0.05$, $*p < 0.01$ and $***p < 0.001$; and from STZ at $^+p < 0.05$, $^{++}p < 0.01$, and $^{+++}p < 0.001$. Values are shown as mean ± SEM for $n = 6$

Similarly, TAU, TTAU, and INS were found to effectively prevent the increases in GSSG levels induced by diabetes in both the plasma (down to only 2–11%) and RBCs (down to ≤7%) (Fig. 27.6). As a result, all the treatments were found to protect against the decreases in GSH/GSSG associated with diabetes, with TAU appearing more protective than TTAU in both the plasma (9% decrease vs. 16% decrease, $p < 0.05$, respectively) and RBCs (7% increase vs. 5% decrease, respectively) (Fig. 27.7). On the other hand a treatment with INS led to plasma and RBC GSH/GSSG ratios that were significantly lower than TAU or TTAU in the plasma (34% decrease, $p < 0.01$) but about equal to TAU and TTAU in the RBCs (13% decrease).

27.3.7 Plasma and RBC Antioxidant Enzymes Activities

The effects of diabetes on enzymatic antioxidant defenses were studied by measuring the activities of plasma and RBC catalase (CAT), glutathione peroxidase (GPx) and superoxide dismutase (SOD). From the results presented in Figs. 27.8–27.10, it is evident that diabetes exerted a reducing effect on these three activities in the plasma (down by 58%, 63%, and 66%, respectively, all at $p < 0.001$ vs. control values) and, to a lesser extent, in RBCs (down by 26%, 35%, and 43%, respectively, all at $p < 0.01$ vs. control values). Without exceptions, all the treat-

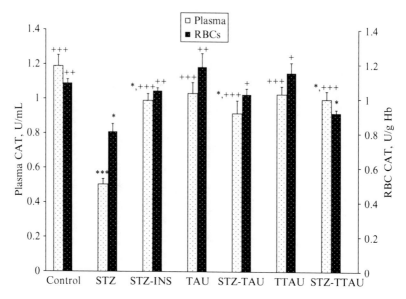

Fig. 27.8 The effects of INS, TAU, and TTAU on the plasma and RBC CAT activity of diabetic rats. Differences were significant from Control at $*p<0.05$ and $***p<0.001$; and from STZ at $^+p<0.05$, $^{++}p<0.01$, and $^{+++}p<0.001$. Values are shown as mean ± SEM for $n=6$

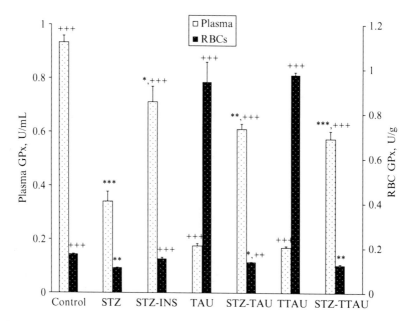

Fig. 27.9 The effects of INS, TAU, and TTAU on the plasma and RBC GPx activity of diabetic rats. Differences were significant from Control at $*p<0.05$, $**p<0.01$, and $***p<0.001$; and from STZ at $^{++}p<0.01$ and $^{+++}p<0.001$. Values are shown as mean ± SEM for $n=6$

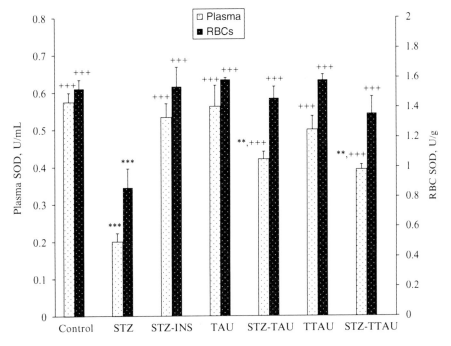

Fig. 27.10 The effects of INS, TAU, and TTAU on the plasma and RBC SOD activity of diabetic rats. Differences were significant from Control at **$p<0.01$ and ***$p<0.001$; and from STZ at +++$p<0.001$. Values are shown as mean $\pm$ SEM for $n=6$

ment compounds were able to reduce these losses to extents that varied according to the particular enzyme. Thus, a 6 week treatment with the test compounds reduced the activity losses of plasma CAT (to 16–23% of control, $p<0.05$) (Fig. 27.8), GPx (to 24–39% of control, $p \leq 0.05$) (Fig. 27.9), and SOD (to 9% with INS; to ~30% with TAU and TTAU, $p<0.01$) (Fig. 27.10) significantly. In RBCs, however, while both INS and TAU were able to virtually abolish the activity losses of CAT and SOD induced by diabetes, only INS was able to have a significant attenuating effect on the losses of SOD. In general, TTAU was insignificantly less effective on RBC enzymes than TAU.

27.3.8 Plasma Cholesterol (CHOL) and Triglycerides (TGs) Levels

From the results presented in Table 27.2, it is evident that diabetes elevated the plasma levels of both CHOL and TGs to a significant extent (by 66% and 190%, respectively, both at $p<0.001$ vs. controls) (Table 27.2). While all the treatment compounds, including INS, kept the CHOL at ~20% above control ($p<0.05$), their

Table 27.2 Effects of INS, TAU, and TTAU on the plasma CHOL and TGs levels of diabetic rats

Control	$59.53 \pm 1.93^{+++}$	$141.15 \pm 10.50^{+++}$
STZ	$99.06 \pm 5.03^{***}$	$405.97 \pm 22.98^{***}$
STZ-INS	$71.08 \pm 3.39^{*,++}$	$144.86 \pm 6.34^{+++}$
TAU	$58.37 \pm 4.88^{+++}$	$142.63 \pm 9.56^{+++}$
STZ-TAU	$70.81 \pm 2.90^{*,++}$	$206.33 \pm 12.82^{***,+++}$
TTAU	$59.27 \pm 6.11^{+++}$	$142.69 \pm 2.74^{+++}$
STZ-TTAU	$69.80 \pm 0.96^{*,++}$	$246.21 \pm 20.44^{***,+++}$

Values are reported as the mean $\pm$ SEM for $n=6$. Differences were significant from Control at $*p<0.05$ and $***p<0.001$; and from STZ at $^{++}p<0.01$ and $^{+++}p<0.001$

Table 27.3 Effects of INS, TAU, and TTAU on the RBC membrane CHOL, PLPs, and CHOL/PLPs ratio of diabetic rats

Treatment groups	CHOL (mg/dL)	PLPs (mg/dL)	CHOL/PLPs ratio
Control	$40.56 \pm 1.19^{+++}$	$62.33 \pm 0.44^{+}$	$0.65 \pm 0.02^{+++}$
STZ	$91.27 \pm 2.64^{***}$	$76.06 \pm 2.87^{*}$	$1.20 \pm 0.08^{***}$
STZ-INS	$72.49 \pm 3.06^{***,+}$	71.35 ± 2.64	$1.02 \pm 0.06^{***}$
TAU	$40.60 \pm 1.00^{+++}$	$61.88 \pm 1.02^{+}$	$0.66 \pm 0.01^{+++}$
STZ-TAU	$72.02 \pm 2.67^{***,+}$	67.70 ± 0.75	$1.06 \pm 0.05^{***}$
TTAU	$43.98 \pm 2.60^{+++}$	$62.80 \pm 0.94^{+}$	$0.70 \pm 0.04^{+++}$
STZ-TTAU	$41.14 \pm 2.01^{***,+}$	$63.36 \pm 2.55^{*}$	$0.65 \pm 0.06^{***,+}$

Values are reported as the mean $\pm$ SEM for $n=6$. Differences were significant from Control at $*p<0.05$ and $***p<0.001$; and from STZ at $^{+}p<0.05$ and $^{+++}p<0.001$

effects on the plasma TGs was quite variable, with TAU providing a greater protection (~46% increase, $p<0.001$) than TTAU (~75% increase, $p<0.001$); and INS returning the TGs content to the baseline value.

27.3.9 RBC Membrane CHOL and Phospholipids (PLPs) Levels

As indicated in Table 27.3, diabetes caused a marked increase in the RBC membrane content of CHOL (by 125%, $p<0.001$) and a mild increase in that of the PLPS (by 22%, $p<0.05$). All the treatment compounds were markedly and uniformly protective in attenuating the increases in RBC membrane CHOL (by 71% with TTAU, by ~79% with TAU and INS, $p<0.001$ vs. diabetes) (Table 27.3). In contrast, the effect of these compounds on the PLPs content was rather weak, with TAU providing a slightly greater stabilizing effect (9% increase) than either TTAU or INS (19% and 14% increases, respectively). While the CHOL/PLPs ratio was elevated by diabetes by 85% ($p<0.001$), a treatment with INS, TAU or TAU lowered the elevations in the ratio by diabetes to only 57%, 63%, and 46%, respectively, above the control value (all at $p<0.001$ vs. control, all at $p\leq0.05$ vs. STZ alone) (Table 27.3).

27.3.10   RBC Morphology and Spectrin Distribution

In comparison with the typical normal concave appearance of RBCs (Fig. 27.11A) from peripheral blood of normal rats, RBCs from diabetic rats exhibited a characteristic echinocytic appearance (Fig. 27.11B) under the scanning electron microscope (SEM). A 6 week treatment of the diabetic animals with INS led to the normalization of the RBC shape (Fig. 27.11C); but when TAU was given as a treatment the RBCs appeared discocytic and aconcave (Fig. 27.11D). In contrast, RBCs from rats receiving TTAU showed a normal appearance (Fig. 27.11E).

RBCs from normal rats showed their cytoskeletal spectrin uniformly distributed throughout (Fig. 27.12A). In contrast, the spectrin of RBCs from diabetic rats became segregated towards the periphery (Fig. 27.12B). A treatment with INS returned the spectrin of diabetic rats to its normal distribution (Fig. 27.12C), but not one with either TAU (Fig. 27.12D) or TTAU (Fig. 27.12E).

27.4   Discussion

This laboratory has previously verified that in Goto–Kakizaki (GK) rats, a substrain of Wistar rats selectively bred over many generations to develop a nonobese type 2 diabetes early in life as a result of β-cell mass deficit (Tourrel et al. 2002), a treatment with TAU for 5 consecutive days protected the circulating RBCs against membrane and biochemical alterations that follow the oxidative stress and cell damage of type 2 diabetes (Gossai and Lau-Cam 2009). Specifically, TAU was able to preserve the integrity of the RBC membrane and, in this manner, lower the leakage of intracellular lactate dehydrogenase and HbA into the circulation. Furthermore, this sulfonate compound was found to reduced LPO and MDA formation, to preserve the intracellular GSH and GSH/GSSG ratio, and to prevent the decreases in CAT, GPx and SOD activities associated with the diabetic state. When RBCs from diabetic rats were examined on a SEM, they displayed an echynocytic appearance rather than the normal discoidal biconcave one; and when put through immunochemical plus microscopic analysis, those from diabetic rats showed their fluorescein-labeled antibody-spectrin conjugates to have a patchy distribution instead of a uniform distribution over the inner surface of the RBC (Ziparo et al. 1978). In the same study, a treatment with the antioxidant *N*-acetylcysteine, but not one with TAU, was able to correct for the morphological change and altered spectrin distribution (Gossai and Lau-Cam 2009).

The present study was undertaken to specifically ascertain if a chronic treatment with TAU will be able to overcome the failure of an acute, 5 day treatment, to normalize the morphology and spectrin distribution changes in RBCs caused by type 2 diabetes. An additional objective was to determine if replacing the sulfonic acid group of TAU for the thiosulfonic group, as present in TTAU, would be able to improve the protective actions of TAU against diabetes-associated alterations in the

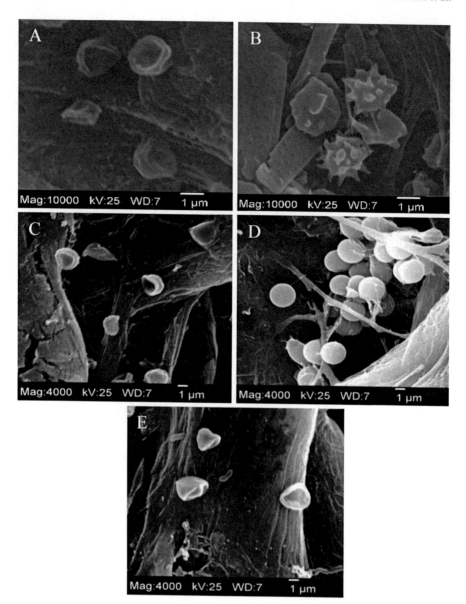

Fig. 27.11 SEM photomicrographs showing the effects of INS, TAU, and TTAU on the morphology of RBCs from diabetic rats: (**A**) Control; (**B**) STZ; (**C**) STZ-INS; (**D**) STZ-TAU; (**E**) STZ-TTAU. (**B**) shows echinocytes and (**D**) shows discocytes. In (**A**), (**C**), and (**E**) the RBCs exhibit a normal appearance

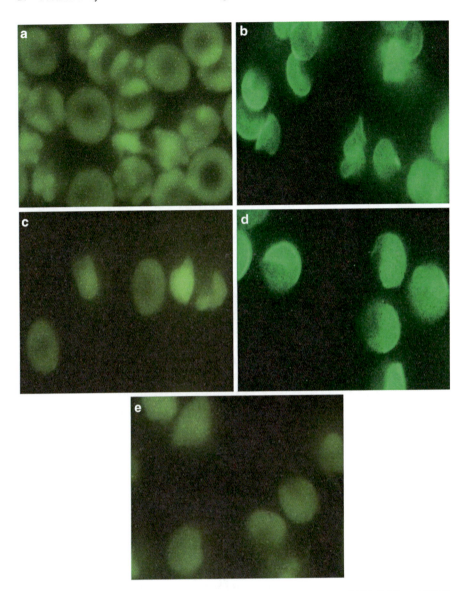

Fig. 27.12 Confocal microscopy photomicrographs showing the effects of INS, TAU, and TTAU on the spectrin distribution in RBCs from diabetic rats: (**A**) Control; (**B**) STZ; (**C**) STZ-INS; (**D**) STZ-TAU; (**E**) STZ-TTAU. In (**B**) and (**D**) there is segregation of spectrin towards the periphery

plasma and RBCs. To attain these objectives, a STZ-induced animal model of diabetes was used taking advantage of the ability of this diabetogen to induce mild to severe types of diabetes, including type 2 diabetes, by adjusting its dose (Arora et al. 2009). The onset of diabetes and its progression was monitored based on weekly measurements of GLC in a sample of tail vein blood. Since the blood GLC level reached a maximum at 2 weeks post-STZ, all the treatments were started on day 15 of the study and they were continued, on a daily basis, until day 56. To more accurately assess the intrinsic potencies of TAU and TTAU, these compounds were given by the same route and at the same molar (2.4 mmol/kg) dose. In parallel, an additional group of diabetic rats was treated with INS to determine the role played by hyperglycemia and hypoinsulinemia on the experimental plasma and RBC parameters evaluated.

While the circulating levels of GLC directly correlated with the formation of MDA, serving as an index of LPO, they were inversely correlated with the ratio of GSH/GSSG and activities of antioxidant enzymes. Interestingly, the increase in blood GLC triggered by a moderate (60 mg/kg) single dose of STZ in the rat was of the same magnitude as that reported for mice by a study using a threefold higher dose (Arora et al. 2009). After 56 days the same animals also manifested a marked hypoinsulinemia, elevated levels of HbA_{1c}, and frank hyperlipidemia. The occurrence of a state of oxidative state was suggested by the ensuing increase in plasma MDA, the drastic drop of the plasma GSH/GSSG ratio, and a concomitant reduction of the activities of CAT, GPx and SOD. In RBCs, an increase in the membrane CHOL/PLPS ratio was accompanied by milder increases in MDA and GSSG levels, and milder decreases in GSH and antioxidant enzymes activities relative to results obtained from plasma samples. In addition, the massive leakage of intracellular HbA into the circulation was suggestive of an extensive change in cell membrane integrity. Some of the explanations accounting for this change have been an increased LPO of the RBC membrane (Jain et al. 1989), a disturbance of aminophospholipids (phosphatidylethanolamine, phosphatidylserine) organization in the membrane bilayer by MDA (Jain 1984), changes in bilayer lipid composition (Jain et al. 1990), and increased oxidative damage due to decreased protection by GSH when this antioxidant has become depleted (Fujiwara et al. 1989). The depletion of GSH under hyperglycemic conditions might have resulted from a direct interaction with reactive aldehydes arising from the metabolism of glucose (Beard et al. 2003), from increased rates of utilization (Darmaun et al. 2005) or from impaired synthesis due either to decreased activity of γ-glutamylcysteine synthetase, the rate-controlling enzyme of GSH synthesis due to excessive ROS levels (Trocino et al. 1995) or to limited availability of the precursor molecules glycine and cysteine (Sekhar et al. 2010).

On the other hand, the high levels of GSSG observed in RBCs from diabetic subjects has been related to both decreased transport of GSSG through the RBC membrane out of the cells and decreased glutathione reductase (GR) activity (Murakami et al. 1989). However, in view of the much higher increase of GSSG in the plasma than in RBCs, it has been postulated that RBCs are responsible for the enhanced amounts of GSSG found in plasma of diabetics as a result of a reduced conversion to GSH by GR, which is a consequence of decreased activity of glucose-6-phosphate dehydrogenase (G6PD). Consequently, G6PD fails to supply GR with optimum concentrations of its cofactor NADPH to carry out the redox cycling of GSH (Costagliola, 1990). An additional contributing factor for the limited availability of NADPH may be the accelerated flux of GLC through the polyol (sorbitol) pathway, which is highly dependent on NADPH (Lee and Chiung 1999).

The trend of changes in the activities of antioxidants enzymes reported by different laboratories for the plasma, RBCs and major organs of diabetic human subjects and experimental animals is surrounded by considerable variability within and between laboratories (Maritim et al. 1999, 2003; Sundaram et al. 1996). In human diabetics, the lack of a uniform trend may also reflect concurrent underlying pathologies, secondary complications, and nutritional deficiencies relevant to enzyme activities (Sundaram et al. 1996). Under the more tightly controlled conditions of studies on animal models of diabetes it has possible to determine that such variability is more apparent during the first weeks after induction of diabetes, at which time the synthesis of some of the enzymes may rise as compensatory mechanism against oxidative stress; a trend that wanes in about 2–3 months of untreated diabetes, and is followed by a generalized decrease in antioxidant enzyme activity (Kędziora-Kornatowska et al. 1998). In agreement with this concept and with the results gathered by other investigators (Hisalkar et al. 2012; Kędziora-Kornatowska et al. 1998), the present work found that the activities of CAT, GPx, and SOD were significantly decreased in both the plasma and RBCs. These decreases may have resulted from their progressive glycation stimulated by hyperglycemia of diabetes since a treatment with INS was found to normalize the blood GLC and HbA_{1c} levels and to reverse the changes in activity of all antioxidant enzymes in the plasma and RBCs (Wohaieb and Godin 1987).

In diabetes, RBCs are subjected to a hyperglycemic environment favoring oxidant stress and the overproduction of ROS through GLC autoxidation and/or protein glycation (Peuchant et al. 1997). As a result, antioxidant defenses and endogenous antioxidants are compromised and LPO is increased as demonstrated by the levels of MDA and changes in enzymatic and nonenzymatic defense systems. In comparison with the results of a previous study conducted on GK rats (Gossai and Lau-Cam 2009), changes of these indicators of oxidative stress in the RBCs from STZ-treated rats followed the same course but they were more pronounced. Chronic treatments with TAU, TTAU or INS resulted in attenuation of MDA formation in both RBCs and plasma, with the potency differences among the three treatments being insignificant. These results were in close agreement with their attenuating effects on changes in blood and plasma GLC, plasma INS and blood HbA_{1c}. INS, followed by TTAU, was the most protective of the three on hyperglycemia and hypoinsulinemia, but all three were equipotent in lowering HbA_{1c} formation. While in the plasma TAU was insignificantly more potent than TTAU and INS was the least potent in counteracting the downward trend of the GSH redox state caused by diabetes, the three treatments were equiprotective in preserving the redox state in RBCs. These findings are in line with those of earlier studies supporting a stimulatory action for TAU on INS secretion in live animals and in cultured pancreatic β-cells (Cherif et al. 1998; Tokunaga et al. 1979) and for an interaction of TAU with the INS receptor (Maturo and Kulakowski, 1988). Furthermore, they confirm the role of antioxidants like TAU in relieving pancreatic β-cells from the inhibitory action of oxidative stress on INS secretion in response to high blood GLC levels (Oprescu et al. 2007). Unfortunately, these effects remain to be realized in diabetic volunteers receiving a TAU supplementation (Brøns et al. 2004; Chauncey et al. 2003).

All the treatment compounds had a marked and about equal decreasing effect on the hypercholesterolemia by diabetes. Also all the treatments attenuated the hyper-triglyceridemic response to diabetes to a significant extent, with INS normalizing this parameter, and TAU providing an almost twofold better protection than TTAU. In contrast, an assessment of the effects of the same treatments on the RBC membrane contents of CHOL and PLPs indicated that TTAU was marginally the most effective treatment in lowering the increase of CHOL/PLPs ratio, with TAU and INS demonstrating equivalent and weaker effects.

In agreement with the results of an earlier short-term (5 days) study by this laboratory with TAU on GK rats (Gossai and Lau-Cam 2009), the present chronic (42 days) treatment with TAU also failed to return the echinocytes present in diabetic rats to the normal biconcave discocytes seen in normal rats. However, by extending the treatment with TAU the RBCs became spherocytes. These findings suggest that in experimental models of diabetes the echinocytic shape of the RBCs is not entirely determined by the lipid composition of the cell membrane and the CHOL/PLPs ratio, which is characteristically elevated in diabetes, since identical treatments with either TTAU or INS were able to normalize the shape of the RBC in spite of reducing the ratio to about the same extent. The effects of the treatment compounds on RBC morphology correlated well with the cytoskeletal spectrin distribution, which was seen segregated to a narrow area of RBCs from untreated and TAU-treated diabetic rats. In contrast, RBCs from rats receiving either TTAU or INS showed the spectrin uniformly distributed throughout the cell. Both spherocytic cells and segregated spectrin may have resulted from changes in cytoskeletal structure and from the restructuring of the PLP bilayer as a result of oxidative stress, changes that may lead to the loss of stability of the membrane bilayer, to the separation of spectrin from the PLP bilayer and to some PLP to leave the membrane and favor the formation of a smaller spherocytic cell (Desouky 2009).

In conclusion, the present results reveal that a chronic treatment of diabetic rats with TTAU, the thiosulfonate analog of TAU, can protect against biochemical changes associated with diabetes in exactly the same manner as the parent compound and INS. However, at variance with TAU, it demonstrated a greater effect on the hyperglycemia, hypoinsulinemia, and decrease in the GSH redox state caused by diabetes. More importantly, it was able to normalize the RBC morphology and cystoskeletal spectrin distribution while TAU was not.

References

Acharya M, Lau-Cam CA (2010) Comparison of the protective actions of N-acetylcysteine, hypo-taurine and taurine against acetaminophen-induced hepatotoxicity in the rat. J Biomed Sci 17:S35

Aebi H (1984) Catalase in vitro. Methods Enzymol 105:121–126

Alciguzel Y, Ozen I, Aslan M, Karayalcin U (2003) Activities of xanthine oxidoreductase and antioxidant enzymes in different tissues of diabetic rats. J Lab Clin Med 142:172–177

Arora S, Ojha SK, Vohora D (2009) Characterisation of streptozotocin induced diabetes mellitus in Swiss albino mice. Global J Pharmacol 3:81–84

Baynes JW, Thorpe SR (1999) Role of oxidative stress in diabetic complications: a new perspective on an old paradigm. Diabetes 48:1–9

Beard KM, Shangari N, Wu B, O'Brien PJ (2003) Metabolism, not autoxidation, plays a role in α-oxoaldehyde- and reducing sugar-induced erythrocyte GSH depletion: relevance for diabetes mellitus. Mol Cell Biochem 252:331–338

Brøns C, Spohr C, Storgaard H, Dyerberg J, Vaag A (2004) Effect of taurine treatment on insulin secretion and action, and on serum lipid levels in overweight men with a genetic predisposition for type II diabetes mellitus. Eur J Clin Nutr 58:1239–1247

Brownlee M (2001) Biochemistry and molecular cell biology of diabetic complications. Nature 414:813–820

Brownlee M (2005) The pathobiology of diabetic complications: a unifying mechanism. Diabetes 54:1615–1625

Ceriello A (2005) Postprandial hyperglycemia and diabetes complications: is it time to treat? Diabetes 54:1–7

Chauncey KB, Tenner TE Jr, Lombardini JB et al (2003) The effect of taurine supplementation on patients with type 2 diabetes mellitus. Adv Exp Med Biol 526:91–96

Cherif H, Reusens B, Ahn MT, Hoet JJ, Remacle C (1998) Effects of taurine on the insulin secretion of rat fetal islets from dams fed a low-protein diet. J Endocrinol 159:341–348

Costagliola C (1990) Oxidative state of glutathione in red blood cells and plasma of diabetic patients: in vivo and in vitro study. Clin Physiol Biochem 8:204–210

Darmaun D, Smith SD, Sweeten S, Sager BK, Welch S, Mauras N (2005) Evidence for accelerated rates of glutathione utilization and glutathione depletion in adolescents with poorly controlled type 1 diabetes. Diabetes 54:190–196

Desco MC, Asensi N, Márquez R et al (2002) Xanthine oxidase is involved in free radical production in type 1 diabetes: protection by allopurinol. Diabetes 51:1118–1124

Desouky OS (2009) Rheological and electrical behavior of erythrocytes in patients with diabetes mellitus. Rom J Biophys 19:239–250

Folch J, Lees M, Stanley GHS (1957) A simple method for the isolation and purification of total lipides from animal tissues. J Biol Chem 226:497–509

Fujiwara Y, Kondo T, Murakami K, Kawakami Y (1989) Decrease of the inhibition of lipid peroxidation by glutathione-dependent system in erythrocytes of non-insulin dependent diabetics. J Mol Med 67:336–341

Goodman HO, Shihabi ZK (1990) Supplemental taurine in diabetic rats: effects on plasma glucose and triglycerides. Biochem Med Metab Biol 43:1–9

Gossai D, Lau-Cam CA (2009) The effects of taurine, taurine homologs and hypotaurine on cell and membrane antioxidative system alterations caused by type 2 diabetes in rat erythrocytes. Adv Exp Med Biol 643:359–368

Günzler WA, Flohé L (1985) Glutathione peroxidase. In: Greenwald RA (ed) CRC handbook of methods for oxygen radical research. CRC, Boca Raton, FL, pp 285–290

Guzik TJ, Mussa S, Gastaldi D et al (2002) Mechanisms of increased vascular superoxide production in human diabetes mellitus: role of NAD(P)H oxidase and endothelial nitric oxide synthase. Circulation 105:1656–1662

Hisalkar PJ, Patne AB, Fawade MM (2012) Assessment of plasma antioxidant levels in type 2 diabetes patients. Int J Biol Med Res 3:1796–1800

Hissin PJ, Hilf R (1976) A fluorometric method for determination of oxidized and reduced glutathione in tissues. Anal Biochem 74:214–226

Jain SK (1984) The accumulation of malonyldialdehyde, a product of fatty acid peroxidation, can disturb aminophospholipid organization in the membrane bilayer of human erythrocytes. J Biol Chem 259:3391–3394

Jain SK, Levine SN, Duetta J, Hollier B (1990) Elevated lipid peroxidation levels in red blood cells of streptozotocin-treated diabetic rats. Metabolism 39:971–975

Jain SK, McVie R, Duett J, Herbst JJ (1989) Erythrocyte membrane lipid peroxidation and glycosylated hemoglobin in diabetes. Diabetes 38:1539–1543

Järvi AE, Karlström BE, Granfeldt YE et al (1999) Improved glycemic control and lipid profile
and normalized fibrinolytic activity on a low-glycemic index diet in type 2 diabetic patients.
Diabetes Care 22:10–18

Kalapos MP, Andrea Littauer A, de Groot H (1993) Has reactive oxygen a role in methylglyoxal
toxicity? A study on cultured rat hepatocytes. Arch Toxicol 67:369–372

Kędziora-Kornatowska KZ, Luciak M, Blaszczyk J, Pawlak W (1998) Lipid peroxidation and
activities of antioxidant enzymes in erythrocytes of patients with non-insulin dependent diabetes with or without diabetic nephropathy. Nephrol Dial Transplant 13:2829–2832

Lee AY, Chiung SS (1999) Contributions of polyol pathway to oxidative stress in diabetic cataract.
FASEB J 13:23–30

Likidlilid A, Patchanans N, Peerapatdit T, Sriratanasathavorn C (2010) Lipid peroxidation and
antioxidant enzyme activities in erythrocytes of type 2 diabetic patients. J Med Assoc Thai
93:682–693

Maritim AC, Moore BH, Sanders RA, Watkins JB III (1999) Effects of melatonin on oxidative
stress in streptozotocin-induced diabetic rats. Int J Toxicol 18:161–166

Maritim AC, Sanders RA, Watkins JB III (2003) Diabetes, oxidative stress, and antioxidants:
a review. J Biochem Mol Toxicol 17:24–38

Maturo J, Kulakowski EC (1988) Taurine binding to the purified insulin receptor. Biochem
Pharmacol 37:3755–3760

Mayfield J (1998) Diagnosis and classification of diabetes mellitus: new criteria. Am Fam Physician
58:1355–1362

Misra HP, Fridovich I (1972) The role of superoxide anion in the autoxidation of epinephrine and
a simple assay for superoxide dismutase. J Biol Chem 247:3170–3175

Moller DE (2001) New drug targets for type 2 diabetes and the metabolic syndrome. Nature
414(6865):821–827

Monnier L, Mas E, Ginet C et al (2006) Activation of oxidative stress by acute glucose fluctuations
compared with sustained chronic hyperglycemia in patients with type 2 diabetes. J Am Med
Assoc 295:1681–1687

Murakami K, Takahito K, Ohtsuka Y et al (1989) Impairment of glutathione metabolism in erythrocytes from patients with diabetes mellitus. Metabolism 38:753–758

Nandhini ATA, Thirunavukkarasu V, Anuradha CV (2004) Stimulation of glucose utilization and
inhibition of protein glycation and AGE products by taurine. Acta Physiol Scand
181:297–303

Nishikawa T, Edelstein D, Brownlee M (2000) The missing link: a single unifying mechanism for
diabetic complications. Kidney Int 58:S26–S30

Obrosova IG, Stavniichuk R, Drel VR (2010) Different roles of 12/15-lipoxygenase in diabetic
large and small fiber peripheral and autonomic neuropathies. Am J Pathol 177:1436–1447

Oprescu AI, Bikopoulos G, Naassan A et al (2007) Fatty acid–induced reduction in
glucose-stimulated insulin secretion: evidence for a role of oxidative stress in vitro and in vivo.
Diabetes 56:2927–2937

Peuchant E, Delmas-Beauvieux MC, Couchouron A et al (1997) Short-term insulin therapy and
normoglycemia. Effects on erythrocyte lipid peroxidation in NIDDM patients. Diabetes Care
20:202–207

Pieczenik SR, Neustadt J (2007) Mitochondrial dysfunction and molecular pathways of disease.
Exp Mol Pathol 83:84–92

Rahimi R, Nikfar S, Larijani B, Abdollahi M (2005) A review on the role of antioxidants in the
management of diabetes and its complications. Biomed Pharmacother 59:365–373

Ricci C, Pastukh V, Leonard J et al (2008) Mitochondrial DNA damage triggers mitochondrial
superoxide generation and apoptosis. Am J Physiol Cell Physiol 294:C413–C422

Schaffer SW, Azuma J, Mozaffari M (2009) Role of antioxidant activity of taurine in diabetes. Can
J Physiol Pharmacol 87:91–99

Sekhar RV, McKay SV, Patel SG et al (2010) Glutathione synthesis is diminished in patients with uncontrolled diabetes and restored by dietary supplementation with cysteine and glycine. Diabetes Care 34:164–167

Sharma R, Premachandra BR (1991) Membrane-bound hemoglobin as a marker of oxidative injury in adult and neonatal red blood cells. Biochem Med Metab Biol 46:33–44

Stewart JCM (1979) Colorimetric determination of phospholipids with ammonium ferrothiocyanate. Anal Biochem 104:10–14

Straface E, Rivabene R, Masella R et al (2002) Structural changes of the erythrocyte as a marker of non-insulin-dependent diabetes: protective effects of N-acetylcysteine. Biochem Biophys Res Commun 290:1393–1398

Sundaram RK, Bhaskar A, Vijayalingam S et al (1996) Antioxidant status and lipid peroxidation in type II diabetes mellitus with and without complications. Clin Sci 90:255–260

Tenner TE Jr, Zhang XJ, Lombardini JB (2003) Hypoglycemic effects of taurine in the alloxan-treated rabbit: a model for type 1 diabetes. Adv Exp Med Biol 526:97–104

Tokunaga H, Yoneda Y, Kuriyama K (1979) Protective actions of taurine against streptozotocin-induced hyperglycemia. Biochem Pharmacol 28:2807–2811

Tourrel C, Bailbe D, Lacorne M, Meile MJ, Kergoat M, Portha B (2002) Persistent improvement of type 2 diabetes in the Goto–Kakizaki rat model by expansion of the β-cell mass during the prediabetic period with glucagon-like peptide-1 or exendin-4. Diabetes 51:1443–1452

Trachtman H, Futterweit S, Maesaka J, Ma C, Valderrama E, Fuchs A, Tarectecan AA, Rao PS, Sturman JA, Boles TH (1995) Taurine ameliorates chronic streptozotocin-induced diabetic nephropathy in rats. Am J Physiol 269:F429–F438

Trocino RA, Akazawa S, Ishibashi M et al (1995) Significance of glutathione depletion and oxidative stress in early embryogenesis in glucose-induced rat embryo culture. Diabetes 44: 992–998

Winiarska K, Szymanski K, Gorniak P, Dudziak M, Bryla J (2009) Hypoglycaemic, antioxidative and nephroprotective effects of taurine in alloxan diabetic rabbits. Biochimie 91:261–270

Wohaieb SA, Godin DV (1987) Alterations in free radical tissue-defense mechanisms in streptozotocin-induced diabetes in rat. Effects of insulin treatment. Diabetes 36:1014–1018

Ziparo E, Lemay A, Marchesi VT (1978) The distribution of spectrin along the membranes of normal and echinocytic human erythrocytes. J Cell Sci 34:91–101

Chapter 28
The Effects of Taurine and Thiotaurine on Oxidative Stress in the Aorta and Heart of Diabetic Rats

Elizabeth Mathew, Michael A. Barletta, and Cesar A. Lau-Cam

Abstract This study has compared the actions of the sulfur-containing compounds taurine (TAU) and thiotaurine (TTAU) with those of insulin (INS) on the oxidative stress that develops in the aorta and heart as a result of diabetes. Diabetes was induced in male Sprague–Dawley rats with streptozotocin (60 mg/kg, i.p.). Starting on day 15, and continuing for the next 41 days, the diabetic rats received each day 2 mL of physiological saline or 2.4 mmol/kg/2 mL of TAU (or TTAU) p.o. or 4 U/kg of isophane INS s.c. Normal rats served as controls. The rats were sacrificed on day 57 to collect blood, heart and thoracic aorta samples. Untreated diabetic rats exhibited a lower body weight gain (by 34%), higher than normal plasma glucose (by ~4-fold), cholesterol (by 66%) and triglycerides (by 188%) levels, and lower INS levels (by 76%). Also there was a marked increase in catalase activity ($\geq$90%); and clear decreases in nitrite ($\geq$40%), glutathione redox status ($\geq$67%), and glutathione peroxidase ($\geq$66%) and superoxide dismutase ($\geq$51%) activities in both the aorta and heart. With only a few isolated instances (plasma lipids), TTAU was either markedly more effective (plasma glucose, plasma INS, aorta and heart glutathione, aorta redox status, and antioxidant enzymes) or marginally more effective (heart redox status) than TAU in attenuating the alterations brought about by diabetes. These results suggest that replacing the sulfonic acid group of TAU by thiosulfonic acid can lead to a greater potency against diabetes-related biochemical changes in the plasma, heart and aorta. However, except for effects on plasma lipids, these sulfur-containing compounds were less effective than INS in counteracting diabetes-related changes.

E. Mathew • M.A. Barletta • C.A. Lau-Cam (✉)
Department of Pharmaceutical Sciences, College of Pharmacy and Allied
Health Professions, St. John's University, Jamaica, NY 11439, USA
e-mail: claucam@usa.net

A. El Idrissi and W.J. L'Amoreaux (eds.), *Taurine 8*, Advances in Experimental
Medicine and Biology 775, DOI 10.1007/978-1-4614-6130-2_28,
© Springer Science+Business Media New York 2013

Abbreviations

STZ	Streptozotocin
TAU	Taurine
TTAU	Thiotaurine
INS	Insulin
GLC	Glucose
CHOL	Cholesterol
TG	Triglycerides
NO_2^-	Nitrite
MDA	Malondialdehyde
GSH	Reduced glutathione
GSSG	Glutathione disulfide
CAT	Catalase
GPx	Glutathione peroxidase
SOD	Superoxide dismutase

28.1   Introduction

Type 2 diabetes is an evolving lifelong disease characterized by clinical manifestations that reflects changes in blood glucose levels and in β-cell mass, phenotype, and function (Weir and Bonner-Weir 2004). One of early stages of type 2 diabetes is glucose intolerance, which is imposed by the insulin resistance of peripheral organs, including liver, fat, and muscle tissues, and by increased insulin secretion by the pancreas as a compensation to maintain normoglycemia (Heather and Clarke 2011; Weir and Bonner-Weir 2004). Depending on the plasticity of the β-cells, the hyperinsulinemic drive may be long lasting or may progress to a state of hypoinsulinemia, hyperglycemia, and insulin-dependent diabetes due to reduction of β-cell mass. As a consequence, diabetes is a highly heterogeneous disease with many systemic effects (Heather and Clarke 2011).

Studies in humans and animals have suggested that type 2 diabetes contributes to the pathogenesis of microvascular (retinopathy, neuropathy, and nephropathy) and macrovascular (accelerated atherosclerosis affecting the heart and brain) diseases (Calles-Escandon and Cipolla 2001) by adversely affecting small (microangiopathy) and large (macroangiopathy) blood vessels. From an epidemiological standpoint, vasculopathy and cardiomyopathy are at the forefront of diabetic complications because of a high incidence and mortality and because of the economic burden associated with them (Sharpe et al. 1998; Gugliucci 2000).

In type 2 diabetes, the etiology of macrovascular disease seems to be multifactorial and to result from the complex interactions of factors such as hyperglycemia, hyperlipidemia, hyperinsulinemia and/or hyperproinsulinemia, and oxidative stress among others (Calles-Escandon et al. 1999). There is also evidence to indicate that endothelial dysfunction is an early marker of atherosclerosis and one that can be detected before structural changes to the vessel wall have become apparent to conventional sonographic or radiologic diagnostic procedures (Davignon and Ganz 2004).

The relevance of hyperglycemia and oxidative stress to the pathogenesis of vascular complications of diabetes has been demonstrated both in vivo and in vitro. For example, a study with cultured porcine aortic vascular smooth muscle cells exposed to different concentrations of glucose revealed that the levels of intracellular malondialdehyde (MDA), serving as a marker of peroxidative damage, and of the intracellular antioxidant reduced glutathione (GSH) were increased and decreased, respectively, in proportion to the concentration of glucose added (Sharpe et al. 1998). Similarly, exposing coronary microvascular endothelial cells to a hyperglycemic environment was found to significantly reduce the levels of GSH and of the vasore-laxant nitric oxide (NO) and to increase the protein expressions and activities of p22-phox, a membrane-bound component of prooxidant NAD(P)H oxidase and anti-oxidant enzymes (Weidig et al. 2004). In the same study, the addition of free radical scavengers like tiron and mercaptopropionylglycine was found to attenuate these effects independently of the increase in medium osmolarity. There is also experimental data derived from rats made diabetic with streptozotocin (STZ) indicating that chronic treatment with insulin (INS), started at the onset of diabetes, was able to preserve the normal relaxation–contraction responses of the aorta (Kobayashi and Kamata 1999); and that a treatment with INS, the antioxidant melatonin or, better with both, restored the normal responses of the isolated aorta to standard relaxing and contracting pharmacological agents (Paskaloglu et al. 2004). The same animal model has been helpful to establish a direct correspondence between the degree of aortic impairment observed in tissue bath preparations and the extent of lipid peroxidation (LPO) and decrease of GSH levels in the aortic tissue (Paskaloglu et al. 2004).

The role of diabetes as a risk factor for the development of cardiovascular complications and cardiovascular disease has also been the subject of extensive investigation. Studies in humans and laboratory animals with type 1 or type 2 diabetes have clearly demonstrated that free radicals play a major role in the development of cardiac remodeling and cardiac dysfunction (Babu et al. 2006), and they are one of the earliest factors triggering other deleterious mechanisms (Potenza et al. 2009). The extreme susceptibility of the heart to oxidative damage and, hence, to changes in morphology and function due to attack by reactive oxygen (ROS) and reactive nitrogen (RNS) species is attributed to its low content of free radical scavengers and antioxidant enzymes (Chen et al. 1994) and/or to compromised antioxidant defense systems (Kędziora-Kornatowska et al. 2003). Under the influence of hyperglycemia, ROS, including superoxide anion, peroxynitrite, hydroxyl radical, and hydrogen peroxide, may originate from glucose autoxidation, protein glycation, oxidative formation of advanced glycation end products, the polyol pathway (Wolff et al. 1991) and, especially, from the mitochondrial electron transport chain (Shen et al. 2006). Based on the close relationship between oxidative stress and cardiovascular disorders in diabetes, numerous antioxidant compounds with different chemical features and biological have been evaluated in STZ- or alloxan-treated animals.

The present study was undertaken in rats with the specific purpose of comparing the actions of taurine (TAU), 2-aminoethanesulfonic acid, with those of its analog thiotaurine (TTAU), 2-aminoethanethiosulfonic acid, to assess the impact of replacing the sulfur-containing functionality of TAU with the thiosulfonic acid group on the protective actions of this compound in the heart and aorta of STZ-treated rats. While the attenuating effects of TAU on hyperglycemia (Brøns et al. 2004; Kaplan et al. 2004; Kulakowski and Maturo 1984), hyperlipidemia (Militante and Lombardini 2004), oxidative stress (Di Leo et al. 2002; Obrosova et al. 2001; Obrosova and Stevens 1999; Shivananjappa and Muralidhara 2012), and functional and biochemical alterations caused by diabetes on the macrovasculature (Fennessy et al. 2003; Moloney et al. 2010; Tan et al. 2007; Wang et al. 2008) and heart (Xu et al. 2008) are well documented, information on the actions of TTAU on diabetes appears to be limited to a single report describing its lowering actions on the blood glucose and triglycerides of alloxan-treated rats (Katsumata et al. 1997). To more accurately gauge the protective potencies of TAU and TTAU, their actions were compared against those of INS.

28.2 Methods

Male Sprague–Dawley rats, 200–250 g in weight, were used in the study. The animals were kept in a room maintained at a constant temperature ($23 \pm 1\,^\circ$C) and relative humidity, with a normal 12 h dark–12 h light cycle. Throughout the study the animals had free access to a commercial rodent diet chow and filtered tap water; and were cared in accordance with guidelines established by the United States Department of Agriculture. The study received the approval of the Institutional Animal Care and Use Committee of St. John's University, Jamaica, New York.

Diabetes was induced with an intraperitoneal, 60 mg/kg/2 mL, dose of streptozotocin (STZ) in citrate buffer pH 4.5, and allowed to progress for 14 days. From day 15 until day 56, groups of six diabetic rats were treated on a daily basis with an oral dose of taurine or thiotaurine (2.4 m mol/kg each) (the treatment groups), a subcutaneous, 4 U/kg, dose of isophane insulin (the reference group), or an oral, 2 mL, volume of physiological saline (the diabetic group). A group of naive rats served as the control group. All the rats were sacrificed by decapitation on day 57 to collect their blood in heparinized tubes and to isolate their thoracic aortae and hearts. The blood samples were centrifuged at $700 \times g$ and $4\,^\circ$C for 10 min to isolate their plasma fraction, which was used for the assay of glucose (GLC), insulin (INS), cholesterol (CHOL) and triglycerides (TG) contents. Aliquots of aorta or heart were homogenized with phosphate buffered saline pH 7.4 (1:30, w/v) to obtain a suspension which, upon centrifugation at $1,000 \times g$ and $4\,^\circ$C, yielded a supernatant suitable for the assay of malondialdehyde (MDA), nitric oxide (measured as the more stable product nitrite, NO_2^-), reduced glutathione (GSH), glutathione disulfide (GSSG), catalase (CAT), glutathione peroxidase (GPx), and superoxide dismutase (SOD).

GLC was assayed using a commercially available colorimetric assay kit from Sigma Chemical Co., St. Louis, MO (Procedure No.510); INS was assayed with an

ELISA kit from Calbiotech Inc., Spring Valley, CA (Item No. IS130D) and CHOL and TG were measured with assay kits from Stanbio Laboratory, Boerne TX (Cholesterol Liquicolor® and Enzymatic Triglycerides Procedure No. 2150, respectively). Remaining biochemical parameter were measured using published methods: MDA according to Buege and Aust (1978), NO_2^- according to Fox et al. (1981), GSH and GSSG according to Akerboom and Sies (1981) supplemented by the method of Guntherberg and Rost (1966) for removing interfering GSH, CAT according to Aebi (1984), GPx according to Flohé and Günzler (1984), and SOD according to Misra and Fridovich (1972).

The experimental results, reported as mean $\pm$ SEM for $n=6$, were analyzed by unpaired Student's t-test followed by one-way analysis of variance and Newman–Keuls multiple comparison test. Differences from the control and diabetic groups were considered to be significant when p was at least 0.05.

28.3   Results

28.3.1   Body Weight Gains and Aortae Weights

At the end of 56 days the body weight of the diabetic rats was 27% lower than that of normal rats ($p<0.05$) (Fig. 28.1). A 6-week treatment with INS allowed the diabetic rats to attain a body weight that was only 4% below that of normal rats. Identical treatments with TAU and TTAU also resulted in steady growths but which were, respectively, 10% and 18% ($p<0.05$) below that of normal rats. However, when the body weights on day 56 were compared with those measured on day 1, the gain of normal rats was ~95% and that of diabetic rats was only 37%. A treatment with INS allowed the weight of diabetic rats to increase by 87%, which was higher than those seen after equal treatments with either TAU (68% gain) or TTAU (56% gain) (Fig. 28.1). Furthermore, a direct proportionality was verified between the body weights and wet weights of the corresponding aortae (Fig. 28.1). Thus, at the end of 56 days aortae from untreated diabetic rats were in the average 37% lighter than aortae from normal rats ($p<0.01$). A treatment with INS, found to normalize the body weight gain, also normalized the aortic weight; and treatments with TAU and TTAU, leading to lower body weight gains, led to aortic weights that were 15% ($p<0.05$) and 23% ($p<0.05$) lower than the weights of control samples.

28.3.2   Plasma GLC and INS

As shown in Figs. 28.2 and 28.3, respectively, diabetes raised the plasma GLC by >300% and lowered the plasma INS by 76% to a significant extent (both at $p<0.001$). Following a chronic treatment with INS, the GLC value was only 23% above

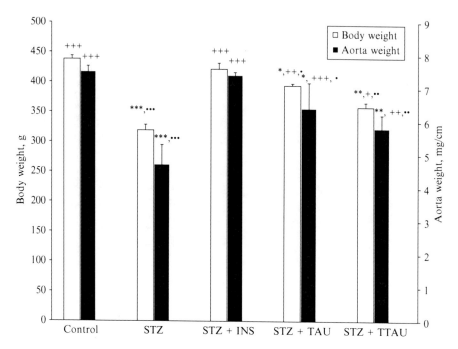

Fig. 28.1 Body weights and aortae weights for normal, diabetic, and treated diabetic rats after 56 days. STZ was administered on day 1 and treatments with INS, TAU, and TTAU were started on day 15. Differences were significant from Control at $*p<0.05$, $**p<0.01$, and $***p<0.001$; from STZ at $^+p<0.05$, $^{++}p<0.01$, and $^{+++}p<0.001$; and from INS at $^•p<0.05$, $^{••}p<0.01$, and $^{•••}p<0.001$. Values are shown as mean ± SEM for $n=6$

control ($p<0.05$) and that of INS exceeded the control value by 21% ($p<0.05$). Both TAU and TTAU also attenuated changes caused by diabetes but to a much lesser extent when compared to INS. With these compounds, the plasma GLC was found to increase by ~220% and ~155%, respectively (both at $p<0.001$) (Fig. 28.2); and the plasma INS was decreased by 52% ($p<0.001$) and 42% ($p<0.01$), respectively (Fig. 28.3).

28.3.3 Plasma CHOL and TG

Figure 28.4 summarizes the results for the plasma CHOL and TG levels, respectively. These lipids were increased in diabetic rats by 66% and 188%, respectively (both at $p<0.001$). A treatment with INS had a weak attenuating effect on both the plasma CHOL (up by ~50% ($p<0.05$)) and TG (up by 181%) relative to diabetic values. While TAU and TTAU were about equipotent with each other (only 20% increase, $p<0.05$), TAU was more effective than TTAU in lowering the elevation in plasma TG (up by 46% and 74%, respectively, both at $p<0.001$ vs. control).

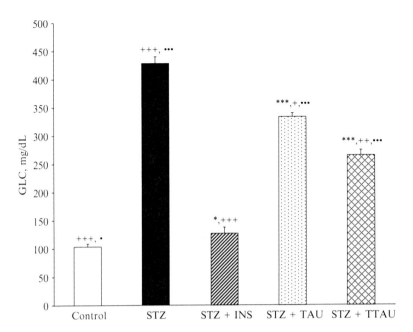

Fig. 28.2 Plasma GLC levels of normal, diabetic, and treated diabetic rats after 56 days. STZ was administered on day 1 and treatments with INS, TAU, and TTAU were started on day 15. Differences were significant from Control at $*p<0.05$ and $***p<0.001$; from STZ at $^+p<0.05$, $^{++}p<0.01$, and $^{+++}p<0.001$; and from STZ+INS at $^•p<0.05$ and $^{•••}p<0.001$. Values are shown as mean±SEM for $n=6$

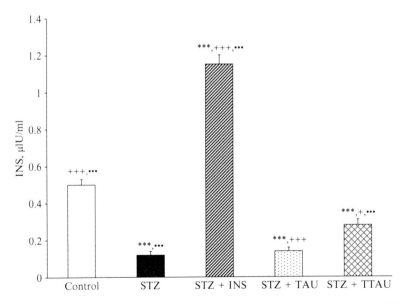

Fig. 28.3 Plasma INS levels of normal, diabetic, and treated diabetic rats after 56 days. STZ was administered on day 1 and treatments with INS, TAU, and TTAU were started on day 15. Comparisons were significantly different from Control at $***p<0.001$; from STZ at $^+p<0.05$ and $^{+++}p<0.001$; and from STZ+INS at $^{•••}p<0.001$. Values are shown as mean±SEM for $n=6$

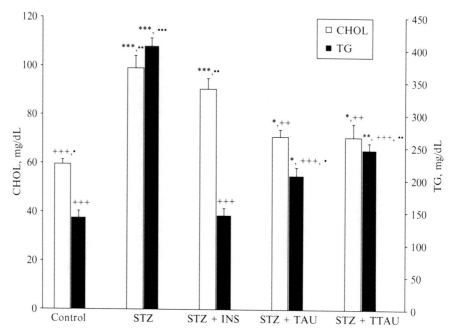

Fig. 28.4 Plasma CHOL and TG levels of normal, diabetic, and treated diabetic rats after 56 days. STZ was administered on day 1 and treatments with INS, TAU, and TTAU were started on day 15. Differences were significant from Control at $*p < 0.05$ and $***p < 0.001$; from STZ at $^{++}p < 0.01$ and $^{+++}p < 0.001$; and from STZ + INS at $^{•}p < 0.05$, $^{••}p < 0.01$, and $^{•••}p < 0.001$. Values are shown as mean ± SEM for $n = 6$

28.3.4 Aorta and Heart MDA and NO_2^-

The results summarized in Fig. 28.5 indicate that STZ-induced diabetes stimulated the production of MDA both in the aorta (up by 233%, $p < 0.001$) and heart (up by 183%, $p < 0.001$), and that a treatment with INS was highly protective against these changes (increases of only 15% and 8%, respectively). TAU and TTAU were also found to provide a good protection but which was not as great as INS in both the aorta (increases of 41% and 29%, respectively, both at $p < 0.01$) and heart (increases of 37% and 27%, respectively, both at $p < 0.01$). The data in Fig. 28.6 indicate that the NO_2^- levels were differently affected by diabetes, with a decrease in the aorta (by 40%, $p < 0.01$) and an increase in the heart (by 103%, $p < 0.001$). Again, INS was highly protective against these changes (4% decrease and 19% increase, $p < 0.05$, respectively). TAU and TTAU were almost equally protective both in the aorta (decreases of 18%, $p < 0.05$, and 12%, respectively) and the heart (increase of only ~49%, $p < 0.001$, with both).

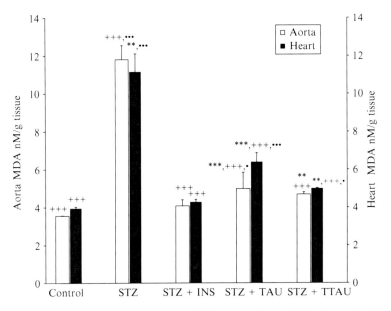

Fig. 28.5 Aorta and heart MDA levels of normal, diabetic, and treated diabetic rats after 56 days. STZ was administered on day 1 and treatments with INS, TAU, and TTAU were started on day 15. Differences were significant from Control at **$p < 0.01$ and ***$p < 0.001$; from STZ at +++$p < 0.001$; and from STZ + INS at •$p < 0.05$ and •••$p < 0.001$. Values are shown as mean ± SEM for $n = 6$

28.3.5   Aorta and Heart GSH and GSSG

Figures 28.7 and 28.8 present the results for the GSH and GSSG contents in the aorta and heart of the various experimental groups. STZ-induced diabetes lowered the intracellular contents of GSH (by 56% in the aorta, $p < 0.001$, by 44% in the heart, $p < 0.01$), but increased those of GSSG (by 72% in the aorta, by 52% in the heart, both at $p < 0.001$). A treatment with INS counteracted all these changes to a significant extent, with the values not being significantly different from those of the control values. While TAU and TTAU were also able to attenuate these changes, they were less effective than INS in preserving the GSH content and somewhat equipotent with INS in lowering the GSSG content. Thus, TTAU was somewhat more potent than TAU in preventing the depletion of GSH in the aorta (21% loss vs. 32%, both at $p < 0.01$) and heart (10% loss vs. 17% loss). While in the aorta, TTAU was found to normalize the GSSG level and TAU to lower the diabetic value by ~20% ($p < 0.05$), in the heart both compounds were about equal in suppressing the GSSG increase. Furthermore, the effects of these treatments were reflected on their respective abilities to prevent the decrease (by 65%, $p < 0.001$) in the GSH/GSSG ratio seen in diabetes (Fig. 28.9). Again INS was the most potent (<10% decreases), TTAU had an intermediate potency (~20% decrease in the aorta, no decrease in the heart), and TAU was the least potent in both the aorta (59% decrease, $p < 0.001$) and heart (15% decrease).

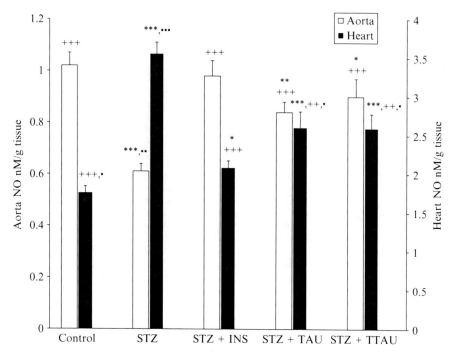

Fig. 28.6 Aorta and heart NO levels of normal, diabetic, and treated diabetic rats after 56 days. STZ was administered on day 1 and treatments with INS, TAU, and TTAU were started on day 15. Differences were significant from Control at *$p < 0.05$, **$p < 0.01$, and ***$p < 0.001$; from STZ at ++$p < 0.01$ and +++$p < 0.001$; and from STZ + INS at •$p < 0.05$, ••$p < 0.01$, and •••$p < 0.001$. Values are shown as mean ± SEM for $n = 6$

28.3.6 Aorta and Heart Antioxidant Enzymes

As shown in Fig. 28.10, the activity of CAT was elevated by diabetes in the aorta (by 127%, $p < 0.001$) and heart (by 90%, $p < 0.001$). These alterations were attenuated by all the treatments, with INS providing the greatest protection (increases of 38%, $p < 0.01$, and 11%, respectively), followed by TTAU (increases of 58%, $p < 0.001$, and 26%, $p < 0.01$, respectively), with TAU demonstrating the weakest effects (increases of 89%, $p < 0.001$, and 33%, $p < 0.01$, respectively). In comparison to the changes in CAT activity, those of GPx and SOD followed a different direction as a result of diabetes. As seen in Fig. 28.11, diabetes lowered the activities of GPx both in the aorta (by 66%, $p < 0.001$) and heart (by 75%, $p < 0.001$). INS was able to keep these decreases to ≤10% below the control values. TAU and TTAU were also able to reduce the losses of GPx activity, with TTAU demonstrating a greater attenuating effect in both the aorta (32%, $p < 0.01$, loss) and heart (20%, $p < 0.05$, loss) than TAU (losses of 40%, $p < 0.01$, and 27%, $p < 0.01$, respectively).

In common with the GPx activities, those for SOD were of comparable magnitude and direction in the aorta (64% loss, $p < 0.001$) and heart (51% loss, $p < 0.001$)

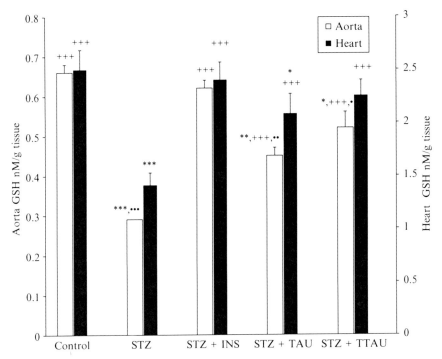

Fig. 28.7 Aorta and heart GSH levels of normal, diabetic, and treated diabetic rats after 56 days. STZ was administered on day 1 and treatments with INS, TAU, and TTAU were started on day 15. Differences were significant from Control at $*p < 0.05$, $**p < 0.01$, and $***p < 0.001$; from STZ at $^{+++}p < 0.001$; and from STZ + INS at $^{\bullet}p < 0.05$, $^{\bullet\bullet}p < 0.01$, and $^{\bullet\bullet\bullet}p < 0.001$. Values are shown as mean ± SEM for $n = 6$

(Fig. 28.12). INS provided a strong protection against these changes (losses of only 26%, $p < 0.01$, and 13%, respectively). A treatment with TTAU was also protective, with the effect being weaker than INS in the aorta (36% loss, $p < 0.01$) and comparable to INS in the heart (17% loss, $p < 0.05$). On the other hand, TAU was less potent than TTAU in both samples (losses equal to 45%, $p < 0.001$, and 27%, $p < 0.01$, respectively).

28.4   Discussion

Information on the actions of TTAU on diabetes appear to be limited to a report by Katsumata et al. (1997) describing its acute hypoglycemic and hypotriglyceridemic effects in alloxan-treated rats when given as a pretreatment. In the present study the actions of this thiosulfonate on experimentally induced diabetes were examined using a chronic treatment approach and rats made diabetic with STZ. This animal

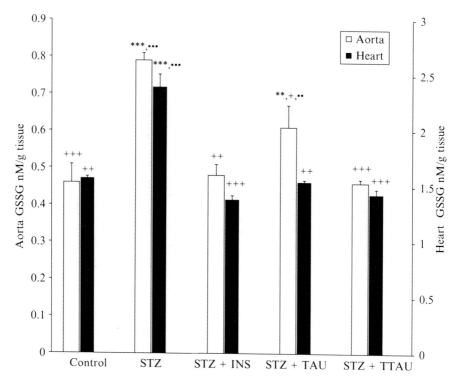

Fig. 28.8 Aorta and heart GSSG levels of normal, diabetic, and treated diabetic rats after 56 days. STZ was administered on day 1 and treatments with INS, TAU, and TTAU were started on day 15. Differences were significant from Control at $*p<0.05$ and $***p<0.001$; from STZ at $^{+}p<0.05$, $^{++}p<0.01$, and $^{+++}p<0.001$; and from STZ + INS at $^{\bullet\bullet}p<0.01$ and $^{\bullet\bullet\bullet}p<0.001$. Values are shown as mean ± SEM for $n=6$

model is quite convenient since it allows the evaluation of exogenous agents for protection against diabetes-induced changes in body weight, biochemical parameters associated with blood GLC and lipids (Montilla et al. 1998; Patel and Goyal 2011), oxidative stress (Coskun et al. 2005; Obrosova et al. 1998; Obrosova and Stevens 1999; Patel and Goyal 2011), energy metabolism (Obrosova et al. 1998) and major complications of diabetes (Banes-Berceli et al. 2007; Bravo-Nuevo et al. 2011; Obrosova et al. 1998; Obrosova and Stevens 1999; Wei et al. 2003). In the present case, STZ-induced diabetes was found to drastically lower the body weights, to elevate the circulating levels of GLC, INS, CHOL, and TG, to promote MDA and GSSG formation, and to lower the NO and GSH contents as well as the activities of antioxidant enzymes in both the aorta and heart.

Under the present experimental conditions, a 6-week treatment with INS, TAU, or TTAU led to a significant increase in the body weights of diabetic rats, with INS demonstrating a reversing effect when administered daily from week 3 of diabetes onwards and which culminated in body weights comparable to those of normal rats by the end of week 8. An identical treatment with TAU and TTAU, administered in equimolar doses, also promoted a steady increase in body weights but not as much

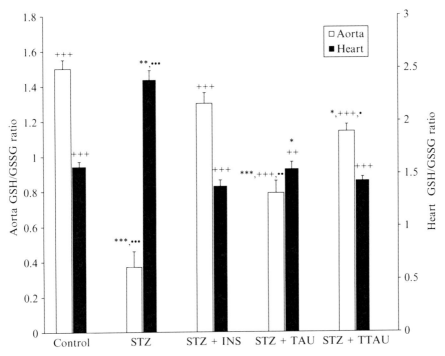

Fig. 28.9 Aorta and heart GSH/GSSG ratios of normal, diabetic, and treated diabetic rats after 56 days. STZ was administered on day 1 and treatments with INS, TAU, and TTAU were started on day 15. Differences were significant from Control at $*p<0.05$, $**p<0.01$, and $***p<0.001$; from STZ at $^{++}p<0.01$ and $^{+++}p<0.001$; and from STZ + INS at $^{•}p<0.05$, $^{••}p<0.01$, and $^{•••}p<0.001$. Values are shown as mean ± SEM for $n=6$

as INS. Moreover, the weights of animals on TTAU were lower than those on TAU when the final body weights were compared with the respective initial weights. This unexplained finding is surprising since the appearance and behavior of these animals was like those of control animals. Moreover, the treatment of diabetic rats either with INS or with a sulfur-containing compound resulted in aortae that were heavier than those of untreated diabetic rats. The weight loss of aorta from diabetic rats has been correlated with a decrease in muscle mass rather than with a decrease in fluid content (Mulhern and Docherty 1989).

TAU and TTAU were also found to lower the plasma GLC and to raise the plasma INS to a significant extent relative to values for untreated diabetic rats, but not as much as INS. In both instances, the effects of TTAU were greater than those from an equal dose of TAU. Interestingly, while this and other studies have repeatedly shown that TAU is a hypoglycemic agent in animal models of diabetes (Kulakowski and Maturo 1984; Nandhini et al. 2005; Tenner et al. 2003), the same consistency has not been observed in human subjects since both positive (Elizarova and Nedosugova 1996) and negative (Chauncey et al. 2003) results have been documented. The hypoglycemic action of TAU has been related to an ability to improve sensitivity to INS (Nakaya et al. 2000), to potentiate the actions of INS

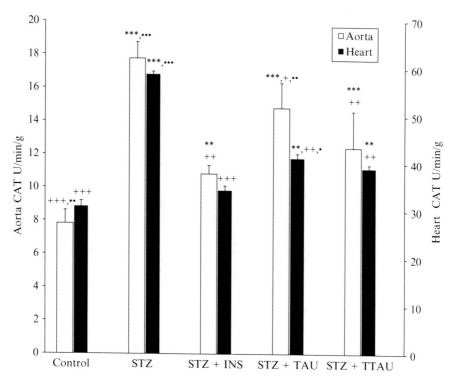

Fig. 28.10 Aorta and heart CAT activities of normal, diabetic, and treated diabetic rats after 56 days. STZ was administered on day 1 and treatments with INS, TAU, and TTAU were started on day 15. Differences were significant from Control at $**p < 0.01$ and $***p < 0.001$; from STZ at $^{++}p < 0.01$ and $^{+++}p < 0.001$; and from STZ + INS at $^{•}p < 0.05$, $^{••}p < 0.01$, and $^{•••}p < 0.001$. Values are shown as mean ± SEM for $n = 6$

(Lampson et al. 1983), to bind to the INS receptor (Maturo and Kulakowski 1988), and to stimulate INS secretion (Cherif et al. 1996). The increase in plasma INS attained with either TAU or TTAU was significant relative to the level of untreated diabetic rats, with the latter compound appearing more potent than the former. Earlier reports on TAU have indicated that this compound can protect the pancreas of rats against damage by STZ when given as a pretreatment (Chang and Kwon 2000), and to have either a stimulatory (Cherif et al. 1996; Hisashi et al. 1979) or inhibitory (Tokunaga et al. 1983) effect on INS secretion. While the actions of TTAU could be of potential benefit in controlling chronic hyperglycemia, a recognized risk factor of diabetic complications such as vascular damage and cardiomyopathy (Militante et al. 2000), it will need to undergo further evaluation in the laboratory and in human subjects since previous studies with TAU in human volunteers failed to produce obvious effects on INS secretion and on GLC levels (Brøns et al. 2004; Chauncey et al. 2003).

Hyperglycemia is recognized as one of the major determinants of the development of microvascular anomalies and of cardiomyopathy in diabetic subjects

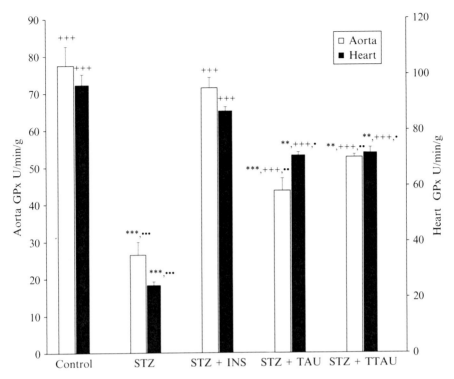

Fig. 28.11 Aorta and heart GPx activities of normal, diabetic, and treated diabetic rats after 56 days. STZ was administered on day 1 and treatments with INS, TAU, and TTAU were started on day 15. Differences were significant from Control at $**p < 0.01$ and $***p < 0.001$; from STZ at $+++p < 0.001$; and from STZ + INS at $•p < 0.05$, $••p < 0.01$, and $•••p < 0.001$. Values are shown as mean ± SEM for $n = 6$

(Hayat et al. 2004). However, the occurrence of oxidative stress appears not to be strictly dependent on the prevailing high levels of glucose since even brief increases can significantly elevate the production of ROS in vascular cells (Yano et al. 2004); and stimulation of oxidative stress have shown a better correlation with acute fluctuations of GLC in the postprandial period than with chronic sustained hyperglycemia (Monnier et al. 2006). In the hyperglycemic state, ROS may originate from the formation of advanced glycation end products (AGEs), increased flux of GLC through the polyol pathway, activation of protein kinase C isoforms and the hexosamine pathway, impairment of the electron transport chain (Brownlee 2001), and activation of membrane-bound NAD(P)H oxidase (Mohazzab et al. 1994). The overproduction of ROS can lead to cell injury through peroxidation of cell membrane lipids to yield MDA and 4-hydroxynonenal (4-HNE) (Haber et al. 2003), to the cleavage of DNA (Pan et al. 2010), and to the oxidation of proteins to yield protein carbonyl derivatives and nitrotyrosine (Adams et al. 2001). In addition, ROS can inhibit enzymes in complexes of the mitochondrial electron transport chain to cause blockade of electron flow, reduction of oxidative phosphorylation

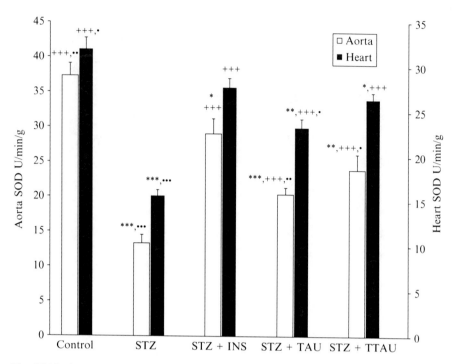

Fig. 28.12 Aorta and heart SOD activities of normal, diabetic, and treated diabetic rats after 56 days. STZ was administered on day 1 and treatments with INS, TAU, and TTAU were started on day 15. Differences were significant from Control at *$p < 0.05$, **$p < 0.01$, and ***$p < 0.001$; and from STZ at $^{+++}p < 0.001$. Values are shown as mean ± SEM for $n = 6$

and the diversion of oxygen towards the formation of superoxide anion (Brownlee 2001). Oxidative stress has also been associated with the activation of oxidative stress pathways related to the development of inflammation and of late complications of diabetes and of signaling pathways associated with the development of INS resistance (Hesselink et al. 2007).

Based on the measurement of MDA levels, the present study verified that LPO was markedly increased by an acute treatment with STZ, more in the aorta than in the heart. INS was found to effectively reduce these elevations to values that were comparable to the respective control values. TAU and TTAU were also effective in attenuating LPO but to a lesser extent than INS. Furthermore, TTAU was found to be roughly twice as potent as TAU. Based on these results, it would appear that in the present animal model LPO is directly dependent on the ability of the test compounds to influence the circulating levels of GLC and INS (Nandhini et al. 2005). Further evidence on a protective antioxidant action by TAU against hyperglycemia-induced oxidative stress and, thereby, against INS resistance, was investigated by infusing rats with GLC in the presence or absence of TAU, followed by a hyperinsulinemic-euglycemic clamp. Changes such as the increased formation of protein carbonyl, MDA and 4-HNE, serving as markers of oxidative stress, and a decrease in INS-stimulated uptake of labeled GLC into the soleus muscle were all normalized by TAU (Haber et al. 2003).

Diabetes mellitus is associated with altered myocardial function and increased susceptibility to cardiovascular complications that are detectable by echocardiography, hemodynamic assessment, isolated heart perfusion techniques, histological evidence of necrosis, and measurement of NO-related parameters (Joffe et al. 1999). Considerable evidence has accumulated to link diabetes to the accelerated development of atherosclerosis, with hyperglycemia, hyperlipidemia, and oxidative stress representing prominent risk factors of this macrovascular complication of diabetes (Ganda and Arkins 1992). Hyperglycemia may cause endothelial dysfunction by lowering the availability of endothelial NO, which protects against vascular disease through actions that include vasodilation, inhibition of platelet aggregation and adhesion to the vascular wall, lowering of the expression of chemoattractant and surface adhesion proteins involved in atherogenesis, reduction of vascular permeability and of the rate of oxidation of LDL to its proatherogenic form, and inhibition of the proliferation of vascular smooth muscle cells (Du et al. 2001). In vitro experiments with bovine aortic endothelial cells (BAECs) have determined that an exposure to either GLC or glucosamine, a product of the hexosamine pathway, can inhibit the endothelial NO synthase (eNOS) through O-linked N-acetylglucosamine modification of eNOS and a reciprocal decrease in O-linked serine phosphorylation, an effect that was enhanced by superoxide anion of mitochondrial origin, found to accumulate as a result of impairment of the mitochondrial electron transport chain and mitochondrial Mn SOD by glucose and glucosamine (Du et al. 2001; Guzik et al. 2002). A chronic, but not an acute, exposure of BAECs to high GLC levels was also accompanied by a reduction of the total NO_2^- content and by the decreased expression of eNOS mRNA and eNOS protein (Srinivasan et al. 2004). Also a reduction of eNOS activity of the same magnitude as that seen in cultured BAECs has been verified in the aorta of STZ-treated rats (Du et al. 2001).

In the present study the level of NO, measured as NO_2^-, was found to be reduced in the aorta and but increased in the heart. This differing trend is in agreement with the results of a study demonstrating that in diabetes there is a decreased expression of eNOS with a concomitant increase in the expression of cardiac iNOS and nitrotyrosine production as the disease progresses (Nagareddy et al. 2005). The dramatic attenuation of these changes by INS not only suggests a role for hyperglycemia in endothelial dysfunction but also for the hormone itself on NO release (Scherrer et al. 1994) and on the expression of eNOS gene (Kuboki et al. 2000) and eNOS protein and mRNA (Fisslthaler et al. 2003; Kobayashi and Kamata 2001) in endothelial cells and microvessels, but without involving direct activation of eNOS (Fisslthaler et al. 2003). TAU and TTAU were also highly and equally effective in preserving NO production, more in the aorta than in the heart, in spite of being less potent than INS. These findings are in agreement with those from an earlier study showing that chronically administered TAU can increase the secretion of NO_2^- (Nakaya et al. 2000) and restore aortic relaxation impaired by CHOL feeding to STZ-treated mice (Kamata et al. 1996). Furthermore, TAU may also protect against endothelial dysfunction and macrovascular disease by reducing the levels of plasma

CHOL since its chronic administration to STZ-treated mice led to a reversal of the attenuation of endothelium-dependent relaxation observed when the circulating levels of CHOL and LDL were elevated (Kamata et al. 1996). Moreover, taking into account the inability of INS to lower the plasma CHOL and TG levels, the effects of TAU on NO and on plasma lipids must occur independently of its hypoglycemic action. At the clinical level, supplementation of young male patients with type 1 diabetes mellitus and statistically significant abnormalities of conduit vessel function with oral TAU (500 mg/day for 2 weeks) was able to normalize endothelial function (Moloney et al. 2010). Further proof of the beneficial impact of TAU on vascular function has come from a study in which young smokers without a history of hyperlipidemia, a family history of premature vascular disease, or any systemic disease predisposing them to endothelial dysfunction, but with demonstrable impaired endothelial-dependent vasodilation were returned to normal function by an oral daily supplementation with TAU (1.5 g/day/5 days) (Fennessy et al. 2003). In the same study, evidenced of upregulation of eNOS expression was obtained in vitro by culturing endothelial cells with monocytes taken from TAU-treated impaired smokers but not with monocytes from untreated smokers.

Diabetes is associated with changes in plasma lipids and lipoproteins pointing in the atherogenic direction. In this regard, the present study found TTAU to reduce the plasma CHOL and TG to the same extents as TAU in spite of demonstrating a greater effect on the plasma GLC and INS. In terms of the plasma TG levels, however, the attenuating action of TTAU was much less than that of TAU. In this regard, the hypolipidemic actions of TAU have been extensively demonstrated in experimental animals. For example, feeding normolipidemic rats a high CHOL diet supplemented with TAU for 5 weeks was found to significantly lower the plasma levels of total CHOL, LDL-CHOL and TG and hepatic CHOL and TG in rats fed a high CHOL diet and to lower the plasma TG and elevate hepatic free fatty acids in rats fed a CHOL free diet (Park and Lee 1998). The lowering effect of TAU on CHOL was ascribed to a stimulatory action on hepatic cholesterol 7-α-hydroxylase (CYP7A1), the enzyme that converts CHOL to bile acids for fecal excretion (Murakami et al. 1999; Yokogoshi et al. 1999). TAU has demonstrated the same stimulatory effect in rats and guinea pigs maintained on a normal diet (Militante and Lombardini 2004). Furthermore, inhibition of intestinal cholesterol absorption through inhibition of the intestinal activity of acyl CoA:cholesterol acyltransferase activity may represent an additional mechanism for the anticholesterolemic action of TAU (Murakami et al. 1996). The potential benefits of TAU in preventing or attenuating macrovascular derangements has also been investigated in young obese or overweight nondiabetic adult patients. A supplementation with TAU produced a significant reduction of the serum TG, atherogenic index, and body weight when compared to individuals not receiving TAU (Zhang et al. 2004). In addition, mice consuming a high CHOL diet along with TAU as part of the drinking water showed significantly less serum LDL and VLDL CHOL, more serum HDL CHOL, and less accumulation of arterial lipids at the end of 6 months (Murakami et al. 2000).

Adhesion molecules have been associated with cardiovascular diseases. Various studies have demonstrated the potential role of adhesion molecules as initiating factors of diabetic vasculopathy and atherogenesis by promoting the recruitment of circulating monocytes, neutrophils and T-lymphocytes at vascular walls (Peschel and Niebauer 2003; Urso et al. 2010). Elevated levels of circulating adhesion molecules have been reported for diabetic patients, possibly as a result of hypertriglyceridemia and hyperglycemia (Ceriello et al. 2004). Acute and chronic hypertriglyceridemia can upregulate the levels of adhesion molecules by activating endothelial cells (Peschel and Niebauer 2003). An accelerating influence of hyperglycemia in the development of atherosclerosis in diabetes has been based on the results of an experiment in which human aortic endothelial cells exposed to a high GLC concentrations for several hours to days exhibited increased binding by monocytes (Kim et al. 1994) and increased expression of adhesion molecule mRNA (Kado et al. 2001). Furthermore, endothelial cells exposed to sera taken from diabetic subjects exhibited a greater expression of adhesion molecules than when exposed to sera from nondiabetics due to a content of component(s), capable of inducing adhesion molecule expression independently of hyperglycemia (Rasmussen et al. 2002). Since the expression of proatherogenic adhesion molecules on endothelial cells appears to be sensitive to the action of antioxidant agents (Sampson et al. 2002), compounds like TAU and TTAU could be of benefit in lowering the expression of adhesion molecule owing to their hypoglycemic, hypotriglyceridemic and antioxidant properties. Indeed, TAU has been found to down regulate the hyperglycemia-induced expression of adhesion molecules and to counteract oxidized LDL-mediated apoptosis in human umbilical cord venous endothelial cells (Ulrich-Merzenich et al. 2007).

In diabetic patients an inverse relationship between the circulating levels of GLC and glycated hemoglobin and the levels of GSH in erythrocytes and the plasma has been established (Jain and McVie 1994). High GLC levels are known to reduce the intracellular GSH content and the uptake of the GSH precursor L-cysteine in cultured vascular smooth muscle cells, and to reduce GSH in the aorta of STZ-treated rats, an effect that has been related to impaired redox cycling of GSH (Tachi et al. 2001). Additional mechanisms that have been invoke to account for the reduction of GSH in diabetes have been its increased oxidation, decreased synthesis and increased degradation (Furfaro et al. 2012). A role for hyperglycemia and for ROS in these reductions was inferred from the normalizing effect that a treatment of diabetic rats with INS had on aortic GSH levels; and from the addition of N-acetylcysteine to isolated ventricular cardiac myocytes cultured with GLC (Fiordaliso et al. 2004). Furthermore, supplementation of type 2 diabetics with vitamin E resulted in a significant reduction of LPO, based on MDA formation, and in higher serum (Nweke et al. 2009) and erythrocyte GSH levels (Jain et al. 2000) relative to diabetics not receiving the vitamin. In diabetes a reduction in GSH is accompanied by an increase in GSSG and a reduction of the redox state, changes that limit the protection of endothelial and cardiac cells from oxidative injuries, cell signaling disturbances and apoptosis (Yang et al. 2006). In the present study, INS exerted a normalizing effect on GSH depletion and GSSG accumulation in the aorta and heart and, hence, on the redox status. When taken together, these results point to a drastic reduction in GLC-driven oxidative stress observed in STZ-treated animals. A similar correlation was confirmed for both TTAU and TAU which, like INS, also reduced the changes in GSH and GSSG to a significant, although lesser, extent. While both compounds were more effective in the heart than in the aorta, the effects of the former were, in both instances, greater than those of the latter. There is also at least one report indicating that the feeding of TAU, as a 5% diet supplement, to STZ-diabetic rats for 3-weeks after the induction of diabetes was able to ameliorate LPO and changes in GSSG/GSH but without affecting the GSH content in the precataractous lens (Obrosova and Stevens 1999).

Hyperglycemia increases the formation of free radicals and other ROS that can impair INS-dependent signaling pathways, damage the pancreas, and impair the vascular endothelium. To counteract the harmful effects of free radicals, cells are equipped with an integrated group of antioxidant enzyme for the detoxification or degradation of these free radicals (Hisalkar et al. 2012). From a comparison of the enzyme activity values reported by different laboratories for different human populations, animal models of diabetes, duration and severity of the diabetic process, specimens analyzed, and extent of the INS deficiency, among others, it is apparent that they vary rather widely and even contradict one another (Szaleczky et al. 1999). On the other hand, decreases in the activity of CAT, SOD and GPx like those observed here have also been documented for sera from subjects with type 2 diabetes (Hisalkar et al. 2012; Pan et al. 2010) and STZ-treated rats (Coskun et al. 2005). Differences in sensitivity to different oxidants (Wijeratne et al. 2005) and in adaptive response to pro-oxidants (Likidlilid et al. 2010) in the diabetic state are some of the reasons that have been invoked to account for the different trends of the antioxidant enzyme activities observed in diabetes. As shown for the GSH levels, INS reversed the decreases in enzyme activities in the aorta and heart to values approximating those derived from nondiabetic rats. Both TAU and TTAU were also able to significantly attenuate the losses of activity for the three antioxidant enzymes in both the aorta and heart, with the extent of the effect being inversely proportional to the severity of the oxidative stress, and with TAU showing an insignificantly greater potency than TAU.

In conclusion, the present results suggest a protective action for both TAU and TTAU on the body weight loss, hyperglycemia, hyperinsulinemia, hyperlipidemia, and aortic and cardiac oxidative stress fostered by a diabetogen like STZ in rats. The substitution of the sulfonic acid group of TAU by the thiosulfonic acid present in TTAU is found to enhance the protective actions of TAU to a significant extent except for an effect on the body weight gain. Hyperglycemia appears to play a major role in the aforementioned variables since INS was able to return them to values comparable to those of normal rats. Further studies on the actions of TTAU on other parameters of endothelial and cardiac dysfunction are warranted.

References

Adams S, Green P, Claxton R et al (2001) Reactive carbonyl formation by oxidative and non-oxidative pathways. Front Biosci 6:a17–a24
Aebi H (1984) Catalase in vitro. Methods Enzymol 105:121–126
Akerboom TP, Sies H (1981) Assay of glutathione, glutathione disulfide, and glutathione mixed disulfides in biological samples. Methods Enzymol 77:373–382
Babu PV, Sabitha KE, Shyamaladevi CS (2006) Therapeutic effect of green tea extract on oxidative stress in aorta and heart of streptozotocin diabetic rats. Chem Biol Interact 162:114–120

Banes-Berceli AK, Ketsawatsomkron P, Ogbi S et al (2007) Angiotensin II and endothelin-1 augment the vascular complications of diabetes via JAK2 activation. Am J Physiol Heart Circ Physiol 293:H1291–H1299

Brøns C, Spohr C, Storgaard H, Dyerberg J, Vaag A (2004) Effect of taurine treatment on insulin secretion and action, and on serum lipid levels in overweight men with a genetic predisposition for type II diabetes mellitus. Eur J Clin Nutr 58:1239–1247

Brownlee M (2001) Biochemistry and molecular cell biology of diabetic complications. Nature 414:813–820

Bravo-Nuevo A, Sugimoto H, Iyer S et al (2011) RhoB loss prevents streptozotocin-induced diabetes and ameliorates diabetic complications in mice. Am J Pathol 178:245–252

Buege JA, Aust SD (1978) Microsomal lipid peroxidation. Methods Enzymol 52:302–310

Calles-Escandon J, Garcia-Rubi E, Mirza S, Mortensen A (1999) Type 2 diabetes: one disease, multiple cardiovascular risk factors. Coron Artery Dis 10:23–30

Calles-Escandon J, Cipolla M (2001) Diabetes and endothelial dysfunction: a clinical perspective. Endocr Rev 22:36–52

Ceriello A, Quagliaro L, Piconi L et al (2004) Effect of postprandial hypertriglyceridemia and hyperglycemia on circulating adhesion molecules and oxidative stress generation and the possible role of simvastatin treatment. Diabetes 53:701–710

Chang KJ, Kwon W (2000) Immunohistochemical localization of insulin in pancreatic β-cells of taurine-supplemented or taurine-depleted diabetic rats. Adv Exp Med Biol 483:579–587

Chauncey KB, Tenner TE Jr, Lombardini JB et al (2003) The effect of taurine supplementation on patients with type 2 diabetes mellitus. Adv Exp Med Biol 526:91–96

Chen Y, Saari JT, Kang YJ (1994) Weak antioxidant defenses make the heart a target for damage in copper-deficient rats. Free Radic Biol Med 17:529–536

Cherif H, Reusens B, Dahri S, Remacle C, Hoet JJ (1996) Stimulatory effects of taurine on insulin secretion by fetal rat islets cultured in vitro. J Endocrinol 151:501–506

Coskun O, Kanter M, Korkmaz A, Oterb S (2005) Quercetin, a flavonoid antioxidant, prevents and protects streptozotocin-induced oxidative stress and β-cell damage in rat pancreas. Pharmacol Res 51:117–123

Davignon J, Ganz P (2004) Role of endothelial dysfunction in atherosclerosis. Circulation 109:III27–III32

Di Leo AM, Santini SA, Cercone S et al (2002) Chronic taurine supplementation ameliorates oxidative stress and Na$^+$K$^+$ATPase impairment in the retina of diabetic rats. Amino Acids 23:401–406

Du XL, Edelstein D, Dimmeler S et al (2001) Hyperglycemia inhibits endothelial nitric oxide synthase activity by posttranslational modification at the Akt site. J Clin Invest 108:1341–1348

Elizarova EP, Nedosugova LV (1996) First experiments in taurine administration for diabetes mellitus. The effect on erythrocyte membranes. Adv Exp Med Biol 403:583–588

Fennessy FM, Moneley DS, Wang JH, Kelly CJ, Bouchier-Hayes DJ (2003) Taurine and vitamin C modify monocyte and endothelial dysfunction in young smokers. Circulation 107:410–415

Fiordaliso F, Bianchi R, Staszewsky L et al (2004) Antioxidant treatment attenuates hyperglycemia-induced cardiomyocyte death in rats. J Mol Cell Cardiol 37:959–968

Fisslthaler B, Benzing T, Busse R, Fleming I (2003) Insulin enhances the expression of the endothelial nitric oxide synthase in native endothelial cells: a dual role for Akt and AP-1. Nitric Oxide 8:253–261

Flohé L, Günzler WA (1984) Assays of glutathione peroxidase. Methods Enzymol 105:114–121

Fox JB, Zell TE, Wasserman AE (1981) Interaction between sample preparation techniques and colorimetric reagents in nitrite analysis in meat. J Assoc Off Anal Chem 64:1397–1402

Furfaro AL, Maengo NM, Domenicotti C et al (2012) Impaired synthesis contributes to diabetes-induced decrease in liver glutathione. Int J Mol Med 29:899–905

Ganda OP, Arkin CF (1992) Hyperfibrinogenemia. An important risk factor for vascular complications in diabetes. Diabetes Care 15:1245–1250

Gugliucci A (2000) Glycation as the glucose link to diabetic complications. J Am Osteopath Assoc 100:621–634

Guntherberg H, Rost J (1966) The true oxidized glutathione content of red blood cells obtained by new enzymic and paper chromatographic methods. Anal Biochem 15:205–210

Guzik TS, Mussa S, Gastaldi D et al (2002) Mechanisms of increased vascular superoxide production in human diabetes mellitus. Circulation 105:1656–1662

Haber CA, Lam TK, Yu Z et al (2003) N-Aacetylcysteine and taurine prevent hyperglycemia-induced insulin resistance in vivo: possible role of oxidative stress. Am J Physiol Endocrinol Metab 285:E744–E753

Hayat SA, Patel B, Khattar S, Malik RA (2004) Diabetic cardiomyopathy: mechanisms, diagnosis and treatment. Clin Sci 107:539–557

Heather LC, Clarke K (2011) Metabolism, hypoxia and the diabetic heart. J Mol Cell Cardiol 50:598–605

Hesselink MK, Mensink M, Schrauwen P (2007) Lipotoxicity and mitochondrial dysfunction in type 2 diabetes. Immun Endoc Metab Agents Med Chem 7:3–17

Hisalkar PJ, Patne AB, Fawade MM (2012) Assessment of plasma antioxidant levels in type 2 diabetes patients. Int J Biol Med Res 3:1796–1800

Hisashi T, Yukio Y, Kuriyama K (1979) Protective actions of taurine against streptozotocin-induced hyperglycemia. Biochem Pharmacol 28:2807–2811

Jain SK, McVie R (1994) Effect of glycemic control, race (white versus black), and duration of diabetes on reduced glutathione content in erythrocytes of diabetic patients. Metabolism 43:306–309

Jain SK, McVie R, Smith T (2000) Vitamin E supplementation restores glutathione and malondi-aldehyde to normal concentrations in erythrocytes of type 1 diabetic children. Diabetes Care 23:1389–1394

Kado S, Wakatsuki T, Yamamoto M, Nagata N (2001) Expression of intercellular adhesion molecule-1 induced by high glucose concentrations in human aortic endothelial cells. Life Sci 68:727–737

Kamata K, Sugiura M, Kojima S, Kasuya Y (1996) Restoration of endothelium-dependent relaxation in both hypercholesterolemia and diabetes by chronic taurine. Eur J Pharmacol 303:47–53

Kaplan B, Karabay G, Zağyapan RD et al (2004) Effects of taurine in glucose and taurine administration. Amino Acids 27:327–333

Katsumata M, Kiuchi K, Tashiro T, Uchikuga S (1997) Methods for scavenging active oxygen compounds and preventing damage from ultra violet B rays using taurine analogues. US Patent 5,601,806, 11 Feb 1997

Kędziora-Kornatowska K, Szramd S, Kornatowski T et al (2003) Effect of vitamin E and vitamin C supplementation on antioxidative state and renal glomerular basement membrane thickness in diabetic kidney. Nephron Exp Nephrol 95:e134–e143

Kim JA, Berliner JA, Natarajan RD, Nadler JL (1994) Evidence that glucose increases monocyte binding to human aortic endothelial cells. Diabetes 42:1103–1107

Kobayashi T, Kamata K (1999) Effect of insulin treatment on smooth muscle contractility and endothelium-dependent relaxation in rat aortae from established STZ-induced diabetes. Br J Pharmacol 127:835–842

Kobayashi T, Kamata K (2001) Effect of chronic insulin treatment on NO production and endothelium-dependent relaxation in aortae from established STZ-induced diabetic rats. Atherosclerosis 155:313–321

Kuboki K, Jiang ZY, Takahara N et al (2000) Regulation of endothelial constitutive nitric oxide synthase gene expression in endothelial cells and in vivo. A specific vascular action of insulin. Circulation 101:676–681

Kulakowski EC, Maturo J (1984) Hypoglycemic properties of taurine: not mediated by enhanced insulin release. Biochem Pharmacol 33:2835–2838

Lampson WG, Kramer JH, Schaffer SW (1983) Potentiation of the actions of insulin by taurine. Can J Physiol Pharmacol 61:457–463

Likidlilid A, Patchanans N, Peerapatdit T, Sriratanasathavorn C (2010) Lipid peroxidation and antioxidant enzyme activities in erythrocytes of type 2 diabetic patients. J Med Assoc Thai 93:682–693

Maturo J, Kulakowski E (1988) Taurine binding to the purified insulin receptor. Biochem Pharmacol 37:3755–3760

Militante JD, Lombardini JB (2004) Dietary taurine supplementation: hypolipidemic and anti-atherogenic effects. Nutr Res 24:787–801

Militante JD, Lombardini JB, Schaffer SW (2000) The role of taurine in the pathogenesis of the cardiomyopathy of insulin-dependent diabetes mellitus. Cardiovasc Res 46:393–402

Misra HP, Fridovich I (1972) The role of superoxide anion in the autoxidation of epinephrine and a simple assay for superoxide dismutase. J Biol Chem 247:3170–3175

Mohazzab HK, Kaminski PM, Wolin MS (1994) NADH oxidoreductase is a major source of superoxide anion in bovine coronary artery endothelium. Am J Physiol 266:H2568–H2572

Moloney MA, Casey RG, O'Donnell DH et al (2010) Two weeks taurine supplementation reverses endothelial dysfunction in young male type 1diabetics. Diab Vasc Dis Res 7:300–310

Monnier L, Mas E, Ginet C et al (2006) Activation of oxidative stress by acute glucose fluctuations compared with sustained chronic hyperglycemia in patients with type 2 diabetes. JAMA 295:1681–1687

Montilla PL, Vargas JF, Túnez IF et al (1998) Oxidative stress in diabetic rats induced by strepto-zotocin: protective effects of melatonin. J Pineal Res 25:94–100

Mulhern M, Docherty JR (1989) Effects of experimental diabetes on the responsiveness of rat aorta. Br J Pharmacol 97:1007–1012

Murakami S, Kondo Y, Nagate T (2000) Effects of long-term treatment with taurine in mice fed a high-fat diet: improvement in cholesterol metabolism and vascular lipid accumulation by tau-rine. Adv Exp Med Biol 483:177–186

Murakami S, Kondo Y, Ohta Y, Tomisawa K (1999) Improvement in cholesterol metabolism in mice given chronic treatment of taurine and fed a high-fat diet. Life Sci 64:83–91

Murakami S, Yamagishi I, Asami Y et al (1996) Hypolipidemic effect of taurine in stroke-prone spontaneously hypertensive rats. Pharmacology 52:303–313

Nagareddy PR, Xia Z, McNeill JH, MacLeod KM (2005) Increased expression of iNOS is associ-ated with endothelial dysfunction and impaired pressor responsiveness in streptozotocin-induced diabetes. Am J Physiol Heart Circ Physiol 289:H2144–H2152

Nakaya Y, Minami A, Harada N et al (2000) Taurine improves insulin sensitivity in the Otsuka Long-Evans Tokushima Fatty rat, a model of spontaneous type 2 diabetes. Am J Clin Nutr 71:54–58

Nandhini AT, Thirunavukkarasu V, Ravichandran MK, Anuradha CV (2005) Effect of taurine on biomarkers of oxidative stress in tissues of fructose-fed insulin-resistant rats. Singapore Med J 46:82–87

Nweke IN, Ohaeri OC, Ezeala CC (2009) Effect of vitamin on malondialdehyde and glutathione levels in type 2 diabetic nigerians. Internet J Nutr Wellness 7(2)

Obrosova I, Cao X, Greene DA, Stevens MJ (1998) Diabetes- induced changes in lens antioxidant status, glucose utilization and energy metabolism: effect of DL-α-lipoic acid. Diabetologia 41:1442–1450

Obrosova IG, Fathallah L, Stevens MJ (2001) Taurine counteracts oxidative stress and nerve growth factor deficit in early experimental diabetic neuropathy. Exp Neurol 172:211–219

Obrosova IG, Stevens MJ (1999) Effect of dietary taurine supplementation on GSH and NAD(P)-redox status, lipid peroxidation, and energy metabolism in diabetic precataractous lens. Invest Ophthalmol Vis Sci 40:680–688

Pan H, Zhang L, Guo M (2010) The oxidative stress status in diabetes mellitus and diabetic neph-ropathy. Acta Diabetol 47(Suppl 1):71–76

Park T, Lee K (1998) Dietary taurine supplementation reduces plasma and liver cholesterol and triglyceride levels in rats fed a high-cholesterol or a cholesterol-free diet. Adv Exp Med Biol 442:319–325

Paskaloglu K, Sener G, Ayangolu-Dülger G (2004) Melatonin treatment protects against diabetes-induced functional and biochemical changes in rat aorta and corpus cavernosum. Eur J Pharmacol 499:345–354

Patel SS, Goyal RK (2011) Prevention of diabetes-induced myocardial dysfunction in rats using the juice of the *Emblica officinalis* fruit. Exp Clin Cardiol 16:87–91

Peschel T, Niebauer J (2003) Role of pro-atherogenic adhesion molecules and inflammatory cytokines in patients with coronary artery disease and diabetes mellitus type 2. Cytometry B Clin Cytom 53B:78–85

Potenza MA, Gagliardi S, Nacci C, Carratú MR, Montagnani M (2009) Endothelial dysfunction in diabetes: from mechanisms to therapeutic targets. Curr Med Chem 16:94–112

Rasmussen LM, Schmitz O, Ledet T (2002) Increased expression of vascular cell adhesion molecule-1 (VCAM-1) in cultured endothelial cells exposed to serum from type 1 diabetic patients: no effects of high glucose concentrations. Scand J Clin Lab Invest 62:485–493

Sampson MJ, Davies IR, Brown JC, Ivory K, Hughes DA (2002) Monocyte and neutrophil adhesion molecule expression during acute hyperglycemia and after antioxidant treatment in type 2 diabetes and control patients. Arterioscler Thromb Vasc Biol 22:1187–1193

Scherrer U, Randin D, Vollenweider P et al (1994) Nitric oxide release accounts for insulin's vascular effects in humans. J Clin Invest 94:2511–2515

Sharpe PC, Liu WH, Yue KK et al (1998) Glucose-induced oxidative stress in vascular contractile cells: comparison of aortic smooth muscle cells and retinal pericytes. Diabetes 47:801–809

Shen X, Zheng S, Metreveli NS, Epstein PN (2006) Protection of cardiac mitochondria by overexpression of MnSOD reduces diabetic cardiomyopathy. Diabetes 55:798–805

Shivananjappa MM, Muralidhara (2012) Taurine attenuates maternal and embryonic oxidative stress in a streptozotocin-diabetic rat model. Reprod Biomed Online 24:558–566

Srinivasan S, Hatley ME, Bolick DT et al (2004) Hyperglycaemia-induced superoxide production decreases eNOS expression via AP-1 activation in aortic endothelial cells. Diabetologia 47:1727–1734

Szaleczky E, Prechl J, Fehér J, Somogyi A (1999) Alterations in enzymatic antioxidant defence in diabetes mellitus—a rational approach. Postgrad Med J 75:13–17

Tachi Y, Okuda Y, Bannai C et al (2001) Hyperglycemia in diabetic rats reduces the glutathione content in the aortic tissue. Life Sci 69:1039–1047

Tan B, Jiang DJ, Huang H, Jia SJ et al (2007) Taurine protects against low-density lipoprotein-induced endothelial dysfunction by the DDAH/ADMA pathway. Vasc Pharmacol 46:338–345

Tenner TE Jr, Zhang XJ, Lombardini JB (2003) Hypoglycemic effects of taurine in the alloxan-treated rabbit: a model for type 1 diabetes. Adv Exp Med Biol 526:97–104

Tokunaga H, Yoneda Y, Kuriyama K (1983) Streptozotocin- induced elevation of pancreatic taurine content and suppressive effect of taurine on insulin secretion. Eur J Pharmacol 87:237–243

Ulrich-Merzenich G, Zeitler H, Vetter H, Bhonde RR (2007) Protective effects of taurine on endothelial cells impaired by high glucose and oxidized low density lipoproteins. Eur J Nutr 46:431–438

Urso C, Hopps E, Caimi G (2010) Adhesion molecules and diabetes mellitus. Clin Ter 161:e17–e24

Wang LJ, Yu YH, Zhang LG et al (2008) Taurine rescues vascular endothelial dysfunction in streptozocin-induced diabetic rats: correlated with downregulation of LOX-1 and ICAM-1 expression on aortas. Eur J Pharmacol 597:75–80

Wei M, Ong L, Smith MT et al (2003) The streptozotocin-diabetic rat as a model of the chronic complications of human diabetes. Heart Lung Circ 12:44–50

Weidig P, McMaster D, Bayraktutan U (2004) High glucose mediates pro-oxidant and antioxidant enzyme activities in coronary endothelial cells. Diabetes Obes Metab 6:432–441

Weir GC, Bonner-Weir S (2004) Five stages of evolving β-cell dysfunction during progression to diabetes. Diabetes 53(Suppl 3):516–521

Wijeratne SSK, Susan L, Cuppett SL, Schlegel V (2005) Hydrogen peroxide induced oxidative stress damage and antioxidant enzyme response in Caco-2 human colon cells. J Agric Food Chem 53:8768–8774

Wolff SP, Jiang ZY, Hunt JV (1991) Protein glycation and oxidative stress in diabetes mellitus and ageing. Free Radic Biol Med 10:339–352

Xu YJ, Arneja AS, Tappia PS, Dhalla NS (2008) The potential health benefits of taurine in cardio-vascular disease. Exp Clin Cardiol 13:57–65

Yang MS, Chan HW, Yu LC (2006) Glutathione peroxidase and glutathione reductase activities are partially responsible for determining the susceptibility of cells to oxidative stress. Toxicology 226:126–130

Yano M, Hasegawa G, Ishii M (2004) Short-term exposure of high glucose concentration induces generation of reactive oxygen species in endothelial cells: implication for the oxidative stress associated with postprandial hyperglycemia. Redox Rep 9:111–116

Yokogoshi H, Mochizuki H, Nanami K et al (1999) Dietary taurine enhances cholesterol degradation and reduces serum and liver cholesterol concentrations in rats fed a high- cholesterol diet. J Nutr 129:1705–1712

Zhang M, Bi LF, Fang JH et al (2004) Beneficial effects of taurine on serum lipids in overweight or obese non-diabetic subjects. Amino Acids 26:267–271

Chapter 29
Comparative Evaluation of Taurine and Thiotaurine as Protectants Against Diabetes-Induced Nephropathy in a Rat Model

Kashyap G. Pandya, Roshil Budhram, George Clark, and Cesar A. Lau-Cam

Abstract Taking into account the proven effectiveness of antioxidants in preventing experimentally induced diabetes in laboratory animals, this study was carried out with the specific purpose of comparing the effectiveness of two known antioxidants, the β-aminosulfonate taurine (TAU) and β-aminothiosulfonate thiotaurine (TTAU), in preventing biochemical, functional and histological alterations indicative of diabetic nephropathy. In the study, streptozotocin (60 mg/kg, orally) was used to induce type 2 diabetes mellitus in Sprague-Dawley rats. Starting on day 15 and continuing up to day 56, the rats received a daily single 2.4 mmol/kg oral dose of a sulfur-containing compound (TAU or TTAU) or 4 U/kg subcutaneous dose of isophane insulin (INS). Rats not receiving any treatment served as controls. After obtaining a 24 h urine sample, the animals were sacrificed by decapitation on day 57, and their blood and kidneys immediately collected. Diabetic rats exhibited marked hyperglycemia, hypoinsulinemia, hypoproteinemia, hyponatremia, hyperkalemia, azotemia, hypercreatinemia, increased plasma TGF β_1, lipid peroxidation, plasma and kidney nitrite, and urine output; decreased glutathione redox status in plasma and kidney, decreased urine Na^+ and K^+, proteinuria and hypocreatinuria. Without exceptions, all the treatment compounds were found to markedly and variously attenuate these changes. Confirmation of protection by INS, TAU and TTAU was provided by the results of histological examination of kidney sections and which showed a more normal appearance than sections from diabetic animals. In most instances protection by TTAU was about equal to that by INS but greater than that by TAU.

K.G. Pandya • R. Budhram • G. Clark • C.A. Lau-Cam (✉)
Department of Pharmaceutical Sciences, St. John's University, College of Pharmacy
and Health Professions, Jamaica, NY 11439, USA
e-mail: claucam@usa.net; gjcrx@hotmail.com

A. El Idrissi and W.J. L'Amoreaux (eds.), *Taurine 8*, Advances in Experimental
Medicine and Biology 775, DOI 10.1007/978-1-4614-6130-2_29,
© Springer Science+Business Media New York 2013

Abbreviations

TAU	Taurine
TTAU	Thiotaurine
STZ	Streptozotocin
GLC	Glucose
INS	Insulin
HbA_{1c}	Glycosylated hemoglobin
TGF-β1	Transforming growth factor-β1
BUN	Blood urea nitrogen
CRN	Creatinine
K^+	Potassium
Na^+	Sodium
TP	Total protein
MDA	Malondialdehyde
GSH	Reduced glutathione
GSSG	Glutathione disulfide

29.1   Introduction

Diabetic nephropathy is a microvascular complication seen in individuals affected with type 1 and type 2 of diabetes and whose occurrence is associated with increased morbidity and mortality (Flyvbjerg 2000; Flyvbjerg et al. 2002). Typical functional manifestations of this leading cause of end-stage renal failure are a progressive and persistent proteinuria, elevated blood and intraglomerular pressure, and progressive decline in glomerular filtration rate and creatinine clearance (Björck et al. 1986; Mathiesen et al. 1991; Rossing et al. 2002a, b). In addition, there are morphological alterations such as glomerular hypertrophy, basement membrane thickening, mesangial expansion, tubular atrophy, interstitial fibrosis, and arteriolar thickening (Kashihara et al. 2010).

Diabetic nephropathy is the result of complex interactions between hemodynamic and metabolic factors (Arya et al. 2010, Rossing et al. 2002a, b), with a genetic predisposition also having a contributory role (Adler 2004). Hemodynamic factors underlining the development of diabetic nephropathy include increased systemic and intraglomerular pressure and the activation of the renin-angiotensin system (RAS), imbalances in the secretion of endothelial vasoactive factors (endothelin-1, nitric oxide, prostacyclin, urotensin II) (Arya et al. 2010), and of certain growth factors (epidermal growth factor, growth hormone, insulin-like growth factors, transforming growth factor β, vascular endothelial growth factor) (Flyvbjerg 2000). Some of the metabolic factors are hyperglycemia-induced excessive formation of advanced glycosylated end products (AGEs), excessive deposition of extracellular matrix (ECM) proteins in tubular epithelial and glomerular mesangial cells, enhanced mesangial and tubular epithelial oxidative stress (Arya et al. 2010; Mathiesen et al. 1991), and decreased renal bioavailability of nitric oxide (Prabhakar et al. 2007).

Along with hyperlipidemia and hypertension, hyperglycemia is regarded as a major determinant of the initiation and progression of diabetic nephropathy since it can create a state of oxidative stress, both by increasing the formation of reactive oxygen species (ROS) and by lowering the activity of antioxidant enzymes through nonenzymatic glycation (Balakumar et al. 2008; Catherwood et al. 2002). Several macromolecules and pathways have been implicated in hyperglycemia-related increased generation of ROS, including NAD(P)H oxidase, AGEs, enhanced polyol pathway activity and arachidonic acid metabolism, uncoupled nitric oxide synthase (NOS) and mitochondrial respiratory chain (Obrosova, 2005). Overproduction of ROS can, in turn, modulate the activity of protein kinase C (PKC) and mitogen-activated protein kinases (MAPKs) and mobilize various cytokines and transcription factors, changes that can eventually increase the expression of ECM genes with progression to fibrosis and end stage renal disease (Ha and Kim, 1999, Ha and Lee, 2000; Kashihara et al. 2010). Furthermore, activation of RAS further worsens the renal injury induced by ROS in diabetic nephropathy (Kashihara et al. 2010).

In both cultured cell preparations (Derlacz et al. 2007; Tada et al. 1997; Trachtman et al. 1994) and animal models of diabetes (Craven et al. 1997; Koya et al. 2003; Naito et al. 2004; Simşek et al. 2005; Winiarska et al. 2008), antioxidant agents representing a wide range of chemical classes and of pharmacological actions have demonstrated potential for ameliorating oxidative stress and, hence, renal cell injury and loss of renal function. One of the antioxidants most widely examined as a preventive measure against diabetic nephropathy has been taurine (TAU). For example, consumption of this sulfur-containing compound (in the drinking water) by rats made diabetic with streptozotocin (STZ) was found to reduce renal oxidant injury by decreasing lipid peroxidation (LPO), the accumulation of AGEs (Trachtman et al. 1995), increases in glomerular TGF-β1 and expression of fibronectin mRNA (Ha et al. 1999), and the activity of PKC isoforms (Ha et al. 2001) in renal tissue. Additional protective actions by TAU against diabetic nephropathy have included a suppressive effect on early renal injury due to lectin-like oxidized low density lipoprotein-1-mediated expression of cell adhesion molecule-1 in the renal cortex (Wang et al. 2008), on the activity of the polyol pathway (Jiang, 2002), and on the production of proinflammatory cytokines and activity of Na^+-K^+-ATPAse (Das and Sil, 2012). A study with renal tubular epithelial cells has suggested that TAU can also protect against high glucose-induced hypertrophic growth and the production of fibronectin and type IV collagen in renal tissue through an attenuation of the activity of stress-sensitive mitogenic signaling pathways (Huang et al. 2007).

A previous study from this laboratory determined that a sulfonate compound like TAU (2-aminoethanesulfonic acid) was able to provide equal or even greater protection in the rat liver against injury, lipid peroxidation (LPO) and changes in the cellular redox state than hypotaurine (HYTAU) (Acharya and Lau-Cam, 2010) even though this 2-aminosulfinate analog had been previously found in vitro to bind to ROS more readily and to a greater extent than TAU

(Aruoma et al. 1988; Green et al. 1991), in all probability because of its facile oxidation to TAU (Fellman et al. 1987; Green et al. 1991). Because of the apparent importance of the sulfur-functionality in the antioxidant and cytoprotective activities of sulfur-containing 2-aminoethane derivatives, this study was conducted in diabetic rats to assess the impact of replacing the sulfonate group of TAU by a thiosulfonate functionality on the protective actions of TAU against diabetes-induced biochemical, functional, and morphological changes. To this end, the renoprotection by TAU was compared with that by thiotaurine (TTAU) in a rat model of diabetes.

29.2 Methods

29.2.1 Animals

Male Sprague-Dawley rats, 225–250 g, acclimated for 1 week in a room maintained at a constant humidity and temperature ($23 \pm 1\,^\circ C$) and a normal 12 h light–12 h dark cycle room. The rats had free access to a commercial rodent diet (LabDiet® 5001, PMI Nutrition International, Brentwood, MO) and filtered tap water. The study was approved by the Institutional Animal Care and Use Committee of St. John's University, Jamaica, NY, and the animals were cared in accordance with guidelines established by the United States Department of Agriculture.

29.2.2 Treatments and Samples

All the experimental groups consisted of six rats. Diabetes was induced with a single 60 mg/kg intraperitoneal (i.p.) dose of streptozotocin in citrate buffer pH 4.5. Starting on day 15 and continuing for the next 41 days, separate groups received a 2.4 mmol/kg, dose of either taurine (TAU) or thiotaurine (TTAU) in distilled water by oral gavage. Normal (control) rats received citrate buffer pH 4.5 (2 mL, i.p.) on day 1, and physiological saline (2 mL) by oral gavage from day 15 onwards. For comparison purposes, an additional diabetic group received a daily subcutaneous dose of isophane insulin (INS) suspension (Humulin N®) from day 15 onwards. A diabetic group received no additional treatment other than STZ. All treatments were continued for a period of 42 days. On day 56, the animals were placed in individual metabolic cages to collect a 24-h urine sample; and on day 57, the animals were sacrificed by decapitation to collect blood and kidney samples. A portion of the blood sample, mixed with sodium heparin, was processed for the plasma fraction. The kidneys were removed by the freeze-clamp technique of Wollenberger et al. (1960) and divided into two portions. One portion was homogenized with Tris buffer pH 7 containing 1 mg of phenylmethylsulfonyl fluoride (1:20 ratio, w/v) over ice, and the resulting suspension was centrifuged at $8,000 \times g$

and 4°C for 30 min, to obtain a supernatant suitable for biochemical assays. The other portion was immediately fixed in10% phosphate buffered (pH 7) formalin and reserved for histological work.

29.2.3   Plasma Glucose (GLC)

The GLC content was measured using a commercially available colorimetric kit (Procedure No. 510 from Sigma-Aldrich, St. Louis, MO). The results were expressed in mg/dL.

29.2.4   Plasma Insulin (INS)

The concentration of INS in plasma was measured by means of a commercial assay kit (Insulin ELISA kit, Calbiotech Inc. Spring Valley, CA). The results were expressed in μIU/mL.

29.2.5   Blood Glycated Hemoglobin (HbA$_{1c}$)

The concentration of blood HbA$_{1c}$ was measured using a commercial optimized ion-exchange resin procedure (Glycohemoglobin Test, Stanbio Laboratory, Boerne, TX). The results were expressed as a percentage of the total hemoglobin content.

29.2.6   Plasma Transforming Growth Factor-β1 (TGF-β1)

The plasma level of TGF-β1 was measured using a commercially available ELISA kit (Invitrogen™ TGF-β1 Multispecies ELISA kit, Life Technologies, Grand Island, NY). The results were expressed in pg/mL.

29.2.7   Blood Urea Nitrogen (BUN)

The BUN level was measured using a commercially available enzymatic-endpoint assay kit based on the Berthelot method (Nitrogen (BUN) Test, Stanbio Laboratory, Boerne, TX). The results were expressed in mg/dL.

29.2.8 Plasma and Urine Creatinine (CRN)

The contents of plasma and urine CRN were measured with a commercially available colorimetric assay kit (Kinetic Creatinine LiquiColor® Test, Stanbio Laboratory, Boerne, TX). The results were expressed in mg/dL.

29.2.9 Plasma and Urine Total Protein (TP)

The contents of plasma and urine TP were measured with a colorimetric assay kit based on the Biuret reaction (Protein, Total LiquiColor® Test, Stanbio Laboratory, Boerne, TX). The results were expressed in g/dL.

29.2.10 Plasma and Urine Sodium (Na⁺)

The concentrations of Na^+ in the plasma and urine were measured colorimetrically with a commercially available assay kit based on reaction with a reagent containing uranyl acetate-zinc acetate (Sodium Test, Stanbio Laboratory, Boerne, TX). The results were expressed in mmol/L.

29.2.11 Plasma and Urine Potassium (K⁺)

The concentrations of K^+ in the plasma and urine were measured by a turbidimetric assay method after reaction with alkaline sodium tetraphenylboron (Potassium Test, Stanbio Laboratory, Boerne, TX). The results were expressed in mmol/L.

29.2.12 Plasma and Kidney MDA

The concentration of MDA in plasma and kidney was measured as thiobarbituric acid reactive substances (TBARS) by the end point assay method of Buege and Aust (1978). The results were expressed in nmol/mL of plasma or nmol/g of tissue.

29.2.13 Plasma and Kidney GSH and GSSG

Both the kidney homogenate and plasma levels of GSH and GSSG were measured fluorometrically by the method of Hissin and Hilf (1976), which is based on the reaction of GSH with ortho-phtaldehdye (OPT) at pH 8 and of GSSG with OPT at pH 12. Prior to the measurement of GSSG, any interfering GSH is complexed with N-ethylmaleimide according to the method of Guntherberg and Rost (1966) to prevent its interfering effect on the measurement of GSSG. The concentrations of GSH and GSSG were expressed as µmol/mL of plasma or µmol/g of tissue.

29.2.14   Plasma and Kidney NO

NO was measured as the more stable product nitrite with the Griess reagent as described by Doerr et al. (1981). At an acid pH, nitrite converts to nitrous acid for reaction with sulfanilamide to form a diazonium salt. Reaction of the diazonium salt with naphthylenediamine dihydrochloride present in Griess reagent yields a purple azo compound amenable to colorimetric measurement at 540 nm. The results were expressed as nmol/mL plasma or nmol/g of tissue.

29.2.15   Histological Analysis

Buffered formalin-fixed kidney tissue samples were embedded in paraffin, sectioned into 5 µm slices, and stained with hematoxylin and eosin (H&E) stain.

29.2.16   Statistic analysis

Experimental results are presented as mean ± SEM for $n = 6$. Statistical comparisons with normal rats and with untreated diabetic rats were made by one-way analysis of variance (ANOVA) and Neuman-Keuls post-hoc test. Differences were statistically significant when $p \leq 0.05$.

29.3   Results

Table 29.1 summarizes the effects of a 6-week treatment with INS, TAU, or TTAU on body weight gains when given at 2 weeks after the induction of diabetes with STZ. Relative to initial values, body weight gains at the end of 56 days amounted to 97% in normal rats, 35% in diabetic rats, and 87% in rats on INS. On the other hand, body weight gains for rats on TAU was 65% and for those on TTAU 56%.

Table 29.2 summarizes the effects of a 6-week treatment with INS, TAU, or TTAU on the plasma GLC, plasma INS, and blood HbA_{1c} levels of diabetic rats measured on day 56 of the study. While diabetic rats showed a 4-fold increase in plasma GLC relative to the normal value, increases for rats on INS, TAU, and TTAU

Table 29.1 Body weight gains at the end of 56 days

Group	Initial body weight, g	Final body weight, g
Control	225.83 ± 2.39	428 ± 5.58^{+++}
STZ	235.83 ± 3.96	319.17 ± 9.61**
TAU	233.33 ± 6.15	419.17 ± 8.21^{+++}
STZ + TAU	235.00 ± 5.63	394.17 ± 3.74^{++}
TTAU	224.17 ± 5.54	418.23 ± 10.22^{+++}
STZ + TTAU	229.17 ± 6.25	358.33 ± 7.92*$^{,+}$
STZ + INS	225.23 ± 8.89	421.67 ± 10.3^{+++}

Values represent mean ± SEM ($n = 6$). *STZ* streptozotocin, *TAU* Taurine, *TTAU* thiotaurine, *INS* insulin. Comparisons were significantly different from Control at *$p < 0.05$ and **$p < 0.01$ and from STZ at $^{+}p < 0.05$, $^{++}p < 0.01$, and $^{+++}p < 0.001$

Table 29.2 Effect of TAU, TTAU, and INS on the circulating levels of GLC, INS, and HbA$_{1c}$ of diabetic rats

Group	GLC, mg/dL	INS, μIU/mL	HbA$_{1c}$, %
Control	103.58 ± 5.05^{+++}	44.08 ± 2.46^{+++}	6.89 ± 0.11^{+++}
STZ	428.07 ± 21.73***	10.58 ± 1.33***	21.14 ± 1.19***
TAU	112.25 ± 3.62^{+++}	40.62 ± 1.58^{+++}	6.89 ± 0.33^{+++}
STZ + TAU	333.27 ± 6.60***$^{,++}$	20.94 ± 0.83^{+++}	10.96 ± 0.96***$^{,+++}$
TTAU	104.27 ± 2.83^{+++}	41.26 ± 0.99***$^{,+++}$	6.79 ± 0.33^{+++}
STZ + TTAU	265.08 ± 35.11**$^{,+++}$	24.55 ± 2.50***$^{,+++}$	10.98 ± 0.56***$^{,+++}$
STZ + INS	127.15 ± 10.71*$^{,+++}$	53.46 ± 6.97^{+++}	10.46 ± 0.41***$^{,+++}$

Values represent mean ± SEM ($n = 6$). *STZ* streptozotocin, *TAU* taurine, *TTAU* thiotaurine, *INS* insulin. Comparisons were significantly different from Control at *$p < 0.05$, **$p < 0.01$ and ***$p < 0.01$, and from STZ at $^{++}p < 0.01$ and $^{+++}p < 0.001$

were only 1.2-fold, 3.2-fold, and 2.6-fold, respectively. Similarly, the plasma INS was decreased by 76% in diabetic rats, was 21% above normal in the INS-treated group, and 52% and 44% below control in the TAU- and TTAU-treated rats, respectively. The plasma HbA$_{1c}$ was found to be >200% above normal in diabetic rats, a value that was reduced to 51% by INS or to 59% by either TAU or TTAU (All at $p < 0.001$ vs. control).

Figure 29.1 shows the plasma changes in TGF-β1 levels as a result of STZ-induced diabetes. Whereas the untreated diabetic rats showed a marked (~12-fold, $p < 0.001$) increase in TGF-β1, in those receiving either TAU (~3.9-fold increase, $p < 0.001$) or TTAU (4-fold increase, $p < 0.001$) the changes were less obvious. In contrast, animal on INS showed a value that was only 1.3-fold above the control value.

Figures 29.2–29.6 summarize the changes in nonenzymatic indices of oxidative stress measured both in the plasma and kidney. Diabetes elevated the plasma MDA levels by 44% ($p < 0.01$), an effect that was virtually abolished by INS and TTAU (≤3% increase) and markedly reduced by TAU (only 13% increase) (Fig. 29.2). The changes in kidney MDA levels correlated closely with those observed in the plasma, namely a 42% increase in diabetic rats ($p < 0.01$), and only ≤4% above control with

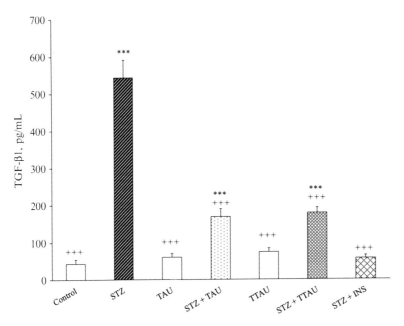

Fig. 29.1 Effects of TAU, TTAU, and INS on the plasma TGF-β1 levels of diabetic rats. *STZ* streptozotocin, *TAU* taurine, *TTAU* thiotaurine, *INS* insulin. Comparisons were significantly different from Control at ***$p < 0.01$; and from STZ at +++$p < 0.001$. Values are shown as mean ± SEM for $n = 6$

all the treatment compounds (Fig. 29.2). Diabetes elevated the plasma nitrite by ~4.4-fold ($p < 0.001$) but lowered the kidney nitrite by 74% ($p < 0.001$) of control values (Fig. 29.3). All the treatment compounds effectively counteracted these changes both in the plasma and kidney, with INS (+10%) appearing more effective than TAU (+27%, $p < 0.01$) and TTAU (+20%, $p < 0.05$) in the plasma, and TTAU (+12%) marginally better than TAU and INS (≤1%) in the kidney (Fig. 29.3). Based on the corresponding GSH (Fig. 29.4) and GSSG (Fig. 29.5) levels, diabetes was found to lower the GSH/GSSG ratios in the plasma and kidney by 75% and 52%, respectively (both at $p < 0.001$) (Fig. 29.6). These changes were effectively attenuated by all the treatment compounds, with TAU and TTAU appearing about equipotent and more effective than INS both in the plasma (<10% decrease vs. 34% decrease, respectively, $p < 0.05$) and kidney (~25% increase vs. 3% decrease, respectively, $p < 0.05$) (Fig. 29.6).

The experimental values for urine output/day, plasma and urine creatinine, and plasma urea nitrogen are presented in Figs. 29.7, 29.8, and 29.9 respectively. While diabetes enhanced the urinary output by ~3.7-fold, neither TAU (3.9-fold) nor TTAU (~3.6-fold) were able to reduce this value as much as INS (1.7-fold) (Fig. 29.7). In diabetic rats, the creatinine and nitrogen urea levels were drastically increased in the plasma (by >300%, $p < 0.001$, and by 63%, $p < 0.001$, respectively) and drastically

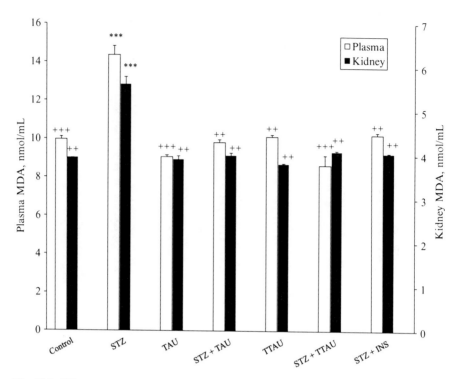

Fig. 29.2 Effects of TAU, TTAU, and INS on the plasma and kidney MDA levels of diabetic rats. *STZ* streptozotocin, *TAU* taurine, *TTAU* thiotaurine, *INS* insulin. Comparisons were significantly different from Control at ***$p < 0.001$; and from STZ at ++$p < 0.01$ and +++$p < 0.001$. Values are shown as mean ± SEM for $n = 6$

decreased in the urine (by ~75%, $p < 0.001$, and 63%, $p < 0.001$, respectively). These changes were significantly attenuated by all the treatment compounds, with the values varying within a narrow range in the plasma (−65% after INS, −55% after TTAU, −58% after TAU, all at $p < 0.001$ vs. control) and within 15% of one another in the urine (−43%, $p < 0.01$, after INS, −128%, $p < 0.001$, after TAU, −144% after TTAU, all at $p < 0.001$ vs. control) (Fig. 29.8). In parallel with the changes in plasma creatinine, that of the urea nitrogen was also markedly increased (+63%, $p < 0.001$) by diabetes. Treatments with either INS or TTAU abolished this increase, and one with TAU markedly reduced the elevation (only 17% increase, $p < 0.05$) (Fig. 29.9).

As shown in Fig. 29.10, diabetes promoted the loss of TP from the plasma (−53%, $p < 0.001$) into the urine (+111%, $p < 0.001$). These changes were effectively counteracted by a treatment with INS (−28%, $p < 0.01$ and −14%, respectively), TAU (−24%, $p < 0.05$ and +6%, respectively) and TTAU (−31%, $p < 0.01$ and −19%, $p < 0.05$, respectively).

The experimental values for plasma and urine Na⁺ and K⁺ levels are presented in Table 29.3. Diabetes promoted the retention of Na⁺ in the plasma (up by 35%, $p < 0.01$) and lowered the urinary excretion of Na⁺ (by 43%, $p < 0.01$). All the treatment compounds were able to attenuate the elevation of the plasma Na⁺ (only

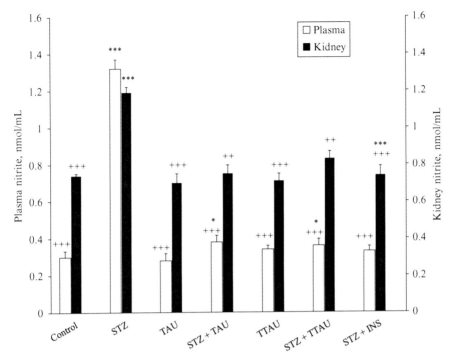

Fig. 29.3 Effects of TAU, TTAU, and INS on the plasma and kidney NO levels of diabetic rats. *STZ* streptozotocin, *TAU* taurine, *TTAU* thiotaurine, *INS* insulin. Comparisons were significantly different from Control at *$p<0.05$ and ***$p<0.001$; and from STZ at ++$p<0.01$ and +++$p<0.001$. Values are shown as mean ± SEM for $n=6$

11–16% above control) and to normalize the urinary output of this electrolyte. On the other hand, changes in K^+ levels followed the trends observed for Na^+, namely an increase (by 60%, $p<0.001$) in the plasma and a decrease (by 33%, $p<0.01$) in the urine. A treatment with INS, TAU or TAU returned the plasma and urine values for K^+ to within normal values, with the potency differences among the three treatments being rather narrow.

Figures 29.11a–e are representative photomicrographs of H&E-stained kidney sections showing the effects of the various treatment agents on the morphological and structural changes caused by STZ-induced diabetes. As seen in Fig. 29.11a, the kidney of a nondiabetic rat displayed well-defined and normal glomeruli, basement membrane, and renal tubules. In contrast, the kidney section of a diabetic rat shown in Fig. 29.11b was characterized by tubular necrosis, thickened glomerular basement membrane, mesangial matrix expansion, extensive vacuolization, and reduced islet size and cell population. Figures 29.11c–e are sections taken from rats treated with INS, TAU, and TTAU, respectively. While the section taken from an INS-treated rat was virtually similar in appearance to that of a normal rat, those from rats receiving either TAU or TTAU showed minimal morphological and architectural changes in comparison to a kidney section from a diabetic rat, especially after a treatment with TTAU.

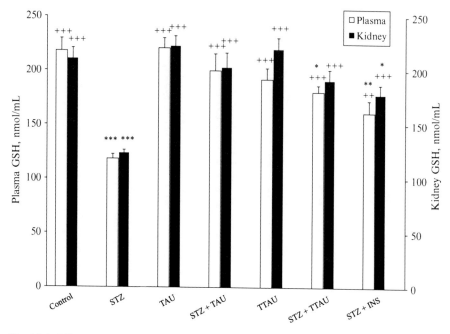

Fig. 29.4 Effects of TAU, TTAU, and INS on the plasma and kidney GSH levels of diabetic rats. *STZ* streptozotocin, *TAU* taurine, *TTAU* thiotaurine, *INS* insulin. Comparisons were significantly different from Control at *$p<0.05$, **$p<0.01$ and ***$p<0.001$; and from STZ at ++$p<0.01$ and +++$p<0.001$. Values are shown as mean ± SEM for $n=6$

29.4   Discussion

TAU is a ubiquitous nonprotein β-aminosulfonic acid possessing intrinsic biological properties that make this compound an attractive candidate for use as a protective agent against diabetic nephropathy. For example, it is found to lower blood GLC in the absence of an effect on INS secretion (Brøns et al. 2004; Kaplan et al. 2004; Kulakowski and Maturo, 1984), to increase cell sensitivity to INS and pancreatic β-cells response to hyperglycemia (Ribeiro et al. 2009), to attenuate oxidative stress and accompanying INS resistance (Haber et al. 2003), to inhibit protein glycosylation and formation of AGEs based on its reactivity towards carbonyl compounds (Hansen, 2001; Nandhini et al. 2004). In the particular case of diabetic nephropathy, TAU was shown to lower hyperglycemia-stimulated secretion of proinflammatory cytokines and to increase NO formation and endothelial NO synthase expression (Das and Sil, 2012), to suppress the activation of high GLC-sensitive signaling cascades like MAPK, signal transducer and activator of transcription 1 and 3 in renal tissue (Das and Sil, 2012; Huang et al. 2007), and the induction of a potential source of ROS such as cytochrome P450 2E1 (Yao et al. 2009). Additional studies have verified the ability of TAU to reduce renal tubular hypertrophy induced by AGEs (Huang et al. 2008), albuminuria, (Jiang, 2002);

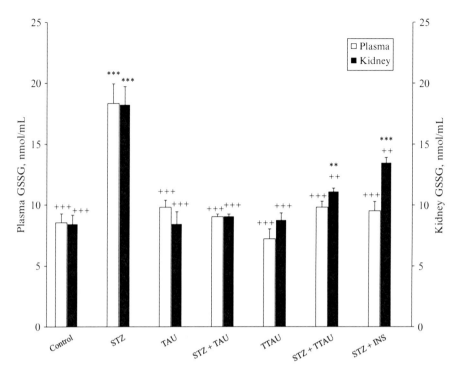

Fig. 29.5 Effects of TAU, TTAU, and INS on the plasma and kidney GSSG levels of diabetic rats. *STZ* streptozotocin, *TAU* taurine, *TTAU* thiotaurine, *INS* insulin. Comparisons were significantly different from Control at **$p < 0.01$ and ***$p < 0.001$; and from STZ at $^{++}p < 0.01$ and $^{+++}p < 0.001$. Values are shown as mean ± SEM for $n = 6$

glomerulonephritis, tubulointerstitial fibrosis (Trachtman et al. 1995), renal glomerular TGF-β1 expression (Higo et al. 2008), LPO and mesangial cell collagen production (Trachtman et al. 1993, 1994), and tubular apoptosis (Verzola et al. 2002) in the face of high GLC and ROS concentrations.

Hyperglycemia is a major determinant of diabetic nephropathy since it is capable of increasing the production of ROS either directly through GLC metabolism and autoxidation or indirectly through the formation of AGEs (Huang et al. 2008). Moreover, ROS are found to cause mesangial expansion by upregulating TGF-β1, plasminogen activator-1 and ECM genes and proteins in glomerular mesangial cells, and to activate signaling transduction cascades (PKC, MAPKs, Janus kinase 2) and signal transducers and activators of transcription factors (nuclear factor-κB, activated protein-1, and specificity protein 1) (Huang et al. 2007; Lee et al. 2003). In spite of the voluminous literature supporting an antioxidant role for TAU in diabetes and other metabolic disorders, the exact mechanism by which this compound acts as an antioxidant is still open to question. Indeed, in vitro evaluation of the rate of interaction of TAU with biologically important ROS, including hydroxyl radical, superoxide anion radical and hydrogen peroxide, appears not to be rapid enough to be protective in vivo (Aruoma et al. 1988). However, TAU is found to inhibit LPO,

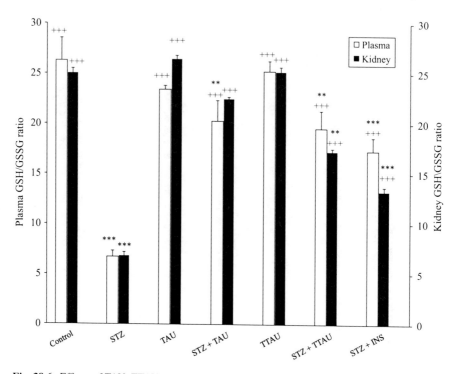

Fig. 29.6 Effects of TAU, TTAU, and INS on the plasma and kidney GSH/GSSG ratio of diabetic rats. *STZ* streptozotocin, *TAU* taurine, *TTAU* thiotaurine, *INS* insulin. Comparisons were significantly different from Control at **$p<0.01$ and ***$p<0.001$; and from STZ at +++$p<0.001$. Values are shown as mean±SEM for $n=6$

to increase the intracellular redox state under conditions of oxidative state (Obrosova and Stevens, 1999; Yao et al. 2009) and to attenuate oxidative damage by decreasing carbonyl group production (Franconi et al. 2004). A recent proposal, based on the observation that in diabetes a decline in intracellular TAU concentration is associated with increased generation of oxidant species and accumulation of reactive carbonyl and AGES, has raised the possibility that low TAU levels might also contribute to oxidant-mediated cellular injury (Franconi et al. 2004; Schaffer et al. 2009). At normal levels, TAU might conjugate to tRNA bind to promote the expression of mitochondrial protein components of the electron transport chain which may become deficient if TAU is limiting. As a result, the electrons flow to generate ATP will not become diverted to oxygen to generate mitochondrial superoxide anion radical (Schaffer et al. 2009). Alternatively, TAU could be exerting an antioxidant action in vivo by maintaining the patency of intracellular nonenzymatic (Obrosova and Stevens, 1999) and enzymatic (Balkan et al. 2001) antioxidant defenses and by helping to modulate the levels of metabolites stimulating relevant metabolic and signaling pathways (Franconi et al. 2004; Hansen, 2001; Jiang, 2002).

TTAU is a sulfane sulfur compound of common occurrence among bacterial endosymbionts with deep-sea vent and seep marine invertebrates, where it may

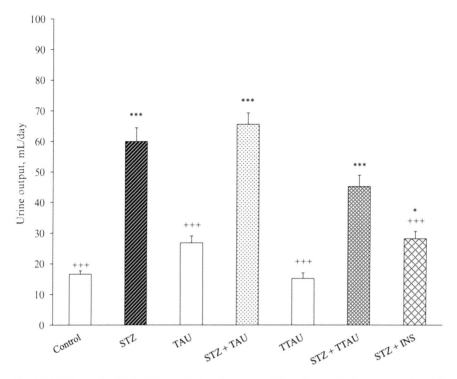

Fig. 29.7 Effects of TAU, TTAU, and INS on the output of diabetic rats. *STZ* streptozotocin, *TAU* taurine, *TTAU* thiotaurine, *INS* insulin. Comparisons were significantly different from Control at *$p < 0.05$ and ***$p < 0.001$; and from STZ at +++$p < 0.001$. Values are shown as mean ± SEM for $n = 6$

arise from the interaction of sulfide and HYTAU (Pruski and Fiala-Médioni, 2003; Yancey et al. 2009). In these organisms TTAU may function in sulfide detoxification by serving as a form of sulfide storage compounds or as a sulfide transport compound between symbiont and host (Joyner et al. 2003, Pruski and Fiala-Médioni, 2003), it may participate in sulfur cycling to provide sulfide to invertebrate mitochondria and symbiont sulfide oxidation pathways (Joyner et al. 2003), and it may contribute to the maintenance of the intracellular osmotic pressure (Yancey, 2005). In vertebrates, TTAU has been detected in rat urine following the oral administration of sulfur-containing compounds such as L-cystamine (Cavallini et al. 1959) and L-cystine (Cavallini et al. 1960). Reports on the biological actions of this compound are scanty, with some describing a protective antioxidant action in the skin (Katsumata et al. 1997; Yoshiyuki and Yoshiki, 2000), in skin sebum (Kono, 1998) and in isolated human skin fibroblasts (Egawa et al. 1999); a hypoglycemic and hypotriglyceridemic effect in alloxan-treated diabetic rats (Katsumata et al. 1997); and a superoxide-dismutase like activity (Katsumata et al. 1997).

This study finds that treating diabetic rats with TAU allowed them to attain body weights that at the end of 8 weeks were about 30% and 10%, respectively, greater

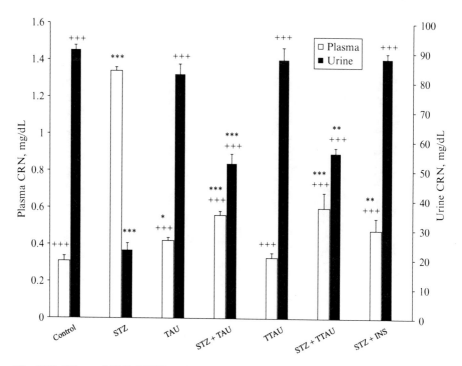

Fig. 29.8 Effects of TAU, TTAU, and INS on the plasma and urine CRN of diabetic rats. *STZ* streptozotocin, *TAU* taurine, *TTAU* thiotaurine, *INS* insulin. Comparisons were significantly different from Control at $*p<0.05$, $**p<0.01$, and $***p<0.001$; and from STZ at $^{+++}p<0.001$. Values are shown as mean ± SEM for $n=6$

than those of untreated diabetic and TTAU-treated rats, but 32% and 22% lower, respectively, than those of normal and INS-treated rats. The lack of a significant difference between diabetic rats treated and nontreated with TAU reported by Li et al. (2004) could be accounted for by the lower dose of STZ (45 mg/kg) and the mode of TAU administration (as a diet supplementation) used by these authors. Similarly, while Li et al. (2004) found no significance difference in plasma GLU between TAU-treated and TAU-untreated diabetic rat, in the present case both TAU and TTAU were able to lower it, with TTAU (-38%, $p<0.01$) appearing more potent than TAU (-22%, $p<0.05$). The same trend existed in terms of the plasma INS, which was higher than diabetic values by 24% and 34%, respectively, after treatments with TAU and TTAU.

Chronic hyperglycemia results in a nonenzymatic glycation of proteins like hemoglobin (HbA) to form HbA_{1c}, the levels of which reflect the status of blood glucose in the preceding 2–3 months (Zheng et al. 2012) and strongly correlate with mean plasma GLC levels at fasting and over postprandial periods in type 1 and type 2 diabetes (Monnier and Colette, 2008). A good correlation also seems to exist between the circulating levels of HbA_{1c} and the incidence of diabetic nephropathy

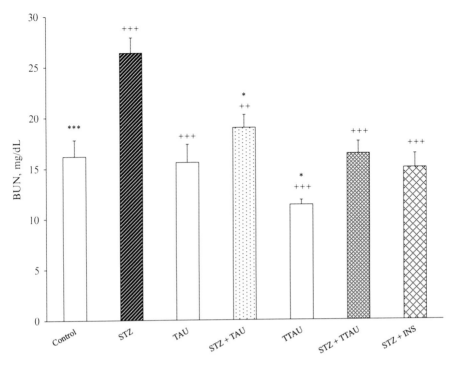

Fig. 29.9 Effects of TAU, TTAU, and INS on the BUN levels of diabetic rats. *STZ* streptozotocin, *TAU* taurine, *TTAU* thiotaurine, *INS* insulin. Comparisons were significantly different from Control at $*p<0.05$ and $***p<0.001$; and from STZ at $^{++}p<0.01$ and $^{+++}p<0.001$. Values are shown as mean $\pm$ SEM for $n=6$

among type 1 diabetics with associated risk factors (Fares et al. 2010). Antioxidants like *N*-acetylcysteine (NAC), oxerutin and TAU, are known to lower the HbA_{1c} levels of diabetic rats (Odetti et al. 2003). In this study, the administration of INS, TAU or TTAU to STZ-treated rats was found to reduce markedly the plasma HbA_{1c} to values that were slightly lower with INS (51% reduction) than with either TAU or TTAU (59% reduction with both) through hypoglycemic and antioxidant mechanisms.

In rats made chronically diabetic with STZ, there is an increase in urine volume output and variable changes in the urine levels of Na^+ and K^+, with one study reporting a substantial increase (Hebden et al. 1986) and another a slight decrease (Vaishya et al. 2009) of both electrolytes. In this work, diabetes drastically enhanced the urine output and reduced the urine excretion of Na^+ and K^+ to a significant extent ($\leq33\%$, $p<0.01$). Conversely, diabetes promoted the retention of these electrolytes in the plasma. In contrast, a treatment with INS, TAU or TTAU had a normalizing effect on both the urinary and plasma electrolytes values.

It is known that moderate hyperglycemia without glucosuria can increase plasma renin activity and mean arterial pressure in young healthy males with early

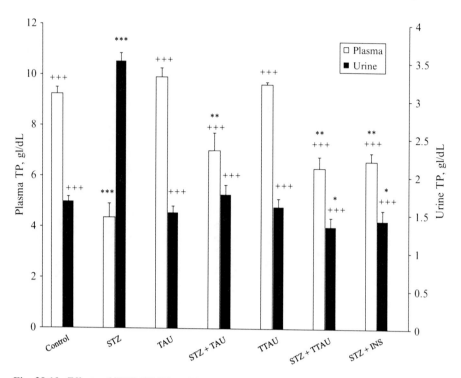

Fig. 29.10 Effects of TAU, TTAU, and INS on the plasma and kidney TP of diabetic rats. *STZ* streptozotocin, *TAU* taurine, *TTAU* thiotaurine, *INS* insulin. Comparisons were significantly different from Control at $*p < 0.05$ and $**p < 0.01$ and $***p < 0.001$; and from STZ at $^{+++}p < 0.001$. Values are shown as mean ± SEM for $n = 6$

Table 29.3 Effect of TAU, TTAU, and INS on the plasma and urine levels of Na^+ and K^+ of diabetic rats

Group	Plasma Na^+ (mmol/L)	Urine Na^+ (mmol/L)	Plasma K^+ (mmol/L)	Urine K^+ (mmol/L)
Control	$144.45 \pm 6.43^{++}$	$119.23 \pm 3.39^{+++}$	$10.02 \pm 0.41^{+++}$	$51.95 \pm 0.49^{+++}$
STZ	$194.83 \pm 1.05^{***}$	$68.34 \pm 2.92^{***}$	$16.01 \pm 0.45^{***}$	$34.61 \pm 1.14^{***}$
TAU	$153.19 \pm 4.30^{++}$	$121.09 \pm 2.89^{+++}$	$9.34 \pm 0.51^{+++}$	$49.17 \pm 0.81^{+++}$
STZ + TAU	$159.81 \pm 2.46^{*,\,++}$	$94.70 \pm 2.23^{**,\,++}$	$8.02 \pm 0.44^{*,\,+++}$	$53.26 \pm 1.71^{+++}$
TTAU	$140.07 \pm 3.41^{++}$	$110.23 \pm 3.02^{+++}$	$7.72 \pm 0.81^{+++}$	$51.20 \pm 0.25^{+++}$
STZ + TTAU	$166.55 \pm 6.59^{*,\,+}$	$118.39 \pm 6.08^{***,\,+++}$	$8.89 \pm 0.55^{+++}$	$48.21 \pm 2.17^{++}$
STZ + INS	$167.10 \pm 2.37^{*,\,+}$	$117.81 \pm 14.12^{+++}$	$10.42 \pm 0.35^{+++}$	$50.00 \pm 3.67^{+++}$

Values represent mean ± SEM ($n = 6$). *STZ* streptozotocin, *TAU* taurine, *TTAU* thiotaurine, *INS* insulin. Comparisons were significantly different from Control at $*p < 0.05$, $**p < 0.01$ and $***p < 0.001$, and from STZ at $^+p < 0.05$, $^{++}p < 0.01$ and $^{+++}p < 0.001$

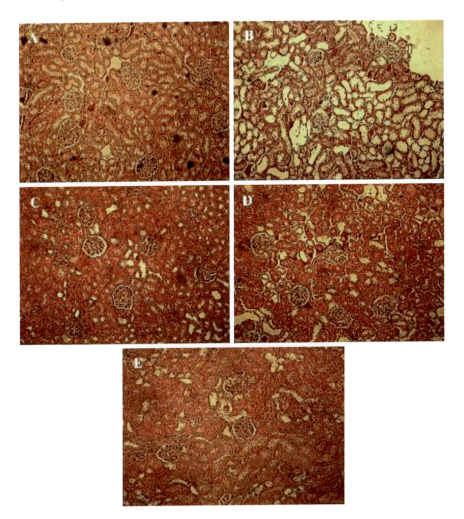

Fig. 29.11 Representative photomicrographs of rat kidney sections stained with H&E from: (a) control (nondiabetic) rat, (b) rat treated with STZ, (c) rat treated with STZ + INS, (d) rat treated with STZ + TAU, (e) rat treated with STZ + TTAU

uncomplicated type 1 diabetes mellitus, and mild to moderate hyperglycemia can affect renal function by increasing the RAS activity in diabetic humans (Miller, 1999). Indeed, activation of the RAS is found to increase arterial blood pressure and albuminuria and to negatively affect kidney function (Rossing et al. 2002a, b). Indeed, STZ-induced diabetes promoted the retention of CRN and urea in the plasma with a corresponding decrease in the urine. In this context, while INS TAU and TTAU were able to provide a significant attenuating action, INS was the most potent and TAU and TTAU were about equipotent. Furthermore, these compounds were able to attenuate the loss of plasma proteins in the urine as a result of diabetes significantly ($p \leq 0.01$) and to similar extents.

The characteristic lesions of diabetic nephropathy may be closely related to the effects of hyperglycemia on various growth factors, among which TGF-β1 has emerged as a key participant in the cascade of events leading to nephrosclerosis. In the early phase of clinical or experimental diabetes, ROS formed under the influence of high ambient glucose, upregulate the formation of TGF-β1 in the kidney to promote glomerular hypertrophy and the accumulation of ECM proteins in mesangial glomerular cells (Goldfarb and Ziyadeh, 2001), in part due to defective extracellular protease activity (Reckelhoff et al. 1993). In the chronic phase, the disease progresses to glomerular mesangial expansion and basement membrane thickening. The close relationship between LPO and expression of TGF-β1 was confirmed by observing their parallel reduction upon the treatment of diabetic rats with antioxidants like TAU, Trolox and NAC (Higo et al. 2008; Lee et al. 2005). In agreement with these results, this study verified that in addition to drastically decreasing then plasma levels of TGF-β1, INS (by 90%), TAU or TTAU (~68% reduction with both) were also able to attenuate LPO, proteinuria, hypercreatininemia and azotemia, and to contribute to the extensive preservation of the normal renal histology in STZ-treated rats. As suggested by Higo et al. (2008), compounds based on the TAU structure could be of value in slowing down the progression of diabetic nephropathy since the onset of their effect on the expression of TGF-β1 appears to start at a time when proteinuria has already set in. In this context, TTAU was somewhat better than TAU.

In view of the relevance of oxidative stress to the development of diabetes nephropathy and to the extent of the ensuing kidney damage and proteinuria, the plasma and kidney levels of MDA, an end product of LPO, the plasma NO, and the plasma and kidney GSH/GSSG ratio were measured. Diabetes was found to enhance the formation of MDA, NO and GSSG, to deplete GSH and to lower the GSH/GSSG ratio. Although all the treatment compounds markedly reduced these changes their potencies varied somewhat with each parameter. For example, TAU and INS were about equipotent and slightly more potent than TTAU in preventing LPO, INS and TTAU were better than TAU in reducing NO levels, and TTAU and TAU were more effective than INS in preserving the GSH/GSSG ratio.

29.5   Conclusion

This work demonstrates the protective actions of TAU against biochemical, functional alterations arising in the kidney as a result of STZ-induced diabetes. In addition, it verifies the protective actions of TTAU, the thiosulfonate analog of TAU, which were qualitatively and quantitatively rather similar to those of the parent compound. In general, however, TTAU demonstrated stronger protective effects than TAU in spite of exerting a weaker promoting effect on the body weight gains of the diabetic rats.

References

Acharya M, Lau-Cam CA (2010) Comparison of the protective actions of N-acetylcysteine, hypotaurine and taurine against acetaminophen-induced hepatotoxicity in the rat. J Biomed Sci 17:S35

Adler S (2004) Diabetic nephropathy: linking histology, cell biology, and genetics. Kidney Int 66:2095–2106

Aruoma OI, Halliwell B, Hoey BM, Butler J (1988) The antioxidant action of taurine, hypotaurine and their metabolic precursors. Biochem J 256:251–255

Arya A, Aggarwal S, Yadav HN (2010) Pathogenesis of diabetic nephropathy. Exp Clin Endocrinol Diabetes 2:24–29

Balakumar P, Chakkarwar VA, Kumar V, Jain A, Reddy J, Singh M (2008) Experimental models for nephropathy. J Renin Angiotensin Aldosterone Syst 9:189–195

Balkan J, Doğru-Abbasoğlu S, Kanbağli O, Cevikbaş U, Aykaç-Toker G, Uysal M (2001) Taurine has a protective effect against thioacetamide-induced liver cirrhosis by decreasing oxidative stress. Hum Exp Toxicol 20:251–254

Björck S, Nyberg G, Mulec H, Granerus G, Herlitz H, Aurell M (1986) Beneficial effects of angiotensin converting enzyme inhibition on renal function in patients with diabetic nephropathy. Br Med J (Clin Res Ed) 293:471–474

Brøns C, Spohr C, Storgaard H, Dyerberg J, Vaag A (2004) Effect of taurine treatment on insulin secretion and action, and on serum lipid levels in overweight men with a genetic predisposition for type II diabetes mellitus. Eur J Clin Nutr 58:1239–1247

Buege JA, Aust SD (1978) Microsomal lipid peroxidation. Methods Enzymol 52:302–310

Catherwood MA, Powell LA, Anderson P, McMaster D, Sharpe PC, Trimble ER (2002) Glucose-induced oxidative stress in mesangial cells. Kidney Int 61:599–608

Cavallini D, De Marco C, Mondovi B (1960) Enzymatic conversion of cystamine into thiotaurine. Boll Soc Ital Biol Sper 36:1915–1918

Cavallinid D, De Marco C, Mondovi B (1959) Chromatographic evidence on the occurrence of thiotaurine in the urine of rats fed with cystine. Biol Chem 234:854–857

Craven PA, DeRubertis FR, Kagan VE, Melhem M, Studer RK (1997) Effects of supplementation with vitamin C or E on albuminuria, glomerular TGF-beta, and glomerular size in diabetes. J Am Soc Nephrol 8:1405–1414

Das J, Sil PC (2012) Taurine ameliorates alloxan-induced diabetic renal injury, oxidative stress-related signaling pathways and apoptosis in rats. Amino Acids 43:1509–1523

Derlacz RA, Sliwinska M, Piekutowska A, Winiarska K, Drozak J, Bryla J (2007) Melatonin is more effective than taurine and 5-hydroxytryptophan against hyperglycemia-induced kidney-cortex tubules injury. J Pineal Res 42:203–209

Doerr RC, Fox JB, Lakritz L, Fiddler W (1981) Determination of nitrite in cured meats by chemi-luminescence. Anal Chem 53:381–384

Egawa M, Kohno Y, Kumano Y (1999) Oxidative effects of cigarette smoke on the human skin. Int J Cosmet Sci 21:83–98

Fares J, Kanaan M, Chaaya M, Azar S (2010) Fluctuations in glycosylated hemoglobin (HbA_{1c}) as a predictor for the development of diabetic nephropathy in type 1 diabetic patients. Int J Diabetes Mellitus 2:10–14

Fellman JH, Green TR, Eicher AL (1987) The oxidation of hypotaurine to taurine: bis-aminoethyl-α-disulfone, a metabolic intermediate in mammalian tissue. Adv Exp Med Biol 217:39–48

Flyvbjerg A (2000) Putative pathophysiological role of growth factors and cytokines in experimental diabetic kidney disease. Diabetologia 43:1205–1223

Flyvbjerg A, Dagnaes-Hansen F, De Vriese AS, Schrijvers BF, Tilton RG, Rasch R (2002) Amelioration of long-term renal changes in obese type 2 diabetic mice by a neutralizing vascular endothelial growth factor antibody. Diabetes 51:3090–3094

Franconi F, Di Leo MA, Bennardini F, Ghirlanda G (2004) Is taurine beneficial in reducing risk factors of diabetes? Neurochem Res 29:143–150

Goldfarb S, Ziyadeh FN (2001) TGF-β: a crucial component of the pathogenesis of diabetic nephropathy. Trans Am Clin Climatol Assoc 112:27–33

Green TR, Fellman JH, Eicher AL, Pratt KL (1991) Antioxidant role and subcellular location of hypotaurine and taurine in human neutrophils. Biochim Biophys Acta 1073:91–97

Güntherberg H, Rost J (1966) The true oxidized glutathione content of red blood cells obtained by new enzymic and paper chromatographic methods. Anal Biochem 15:205–210

Ha H, Kim KH (1999) Pathogenesis of diabetic nephropathy: the role of oxidative stress and kinase C. Diabetes Res Clin Pract 45:147–151

Ha H, Lee HB (2000) Reactive oxygen species as glucose signaling molecules in mesangial cells cultured under high glucose. Kidney Int 58(Suppl 77):S-19–S-25

Ha H, Yu MR, Choi YJ, Lee HB (2001) Activation of protein kinase C-δ and C-ε by oxidative stress in early diabetic rat kidney. Am J Kidney Dis 38:S204–S207

Ha H, Yu MR, Kim KH (1999) Melatonin and taurine reduce early glomerulopathy in diabetic rats. Free Radic Biol Med 26:944–950

Haber CA, Lam TK, Yu Z, Gupta N, Goh T, Bogdanovic E, Giacca A, Fantus IG (2003) N-Acetylcysteine and taurine prevent hyperglycemia-induced insulin resistance in vivo: possible role of oxidative stress. Am J Physiol Endocrinol Metab 285:E744–E753

Hansen SH (2001) The role of taurine in diabetes and the development of diabetic complications. Diabetes Metab Res Rev 17:330–346

Hebden RA, Gardiner SM, Bennett T, MacDonald IA (1986) The influence of streptozotocin-induced diabetes mellitus on fluid and electrolyte handling in rats. Clin Sci (Lond) 70:111–117

Higo S, Miyata S, Jiang QY, Kitazawa R, Kitazawa S, Kasuga M (2008) Taurine administration after appearance of proteinuria retards progression of diabetic nephropathy in rats. Kobe J Med Sci 54:E35–E45

Hissin PJ, Hilf R (1976) A fluorometric method for determination of oxidized and reduced glutathione in tissues. Anal Biochem 74:214–226

Huang JS, Chuang LY, Guh JY, Huang YJ, Hsu MS (2007) Antioxidants attenuate high glucose-induced hypertrophic growth in renal tubular epithelial cells. Am J Physiol 293:F1072–F1082

Huang JS, Chuang LY, Guh JY, Yang YL, Hsu MS (2008) Effect of taurine on advanced glycation end products-induced hypertrophy in renal tubular epithelial cells. Toxicol Appl Pharmacol 233:220–226

Jiang QY (2002) Mechanisms of beneficial effects of taurine on diabetic nephropathy. Med J Kobe Univ 62:119–126

Joyner JL, Peyer SM, Lee RW (2003) Possible roles of sulfur-containing amino acids in a chemoautotrophic bacterium-mollusc symbiosis. Biol Bull 205:331–338

Kaplan B, Karabay G, Zağyapan RD, Ozer C, Sayan H, Duyar I (2004) Effects of taurine in glucose and taurine administration. Amino Acids 27:327–333

Kashihara N, Haruna Y, Kondeti VK, Kanwar YS (2010) Oxidative stress in diabetic nephropathy. Curr Med Chem 17:4256–4269

Katsumata M, Kiuchi K, Tashiro T, Uchikuga S (1997) Methods for scavenging active oxygen compounds and preventing damage from ultra violet B rays using taurine analogues. US Patent Number 5:601,806

Kono Y (1998) New raw materials and new technologies for cosmetics. (Part I). Development and its application of "sebum antioxidant thiotaurine" for cosmetics. Fragr J 26:9–14

Koya D, Hayashi K, Kitada M, Kashiwagi A, Kikkawa R, Haneda M (2003) Effects of antioxidants in diabetes-induced oxidative stress in the glomeruli of diabetic rats. J Am Soc Nephrol 14:S250–S253

Kulakowski EC, Maturo J (1984) Hypoglycemic properties of taurine: not mediated by enhanced insulin release. Biochem Pharmacol 33:2835–2838

Lee EA, Seo JY, Jiang Z, Yu MR, Kwon MK, Ha H, Lee HB (2005) Reactive oxygen species mediate high glucose-induced plasminogen activator inhibitor-1 up-regulation in mesangial cells and in diabetic kidney. Kidney Int 67:1762–1771

Lee HB, Yu MR, Yang Y, Jiang Z, Ha H (2003) Reactive oxygen species-regulated signaling pathways in diabetic nephropathy. J Am Soc Nephrol 14:S241–S245

Li F, Obrosova IG, Abatan O, Tian D, Larkin D, Stuenkel EL, Stevens MJ (2004) Taurine replacement attenuates hyperalgesia and abnormal calcium signaling in sensory neurons of STZ-D rats. Am J Physiol Endocrinol Metab 288:E29–E36

Mathiesen ER, Hommel E, Giese J, Parving HH (1991) Efficacy of captopril in postponing nephropathy in normotensive insulin dependent diabetic patients with microalbuminuria. Br Med J 303:81–87

Miller JA (1999) Impact of hyperglycemia on the renin angiotensin system in early human type 1 diabetes mellitus. J Am Soc Nephrol 10:1778–1785

Monnier L, Colette C (2008) Fasting glucose and postprandial glycemia: which is the best target for improving outcomes? The Apollo and 4-T Trials. Expert Opin Pharmacother 9:2857–2865

Naito Y, Uchiyama K, Aoi W, Hasegawa G, Nakamura N, Yoshida N, Maoka T, Takahashi J, Yoshikawa T (2004) Prevention of diabetic nephropathy by treatment with astaxanthin in diabetic db/db mice. Biofactors 20:49–59

Nandhini AT, Thirunavukkarasu V, Anuradha CV (2004) Stimulation of glucose utilization and inhibition of protein glycation and AGE products by taurine. Acta Physiol Scand 181:297–303

Obrosova IG (2005) Increased sorbitol pathway activity generates oxidative stress in tissue sites for diabetic complications. Antioxid Redox Signal 7:1543–1552

Obrosova IG, Stevens MJ (1999) Effect of dietary taurine supplementation on GSH and NAD(P)-redox status, lipid peroxidation, and energy metabolism in diabetic precataractous lens. Invest Ophthalmol Vis Sci 40:680–688

Odetti P, Pesce C, Traverso N, Menini S, Maineri EP, Cosso L, Valentini S, Patriarca S, Cottalasso D, Marinari UM, Pronzato MA (2003) Comparative trial of N-acetyl-cysteine, taurine, and oxerutin on skin and kidney damage in long-term experimental diabetes. Diabetes 52:499–505

Prabhakar S, Starnes J, Shi S, Lonis B, Tran R (2007) Diabetic nephropathy is associated with oxidative stress and decreased renal nitric oxide production. J Am Soc Nephrol 18:2945–2952

Pruski AM, Fiala-Médioni A (2003) Stimulatory effect of sulphide on thiotaurine synthesis in three hydrothermal-vent species from the East Pacific Rise. J Exp Biol 206:2923–2930

Reckelhoff JF, Tygart VL, Mitias MM, Walcott JL (1993) STZ-induced diabetes results in decreased activity of glomerular cathepsin and metalloprotease in rats. Diabetes 42:1425–1432

Ribeiro RA, Bonfleur ML, Amaral AG, Vanzela EC, Rocco SA, Boschero AC, Carneiro EM (2009) Taurine supplementation enhances nutrient-induced insulin secretion in pancreatic mice islets. Diabetes Metab Res Rev 25:370–379

Rossing K, Christensen PK, Jensen BR, Parving HH (2002a) Dual blockade of the renin-angiotensin system in diabetic nephropathy: a randomized double-blind crossover study. Diabetes Care 25:95–100

Rossing P, Hougaard P, Parving HH (2002b) Risk factors for development of incipient and diabetic nephropathy in type 1 diabetic patients: a 10-year prospective observational study. Diabetes Care 25:859–864

Schaffer SW, Azuma J, Mozaffari M (2009) Role of antioxidant activity of taurine in diabetes. Can J Physiol Pharmacol 87:91–99

Simşek M, Naziroğlu M, Erdinç A (2005) Moderate exercise with a dietary vitamin C and E combination protects against streptozotocin-induced oxidative damage to the kidney and lens in pregnant rats. Exp Clin Endocrinol Diabetes 113:53–59

Tada H, Ishii H, Isogai S (1997) Protective effect of D-alpha-tocopherol on the function of human mesangial cells exposed to high glucose concentrations. Metabolism 46:779–784

Trachtman H, Futterweit S, Bienkowski R (1993) Taurine prevents glucose-induced lipid peroxidation and increased collagen production in cultured rat mesangial cells. Biochem Biophys Res Commun 191:759–765

Trachtman H, Futterweit S, Maesaka J, Ma C, Valderrama E, Fuchs A, Tarectecan AA, Rao PS, Sturman JA, Boles TH (1995) Taurine ameliorates chronic streptozocin-induced diabetic nephropathy in rats. Am J Physiol 269:F429–F438

Trachtman H, Futterweit S, Prenner J, Hanon S (1994) Antioxidants reverse the antiproliferative effect of high glucose and advanced glycosylation end products in cultured rat mesangial cells. Biochem Biophys Res Commun 199:346–352

Vaishya R, Singh J, Lal H (2009) Biochemical effects of irbesartan in experimental diabetic nephropathy. Indian J Pharmacol 41:252–254

Verzola D, Bertolotto MB, Villaggio B, Ottonello L, Dallegri F, Frumento G, Berruti V, Gandolfo MT, Garibotto G, Deferran G (2002) Taurine prevents apoptosis induced by high ambient glucose in human tubule renal cells. J Investig Med 50:443–451

Wang LJ, Zhang L, Yu Y, Wang Y, Niu N (2008) The protective effects of taurine against early renal injury in STZ-induced diabetic rats, correlated with inhibition of renal LOX-1-mediated ICAM-1 expression. Ren Fail 30:763–771

Winiarska K, Szymanski K, Gorniak P, Dudziak M, Bryla J (2008) Hypoglycaemic, antioxidative and nephroprotective effects of taurine in alloxan diabetic rabbits. Biochimie 91:261–270

Wollenberger A, Ristau O, Schoffa G (1960) Eine einfache Technik der extrem schneller Abkühlung grösserer Gewebstücke. Pflügers Arch. Gesamte Physiol. Menschen Tiere 270:399–412

Yancey PH (2005) Organic osmolytes as compatible, metabolic and counteracting cytoprotectants in high osmolarity and other stresses. J Exp Biol 208:2819–2830

Yancey PH, Ishikawa J, Meyer B, Girguis PR, Lee RW (2009) Thiotaurine and hypotaurine contents in hydrothermal-vent polychaetes without thiotrophic endosymbionts: correlation with sulfide exposure. J Exp Zool A Ecol Genet Physiol 311A:439–447

Yao HT, Lin P, Chang YW, Chen CT, Chiang MT, Chang L, Kuo YC, Tsai HT, Yeh TK (2009) Effect of taurine supplementation on cytochrome P450 2E1 and oxidative stress in the liver and kidneys of rats with streptozotocin-induced diabetes. Food Chem Toxicol 47:1703–1709

Yoshiyuki K, Yoshiki M (2000) Peroxidation in the skin and its prevention. Jpn J Inflamm 20:119–129

Zheng CM, Ma WY, Wu CC, Lu KC (2012) Glycated albumin in diabetic patients with chronic kidney disease. Clin Chim Acta 413:1555–1561

Chapter 30
Taurine May Not Alleviate Hyperglycemia-Mediated Endoplasmic Reticulum Stress in Human Adipocytes

Kyoung Soo Kim, Hye-In Ji, and Hyung-In Yang

Abstract In obesity and diabetes, adipocytes show significant endoplasmic reticulum (ER) stress. Hyperglycemia-induced ER stress has not been studied in adipocyte differentiation and adipokine expression. Taurine has been known to protect the cells against ER stress. This study examined the effect of taurine on ER stress-induced adipocyte differentiation and adipokine expression to explain the therapeutic effect of taurine on diabetes and obesity. To do this, human preadipocytes were differentiated into adipocytes, in the presence or absence of taurine, under ER stress conditions. Human preadipocytes were treated with thapsigargin (10 nM) or high glucose concentrations (100 mM) as ER stress inducers during differentiation into adipocytes. Thapsigargin inhibited the differentiation of adipocytes in a dose-dependent manner, but the high glucose concentration treatment did not. Taurine 100 mM treatment did not block the inhibition of differentiation of preadipcytes into adipocytes. Furthermore, the high glucose concentration treatment inhibited the expression of adiponectin and increased the expression of leptin in human adipocytes. However, taurine treatment did not affect the expression of two adipokines. In conclusion, the therapeutic mechanism of taurine in diabetes and obesity does not appear to occur by alleviating hyperglycemia-mediated ER stress. To clarify the molecular mechanism by which taurine improves diabetic symptoms and obesity in animal models, the protective effect of taurine against hyperglycemia- or overnutrition-mediated ER stress should be further evaluated under various conditions or types of ER stress.

K.S. Kim, (✉) • H.-I Ji • H.-I Yang
East-West Bone & Joint Disease Research Institute, Kyung Hee University
Hospital at Gangdong, 149 Sangil-dong, Gangdong-gu, Seoul 134-727, Korea
e-mail: labrea46@yahoo.co.kr

A. El Idrissi and W.J. L'Amoreaux (eds.), *Taurine 8*, Advances in Experimental
Medicine and Biology 775, DOI 10.1007/978-1-4614-6130-2_30,
© Springer Science+Business Media New York 2013

Abbreviations

ER	Endoplasmic reticulum
TG	Thapsigargin
T2DM	Type 2 diabetes mellitus
TUDCA	Tauroursodeoxycholic acid
NAC	N-Acethylcysteine

30.1 Introduction

Obesity is a well-known risk factor for the development of type 2 diabetes mellitus (T2DM). T2DM is a multifactorial chronic metabolic disease characterized by hyperglycemia (Kahn et al. 2006). A relatively new player in the diabetes mellitus (DM) field is endoplasmic reticulum (ER) stress (Cnop et al. 2010). The ER is where the synthesis, assembly, and/or modification of transmembrane and secreted proteins occur. Importantly, obesity impairs the proper functioning of the ER in fat and other cell types (Ozcan et al. 2004). ER stress and/or ER stress-induced apoptosis are increasingly acknowledged as important mechanisms in the development of DM, not only for β-cell loss but also for the development of insulin resistance. Recently, the importance of ER stress has been noted in adipocytes in obesity because ER stress affects the expression of adipokines in adipocytes (Xu et al. 2010). Adipokines, particularly adiponectin and leptin, contribute to the maintenance of whole-body glucose homeostasis by modulating gluconeogenesis in the liver (Daval et al. 2006). Meanwhile, ER stress can be induced by hyperlipidemia and hyperglycemia (Alhusaini et al. 2010). Recent studies showed that ER stress induces the downregulation of adiponectin, leptin, and resistin, but increases the expression of IL-6 in adipocytes (Xu et al. 2010). Furthermore, ER stress inhibits the differentiation of preadipocytes into adipocytes during adipogenesis (Shimada et al. 2007; Zha and Zhou 2012).

Taurine (2-aminoethylsulphonic acid) is a non-protein amino acid present in almost all animal tissues, and is abundantly present as a free intracellular amino acid in human cells (Kim et al. 2007). Due to its unique chemical structure, taurine is involved in numerous biological and physiological functions that confer important health benefits. Thus, taurine is a cytoprotective agent in a variety of tissues. It also modulates a variety of cellular functions, including antioxidation, modulation of ion movement, osmoregulation, modulation of neurotransmitters, and conjugation of bile acids (Lourenco and Camilo 2002). Taurine supplementation improved diabetic symptoms in diabetic animal models and showed anti-obesity effects in animal models (Ito et al. 2012). In addition, taurine had a protective effect in PC 12 cells against ER stress induced by oxidative stress and in cortical neurons against ER stress induced by glutamate (Pan et al. 2011). It also reduced ER stress in *Caenorhabditis elegans* (Kim et al. 2010). In this study, to explain the therapeutic effect of taurine on diabetes and obesity as an ER stress reducer, we investigated whether taurine blocks ER stress-mediated inhibition of adipogenesis and whether it reverses the change in adiponectin and leptin expression in human adipocytes under hyperglycemia-mediated ER stress.

30.2   Methods

30.2.1   Preadipocyte Cell Culture and Differentiation into Adipocytes

Human preadipocytes were purchased from Cell Applications (San Diego, CA, USA) and maintained in the Preadipocyte Growth Medium Kit (Cell Applications). Preadipocytes were seeded into 6-well plates (1.5×10^5 cells/well in 2 ml of medium) or 60-mm dishes (2.5×10^5 cells/60-mm dish in 2 ml of medium) and cultured until confluent. For differentiation, the culture medium was changed to Adipocyte Differentiation Medium (Cell Applications) and cells were cultured for 2 weeks, with changes in the medium every 2 days, and grown in the presence or absence of taurine at different concentrations under ER stress conditions induced by thapsigargin (TG) or high glucose concentration.

30.2.2   Oil Red O Staining

Lipid accumulation was examined with Oil red O staining (Ramirez-Zacarias et al. 1992). Cultured cells were rinsed twice with phosphate-buffered saline (PBS) and fixed in 10% (v/v) formaldehyde for 1 h. After the formaldehyde was removed, the cells were rinsed three times with deionized water and stained with a saturated solution of Oil red O in 60% isopropanol solution for 2 h at room temperature. Microscopic images (Olympus, Tokyo, Japan) of the stained cells were obtained after removing the staining solution. Finally, the dye retained in the cells was eluted with isopropanol and quantified by measuring the optical absorbance at 500 nm.

30.2.3   Enzyme-Linked Immunosorbent Assay (ELISA)

The differentiated adipocytes (2.5×10^5 cells/60-mm dish in 2 ml medium) were further cultured in serum-free medium in the absence and presence of tauroursodeoxycholic acid (TUDCA), taurine, or N-acethylcysteine (NAC) under hyperglycemia-induced ER stress. The culture supernatants were collected after 7 days in culture. The supernatants were collected and analyzed for adiponectin and leptin with an ELISA kit (R&D Systems Inc., Minneapolis, MN, USA).

30.2.4   Statistical Analysis

The in vitro experimental data are expressed as the mean±standard error of the mean (SEM) of quadruplicate samples. Differences between groups were compared with the Mann–Whitney test. Prism software 4 (Graphpad Software, San Diego, CA) was used for statistical analysis and graphing. Differences were considered significant at $P<0.05$.

30.3   Results

30.3.1   Effect of ER Stress on the Differentiation of Human Preadipocytes to Adipocytes

To study the physiological effects of ER stress on human adipocyte differentiation, preadipocytes were cultured under differentiation conditions for 14 days in the presence or absence of the ER stress inducers TG or high glucose concentration. As shown in Fig. 30.1a, the preadipocytes differentiated into adipocytes (adipogenesis) in the absence of TG, whereas differentiation was inhibited by TG in a dose-dependent manner. However, high glucose concentration did not inhibit adipogenesis (data not shown). Meanwhile, the differentiated adipocytes showed intracellular lipid accumulation. The accumulated lipid droplets were examined by oil red O staining. In accordance with the degree of differentiation, the degree of oil red O staining was dose-dependently decreased by TG (Fig. 30.1b). To test whether taurine could reverse the TG-mediated inhibition of adipogenesis, taurine was added to the culture medium at concentrations up to 100 mM. The inhibition of adipogenesis was not blocked by taurine (data not shown). In contrast, high concentrations of glucose (100 mM) added to the culture media to mimic hyperglycemia-induced ER stress did not inhibit adipocyte differentiation (data not shown).

30.3.2   Effect of TUDCA, Taurine, or NAC on ER Stress-Mediated Change of Adiponectin and Leptin Production in Differentiated Human Adipocytes

To examine whether hyperglycemia-induced ER stress modulates the expression of adipokines in adipocytes, glucose was added to fully-differentiated adipocytes at concentrations ranging from 0 to 40 mM and cultured for 7 days. After 7 days, the culture supernatants were assayed for the expression of adiponectin and leptin. The supernatant from differentiated adipocytes had greatly increased levels of adiponectin and leptin compared with the preadipocytes. Glucose treatment for 7 days significantly decreased the production of adiponectin by differentiated human

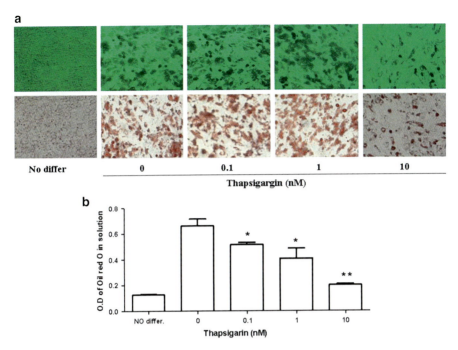

Fig. 30.1 Effect of the ER stress inducer, thapsigargin, on the differentiation of preadipocytes into adipocytes. (**a**) Microscopic image of differentiated adipocytes before (*top row*) and after (*bottom row*) oil red O staining. (**b**) Optical absorbance at 500 nm of the dye retained in adipocytes. Three independent experiments were performed in triplicate. Values are expressed as the mean ± standard error of the mean (SEM). *$P<0.05$, **$P<0.01$ versus no treatment with thapsigargin

adipocytes, but leptin production was significantly increased in a dose-dependent manner (Fig. 30.2a). Next, to determine if the ER stress reducer, TUDCA can block the change in adipokine production in adipocytes, TUDCA was added to adipocytes under hyperglycemia-induced ER stress. TUDCA reversed the change in adipokine production even at the low concentration of 0.01 mM (Fig. 30.2b). Next, we tested whether taurine and NAC have the same effect as TUDCA on the change in adipokine production in adipocytes under ER stress. Taurine did not modulate the change, even at the highest concentration of 100 mM (Fig. 30.3a). In contrast, NAC slightly modulated the change at the concentration of 500 mM (Fig. 30.3b).

30.4 Discussion

In this study, we investigated whether taurine could reverse the ER stress-mediated inhibition of adipogenesis or if taurine could modulate the expression of adiponectin and leptin in adipocytes under ER stress. The ER stress inducer TG blocked

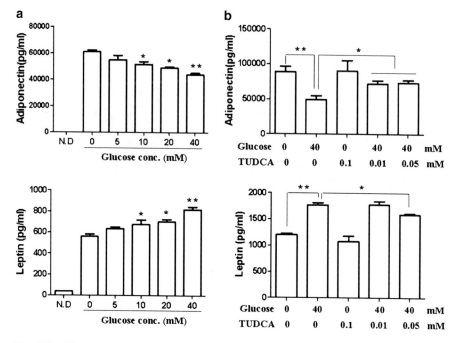

Fig. 30.2 Effect of the ER stress blocker, TUDCA on the modulation of adipokine production in adipocytes treated with high glucose concentrations. (**a**) The modulation of adipokines according to glucose concentration. (**b**) Recovery of the modulation of adipokines in adipocytes high glucose concentration by TUDCA. Values are expressed as the mean ± standard error of the mean (SEM). *$P < 0.05$, **$P < 0.01$ versus no treatment with glucose or TUDCA

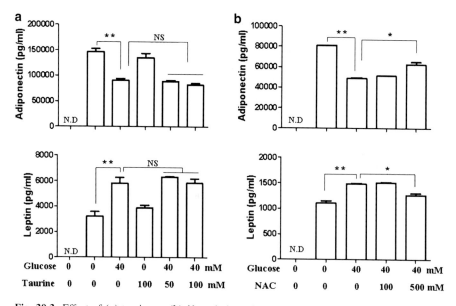

Fig. 30.3 Effect of (**a**) taurine or (**b**) *N*-acethylcysteine (NAC) on the modulation of adipokine production in adipocytes treated with high glucose concentrations. Values are expressed as the mean ± standard error of the mean (SEM). *$P < 0.05$, **$P < 0.01$ versus no treatment with taurine or NAC. *N.D.* No Differentiation

adipogenesis, but high concentrations of glucose (100 mM) did not. In contrast, high glucose concentrations (40 mM) downregulated the expression of adiponectin, but increased leptin expression in differentiated human adipocytes. TUDCA, an ER stress reducer, reversed the change in adiponectin and leptin expression under hyperglycemia-induced ER stress. However, taurine did not modulate the expression of these adipokines in this in vitro system.

These results differ from those of previous reports in some aspects. First, in previous reports the ER stress inducer reduced the expression of adiponectin, leptin, and resistin (Xu et al. 2010). However, in our system the ER stress inducer, high glucose concentration increased leptin expression and decreased adiponectin expression in differentiated human adipocytes. In contrast, the ER stress inducer TG did not modulate the expression of either adiponectin or leptin in this system (data not shown). Second, TG inhibited the differentiation of preadipocytes into adipocytes, but high glucose concentration did not. Inconsistent with our result, TG or other types of ER stress inducers like A23187 (calcium ionophore) inhibited the adipogenesis process (Shimada et al. 2007). However, a high glucose concentration (100 mM) did not block adipogenesis in this system. Thus, we hypothesize that types of ER stress inducers can differentially affect the expression of adipokines and the adipogenesis process. In support of this hypothesis, TG decreased extracellular-superoxide dismutase (EC-SOD) expression, whereas the expression of Cu/Zn-SOD and Mn-SOD was unchanged. On the other hand, another ER stress inducer, tunicamycin did not affect the expression of EC-SOD. Furthermore, TG has the ability to activate extracellular-signal regulated kinase (ERK), but tunicamycin does not (Kamiya et al. 2011). Furthermore, high glucose concentrations amplify fatty acid-induced ER stress in pancreatic β-cells via activation of mTORC1 (Bachar et al. 2009). In addition, cells respond differentially to different degrees of ER stress (D'Hertog et al. 2010). For example, insulin-producing INS-1E cells were exposed in vitro to the ER-stress inducer, cyclopiazonic acid (CPA) at two concentrations. CPA 25 μM led to massive apoptosis accompanied by a near complete shut-down of protein translation, but CPA 6.25 μM led to adaptation of the β-cells to ER stress. This difference in response may be due to different defense pathways against ER stress.

Deterioration of ER homeostasis through a variety of biochemical or pathophysiological stimuli can impair protein folding processes in the ER by disrupting protein glycosylation, disulfide bond formation, or the ER calcium pool. These disruptions can cause the accumulation of unfolded or misfolded proteins in the ER lumen, a condition termed "ER stress"(van der Kallen et al. 2009) that can be sensed by three ER transmembrane protein sensors; PERK (protein kinase-like ER kinase), IRE1α (insositol-requiring ER-to-nucleus signal kinase 1), and ATF6α (activating transcription factor 6). These sensors activate a cellular emergency program to re-establish homeostasis called the unfolded protein response (UPR), resulting in attenuation of protein translation and transcriptional activation of UPR genes. Another pathway termed "ER-associated degradation (ERAD)" is also activated to dispose of misfolded proteins. When these two systems are imperfect, terminally misfolded proteins can be cleared from the ER by an additional process like autophagy.

Inconsistent with other previous studies reporting that taurine protects cells from ER stress (Kim et al. 2010; Pan et al. 2011), taurine did not reverse the effect of ER stress on the inhibition of adipocyte differentiation or the change in adipokine expression in adipocytes, whereas TUDCA could reverse these changes in this system. This discrepant result on the effects of taurine may be due to the differential activation of defense signal pathways induced by ER stress. Therefore, the protective effect of taurine against ER stress should be evaluated under different degrees of ER stress or with different types of ER stress inducers.

Meanwhile, deregulated adipokine expression in adipocytes under hyperglycemia-mediated ER stress is closely linked to diabetes; thus, ER stress induced by hyperglycemia may be a therapeutic target in diabetes. In other experiments in our lab, taurine supplementation downregulated blood glucose levels in a diabetic animal model, the OLETF rat (Kim et al. 2012). In addition, taurine supplementation reduced the serum leptin levels in this animal model, whereas it did not increase serum adiponectin levels. We do not understand the molecular mechanisms by which taurine improved the diabetic symptoms, but taurine may reduce ER stress through the downregulation of blood glucose levels. Thus, leptin levels may be decreased by taurine supplementation.

30.5   Conclusion

Taurine did not reverse the inhibition of adipocyte differentiation and the change in adipokine expression in human adipocytes under ER stress, although taurine has a protective effect in cells in which ER stress has been induced. Thus, the therapeutic mechanism of taurine on diabetes cannot be explained by its ability to alleviate hyperglycemia-mediated ER stress. The protective effect of taurine against hyperglycemia-mediated ER stress should be further evaluated under various degrees of ER stress.

Acknowledgments This research was supported by the Basic Science Research Program through the National Research Foundation of Korea (NRF) funded by the Ministry of Education, Science and Technology (2012-0002659).

References

Alhusaini S, McGee K, Schisano B, Harte A, McTernan P, Kumar S, Tripathi G (2010) Lipopolysaccharide, high glucose and saturated fatty acids induce endoplasmic reticulum stress in cultured primary human adipocytes: salicylate alleviates this stress. Biochem Biophys Res Commun 397:472–478

Bachar E, Ariav Y, Ketzinel-Gilad M, Cerasi E, Kaiser N, Leibowitz G (2009) Glucose amplifies fatty acid-induced endoplasmic reticulum stress in pancreatic beta-cells via activation of mTORC1. PLoS One 4:e4954

Cnop M, Ladriere L, Igoillo-Esteve M, Moura RF, Cunha DA (2010) Causes and cures for endoplasmic reticulum stress in lipotoxic beta-cell dysfunction. Diabetes Obes Metab 12(Suppl 2):76–82

D'Hertog W, Maris M, Ferreira GB, Verdrengh E, Lage K, Hansen DA, Cardozo AK, Workman CT, Moreau Y, Eizirik DL, Waelkens E, Overbergh L, Mathieu C (2010) Novel insights into the global proteome responses of insulin-producing INS-1E cells to different degrees of endoplasmic reticulum stress. J Proteome Res 9:5142–5152

Daval M, Foufelle F, Ferre P (2006) Functions of AMP-activated protein kinase in adipose tissue. J Physiol 574:55–62

Ito T, Schaffer SW, Azuma J (2012) The potential usefulness of taurine on diabetes mellitus and its complications. Amino Acids 42:1529–1539

Kahn SE, Hull RL, Utzschneider KM (2006) Mechanisms linking obesity to insulin resistance and type 2 diabetes. Nature 444:840–846

Kamiya T, Obara A, Hara H, Inagaki N, Adachi T (2011) ER stress inducer, thapsigargin, decreases extracellular-superoxide dismutase through MEK/ERK signalling cascades in COS7 cells. Free Radic Res 45:692–698

Kim HM, Do CH, Lee DH (2010) Taurine reduces ER stress in C. elegans. J Biomed Sci 17 (Suppl 1):S26

Kim KS, Oh DH, Kim JY, Lee BG, You JS, Chang KJ, Chung HJ, Yoo MC, Yang HI, Kang JH, Hwang YC, Ahn KJ, Chung HY, Jeong IK (2012) Taurine ameliorates hyperglycemia and dyslipidemia by reducing insulin resistance and leptin level in Otsuka Long-Evans Tokushima fatty (OLETF) rats with long-term diabetes. Exp Mol Med 44:665–673

Kim SJ, Gupta RC, Lee HW (2007) Taurine-diabetes interaction: from involvement to protection. Curr Diabetes Rev 3:165–175

Lourenco R, Camilo ME (2002) Taurine: a conditionally essential amino acid in humans? An overview in health and disease. Nutr Hosp 17:262–270

Ozcan U, Cao Q, Yilmaz E, Lee AH, Iwakoshi NN, Ozdelen E, Tuncman G, Gorgun C, Glimcher LH, Hotamisligil GS (2004) Endoplasmic reticulum stress links obesity, insulin action, and type 2 diabetes. Science 306:457–461

Pan C, Prentice H, Price AL, Wu JY (2011) Beneficial effect of taurine on hypoxia- and glutamate-induced endoplasmic reticulum stress pathways in primary neuronal culture. Amino Acids 43(2):845–855, Epub ahead of print

Ramirez-Zacarias JL, Castro-Munozledo F, Kuri-Harcuch W (1992) Quantitation of adipose conversion and triglycerides by staining intracytoplasmic lipids with Oil red O. Histochemistry 97:493–497

Shimada T, Hiramatsu N, Okamura M, Hayakawa K, Kasai A, Yao J, Kitamura M (2007) Unexpected blockade of adipocyte differentiation by K-7174: implication for endoplasmic reticulum stress. Biochem Biophys Res Commun 363:355–360

van der Kallen CJ, van Greevenbroek MM, Stehouwer CD, Schalkwijk CG (2009) Endoplasmic reticulum stress-induced apoptosis in the development of diabetes: is there a role for adipose tissue and liver? Apoptosis 14:1424–1434

Xu L, Spinas GA, Niessen M (2010) ER stress in adipocytes inhibits insulin signaling, represses lipolysis, and alters the secretion of adipokines without inhibiting glucose transport. Horm Metab Res 42:643–651

Zha BS, Zhou H (2012) ER stress and lipid metabolism in adipocytes. Biochem Res Int 2012:312943

Part IV
Function of Taurine in the Cardiovascular System

Chapter 31
Taurine Regulation of Blood Pressure and Vasoactivity

Abdeslem El Idrissi, Evelyn Okeke, Xin Yan, Francoise Sidime, and Lorenz S. Neuwirth

Abstract Taurine plays an important role in the modulation of cardiovascular function by acting not only within the brain but also within peripheral tissues. We found that IV injection of taurine to male rats caused hypotension and tachycardia. A single injection of taurine significantly lowered the systolic, diastolic, and mean arterial blood pressure in freely moving Long–Evans control rats. We further confirm the vasoactive properties of taurine using isolated aortic ring preparations. Mechanical responses of circular aortic rings to pharmacological agents were measured by an isometric force transducer and amplifier. We found that bath application of taurine to the aortic rings caused vasodilation which was blocked by picrotoxin. Interestingly, picrotoxin alone induced a constriction of the aortic ring in the absence of exogenously added taurine, suggesting a tonic activation of $GABA_A$ receptors by circulating either taurine or GABA. Additionally, we found that the endothelial cells express high levels of taurine transporters and $GABA_A$ receptors. We have previously shown that taurine activates $GABA_A$ receptors and thus we suggest that the functional implication of $GABA_A$ receptor activation is the relaxation of the arterial muscularis, vasodilation, and a decrease in blood pressure. Interestingly however, the effects of acute taurine injection were very different than chronic supplementation of taurine. When rats were supplemented taurine (0.05%, 4 weeks) in their

A. El Idrissi (✉)
Department of Biology, Center for Developmental Neuroscience,
College of Staten Island, City University of New York Graduate School,
Building 6S, Room 320, 2800 Victory Boulevard, Staten Island, NY 10314, USA
e-mail: abdeslem.elidrissi@csi.cuny.edu

E. Okeke
Department of Biology, City University of New York Graduate School,
Staten Island, NY 10314, USA

X. Yan • F. Sidime • L.S. Neuwirth
Center for Developmental Neuroscience, City University of New York Graduate School,
Staten Island, NY 10314, USA

A. El Idrissi and W.J. L'Amoreaux (eds.), *Taurine 8*, Advances in Experimental
Medicine and Biology 775, DOI 10.1007/978-1-4614-6130-2_31,
© Springer Science+Business Media New York 2013

drinking water, taurine has significant hypertensive properties. The increase in blood pressure was observed however only in females; males supplemented with taurine did not show an increase in systolic, diastolic, or mean arterial pressure. In both genders however, taurine supplementation caused a significant tachycardia. Thus, we suggest that acute administration of taurine may be beneficial to lowering blood pressure. However, our data indicate that supplementation of taurine to females caused a significant increase in blood pressure. The effect of taurine supplementation on hypertensive rats remains to be seen.

Abbreviations

Epi   Epinephrine
GAD   Glutamic acid decarboxylase
GABA   γ-Aminobutyric acid

31.1   Introduction

γ-Aminobutyric acid (GABA) is one of the major inhibitory neurotransmitters in the central nervous system and is also found in many peripheral tissues. GABA has been shown to play an important role in the modulation of cardiovascular function by acting not only within the central nervous system but also within peripheral tissues. GABA has been reported to reduce blood pressure in experimental animals (Takahashi et al. 1955) and humans (Elliott and Hobbiger 1959) following its systemic or central administration, and it has been suggested that the depressor effect induced by systemic administration of GABA is due to the blockade of sympathetic ganglia. The blood–brain barrier is impermeable to GABA, and its concentration in the brain is not changed following i.v. injection (Roberts, Tsukada, and Gelder). Thus, the antihypertensive effects seen following i.p. or i.v. administration of GABA are due to its actions within the peripheral tissues presumably, blood vessels, or autonomic nervous system. It has been reported that GABA can modulate the vascular tone by suppressing the noradrenaline release in the isolated rabbit ear artery and rat kidney (Manzini, Monasterolo, and Fujimura). The effects produced by GABA in many kinds of peripheral tissues as well as within the central nervous system are mediated by at least two distinct receptor types, $GABA_A$ and $GABA_B$. It has been reported that GABA inhibits sympathetic neurotransmission in the rabbit ear artery through the stimulation of a presynaptic of $GABA_B$ receptor subtype (Manzini et al. 1985), and that GABA acts on presynaptic $GABA_B$ receptors to suppress neurotransmitter release (and thereby attenuate renal vasoconstriction) during the activation of the sympathetic nervous supply to the rat kidney (Monasterolo and Fujimura). Baclofen, a selective $GABA_B$ receptor agonist, attenuated the perivascular nerve stimulation-induced increase in perfusion pressure and noradrenaline release to the same extent as did GABA itself (Bowery 1993). Consistent with this, it has been reported that baclofen has hypertensive properties after systemic or intracerebro-ventricular administration in rats (Persson and Crambes). Furthermore, these inhibitory effects of GABA were completely antagonized by the selective $GABA_B$ receptor antagonist, saclofen (Bowery 1993), but not by the selective $GABA_A$ receptor antagonist, bicuculline (Curtis and Kwan). These results strongly suggest that GABA acts on presynaptic $GABA_B$ receptors to inhibit noradrenaline release, and thus the increase in perfusion pressure, induced by perivascular nerve stimulation. Because taurine has been shown to act as an agonist for $GABA_A$ receptors (El Idrissi et al. 2003), we tested the effects of taurine on cardiovascular function, specifically on blood pressure and heart rate. But unlike GABA, taurine crosses the blood–brain barrier. Thus the effects of taurine on cardiovascular function could be mediated either centrally or peripherally. Based on the results of the present study, we show that taurine injection significantly lowered systemic blood pressure in fully awake rats. Using aortic ring preparation, we further confirmed that taurine acts as a vasodilator. In the aortic preparations taurine-induced vasorelaxation may be due primarily to the activation of $GABA_A$ receptors expressed on smooth muscle. Interestingly however, the chronic supplementation of taurine to rats resulted in gender-specific increase in blood pressure. This increase in blood pressure was observed only in females; males supplemented with taurine did not show any increase in systolic, diastolic, or mean arterial pressure. In both genders however, taurine supplementation caused a significant tachycardia.

Taurine is usually described as a free amino acid and does not participate in protein synthesis. Most animals (but not cats) can synthesize taurine from cysteine in a reaction pathway that involves decarboxylation and multiple oxidations of the sulfhydryl group (Huxtable 1992a, b). However, capacity for endogenous synthesis is limited in humans and the majority of body taurine stores are usually derived from food sources. The neonatal brain contains high levels of taurine (Huxtable 1989, 1992a, b; Sturman 1993). As the brain matures its taurine content declines and reaches stable adult concentrations that are second to those of glutamate, the principal excitatory neurotransmitter in the brain. Taurine levels in the brain significantly increase under stressful conditions (Wu et al. 1998), suggesting that taurine may play a vital role in neuroprotection. A possible mechanism of taurine's neuroprotection lies in its calcium modulatory effects. We have shown that taurine modulates both cytoplasmic and intra-mitochondrial calcium homeostasis. (El Idrissi and Trenkner 1999, 2004a, b). Furthermore, taurine acts as an agonist of $GABA_A$ receptors (Quinn and Harris 1995; Wang et al. 1998; del Olmo et al. 2000; Mellor et al. 2000; El Idrissi et al. 2003; El Idrissi and Trenkner 2004a, b). The effect of taurine on excitable tissues has been well studied with the exception of smooth muscle cells where not much attention has been devoted. Ristori and Verdetti have shown that perfusion of aortic rings from rats with taurine perfusion (1–10 mM) reduced basal tone and had a relaxant effect on rings preconstricted with KCl or norepinephrine (Ristori and Verdetti 1991). This effect was not mediated by endothelium or by muscarinic or adrenoreceptors, and thus probably represented a direct effect of taurine on vascular smooth muscle cells. Although the mechanism of vasorelaxation mediated by taurine was not elucidated, taurine may be minimizing $[Ca^{2+}]_i$ by enhancing the activity of calcium-transporting enzymes. Consistent with this, taurine has been shown to stimulate Ca-ATPase and Na/Ca-antiport in cardiac sarcolemma. In cardiac myocytes, taurine inhibits the rise in $[Ca^{2+}]_i$ induced by beta adrenergic receptor stimulation (Failli et al. 1992). Clearly, more work is required to define the impact of taurine on calcium transport mechanisms in vascular smooth muscle; however, the net effect of these actions appears to be a reduction of $[Ca^{2+}]_i$. The calcium modulatory role of taurine has been well established (El Idrissi and Trenkner 1999; El Idrissi and Trenkner 2004a, b). In this study, we found that the effect of taurine on blood pressure was dependent on the duration of treatment. Acute taurine injection induced hypotension whereas chronic supplementation proved hypertensive but interestingly, only in females. Several clinical studies indicate that chronic oral administration of taurine reduces elevated blood pressures (Meldrum et al. 1994). Therefore, it seems that the effects of taurine on blood pressure are not only gender specific but also depend on the level of blood pressure prior to taurine supplementation. Thus, the findings of this study cast some light on the ability of dietary taurine to regulate blood pressure. The benefit of dietary supplementation may depend on the model examined (Failli et al. 1992; Nara et al. 1978; Nakagawa et al. 1994; Meldrum et al. 1994).

31.2 Methods

31.2.1 Animals

All rats used in this study were 2- to 3-month-old Long–Evans. All rats were housed in groups of three in a pathogen-free room maintained on a 12-h light/dark cycle and given food and water ad libitum. All procedures were approved by the Institutional Animal Care and Use Committee of the College of Staten Island/CUNY, and were in conformity with National Institutes of Health Guidelines. The number of rats sufficient to provide statistically reliable results was used in these studies.

31.2.2 Immunohistochemistry

Cryosections of thoracic aorta were placed onto gelatin-subbed slides. Nonspecific binding sites were blocked using 4% bovine serum albumin (BSA), 2% normal goat serum (NGS), and 0.05% Triton X-100 in 0.01 M phosphate-buffered saline (pH 7.2). Following the blocking step, the slides were rinsed in an antibody dilution cocktail (ABD) consisting of 2% BSA and 1% NGS in 0.01 M PBS. Primary antibodies (Chemicon International) employed were directed against β subunit of the $GABA_A$ receptors (mouse host) and taurine transporter (rabbit host) and diluted 1:500 in ABD. For these studies, the mouse anti-$GABA_A$ receptors were paired with rabbit taurine transporter. The primary antibodies were incubated overnight at 40°C and then unbound antibodies rinsed with ABD. Secondary antibodies were all raised in goat and directed against appropriate primary antibody type. The anti-mouse IgG was conjugated to Alexa Fluor 488 (Invitrogen/Molecular probes) and the anti-rabbit IgG was conjugated to Cy5 (Jackson Immunological). Images were obtained by confocal microscopy (Leica SP2 AOBS). To determine relative changes in protein expression, the gain and offset were identical for all comparisons. Images were reconstructed from Z stack using Leica software (Fig. 31.9a) or Imaris x64 software (Fig. 31.9b).

31.2.3 Blood Pressure Measurements

For indirect blood pressure and heart rate readings, the rats were placed in a chamber at 37°C for 10 min, and then transferred to a standard setup with heating pad and acrylic restrainer, tail cuff, and pulse sensor (CODA monitor, Kent Scientific, Torrington, CT). The tail cuff was connected to a blood pressure monitor that through an arrangement of inlet and outlet valves permitted inflation and deflation of the cuff at a constant rate. The tail cuff pressure was continuously recorded with a solid-state pressure sensor (Kent Scientific). The signals from the pulse and pressure sensors were conveniently amplified and then digitized with an analog–digital board directly on the blood pressure monitor. For each indirect BP determination the inflation and deflation readings were always recorded, as well as the compression interval. The indirect measurements were all performed by the same person, who was kept blind about the purpose of the study. The animals quickly became familiar with the procedure and remained calm within the restrainer. In the rare cases when signs of discomfort were present the procedure was interrupted and the animal was disqualified from the study.

31.2.4 Ex Vivo Measurements of Vascular Response

Long–Evans rats (250–350 g) were anesthetized with intraperitoneal injection of 50 mg/kg pentobarbitol. After opening the chest, descending aorta was removed and immediately placed in a Petri dish containing ice-cold physiological Krebs solution (in mM: 119.9 NaCl, 6 KCl, 15.6 $NaHCO_3$, 1.2 $MgCl_2\cdot6H_2O$, 11.7 glucose, 2.5 $CaCl_2$, $2H_2O$, pH 7.4). Rats were killed by an overdose of anesthetic. The peri-aortic fibroadipose tissue was carefully removed with fine microdissecting forceps and iridectomy scissors paying special attention not to damage the aortic wall. The thoracic aorta was then cut into rings (3 mm in length) with the help of a paper ruler placed under the Petri dish. After dissection, the rings were calibrated at room temperature for 45 min in aerated (95% O_2/5% CO_2) Krebs solution. Each of the rings was suspended horizontally in the same organ chamber (volume 4 ml) between two stainless steel hooks. One of the hooks was fixed to the chamber wall whereas the other was attached to an isometric force transducer (Refined Myograph Systems, Kent Scientific, Torrington, CT). The rings were continuously superfused with prewarmed (37°C), aerated (95% O_2/5% CO_2) Krebs solution. The rings were initially stretched until resting tension reached 2 g and allowed to equilibrate for 15 min. During this period the resting tension was continuously monitored and, if needed, readjusted to 2 g by further stretching. All subsequent measurements represent force generated above this baseline. Mechanical responses to pharmacological stimulation of circular preparations were measured by means of an isometric force transducer (Refined Myograph Systems, Kent Scientific, Torrington, CT) and amplifier (PowerLab, ADinstruments, Colorado Springs, CO) and were visualized using a graphic recorder (ADInstruments LabChart 7, Colorado Springs, CO). Data was sampled at 100 kHz and band-pass filtered between 0.3 Hz and 3 kHz.

31.2.5 Statistical Analysis

Statistical significance was determined by Student's t-test. Each value was expressed as the mean ±SEM. Differences were considered statistically significant when the calculated P value was less than 0.05.

31.3 Results

31.3.1 Taurine Supplementation Increases Systemic Blood Pressure

To investigate the effects of taurine on the regulation of hemodynamics and peripheral resistance, we supplemented taurine (0.05%) to rats in their drinking water for 4 weeks and monitored their blood pressure. We found that the effects of taurine on blood pressure were gender specific. While the blood pressure of adult male rats was not affected by taurine supplementation (Fig. 31.1), females on the other hand showed a significant increase in systolic, diastolic, and mean arterial pressure (Fig. 31.2).

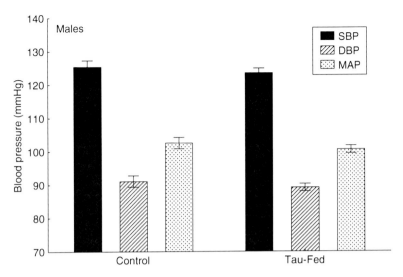

Fig. 31.1 Blood pressure measurements from male rats supplemented with taurine (0.05%) for 4 weeks. Taurine induced a slight but not significant decrease in systolic (SBP), diastolic (DBP), and mean arterial pressure (MAP). Rats were 2 months old and 15 rats were used for this experiment

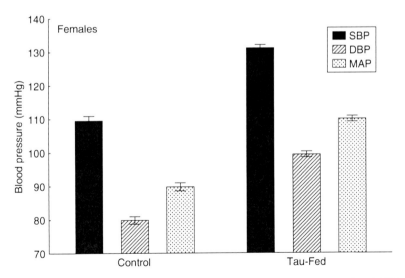

Fig. 31.2 Blood pressure measurements from female rats supplemented with taurine (0.05%) for 4 weeks. Taurine induced a significant ($p < 0.01$) increase in systolic (SBP), diastolic (DBP), and mean arterial pressure (MAP). Rats were 2 months old and 15 rats were used for this experiment

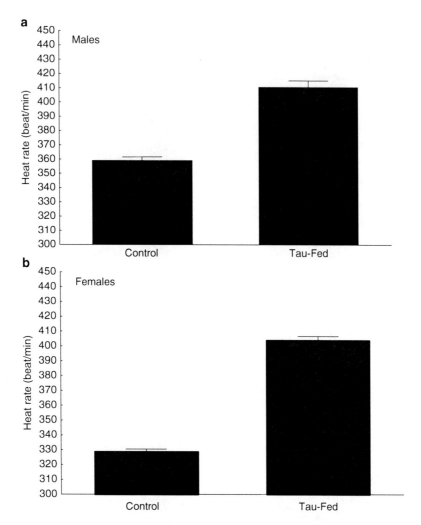

Fig. 31.3 Heart rate measurements from male female rats supplemented with taurine (0.05%) for 4 weeks. Taurine induced a significant ($p < 0.01$) increase in heart rates in both genders. Rats were 2 months old and 15 male and 15 female rats were used for this experiment. These measurements were taken from the same rats used in Figs. 31.1 and 31.2. Recording of heart rate was simultaneous with blood pressure

31.3.2   Taurine Effects on Cardiac Function

While the effects of taurine on peripheral resistance were gender specific with only females being affected, the effect of taurine supplementation on cardiac function was observed in both males and females. In response to 4 weeks of taurine supplementation, rats of both gender showed a drastic increase in heart rate (Fig. 31.3). The increase in heart rate in response to the increase in vascular resistance in female

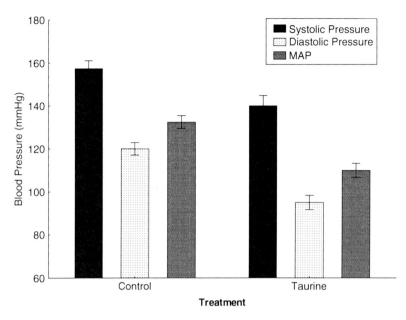

Fig. 31.4 Blood pressure measurements from rats injected with taurine (43 mg/kg). Rats received an intravenous injection of taurine into the tail vein and blood pressure was measured 15 min post injection. Taurine induced a significant ($p<0.01$) lowered systolic (SBP), diastolic (DBP), and mean arterial pressure (MAP). Rats were 2 months old and 18 rats were used for this experiment

could be a mechanism to increase tissue perfusion. However, the tachycardia observed in males in the absence of an effect on vascular resistance may suggest a direct effect of taurine on heart physiology. This could be mediated at the myocardiocytes level (e.g., regulation of calcium homeostasis and contractile properties) or through interaction of taurine with the autonomic nervous system innervating the heart.

31.3.3 Taurine Injection Lowered Peripheral Resistance

To further characterize the vasoactive properties of taurine, we measured blood pressure in response to acute injection of taurine (43 mg/kg). Taurine was injected in the tail vein and blood pressure was monitored 15 min post injection. In response to taurine injection rats showed a drastic decrease in systolic, diastolic, and mean arterial pressure (Fig. 31.4). The observed effects of acute taurine injection on vascular resistance could be mediated by direct interaction of taurine with endothelial or smooth muscle cells, or alternatively through modulation of the nervous control of the cardiovascular function. The effect of taurine on peripheral resistance after acute injection was not gender specific. Both males and females responded by a drop in blood pressure.

31.3.4   Taurine Causes Vasorelaxation of Aortic Rings

Taurine has been shown to regulate the intracellular calcium homeostasis. Within the context of cardiac and smooth muscle physiology, calcium ions are very important in the regulation of the contractility of these muscle cells and thus regulate both peripheral resistance and cardiac output. Taurine also has been shown to be a potent agonist of $GABA_A$ receptors and activation of these receptors has been shown to affect cardiovascular function and peripheral resistance. Thus, we used aortic rings to further elucidate the mechanisms by which taurine mediates its vasoactive properties. Freshly prepared aortic rings from the thoracic aorta were equilibrated for 45 min and isometric contractions were monitored in the presence of bath application of taurine. We first tested the effects of epinephrine (Epi), a well-established and potent vasoconstrictor on the contractibility of the aortic rings (Fig. 31.5). Bath application of 1 μM Epi resulted in a potent vasoconstriction of the aortic rings. The kinetics of the constriction elicited by Epi was consistent with the mechanisms of action of this hormone/neurotransmitter. Upon application of Epi, it took approximately two minutes to observe the beginning of smooth muscle cell contraction, as Epi activates adrenergic receptors that are G protein-coupled receptors. The peak tension was reached 15 min post bath application of Epi (Fig. 31.5). Similar to its slow onset of action, the effects of Epi persisted long after its removal from the bath. The persistent vasoconstriction in the absence of extracellular Epi indicates the continual activation of intracellular pathways and presence of second messenger systems triggered by Epi long after the removal of the agonist (Fig. 31.5).

31.3.5   Taurine Activates GABA_A Receptors in Aortic Preparations

In search of the potential cellular mechanism by which taurine mediates vasodilation, we used aortic rings and pharmacologically characterized the vasoactive properties of taurine on smooth muscle cells. Addition of taurine (10 μM) to the tissue bath resulted in a vasorelaxation of the aortic rings (Fig. 31.6). The onset and offset of taurine action were much faster than those observed with Epi application, suggesting that taurine may activate an ionotropic receptor.

Since taurine is a $GABA_A$ receptor agonist, we tested the effects of taurine in the presence of $GABA_A$ receptor antagonist to determine if the effects were mediated through activation of $GABA_A$ receptors. Application of taurine in the presence of picrotoxin resulted in a rapid vasoconstriction as measured by the increased tension developed by the aortic ring (Fig. 31.7). The onset of vasoconstriction was very rapid within 30 s and reached its plateau within 2 min. 90% of the tension was produced within 1 min of taurine and picrotoxin application (Fig. 31.7). Removing picrotoxin and taurine from the bath resulted in a relatively rapid vasodilation and the tension returned to baseline levels (Fig. 31.7) relatively quickly.

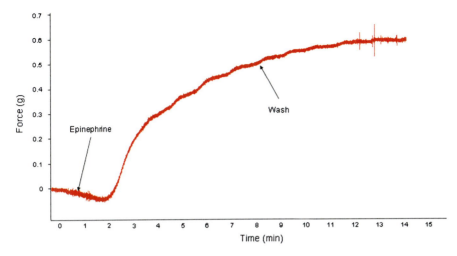

Fig. 31.5 Pharmacological response of aortic rings to epinephrine (Epi). Addition of 1 μM Epi to the tissue bath resulted in a delay and sustained contraction of smooth muscle cells of the aorta. After removal of Epi from the bath, the rings remained constricted for a long time indicating the slow onset and offset action of Epi

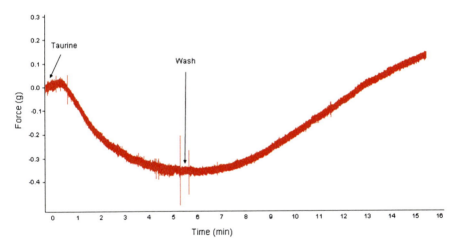

Fig. 31.6 Pharmacological response of aortic rings to taurine. Addition of 10 μM taurine to the tissue bath resulted in a rapid relaxation of smooth muscle cells of the aorta. After removal of taurine from the bath, the rings regained their pre-taurine contraction state relatively quickly, indicating that the fast onset and offset action of taurine could be mediated through an ionotropic receptor

Picrotoxin is a $GABA_A$ receptor competitive antagonist and competes with GABA to the binding site in an open-channel state. Since taurine is a $GABA_A$ receptor agonist, application of taurine would open a $GABA_A$ receptor and in the presence of picrotoxin, the effects of taurine on $GABA_A$ receptor would be antagonized. Interestingly however, application of picrotoxin alone resulted in a vasoconstriction (Fig. 31.8). The kinetics of tension development by the aortic ring in the presence of

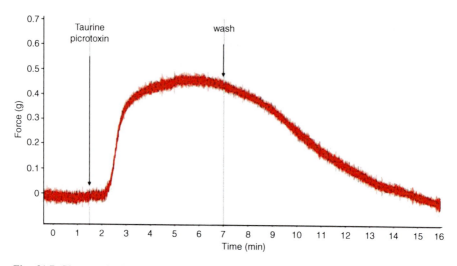

Fig. 31.7 Pharmacological response of aortic rings to taurine in the presence of pictrotoxin. Addition of 10 μM taurine in the presence of 5 μM picrotoxin to the tissue bath resulted in a rapid vasoconstriction of smooth muscle cells of the aorta to reach a plateau within 2–3 min. After removal of taurine and picrotoxin from the bath, the rings regained their pre-taurine contraction state relatively quickly, indicating that taurine mediates its actions on the smooth muscle cells through activation of GABA$_A$ receptors

picrotoxin alone was different than that in combination with taurine. When picrotoxin alone was added to the bath, the tension was slow to develop (max tension was produced within 5 min) and the amount of tension was lower than when picrotoxin was combined with taurine (compare Figs. 31.7 and 31.8; 0.5 g vs. 0.3 g). We infer from these data the following: Taurine exerts its vasoactive properties through activation of the GABA$_A$ receptors with subsequent hyperpolarization of smooth muscle cells and relaxation. The fact that picrotoxin application resulted in a relaxation of the muscularis and a drop in tension suggests that there is a tonic activation of GABA$_A$ receptors that could be mediated through circulating levels of GABA or taurine. However, the aortic rings are in an incubation chamber with controlled environment. The source of taurine or GABA in such a milieu could arise from release of these substances from the smooth muscle tissue itself. Smooth muscle cells have been shown to contain taurine (Lobo et al. 2001). Therefore, taurine or potentially GABA could be released from the tissue and causes a tonic relaxation of the smooth muscle cell within the muscularis of the aorta.

31.3.6 GABA$_A$ Receptors Are Expressed in the Aorta

As further evidence for the activation of GABA$_A$ receptors by taurine, we examined immunohistochemically the presence of GABA$_A$ receptors on the aortic wall. We found that the muscularis contains high levels of immunoreactivity for the β subunit

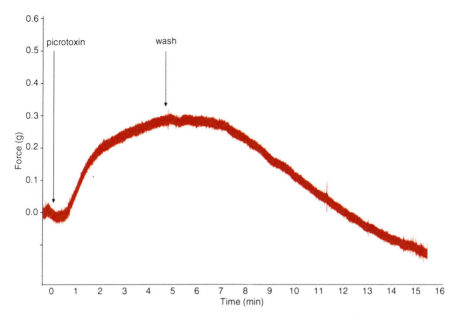

Fig. 31.8 Pharmacological response of aortic rings to pictrotoxin. Addition of 5 μM picrotoxin to the tissue bath resulted in a rapid vasoconstriction of smooth muscle cells of the aorta to reach a plateau within 4–5 min. The tension developed by picrotoxin alone was much smaller than when taurine was present. After removal of picrotoxin from the bath, the rings regained their pre-taurine contraction state relatively quickly, indicating a tonic activation of GABA$_A$ receptors in the smooth muscle cells of the aorta

of the GABA$_A$ receptors (Fig. 31.9). Most GABA$_A$ immunoreactivity was localized to the outer layers of the muscularis of the aorta (Fig. 31.9). GABA$_A$ immunoreactivity was found in both cerebral and extracerebral blood vessel vasculature (Fig. 31.9a, b). We also examined the presence of taurine transporter on the wall of blood vessels. Interestingly, we found that the localization of taurine transporter in both cerebral and extracerebral vasculature is confined mostly to the endothelium layer (Fig. 31.9a, b). Low level of taurine transporter immunoreactivity was observed on the smooth muscle cell of blood vessels.

31.4   Discussion

GABA has been shown to play an important role in the modulation of cardiovascular function and hemodynamics. GABA mediates these actions by acting not only within the central nervous system but also within peripheral tissues. Since taurine is an agonist for GABA$_A$ receptors, we sought to determine the vasoactive properties of taurine, presumably through activation of GABA$_A$ receptors.

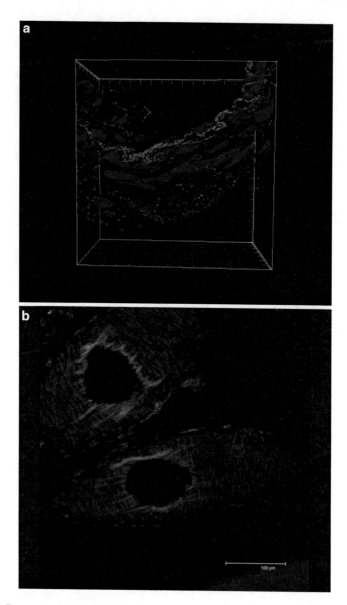

Fig. 31.9 Representative images of GABA$_A$ receptors and taurine transporter immunoreactivity in the wall of the aorta (**a**) and middle cerebral artery. (**b**) The image in (**a**) was reconstructed from a Z stack obtained with a confocal microscope and processed using Imaris software. (**b**) is a maximum projection of a Z stack images. GABA$_A$ immunoreactivity is localized to the outer muscularis, whereas taurine transporter is expressed in the apical side, presumably within the endothelium layer. Images captured with a 60× oil objective

Taurine (2-aminoethanesulfonic acid) is a sulfur-containing amino acid. It is one of the most abundant free amino acids in many excitable tissues, including the brain, skeletal, and cardiac muscles. Physiological actions of taurine are widespread and include bile acid conjugation, detoxification, membrane stabilization, osmoregulation, neurotransmission, and modulation of cellular calcium levels (Foos and Wu 2002; Lombardini 1985; Saransaari and Oja 2000; Schaffer et al. 2000; Solis et al. 1988). Furthermore, taurine plays an important role in modulating glutamate and GABA neurotransmission (El Idrissi and Trenkner 1999; El Idrissi and Trenkner 2004a, b; Militante and Lombardini 1998). We have previously shown that taurine prevents excitotoxicity in vitro primarily through modulation of intracellular calcium homeostasis (El Idrissi and Trenkner 1999). In neurons, calcium plays a key role in mediating glutamate excitotoxicity.

Outside of the central nervous system, taurine also is essential during developmental processes. Taurine is added to milk formula and in solution for parenteral nutrition of premature babies to prevent retinal degeneration and cholestasis (Huxtable 1992a, b; Lourenco and Camilo 2002). Taurine is found at high concentrations in pancreatic islets (Huxtable 1992a, b). Taurine is able to prevent pancreatic alterations induced by gestational malnutrition, especially low-protein diet (Boujendar et al. 2002; Cherif et al. 1996; Dahri et al. 1991; Merezak et al. 2001).

The antihypertensive effects of taurine have been demonstrated in several experimental models (Fujita and Sato 1984; Harada et al. 2000; Nara et al. 1978). Studies in vitro showed that taurine relaxed pre-constricted rabbit ear artery (Franconi et al. 1982), rat aorta (Ristori and Verdetti 1991), and rat mesenteric artery (Li et al. 1996). Thoracic aortic rings isolated from rats that were chronically given beta-alanine to deplete internal taurine showed enhanced contractile responses to norepinephrine and high potassium, and reduced relaxant responses to sodium nitroprusside and acetylcholine (Abebe and Mozaffari 2003). Thoracic aortic rings isolated from rats that were chronically given taurine showed reduced contractile responses to norepinephrine and high potassium nonspecifically (Abebe and Mozaffari 2000). These experiments suggest that taurine plays an important role in the maintenance and regulation of vascular tone in normal and pathological situations.

We found that acute injection of taurine (43 mg/kg) to adult rats resulted in a significant decrease in peripheral resistance with no effects on heart rate (Fig. 31.4). We further confirmed the vasorelaxant action of taurine using aortic ring preparations. Addition of taurine to aortic ring preparation resulted in a rapid decrease in tension attributed to the relaxation of the arterial muscularis (Fig. 31.6). The effects of taurine on smooth muscle were mediated through activation of $GABA_A$ receptors. Bath application of picrotoxin, a $GABA_A$ receptor antagonist, resulted in a vasoconstriction of the aortic rings (Fig. 31.7). Interestingly, picrotoxin alone induced a constriction of the aortic ring in the absence of exogenously added taurine (Fig. 31.8), suggesting a tonic activation of $GABA_A$ receptors by circulating either taurine or GABA. Picrotoxin is a $GABA_A$ receptor competitive antagonist and binds to $GABA_A$ in the open state. The finding that picrotoxin caused a vasoconstriction in the absence of exogenously added agonist (GABA or taurine), coupled with the kinetic of constriction, suggests that there is a tonic low level of activation of $GABA_A$ receptors. This could be mediated by release of taurine or GABA from the aortic ring tissue under stretch conditions. Additionally, we found that the endothelial cells express high levels of taurine transporters and $GABA_A$

receptors (Fig. 31.9). The presence of high-level expression of taurine transporter in the endothelial cells suggests a high-affinity uptake mechanism for taurine by the endothelial cells. Once taurine is removed from the plasma it would activate the $GABA_A$ receptors that are abundantly expressed on smooth muscle cells of aortic wall (Fig. 31.9).

Peripheral resistance within the large arteries is predominantly controlled by the level of tonic activity of the sympathetic nervous system and the level of adrenergic receptor activation. Thus one could suggest an antagonistic system to the sympathetic innervation of the vasculature. While the sympathetic nervous system causes vasoconstriction proportional to the level of activation of adrenergic receptors, the GABAergic system opposes that by mediating vasodilation. The GABAergic system mediates its vasoactive properties through activation of $GABA_A$ receptors expressed throughout the length of the muscularis of the vasculature. $GABA_A$ receptors can be activated either with taurine or GABA, both of which are found at relatively high levels in the plasma.

The finding that acute taurine had an opposite effect on peripheral resistance than chronic suggests that the chronic supplementation of taurine in drinking water may cause alterations to the mechanisms responsible for taurine regulation of blood pressure and peripheral resistance. Consistent with this observation, we found that the effects of taurine on the GABAergic system in the brain are dependent on the duration of treatment.

We have previously shown that taurine-fed mice have reduced expression of $GABA_A$ receptors in the hippocampus (El Idrissi 2006). We suggested that a down-regulation of $GABA_A$ receptor expression was due to the sustained interaction of taurine with $GABA_A$ receptors which causes a change in subunit composition of the $GABA_A$ receptors and concomitant decrease in the efficacy of the inhibitory system (El Idrissi 2006). Similar observations were noted in peripheral tissues, mainly the pancreas (El Idrissi et al. 2008). Therefore, we suggest that a potential decrease in $GABA_A$ receptor expression in the muscularis aortic wall in response to chronic treatment with taurine would result in a reduced efficiency of vasodilative properties of GABA on peripheral resistance. This would lead to hypertensive effect when taurine is chronically supplemented to rats. However, the hypertensive properties of taurine were gender specific. Only females showed a significant decrease in blood pressure when chronically fed taurine. The gender-specific effects of taurine on peripheral resistance are intriguing are require further investigation of the mechanism of action. We suggest that $GABA_A$ could mediate this gender specificity. Steroid hormones have been shown to act as allosteric modulators of $GABA_A$ receptors. Thus, the gender-specific hormonal phenotype could underlie the selective modulation of $GABA_A$ receptor conductance by taurine. This however remains to be elucidated. Alternatively, taurine may mediate its vasoactive properties, in addition to activating $GABA_A$ receptors, through other known vasoactive substances. These include prostaglandins, nitric oxide, opening of K+ channels, reduced Ca^{2+} availability or yet other mechanisms that could be involved in the vasorelaxant effects of taurine (Félétou and Vanhoutte 2006).

31.5 Conclusion

In summary, this study shows that taurine could have both hypo- and hypertensive properties. If chronically administered, taurine induces hypertension in female and tachycardia in both female and male rats. Intravenous injection of taurine causes a rapid decrease in blood pressure. This hypotensive effect was mediated through activation of $GABA_A$ receptors expressed on the muscularis of the aorta and cerebral blood vessels. Taurine, therefore, could act as a vasorelaxant when acutely injected. Furthermore, this study shows that GABA plays an important role in the regulation of cardiovascular function both centrally and peripherally. Drugs that target $GABA_A$ in the CNS would affect peripheral resistance in addition to the intended central effects.

Acknowledgments This study was funded by a fellowship to E.O. from CUNY Summer Undergraduate Research Program (C-SURP) and CSI.

References

Abebe W, Mozaffari MS (2000) Effects of chronic taurine treatment on reactivity of the rat aorta. Aminosan 19:615–623

Abebe W, Mozaffari MS (2003) Taurine depletion alters vascular reactivity in rats. Can J Physiol Pharmacol 81:903–909

Boujendar S, Reusens B, Merezak S et al (2002) Taurine supplementation to a low protein diet during foetal and early postnatal life restores a normal proliferation and apoptosis of rat pancreatic islets. Diabetologia 45:856–866

Bowery GN (1993) GABAB receptor pharmacology. Annu Rev Pharmacol 33:109–147

Cherif H, Reusens B, Dahri S et al (1996) Stimulatory effects of taurine on insulin secretion by fetal rat islets cultured in vitro. J Endocrinol 151:501–506

Dahri S, Snoeck A, Reusens-Billen B et al (1991) Islet function in offspring of mothers on low-protein diet during gestation. Diabetes 40(Suppl 2):115–120

del Olmo N, Bustamante J, del Rio RM, Soli J (2000) Taurine activates GABA(A) but not GABA(B) receptors in rat hippocampal CA1 area. Brain Res 864:298–307

Elliott CAK, Hobbiger F (1959) Gamma aminobutyric acid: circulatory and respiratory effects in different species: re-investigation of the anti-strychine action in mice. J Physiol 146:70–84

El Idrissi A (2006) Taurine and brain excitability. Adv Exp Med Biol 583:315–322. PMID: 17153616

El Idrissi A, Trenkner E (1999) Growth factors and taurine protect against excitotoxicity by stabilizing calcium homeostasis and energy metabolism. J Neurosci 19:9459–9468

El Idrissi A, Trenkner E (2004b) Taurine as a modulator of excitatory and inhibitory neurotransmission. Neurochem Res 29:189–197

El Idrissi A, Trenkner E (2004a) Taurine as a modulator of excitatory and inhibitory neurotransmission. Neurochem Res 29:189–197

El Idrissi A, Messing J, Scalia J, Trenkner E (2003) Prevention of epileptic seizures through taurine. Adv Exp Med Biol 526:515–525

El Idrissi A, Boukarrou L, L'Amoreaux W (2008) Taurine supplementation and pancreatic remodeling. Adv Exp Med Biol 643(7):353–358. PMID:1923916

Failli P, Fazzini A, Franconi F et al (1992) Taurine antagonizes the increases in intracellular calcium concentration induced by α-adrenergic stimulation in freshly isolated guinea-pig cardiomyocytes. J Mol Cell Cardiol 24:1253–1265

Félétou M, Vanhoutte PM (2006) Endothelium-derived hyperpolarizing factor where are we now? Arterioscler Thromb Vasc Biol 26:1215–1225

Foos TM, Wu JY (2002) The role of taurine in the central nervous system and the modulation of intracellular calcium homeostasis. Neurochem Res 27:21–26

Franconi F, Giotti A, Manzini S, Martini F, Stendardi I, Zilletti L (1982) The effect of taurine on high potassium- and noradrenaline-induced contraction in rabbit ear artery. Br J Pharmacol 75:605–612

Fujita T, Sato Y (1984) The antihypertensive effect of taurine in DOCA-salt rats. J Hypertens Suppl 2:S563–S565

Harada H, Kitazaki K, Tsujino T, Watari Y, Iwata S, Nonaka H, Hayashi T (2000) Oral taurine supplementation prevents the development of ethanol-induced hypertension in rats. Hypertens Res 23:277–284

Huxtable RJ (1989) Taurine in the central nervous system and the mammalian action actions of taurine. Prog Neurobiol 32:471–533

Huxtable RJ (1992a) The physiological actions of taurine. Physiol Rev 72:101–163

Huxtable RJ (1992b) Physiological actions of taurine. Physiol Rev 72:101–163

Li N, Sawamura M, Nara Y, Ikeda K, Yamori Y (1996) Direct inhibitory effects of taurine on norepinephrine-induced contraction in mesenteric artery of stroke-prone spontaneously hypertensive rats. Adv Exp Med Biol 403:257–262

Lobo MV, Alonso FJ, Latorre A, del Río RM (2001) Immunohistochemical localization of taurine in the rat ovary, oviduct, and uterus. J Histochem Cytochem 49(9):1133–1142

Lombardini JB (1985) Effects of taurine on calcium ion uptake and protein phosphorylation in rat retinal membrane preparations. J Neurochem 45:268–275

Lourenco R, Camilo ME (2002) Taurine: a conditionally essential amino acid in humans? An overview in health and disease. Nutr Hosp 17:262–270

Manzini S, Maggi CA, Meli A (1985) Inhibitory effect of GABA on sympathetic neurotransmission in rabbit ear artery. Arch Int Pharmacodyn Ther 273:100–109

Meldrum MJ, Tu R, Patterson T et al (1994) The effect of taurine on blood pressure, and urinary sodium, potassium and calcium excretion. In: Huxtable R, Michalk DV (eds) Taurine in health and disease. Plenum Press, New York, pp 207–215

Mellor JR, Gunthorpe MJ, Randall AD (2000) The taurine uptake inhibitor guanidinoethyl sulphonate is an agonist at gamma-aminobutyric acid (A) receptors in cultured murine cerebellar granule cells. Neurosci Lett 286:25–28

Merezak S, Hardikar AA, Yajnik CS et al (2001) Intrauterine low protein diet increases fetal beta-cell sensitivity to no and il-1 beta: The protective role of taurine. J Endocrinol 171:299–308

Militante JD, Lombardini JB (1998) Pharmacological characterization of the effects of taurine on calcium uptake in the rat retina. Amino Acids 15:99–108

Nakagawa M, Takeda K, Yoshitomi T et al (1994) Antihypertensive effect of taurine on salt-induced hypertension. In: Huxtable R, Michalk DV (eds) Taurine in health and disease. Plenum Press, New York, pp 197–206

Nara Y, Yamori Y, Lovenberg W (1978) Effect of dietary taurine on blood pressure in spontaneously hypertensive rats. Biochem Pharmacol 27:2689–2692

Quinn MR, Harris CL (1995) Tautine allosterically inhibits binding of [35S]-tbutylbicyclophosphorothionate (TBPS) to rat brain synaptic membranes. Neuropharmacol 34:1607–1613

Ristori MT, Verdetti J (1991) Effects of taurine on rat aorta in vitro. Fundam Clin Pharmacol 5:245–258

Saransaari P, Oja SS (2000) Taurine and neural cell damage. Amino Acids 19:509–526

Schaffer S, Takahashi K, Azuma J (2000) Role of osmoregulation in the actions of taurine. Amino Acids 19:527–546

Solis JM, Herranz AS, Herreras O et al (1988) Does taurine act as an osmoregulatory substance in the rat brain? Neurosci Lett 91:53–58

Sturman JA (1993) Taurine in development. Physiol Rev 73:119–147

Takahashi H, Tiba M, Iino M, Takayasu T (1955) The effect of gamma-aminobutyric acid on blood pressure. Jpn J Physiol 5(4):334–341

Wu JY, Tang XW, Schloss JV, Faiman MD (1998) Regulation of taurine biosynthesis and its physiological significance in the brain. Adv Exp Med Biol 442:339–345

Wang DS, Xu TL, Pang ZP, Li JS, Akaike N (1998) Taurine-activated chloride currents in the rat sacral dorsal commissural neurons. Brain Res 792:41–47

Chapter 32
Synergistic Effects of Taurine and L-Arginine on Attenuating Insulin Resistance Hypertension

Ying Feng, Jitao Li, Jiancheng Yang, Qunhui Yang, Qiufeng Lv, Yongchao Gao, and Jianmin Hu

Abstract To elucidate the synergistic effects of taurine and L-arginine on hypertension, 25% fructose were administered to male Wistar rats for 3 months to establish insulin resistance hypertensive models. Rats with the systolic blood pressure (SBP) higher than 150 mmHg were considered as model rats. Forty-two model rats were randomly divided into six groups and administered with 3% taurine, 2.7% taurine + 0.3% L-arginine, 2.1% taurine + 0.9% L-arginine, 1.5% taurine + 1.5% L-arginine and 3% L-arginine in drinking water respectively. The results showed that coadministration of taurine (1.5%) and L-arginine (1.5%) could bring the levels of SBP, blood glucose, and insulin down to normal levels after 4 weeks. The thickness of blood vessels increased significantly in model group, which could be reversed by taurine and L-arginine. Serum NO, cGMP, and ET levels could return to normal levels. These data indicated that both taurine and L-arginine could ameliorate vascular remodeling and showed obvious antihypertensive effects, and taurine (1.5%) and L-arginine (1.5%) in the drinking water showed a better result in the cure of hypertension.

Abbreviations

SBP	Systolic blood pressure
Ins	Insulin
ET	Endothelin
cGMP	Cyclic guanosine monophosphate

Y. Feng • J. Li • J. Yang • Q. Yang • Q. Lv • Y. Gao • J. Hu(✉)
College of Animal Science & Veterinary Medicine, Shenyang
Agricultural University, Shenyang 110866, P.R. China
e-mail: hujianmin59@163.com

A. El Idrissi and W.J. L'Amoreaux (eds.), *Taurine 8*, Advances in Experimental
Medicine and Biology 775, DOI 10.1007/978-1-4614-6130-2_32,
© Springer Science+Business Media New York 2013

32.1 Introduction

Hypertension is a systemic disease associated with heart, blood vessels, brain, kidneys, and other organs. Taurine, a sulfur-containing amino acid, which is present in high concentrations in mammalian plasma and cells, has been reported to have antihypertensive effects in both rat and human studies (Militante and Lombardini 2002; Hu et al. 2009; Rahman et al. 2011; Harada et al. 2004; Anuradha and Balakrishnan 1999). Taurine can scavenge oxygen free radicals, act as a membrane stabilizer, increase renal kallikrein synthesis, blockade angiotensin II, and also have a sympatholytic effect (Hagar et al. 2006). L-Arginine is a precursor to nitric oxide which dilates blood vessels and lowers blood pressure. Dietary L-arginine supplementation has been proposed to reverse endothelial dysfunction under certain conditions, including hypercholesterolemia, coronary heart disease, and some forms of animal hypertension. Chronic oral administration of L-arginine prevented the increase of blood pressure induced by sodium chloride loading in salt-sensitive rats. In addition, creatinine clearance was improved and fasting blood sugar was decreased by the addition of L-arginine (Siani et al. 2000). However, we still have no idea about the synergistic antihypertensive effects of these two amino acids. In the present study, 25% fructose was administered in the drinking water to establish insulin resistance hypertensive models, and taurine and L-arginine were administered alone or together in the drinking water in different proportions.

32.2 Methods

32.2.1 Animals

Six-week-old male Wistar rats weighing 140–180 g were maintained under a controlled condition of light (12 h of light, 12 h of dark) and temperature ($23 \pm 2°C$), and were given free access to food (commercial standard rat chow) and water. To establish insulin resistance hypertensive models, rats were given 25% fructose in drinking water for 12 weeks.

32.2.2 Experimental Design

Rats with the systolic blood pressure (SBP) higher than 150 mmHg were considered as hypertensive model rats. Forty-two hypertensive rats were randomly divided into six groups, and were all given drinking water administered with 25% fructose for four consecutive weeks. The rats were designated as model group (M) and treatment groups (T) with subscript 1, 2, and 3. That is, treatment groups were coadministered with 25% fructose and 3% taurine (T1), 2.7% taurine+0.3% L-arginine (T2), 2.1% taurine+0.9% L-arginine (T3), 1.5% taurine+1.5% L-arginine (T4), and 3% L-arginine (T5) in drinking water, respectively ($n=7$).

32.2.3   Measurement of Systolic Blood Pressure

Systolic blood pressure (SBP) was measured weekly in conscious rats with a noninvasive blood pressure measurement system (BP-98A softron, Japan) using the tail-cuff method. Rats were placed in a plastic holder mounted on a thermostatically controlled warm plate that was maintained at 37°C during measurements. An average value from three blood pressure readings (that differed by no more than 2 mmHg) was determined for each animal after they became acclimated to the environment. All blood pressure measurements were made between 09:00 and 12:00 h once a week.

32.2.4   Histological Analysis

Thoracic aortas were fixed in 40 g/L paraformaldehyde for 24 h at room temperature, followed by dehydration in 300 g/L sucrose overnight at 4°C. After embedding with OCT, the tissues were frozen in liquid nitrogen, then stored at −80°C. Cryostat-cut sections (6 µm) were picked up onto gelatin-coated glass slides and then stained with hematoxylin and eosin.

32.2.5   Sample Collection and Biochemical Analysis

Blood samples were collected, and serum was separated by centrifuging at 1,500 rpm for 15 min at 4°C, stored at −20°C. Serum nitric oxide (NO) levels were detected by nitrate reductase method according to the instruction of reagent kits. Insulin (Ins), endothelin (ET), and cyclic guanosine monophosphate (cGMP) were detected by enzyme-linked immunosorbent assay. All the reagent kits were purchased from Nanjing Jiancheng Bioengineering Institute (China).

32.2.6   Statistical Analysis

Data were presented as the mean±SD and significant differences were determined by Duncan's multiple range tests using SPSS 16.0 statistical analysis software. p values less than 0.05 were considered as significant.

32.3 Results

32.3.1 Assessments of Systolic Blood Pressure

As expected, consumption of fructose-containing water resulted in significant increase in systolic blood pressure (SBP) after 12 weeks. SBP of treatment groups showed a gradual downward tendency during the trial period. Only after 1 week treatment, SBP levels had been significantly lower than M group ($p<0.05$). Four weeks later, SBP of treatment groups were all approaching to normal levels, especially in T4 group, which showed no significant difference compared with control group ($p>0.05$) (Table 32.1).

32.3.2 Detection of Blood Glucose

As shown in Table 32.2, fructose-fed rats had higher blood glucose levels compared with the control group. Blood glucose levels showed a gradual downward tendency in the treatment groups. Four weeks later, there were no significant differences between T4 group and control group.

32.3.3 Serum Insulin Levels

The results of Fig. 32.1 showed that serum insulin levels in model group significantly increased compared with the control group, and the insulin levels of treatment groups were all significantly lower than the model group ($p<0.05$).

Table 32.1 Effect of taurine and L-arginine on systolic blood pressure (SBP) of insulin resistance hypertensive rats

Groups	Before treatment	Treatment			
		1 week	2 weeks	3 weeks	4 weeks
Control group	128.20 ± 3.03^a	126.71 ± 2.54^a	127.34 ± 1.59^a	129.20 ± 3.32^a	128.20 ± 1.94^a
Model group	158.29 ± 2.56^b	156.43 ± 4.07^c	157.86 ± 6.09^c	155.14 ± 4.41^d	158.14 ± 3.89^c
T1 group	152.86 ± 8.23^b	149.86 ± 8.23^b	145.71 ± 4.61^b	141.71 ± 1.98^c	135.71 ± 2.06^b
T2 group	155.86 ± 6.09^b	151.29 ± 4.54^b	147.14 ± 5.24^b	$139.29 \pm 4.99^{b,c}$	134.71 ± 3.59^b
T3 group	156.71 ± 8.65^b	149.57 ± 3.15^b	142.14 ± 2.12^b	$138.29 \pm 2.3c^{b,c}$	134.14 ± 2.41^b
T4 group	156.43 ± 3.99^b	148.43 ± 2.76^b	143.14 ± 4.02^b	135.57 ± 3.36^b	129.14 ± 3.07^a
T5 group	155.43 ± 3.90^b	149.86 ± 3.34^b	145.29 ± 1.80^b	141.57 ± 1.99^c	136.57 ± 2.82^b

The rats were designated as model group (M) and treatment groups (T) with subscript 1, 2, and 3. That is, treatment groups were coadministered with 25% fructose and 3% taurine (T1), 2.7% taurine + 0.3% L-arginine (T2), 2.1% taurine + 0.9% L-arginine (T3), 1.5% taurine + 1.5% L-arginine (T4), and 3% L-arginine (T5) in drinking water, respectively, for four consecutive weeks. SBP were measured once a week. Data are the mean ± SD ($n=7$). Values with different letters on the shoulder represent significant differences ($p<0.05$)

Table 32.2 Effect of taurine and L-arginine on blood glucose of insulin resistance hypertensive rats

Groups	Before treatment	Treatment			
		1 week	2 weeks	3 weeks	4 weeks
Control group	$3\,03 \pm 0.35^a$	3.33 ± 0.28^a	3.43 ± 0.37^a	3.06 ± 0.39^a	3.16 ± 0.37^a
Model group	5.41 ± 0.50^b	5.21 ± 0.33^b	5.63 ± 0.28^c	5.79 ± 0.21^d	6.03 ± 0.38^d
Tl group	5.79 ± 0.60^b	5.34 ± 0.49^b	4.87 ± 0.51^b	4.61 ± 0.38^c	4.06 ± 0.28^b
T2 group	5.81 ± 0.95^b	5.31 ± 0.95^b	4.86 ± 0.85^b	$4.39 \pm 0.67^{b,c}$	3.94 ± 0.54^b
T3 group	6.02 ± 0.39^b	5.63 ± 0.39^b	$5.13 \pm 0.39^{b,c}$	$4.43 \pm 0.39^{b,c}$	4.13 ± 0.39^b
T4 group	5.67 ± 0.62^b	5.23 ± 0.55^b	4.57 ± 0.35^b	4.06 ± 0.28^b	3.214 ± 0.35^a
T5 group	5.71 ± 0.80^b	5.18 ± 0.65^b	4.84 ± 0.59^b	$4.53 \pm 0.59^{b,c}$	4.19 ± 0.46^b

The rats were designated as model group (M) and treatment groups (T) with subscript 1, 2, and 3. That is, treatment groups were given 25% fructose coadministered with 3% taurine (T1), 2.7% taurine + 0.3% L-arginine (T2), 2.1% taurine + 0.9% L-arginine (T3), 1.5% taurine + 1.5% L-arginine (T4), and 3% L-arginine (T5) in drinking water, respectively, for four consecutive weeks. Measurements of blood glucose were made once a week. Data are the mean $\pm$ SD ($n = 7$). Values with different letters on the shoulder represent significant differences ($p < 0.05$)

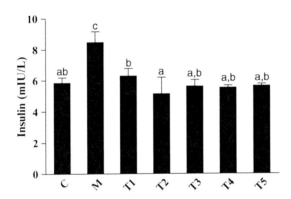

Fig. 32.1 Effect of taurine and L-arginine on serum insulin levels of insulin resistance hypertensive rats. The rats were designated as model group (*M*) and treatment groups (*T*) with subscript 1, 2, and 3. That is, treatment groups were given 25% fructose coadministered with 3% taurine (*T1*), 2.7% taurine + 0.3% L-arginine (*T2*), 2.1% taurine + 0.9% L-arginine (*T3*), 1.5% taurine + 1.5% L-arginine (*T4*), and 3% L-arginine (*T5*) in drinking water, respectively, for four consecutive weeks. Measurements were made after 4 weeks treatment. Data are the mean $\pm$ SD ($n = 7$). Values with different letters on the shoulder represent significant differences ($p < 0.05$)

32.3.4   Serum NO/cGMP and ET

As shown in Fig. 32.2, NO levels of model group decreased significantly compared with the control group ($p < 0.05$). After treatment, T3 and T4 groups showed no significant difference compared with control group ($p > 0.05$). cGMP levels showed no significant difference among groups, but it could be seen that the treatment groups have higher cGMP levels than the model group, and cGMP

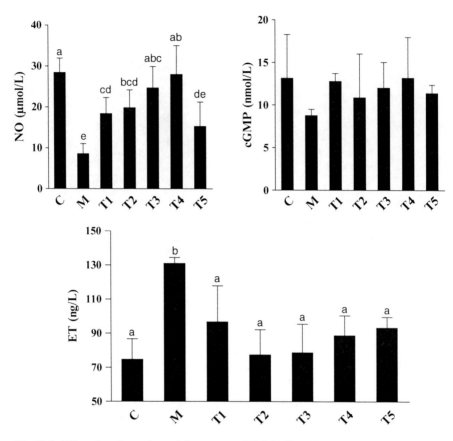

Fig. 32.2 Effect of taurine and L-arginine on serum NO/cGMP and ET levels of insulin resistance hypertensive rats. The rats were designated as model group (*M*) and treatment groups (*T*) with subscript 1, 2, and 3. That is, treatment groups were given 25% fructose coadministered with 3% taurine (*T1*), 2.7% taurine + 0.3% L-arginine (*T2*), 2.1% taurine + 0.9% L-arginine (*T3*), 1.5% taurine + 1.5% L-arginine (*T4*), and 3% L-arginine (*T5*) in drinking water, respectively, for four consecutive weeks. Measurements were made after 4 weeks treatment. Data are the mean ± SD (*n* = 7). Values with different letters on the shoulder represent significant differences ($p < 0.05$)

level of T4 group was very close to the control group. Serum ET level increased significantly in the model group, and all the treatment groups showed no significant difference compared with the control group. ET levels of T2 and T3 groups were close to the control group.

32.3.5   Histological Analysis

As shown in Fig. 32.3, the abnormal proliferation of vascular smooth muscle cells were observed in model group, and the thickness of blood vessels wall significantly increased in model group, which could be reversed by taurine and L-arginine.

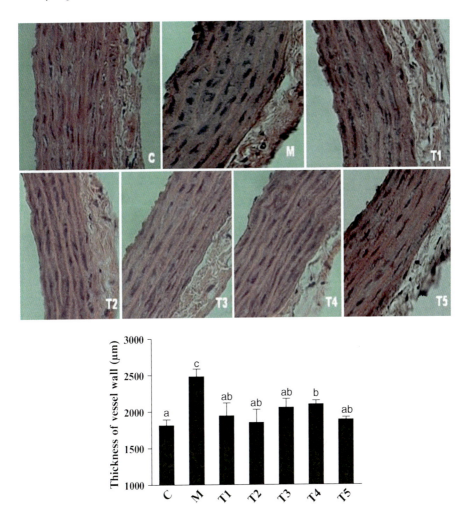

Fig. 32.3 Effect of taurine and L-arginine on vascular remodeling of insulin resistance hypertensive rats. The rats were designated as model group (*M*) and treatment groups (*T*) with subscript 1, 2, and 3. That is, treatment groups were given 25% fructose coadministered with 3% taurine (*T1*), 2.7% taurine + 0.3% L-arginine (*T2*), 2.1% taurine + 0.9% L-arginine (*T3*), 1.5% taurine + 1.5% L-arginine (*T4*), and 3% L-arginine (*T5*) in drinking water, respectively, for four consecutive weeks. Measurements were made after 4 weeks treatment. Data are the mean ± SD (*n* = 7). Values with different letters on the shoulder represent significant differences ($p < 0.05$). Images were acquired with 200× magnification

32.4 Discussion

Taurine and L-arginine have both been confirmed to have antihypertensive effects in the previous studies. Treatment with 2% taurine in drinking water prevented the blood pressure elevation and attenuated the hyperinsulinemia in fructose-fed rats. Taurine supplementation could be beneficial in circumventing metabolic alterations in insulin resistance (Anuradha and Balakrishnan 1999). Numerous studies, though not uniformly, demonstrated a beneficial effect of acute and chronic L-arginine supplementation on endothelium-derived nitric oxide production and endothelial function, and L-arginine has been shown to reduce blood pressure in some forms of experimental hypertension (Gokce 2004). Our results showed that coadministration of the two amino acids (1.5% taurine + 1.5% L-arginine) in drinking water could reverse the SBP of insulin-resistance hypertensive rats to normal levels.

Insulin resistance appears to be a common feature and a possible contributing factor to hypertension. Abnormalities in glucose, insulin, and lipoprotein metabolism are common in hypertension, and insulin resistance may represent a unifying pathophysiological mechanism. Anuradha and Balakrishnan (1999) reported that adding taurine to the diet of fructose-fed rats moderated the fructose-induced exaggerated glucose levels and hyperinsulinemia. Nakaya et al. (2000) showed that administration of taurine resulted in significantly less abdominal fat accumulation, hyperglycemia, and insulin resistance. L-arginine (0.52 mg/kg/min) administered intravenously could restore the impaired insulin-mediated vasodilation observed in patients with obesity and type 2 diabetes and improved insulin sensitivity (Wascher et al. 1997). In the present study, model rats showed a higher blood glucose and fasting insulin levels and coadministration of the two amino acids (1.5% taurine + 1.5% L-arginine) had better effects than the other treatment groups in increasing insulin sensitivity.

Increased vascular smooth muscle cell (VSMC) hypertrophy, migration and proliferation are among key events that contribute to remodeling of the vasculature associated with cardiovascular diseases. Reversal of vascular remodeling is very important in the cure of hypertension. Taurine and L-arginine both could reverse the vascular remodeling, and T4 group was not as good as T2 group in reversing vascular remodeling.

Nitric oxide is an important vasoprotective molecule that serves not only as a vasodilator but also exerts antihypertrophic and antiproliferative effects in vascular smooth muscle cells (VSMC). Endothelin-1 is a powerful vasoconstrictor peptide with mitogenic and growth stimulatory properties. Both cGMP-dependent and independent events have been reported to mediate the effect of NO on these pathways leading to its vasoprotective response (Kapakos et al. 2010). ET-1 plays an important vasoconstrictor and growth-promoting role in peripheral resistance vessels and may contribute to BP elevation in some animal models of hypertension. In the present study, it could be seen that NO/cGMP levels significantly decreased and ET-1 levels elevated in insulin resistance hypertensive model rats, while taurine and L-arginine could bring them to normal levels, especially when they were coadministered in drinking water.

32.5 Conclusion

In summary, the present study showed that supplementation of taurine and L-arginine together in drinking water had better antihypertensive effects than when taurine was administered alone.

References

Anuradha CV, Balakrishnan SD (1999) Taurine attenuates hypertension and improves insulin sensitivity in the fructose-fed rat, an animal model of insulin resistance. Can J Physiol Pharmacol 77:749–754

Gokce N (2004) L-arginine and hypertension. J Nutr 134(10 Suppl):2807S–2811S, discussion 2818S-2819S

Harada H, Tsujino T, Watari Y, Nonaka H, Emoto N, Yokoyama M (2004) Oral taurine supplementation prevents fructose-induced hypertension in rats. Heart Vessels 19(3):132–136

Hagar HH, El Etter E, Arafa M (2006) Taurine attenuates hypertension and renal dysfunction induced by cyclosporine A in rats. Clin Exp Pharmacol Physiol 33(3):189–196

Hu J, Xu X, Yang J, Wu G, Sun C, Lv Q (2009) Antihypertensive effect of taurine in rat. Adv Exp Med Biol 643:75–84

Kapakos G, Bouallegue A, Daou GB, Srivastava AK (2010) Modulatory role of nitric oxide/cGMP System in endothelin-1-induced signaling responses in vascular smooth muscle cells. Curr Cardiol Rev 6(4):247–254

Militante JD, Lombardini JB (2002) Treatment of hypertension with oral taurine: experimental and clinical studies. Amino Acids 23(4):381–393

Nakaya Y, Minami A, Harada N et al (2000) Taurine improves insulin sensitivity in the Otsuka Long-Evans Tokushima fatty rat, a model of spontaneous type 2 diabetes. Am J Clin Nutr 71:54–58

Rahman MM, Park HM, Kim SJ, Go HK, Kim GB, Hong CU, Lee YU, Kim SZ, Kim JS, Kang HS (2011) Taurine prevents hypertension and increases exercise capacity in rats with fructose-induced hypertension. Am J Hypertens 24(5):574–581

Siani A, Pagano E, Iacone R et al (2000) Blood pressure and metabolic changes during dietary L-arginine supplementation in humans. Am J Hypertens 13:547–551

Wascher TC, Graier WF, Dittrich P et al (1997) Effects of low-dose L-arginine on insulin mediated vasodilatation and insulin sensitivity. Eur J Clin Invest 27:690–695

Chapter 33
High Sugar Intake Blunts Arterial Baroreflex via Estrogen Receptors in Perinatal Taurine Supplemented Rats

Atcharaporn Thaeomor, J. Michael Wyss, Stephen W. Schaffer, Wiyada Punjaruk, Krissada Vijitjaroen, and Sanya Roysommuti

Abstract In adult rats, perinatal taurine depletion followed by high sugar intake alters neural and renal control of arterial pressure via the renin–angiotensin system. This study tests the hypothesis that perinatal taurine supplementation predisposes adult female rats to the adverse arterial pressure effect of high sugar intake via the renin–angiotensin system, rather than via estrogen. Female Sprague-Dawley rats were fed normal rat chow with 3% taurine (taurine supplementation, TS) or water alone (control, C) from conception to weaning. Their female offspring were fed normal rat chow with either 5% glucose in tap water (TSG, CG) or tap water alone (TSW, CW). At 7–8 weeks of age, the female offspring's renin–angiotensin system or estrogen receptors were inhibited by captopril or tamoxifen, respectively. Body weight, heart weight, kidney weight, mean arterial pressures (MAP), and heart rates were not significantly different among groups without captopril or tamoxifen. Captopril (but not tamoxifen) decreased MAP but not heart rates in all groups. In TSG compared to TSW, CW, and CG groups, baroreflex sensitivity of heart rate

A. Thaeomor
Department of Physiology, Faculty of Medicine, Khon Kaen University,
Khon Kaen 40002, Thailand

School of Physiology, Institute of Science, Suranaree University of Technology,
Nakhonratchasima 30000, Thailand

J.M. Wyss
Department of Cell, Developmental and Integrative Biology, School of Medicine,
University of Alabama at Birmingham, Birmingham, AL 35294, USA

S.W. Schaffer
Department of Pharmacology, College of Medicine, University of South Alabama,
Mobile, AL 36688, USA

W. Punjaruk • K. Vijitjaroen • S. Roysommuti (✉)
Department of Physiology, Faculty of Medicine, Khon Kaen University,
Khon Kaen 40002, Thailand
e-mail: sanya@kku.ac.th

A. El Idrissi and W.J. L'Amoreaux (eds.), *Taurine 8*, Advances in Experimental
Medicine and Biology 775, DOI 10.1007/978-1-4614-6130-2_33,
© Springer Science+Business Media New York 2013

437

(BSHR) and renal nerve activity (BSRA) were significantly decreased. Neither captopril nor tamoxifen altered BSHR in TSG, but tamoxifen (but not captopril) restored TSG BSRA to CW or CG control levels. Perinatal taurine supplementation did not disturb sympathetic and parasympathetic nerve activity in the adult rats on high or basal sugar intake. Compared to its effect in CW and CG groups, tamoxifen increased sympathetic but decreased parasympathetic activity less in TSG and TSW groups. Inhibition of the renin–angiotensin system did not affect autonomic nerve activity in any group. These data suggest that in adult female rats that are perinatally supplemented with taurine, high sugar intake after weaning blunts arterial baroreflex via an estrogen (but not renin–angiotensin) mechanism.

Abbreviations

CW	Control with water intake alone
CW + Cap	CW plus captopril treatment
CW + Tam	CW plus tamoxifen treatment
CG	Control with high sugar intake
CG + Cap	CG plus captopril treatment
CG + Tam	CG plus tamoxifen treatment
TSW	Perinatal taurine supplementation with water intake alone
TSW + Cap	TSW plus captopril treatment
TSW + Tam	TSW plus tamoxifen treatment
TSG	Perinatal taurine supplementation with high sugar intake
TSG + Cap	TSG plus captopril treatment
TSG + Tam	TSG plus tamoxifen treatment
SD	Sprague-Dawley
HF	High frequency
LF	Low frequency

33.1 Introduction

The perinatal environment influences not only growth and development of the fetus and newborn, but the phenotypic expression of the adult offspring. For examples, inhibition of the renin–angiotensin system from conception throughout life or during the perinatal period alone attenuates or prevents hypertension and other related damage in adult spontaneously hypertensive rats (Wyss et al. 1994). However, such inhibition produces salt-sensitive hypertension in adult normotensive rats. Several lines of evidence have reported that perinatal over or under dietary exposure programs adult function and dysfunction (Barnes and Ozanne 2011; Marino et al. 2011; Zhang et al. 2011). Previous experiments indicate that either taurine depletion or supplementation in early life alters renal function in adult male and female rats (Roysommuti et al. 2009a; Roysommuti et al. 2010a; Roysommuti et al. 2010b) while perinatal taurine depletion paired with a high sugar diet after weaning blunts baroreflex function and depresses autonomic nervous system control of arterial pressure (Roysommuti et al. 2009b; Thaeomor et al. 2010). These abnormalities are improved by acute inhibition of renin–angiotensin system (Thaeomor et al. 2010).

The effect of perinatal taurine exposure on adult function is sex dependent (Roysommuti et al. 2009c), suggesting the possible role of sex hormones particularly estrogen in these effects. Without high sugar intake after weaning, perinatal taurine depletion only slightly but significantly blunts baroreflex sensitivity in male but not female rats. In contrast, a high sugar diet markedly depresses baroreflex function and increases sympathetic nerve activity in both sexes (Roysommuti et al. 2009b; Thaeomor et al. 2010). The significant role of estrogen on arterial pressure control is also supported by studies in spontaneously hypertensive rats (Bonacasa et al. 2008) and salt-induced hypertensive and ovariectomized rats (Fang et al. 2001). The much lower incidence of hypertension in premenopausal compared to postmenopausal women also supports a role for estrogen in cardiovascular disorders (Leuzzi and Modena 2011; Yanes and Reckelhoff 2011).

Perinatal taurine supplementation is common in pregnant women and as a dietary supplement in newborns (McPherson and Hardy 2011; Wu 2009; Yamori et al. 2010). Although adverse effects of taurine excess have not been definitively demonstrated in humans, taurine supplementation during late pregnancy in rats stimulates postnatal growth and induces obesity and insulin resistance in adult offspring (Hultman et al. 2007). However, perinatal taurine supplementation's effect and possible mechanisms of action on arterial pressure control have not been established. The present study tests the hypothesis that perinatal taurine supplementation predisposes female adult rats to adverse effects of high sugar intake on arterial pressure control by a renin–angiotensin (rather than estrogen) mechanism.

33.2 Methods

33.2.1 Experimental Protocol

Sprague-Dawley (SD) rats were bred at the animal unit of Faculty of Medicine, Khon Kaen University and maintained at constant humidity ($60 \pm 5\%$), temperature ($24 \pm 1°C$), and light cycle (06.00–18.00 h). Female SD rats were fed normal rat chow with 3% taurine (taurine supplementation, TS) or water alone (control, C) from conception to weaning. Their female offspring were fed with the normal rat chow with either 5% glucose in tap water (TS with glucose, TSG; C with glucose, CG) or tap water alone (TSW, CW) throughout the experiment. All experimental procedures were approved by the Animal Ethics Committee of Khon Kaen University (Khon Kaen, Thailand) and were conducted in accordance with the US National Institutes of Health guidelines.

To test the possible role of renin–angiotensin system, another six separated groups were treated with captopril in drinking water (an angiotensin-converting enzyme inhibitor, 400 mg/l) from 7 days before parameter measurements until the end of experiment (CW + Cap, CG + Cap, TSW + Cap, TSG + Cap). In addition to test the possible role of estrogen, more six separated groups were treated with an estrogen receptor antagonist (tamoxifen, 10 mg/kg/day, oral) since 7 days before the study (CW + Tam, CG + Tam, TSW + Tam, TSG + Tam).

At 7–8 weeks of age, under thiopental anesthesia (50 mg/kg body weight, i.p.), all female rats were implanted with femoral arterial and venous catheters. Three days later, arterial pressure, heart rate, arterial baroreflex sensitivity of heart rate (BSHR) were measured in conscious freely moving rats before and during infusion of phenylephrine (BSHR-phenylephrine) or sodium nitroprusside (BSHR-nitroprusside). After 24 h resting, rats were anesthetized with thiopental sodium, were tracheostomized, and their arterial pressures were recorded continuously, respectively. Renal sympathetic nerve activity was continuously recorded by using stainless steel electrodes (12 MΩ, 0.01 Taper, A-M System, Sequim, Washington, USA) connected to DAM-80 amplifier (World Precision Instruments, Sarasota, Florida, USA) and BioPac Systems (Goleta, California, USA), respectively. Multiunit recording of renal nerve activity was conducted only on nerve units that responded to changes in arterial pressure following nitroprusside or phenylephrine infusion. Body temperature was servo-control at $37 \pm 0.5°C$ by a rectal probe connected to a temperature regulator controlling an overhead heating lamp. At the end of experiment, all animals were terminated by a high dose of thiopental anesthesia and kidney and heart weights were then collected.

33.2.2 Data Analyses

All data were analyzed by Acknowledge software version 3.9.1 (BioPac Systems, Goleta, California, USA). Frequency domains of arterial pressure pulse spectrum were analyzed using the Acknowledge software to indirectly estimate the sympathetic (low frequency 0.3–0.5 Hz, LF) and the parasympathetic nerve activities (high frequency 0.5–4.0 Hz, HF). The percent power spectral densities of LF or HF to the total power of LF and HF indicate sympathetic and parasympathetic nerve activity, respectively (Roysommuti et al. 2009b; Thaeomor et al. 2010). Baroreflex sensitivity values were calculated from changes in heart rates (BSHR) or renal nerve activity (BSRA) per changes in mean arterial pressures.

33.2.3 Statistical Analyses

All data are expressed as mean $\pm$ SEM and were statistically analyzed using one-way ANOVA and a post hoc Duncan's multiple range test. Statistically significant difference is p-values < 0.05.

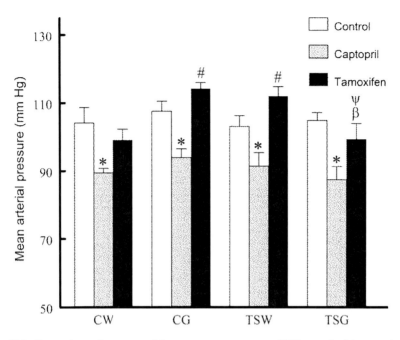

Fig. 33.1 Comparison of mean arterial pressure among groups (*CW* control with water intake alone, *CW + Cap* CW plus captopril treatment, *CW + Tam* CW plus tamoxifen treatment, *CG* control with high sugar intake, *CG + Cap* CG plus captopril treatment, *CG + Tam* CG plus tamoxifen treatment, *TSW* perinatal taurine supplementation with water intake alone, *TSW + Cap* TSW plus captopril treatment, *TSW + Tam* TSW plus tamoxifen treatment, *TSG* perinatal taurine supplementation with high sugar intake, *TSG + Cap* TSG plus captopril treatment, *TSG + Tam* TSG plus tamoxifen treatment, *$p < 0.05$ vs. control, #$p < 0.05$ vs. CW of same treatment, β$p < 0.05$ vs. CG of same treatment, ψ$p < 0.05$ vs. TSW of same treatment)

33.3   Results

At 7–8 weeks of age, all groups displayed similar body, heart, and kidney weights as previously reported (Thaeomor et al. 2010). Mean arterial pressures were not significantly different among untreated control groups and acute inhibition of renin–angiotensin system by captopril significantly decreased mean arterial pressures to a similar extent (about 15 mm Hg decreases). Compared to control treatment, inhibition of the estrogen receptors by tamoxifen did not significantly alter mean arterial pressures in any group, but the values of CG and TSW were slightly and significantly higher than those of CW and TSG (Fig. 33.1).

Heart rates were not significantly different among groups (Fig. 33.2). Baroreflex sensitivity of heart rate was not significantly different among control CW, CG, and TSW, but it was impaired significantly in control TSG (Fig. 33.3). While tamoxifen treatment significantly decreased baroreflex sensitivity in CW, CG, and TSW, captopril treatment significantly decreased only in TSW. Neither captopril nor tamoxifen altered the baroreflex sensitivity of heart rate in TSG. Compared to CW and

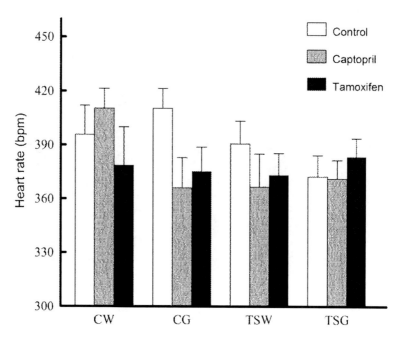

Fig. 33.2 Comparison of heart rate among groups (*CW* control with water intake alone, *CW+Cap* CW plus captopril treatment, *CW+Tam* CW plus tamoxifen treatment, *CG* control with high sugar intake, *CG+Cap* CG plus captopril treatment, *CG+Tam* CG plus tamoxifen treatment, *TSW* perinatal taurine supplementation with water intake alone, *TSW+Cap* TSW plus captopril treatment, *TSW+Tam* TSW plus tamoxifen treatment, *TSG* perinatal taurine supplementation with high sugar intake, *TSG+Cap* TSG plus captopril treatment, *TSG+Tam* TSG plus tamoxifen treatment, no significant difference was observed among groups)

TSW, baroreflex sensitivity of renal nerve activity was impaired significantly in CG and TSG rats (Fig. 33.4). While tamoxifen treatment (compared to control) significantly decreased baroreflex sensitivity in CW, CG and TSW, captopril treatment significantly decreased baroreflex sensitivity in CW and CG (only BSRA-nitroprusside). Tamoxifen but not captopril treatment restored baroreflex sensitivity of TSG compared to TSW and CG.

All groups with or without captopril treatment displayed similar sympathetic and parasympathetic nerve activity while those with tamoxofen treatment showed significantly increased sympathetic and decreased parasympathetic nerve activity (Fig. 33.5). However, the autonomic nerve activity was less in TSW+Tam and TSG+Tam, compared to CW+Tam and CG+Tam, respectively.

33.4   Discussion

Perinatal taurine depletion followed by high sugar intake after weaning blunts baroreceptor reflex function and increases sympathetic nerve activity in adult female rats, and this effect is abolished by short-term inhibition of the renin–angiotensin

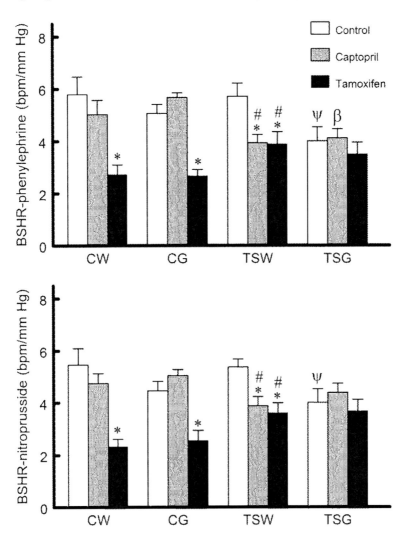

Fig. 33.3 Comparison of baroreflex sensitivity of heart rate (BSHR) among groups (*BSHR-phenylephrine* BSHR tested by phenylephrine infusion, *BSHR-nitroprusside* BSHR tested by sodium nitroprusside infusion, *CW* control with water intake alone, *CW + Cap* CW plus captopril treatment, *CW + Tam* CW plus tamoxifen treatment, *CG* control with high sugar intake, *CG + Cap* CG plus captopril treatment, *CG + Tam* CG plus tamoxifen treatment, *TSW* perinatal taurine supplementation with water intake alone, *TSW + Cap* TSW plus captopril treatment, *TSW + Tam* TSW plus tamoxifen treatment, *TSG* perinatal taurine supplementation with high sugar intake, *TSG + Cap* TSG plus captopril treatment, *TSG + Tam* TSG plus tamoxifen treatment, *$p < 0.05$ vs. control, #$p < 0.05$ vs. CW of same treatment, $^\beta p < 0.05$ vs. CG of same treatment, $^\psi p < 0.05$ vs. TSW of same treatment)

system (Thaeomor et al. 2010), but not by inhibition of estrogen receptors (Roysommuti et al. 2011). The present data indicate that in adult female rats that were perinatally taurine supplemented, estrogen (but not the renin–angiotensin system) contributes to the ability of high sugar intake to depress baroreflex renal nerve

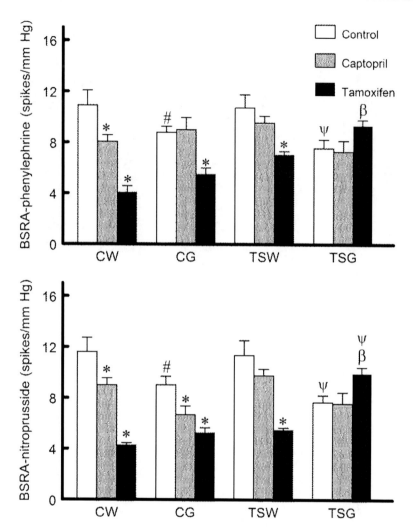

Fig. 33.4 Comparison of baroreflex sensitivity of renal nerve activity (BSRA) among groups (*BSRA-phenylephrine* BSRA tested by phenylephrine infusion, *BSRA-nitroprusside* BSRA tested by sodium nitroprusside infusion, *CW* control with water intake alone, *CW+Cap* CW plus captopril treatment, *CW+Tam* CW plus tamoxifen treatment, *CG* control with high sugar intake, *CG+Cap* CG plus captopril treatment, *CG+Tam* CG plus tamoxifen treatment, *TSW* perinatal taurine supplementation with water intake alone, *TSW+Cap* TSW plus captopril treatment, *TSW+Tam* TSW plus tamoxifen treatment, *TSG* perinatal taurine supplementation with high sugar intake, *TSG+Cap* TSG plus captopril treatment, *TSG+Tam* TSG plus tamoxifen treatment, *$p < 0.05$ vs. control, #$p < 0.05$ vs. CW of same treatment, $\beta$$p < 0.05$ vs. CG of same treatment, $\psi$$p < 0.05$ vs. TSW of same treatment)

control but not baroreflex control of heart rate. In contrast, the renin–angiotensin system (but not estrogen) plays a more important role in baroreflex impairments of heart rate control in TSW when compared to control rats. This phenomenon

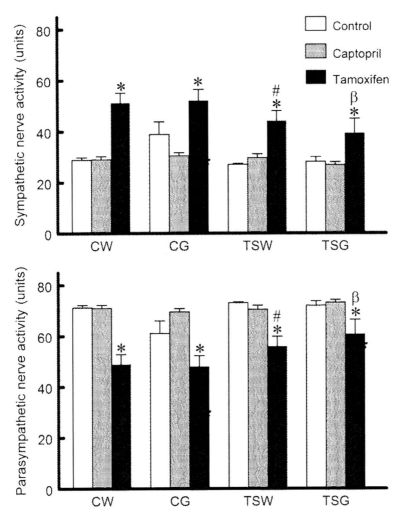

Fig. 33.5 Comparison of sympathetic and parasympathetic nerve activity among groups (*CW* control with water intake alone, *CW + Cap* CW plus captopril treatment, *CW + Tam* CW plus tamoxifen treatment, *CG* control with high sugar intake, *CG + Cap* CG plus captopril treatment, *CG + Tam* CG plus tamoxifen treatment, *TSW* perinatal taurine supplementation with water intake alone, *TSW + Cap* TSW plus captopril treatment, *TSW + Tam* TSW plus tamoxifen treatment, *TSG* perinatal taurine supplementation with high sugar intake, *TSG + Cap* TSG plus captopril treatment, *TSG + Tam* TSG plus tamoxifen treatment, $*p < 0.05$ vs. control, $^{\#}p < 0.05$ vs. CW of same treatment, $^{\beta}p < 0.05$ vs. CG of same treatment)

disappears after a high sugar diet. In addition, these changes may not relate to autonomic nervous system dysregulation.

The effect of estrogen on baroreceptor function is complex. Direct infusion of phytoestrogen into isolated carotid sinus depresses the sinus nerve activity via inhibition of protein tyrosine kinase and decreased Ca^{2+} influx to stretch activated channels (Ma et al. 2005). In contrast, estrogen enhances baroreflex sensitivity induced by phenylephrine via interaction with central projections of the aortic nerve (Mohamed et al. 1999). In addition, central estrogen enhances baroreceptor mediated increases in adrenal hormone secretion induced by hypotension in sheep (Purinton and Wood 2002). However, many brain areas respond differently to estrogen injection. Estrogen injection into the insular cortex increases renal sympathetic (but not vagal) nerve activity while injection into the central nucleus of amygdala decreases sympathetic nerve activity (Saleh and Connell 2003). Although estrogen can modify baroreflex sensitivity and autonomic tone, its effect on arterial pressure is minimal (Saleh et al. 2005; Saleh and Connell 2003). This may explain why in the present study, inhibition of estrogen receptors alters autonomic nerve activity and baroreflex sensitivity without any effect on arterial pressure and heart rate.

Perinatal taurine exposure influences brain growth and development and programs adult function and abnormalities (Aerts and Van Assche 2002; Huxtable 1992; Sturman 1993). It is likely that some brain areas in perinatal taurine supplemented rats respond to estrogen differently from control taurine treated animals, in such a manner that their estrogen-dependent baroreflex functions are depressed. This notation is confirmed by baroreflex sensitivity data in TSW + Tam (about 50% increases) compared to CW + Tam (about 30% increases) rats. This effect disappears after high sugar intake after weaning in TSG but not CG groups, suggesting that perinatal taurine supplementation changes the estrogen-mediated baroreflex interaction, making it more sensitive to some mechanisms related to glucose–insulin regulation.

The renin–angiotensin system underlies sugar-induced hypertension in many animal models (Erlich and Rosenthal 1995; Freitas et al. 2007; Rosenthal et al. 1995). High sugar intake (similar to that in the present study) induces renal dysregulation prior to its effects on hypertension and diabetes mellitus, and these changes are related to renin–angiotensin system overactivity (Roysommuti et al. 2002). The role of the renin–angiotensin system on autonomic nerve function, arterial pressure, and heart rate is completely preserved in perinatal taurine supplemented rats, even with high sugar intake after weaning. Its inhibition decreased baroreflex control of heart rate in TSW but not in CW and CG, but it decreased baroreflex control of renal nerve in CW and CG but not in TSW. In addition, this effect disappears when these animals are treated with a high sugar diet after weaning. The present data suggest that perinatal taurine supplementation alters the interplay of renin–angiotensin system on baroreflex function but not autonomic tone. Angiotensin II generally affects baroreflex sensitivity by acting via the brain, especially the medulla. This study thus confirms that perinatal taurine supplementation programs baroreflex function by altering central nervous system.

33.5 Conclusion

In summary, perinatal taurine exposure programs adult health and disease. Many lines of evidence report that perinatal taurine depletion induces disorders in adult offspring, especially hypertension, renal dysfunction, and metabolic disorders. In contrast, taurine supplementation, either during perinatal life or in the elderly, has been accepted as beneficial for health. The present study suggests that although perinatal taurine excess alone may have minimal effect on cardiovascular control, high sugar intake after weaning can modify the baroreflex function mediated by estrogen and renin–angiotensin system. In the female rat, the estrogen rather than the renin–angiotensin system plays a protective role on the adverse effects of a high sugar diet.

Acknowledgment This study was supported by a grant from the Faculty of Medicine, Khon Kaen University, Khon Kaen 40002, Thailand and by the US National Institutes of Health (NIH) grants AT 00477 and NS057098 (JMW).

References

Aerts L, Van Assche FA (2002) Taurine and taurine-deficiency in the perinatal period. J Perinat Med 30:281–286
Barnes SK, Ozanne SE (2011) Pathways linking the early environment to long-term health and lifespan. Prog Biophys Mol Biol 106:323–336
Bonacasa B, Sanchez ML, Rodriguez F, Lopez B, Quesada T, Fenoy FJ, Hernandez I (2008) 2-Methoxyestradiol attenuates hypertension and coronary vascular remodeling in spontaneously hypertensive rats. Maturitas 61:310–316
Erlich Y, Rosenthal T (1995) Effect of angiotensin-converting enzyme inhibitors on fructose induced hypertension and hyperinsulinaemia in rats. Clin Exp Pharmacol Physiol Suppl 22:S347–S349
Fang Z, Carlson SH, Chen YF, Oparil S, Wyss JM (2001) Estrogen depletion induces NaCl-sensitive hypertension in female spontaneously hypertensive rats. Am J Physiol Regul Integr Comp Physiol 281:R1934–R1939
Freitas RR et al (2007) Sympathetic and renin-angiotensin systems contribute to increased blood pressure in sucrose-fed rats. Am J Hypertens 20:692–698
Hultman K, Alexanderson C, Manneras L, Sandberg M, Holmang A, Jansson T (2007) Maternal taurine supplementation in the late pregnant rat stimulates postnatal growth and induces obesity and insulin resistance in adult offspring. J Physiol 579:823–833
Huxtable RJ (1992) Physiological actions of taurine. Physiol Rev 72:101–163
Leuzzi C, Modena MG (2011) Hypertension in postmenopausal women: pathophysiology and treatment. High Blood Press Cardiovasc Prev 18:13–18
Ma HJ, Liu YX, Wang FW, Wang LX, He RR, Wu YM (2005) Genistein inhibits carotid sinus baroreceptor activity in anesthetized male rats. Acta Pharmacol Sin 26:840–844
Marino M, Masella R, Bulzomi P, Campesi I, Malorni W, Franconi F (2011) Nutrition and human health from a sex-gender perspective. Mol Aspects Med 32:1–70
McPherson RA, Hardy G (2011) Clinical and nutritional benefits of cysteine-enriched protein supplements. Curr Opin Clin Nutr Metab Care 14:562–568
Mohamed MK, El-Mas MM, Abdel-Rahman AA (1999) Estrogen enhancement of baroreflex sensitivity is centrally mediated. Am J Physiol 276:R1030–R1037
Purinton SC, Wood CE (2002) Oestrogen augments the fetal ovine hypothalamus- pituitary-adrenal axis in response to hypotension. J Physiol 544:919–929
Rosenthal T, Erlich Y, Rosenmann E, Grossman E, Cohen A (1995) Enalapril improves glucose tolerance in two rat models: a new hypertensive diabetic strain and a fructose-induced hyperinsulinaemic rat. Clin Exp Pharmacol Physiol Suppl 22:S353–S354
Roysommuti S, Khongnakha T, Jirakulsomchok D, Wyss JM (2002) Excess dietary glucose alters renal function before increasing arterial pressure and inducing insulin resistance. Am J Hypertens 15:773–779

Roysommuti S, Lerdweeraphon W, Malila P, Jirakulsomchok D, Wyss JM (2009a) Perinatal taurine alters arterial pressure control and renal function in adult offspring. Adv Exp Med Biol 643:145–156

Roysommuti S, Malila P, Jirakulsomchok D, Wyss JM (2010a) Adult renal function is modified by perinatal taurine status in conscious male rats. J Biomed Sci 17(Suppl 1):S31

Roysommuti S, Malila P, Lerdweeraphon W, Jirakulsomchok D, Wyss JM (2010b) Perinatal taurine exposure alters renal potassium excretion mechanisms in adult conscious rats. J Biomed Sci 17(Suppl 1):S29

Roysommuti S, Suwanich A, Jirakulsomchok D, Wyss JM (2009b) Perinatal taurine depletion increases susceptibility to adult sugar-induced hypertension in rats. Adv Exp Med Biol 643:123–133

Roysommuti S, Suwanich A, Lerdweeraphon W, Thaeomor A, Jirakulsomchok D, Wyss JM (2009c) Sex dependent effects of perinatal taurine exposure on the arterial pressure control in adult offspring. Adv Exp Med Biol 643:135–144

Roysommuti S, Thaewmor A, Lerdweeraphon W, Khimsuksri S, Jirakulsomchok D, Schaffer SW (2011) Perinatal taurine exposure alters neural control of arterial pressure via the renin-angiotensin system but not estrogen in rats. Amino Acids 41:S84

Saleh TM, Connell BJ (2003) Central nuclei mediating estrogen-induced changes in autonomic tone and baroreceptor reflex in male rats. Brain Res 961:190–200

Saleh TM, Connell BJ, Cribb AE (2005) Sympathoexcitatory effects of estrogen in the insular cortex are mediated by GABA. Brain Res 1037:114–122

Sturman JA (1993) Taurine in development. Physiol Rev 73:119–147

Thaeomor A, Wyss JM, Jirakulsomchok D, Roysommuti S (2010) High sugar intake via the renin-angiotensin system blunts the baroreceptor reflex in adult rats that were perinatally depleted of taurine. J Biomed Sci 17(Suppl 1):S30

Wu G (2009) Amino acids: metabolism, functions, and nutrition. Amino Acids 37:1–17

Wyss JM, Roysommuti S, King K, Kadisha I, Regan CP, Berecek KH (1994) Salt-induced hypertension in normotensive spontaneously hypertensive rats. Hypertension 23:791–796

Yamori Y, Taguchi T, Hamada A, Kunimasa K, Mori H, Mori M (2010) Taurine in health and diseases: consistent evidence from experimental and epidemiological studies. J Biomed Sci 17(Suppl 1):S6

Yanes LL, Reckelhoff JF (2011) Postmenopausal hypertension. Am J Hypertens 24:740–749

Zhang S, Rattanatray L, McMillen IC, Suter CM, Morrison JL (2011) Periconceptional nutrition and the early programming of a life of obesity or adversity. Prog Biophys Mol Biol 106:307–314

Index

A. El Idrissi and W.J. L'Amoreaux (eds.), *Taurine 8*, Advances in Experimental
Medicine and Biology 775, DOI 10.1007/978-1-4614-6130-2,
© Springer Science+Business Media New York 2013

Printed by Publishers' Graphics LLC
DBT140405.15.14.23